Satoshi Horikoshi · Robert F. Schiffmann
Jun Fukushima · Nick Serpone

Microwave Chemical and Materials Processing

A Tutorial

Springer

Satoshi Horikoshi
Department of Materials and Life Sciences and Microwave Science Research Center (MSRC)
Sophia University
Tokyo
Japan

Robert F. Schiffmann
R.F. Schiffmann Associates, Inc.
New York, NY
USA

Jun Fukushima
Department of Applied Chemistry
Tohoku University
Sendai
Japan

Nick Serpone
PhotoGreen Laboratory, Dipartimento di Chimica
Università di Pavia
Pavia
Italy

ISBN 978-981-13-4885-3 ISBN 978-981-10-6466-1 (eBook)
https://doi.org/10.1007/978-981-10-6466-1

Printed on acid-free paper

This Springer imprint is published by Springer Nature
The registered company is Springer Nature Singapore Pte Ltd.
The registered company address is: 152 Beach Road, #21-01/04 Gateway East, Singapore 189721, Singapore

Microwave Chemical and Materials Processing

Preface

Research on human evolution reports that our ancestors learned how to *cook using fire*; it became a developmental turning point, enlarging their brains. Secondary benefits of fire included light, warmth, and protection from predators. In other words, civilization likely was born out of those initial uses of fire, after which it fulfilled a big role in human life. Securing fuel for its citizenry has been at the core of politics for centuries. Currently, humankind's main fuel is petroleum. Changes in oil prices have a dramatic impact on the world's economy. Unfortunately, the generation of heat by the combustion of wood (biomass) and fossil fuels inevitably generates greenhouse gases such as carbon dioxide (CO_2), a gas which has been designated as being partly responsible for global warming. However, energy released from the combustion process cannot be sustained for very long, therefore further development of sustainable energy is necessary.

Chemical syntheses and materials sintering must be tempered to avoid the unnecessary production of CO_2. Today's chemical industry has changed from processes that use fire to processes that use steam or Joule heating—which can be easily controlled. However, more than 100 years have passed since they were initially used, and further innovation is desired. When sintering ceramics in an electric furnace, the refractory material in the furnace is first heated with an electric heater, and then the sample is heated by heat transfer. Consequently, the temperature of the sample is lower than the temperature in the furnace. Approximately 20–30% of the available energy is used in the heating process. Chemical engineering and heat transfer engineering has been used to optimize the relationship between substrate and heating—these two disciplines have witnessed remarkable development. Nonetheless, more advanced heating methods are required today.

Microwave radiation has been used for more than 30 years as a heat source for materials synthesis and sintering. Accordingly, the microwave energy must be absorbed directly by the materials. Energy savings represent one of the advantages of using microwave radiation. With microwave heating, over 60% of the available power can be used to heat the sample. Pioneering microwave scientists initially focused their efforts on developing the domestic microwave oven—a device which can be obtained at low cost, unlike other chemical heating sources. It was only natural for such ovens to find their way into the research laboratory. The convenience of microwaves meant scientist felt that new processes might be developed for rapid heating and cooling, as well as to produce high-quality materials. Several excellent review articles and books that describe the vast results of microwave chemistry and materials processing have been published in recent years. The presence of a microwave heating device in a laboratory is no longer unusual. At present, researchers use microwaves as a tool to perform chemical reactions and materials sintering. Regrettably, this situation has increased the number of researchers who have little understanding of the principles of microwaves as an energy source and little knowledge of microwave equipment. Such a situation will likely slow down innovative research using microwave heating in chemical and materials processing, unless of course the fundamentals and possible applications of microwaves are clearly understood.

This tutorial book represents the backbone of the fundamentals and applications of microwaves in chemistry and materials processing. It aims to develop new research ideas into microwave chemistry and materials processing. In addition, when researchers are confronted with problematic issues in performing microwave-assisted chemistry and materials processing, they may find in this book the means to avoid such issues. The contents embodied in this book are based on the authors' experiences and learnings. Thus, it has the characteristics of a textbook aimed not only at students, but also professionals and engineers. Every attempt has been made to avoid the use of complicated equations to describe the principles of microwaves; instead, they have been explained as much as possible through the use of illustrations and examples.

In Chap. 1 the readers will learn about microwave heating; in Chap. 2 they will find a discussion about heat, and in Chap. 3 will learn the basics of electromagnetic waves. Chapters 4 and 5 clarify the unique characteristics of microwave heating and its underlying theory. Chapter 6 introduces the various functions of each component found in microwave equipment, while Chaps. 7 and 9 present various caveats, features, and uses of microwave chemistry. Chaps. 8 and 10 describe points connected to microwave materials processing. Appendix A presents a summary of various data concerning the use of microwave radiation. A unique find by the reader will be the *Coffee Break* paragraphs added at the end of each chapter that are intended to refresh the reader's memory before moving to the next chapter.

We are grateful to the Springer editorial staff for their thorough and professional assistance. The data presented would not have been possible without the fruitful collaboration of many university and industrial researchers, not least without the cooperation of students whose names appear in many of the publications cited. We are indeed very grateful for their effort.

Tokyo, Japan — Satoshi Horikoshi
New York, USA — Robert F. Schiffmann
Sendai, Japan — Jun Fukushima
Pavia, Italy/Cote St. Luc, Canada — Nick Serpone
July 2017

Contents

About the Authors

Satoshi Horikoshi Sophia University, Department of Materials and Life Sciences (Associate Professor).

Microwave Science Research Center MSRC (Director).

Satoshi Horikoshi received his Ph.D. degree in 1999, and was subsequently a postdoctoral researcher at the Frontier Research Center for the Global Environment Science (Ministry of Education, Culture, Sports, Science and Technology) until 2006. He joined Sophia University as Assistant Professor in 2006, moved to Tokyo University of Science as Associate Professor in 2008, after which he returned to Sophia University as Associate Professor in 2011. Currently, he is Director of the Japan Society of Electromagnetic Wave Energy Applications (JEMEA), and is on the Editorial Advisory Board of the Journal of Microwave Power and Electromagnetic Energy as well as three other international journals. His research interests involve new functional materials or nanomaterial synthesis, molecular biology, the formation of sustainable energy, and environmental protection using microwave-energy and/or photo-energy. He has co-authored over 190 scientific publications and has contributed to and edited or co-edited 23 books.

Prof. Robert F. Schiffmann R.F. Schiffmann Associates, Inc. (President).

Bob Schiffmann has been conducting R&D with microwave heating since 1961. He is an independent consultant on microwave-heating projects. His work includes applications research in microwave ovens, industrial microwave systems, medical applications, and more. He has been the President of the International Microwave Power Institute (IMPI) for 18 years; is a Founding Member of AMPERE; Honorary Member of the Microwave Technology Association of the UK; and is a Certified Food Scientist. He is the first recipient of the Metaxas Microwave Pioneer Award. He has been teaching international microwave science and technology courses since 1973 and has chaired over 20 international microwave conferences. He has 28 US patents and over 50 publications related to microwave science and technology. He has served as an expert witness in 40 microwave-related cases. He is a visiting professor at the Kunming University of Science and Technology in China, and received his MS in Physical and Analytical Chemistry from Purdue University.

Jun Fukushima Advanced Inorganic Materials, Department of Applied Chemistry, Tohoku University, 6-6-07, Aoba Aramaki, Aoba-ku, Sendai, Miyagi 980-8579, Japan, E-mail: fukushima@aim.che.tohoku.ac.jp

Jun Fukushima received his Ph.D. degree in 2012 (Nagoya University, Japan), and was subsequently an Assistant Professor at Tohoku University (Department of Applied Chemistry). His research interests involve novel solid-state materials synthesis, microstructure and diffusion control by microwave processing, and the physics of microwave effects. He has co-authored over 50 scientific publications and books.

Nick Serpone, Ph.D., F. EurASc. Visiting Professor, PhotoGreen Laboratory, Dipartimento di Chimica, Universita di Pavia, Italia, nick.serpone@unipv.it; Postal address: 5647 Smart Avenue, Cote St. Luc (QC), Canada H4W-2M4.

Nick Serpone is Professor Emeritus (Concordia University, Montreal, Canada), and since 2002 has been a Visiting Professor at the University of Pavia (Italy). He was also a Visiting Professor at the Universities of Bologna and Ferrara (Italy), École Polytechnique Fédérale de Lausanne (Switzerland), École Centrale de Lyon (France), and Tokyo University of Science (Japan), and a Guest Lecturer at the University of Milan (Italy). He was Program Director at the National Science Foundation (Washington, USA) and consultant to the 3M Company (USA). He has co-edited/co-authored several books, contributed 23 chapters to books, and published over 450 articles. His principal interests have focused on the photophysics and photochemistry of coordination compounds and metal-oxide semiconductors, environmental remediation, and microwave chemistry. He is a Fellow of the European Academy of Sciences (EurASc) where he is currently Head of the Materials Science Division.

Chapter 1
Microwave as a Heat Source

Abstract Microwave dielectric heating has been used for various purposes for well over five decades. This chapter thus introduces historical trends on the use of these electromagnetic waves in not only domestic microwave ovens but also more importantly applied in such fields as microwave chemistry and materials processing, as well as being envisioned in solar power satellite (SPS) systems. Furthermore, the chapter classifies the industrial fields in which microwave heating is used and shows a summary of the application fields of microwave chemistry and materials. In addition, overviews are given of microwaves being used in organic synthesis, polymer synthesis, catalyzed reactions, sintering of ceramics, heating of metals, extraction, and discharge electrodeless lamps. A coffee break talks about the Raytheon Corporation where the microwave ovens were first created.

Keywords Microwave history · Microwave organic synthesis
Microwave polymer synthesis · Microwave-assisted catalyzed reactions
Microwave sintering of ceramics · Microwave heating of metals
Microwave extraction · Raytheon · Solar power satellites (SPS)

1.1 Applications of Microwaves

Microwave radiation is electromagnetic radiation that spans a frequency ranging from 300 GHz to 300 MHz (i.e., from a wavelength of 1 mm to 1 m). It is used widely in communications and heating, especially in the heating of foodstuff (Fig. 1.1). Historically, the strong interaction of microwaves with materials was discovered in 1945 from the melting of a chocolate bar in the discoverer's pocket, which he attributed to microwave heating from a radar system in his work area. The first commercial microwave oven was developed by Spencer and others at the Raytheon Company [1], and the first commercial model was introduced in 1952; this is a good example of a serendipitous discovery. However, the microwave phenomenon as a potential heat source was known long before Spencer's development of the microwave oven. In the years prior to World War II, Japan had

S. Horikoshi et al., *Microwave Chemical and Materials Processing*,
https://doi.org/10.1007/978-981-10-6466-1_1

knowledgeable researchers in technologies necessary for radar development, who were particularly advanced in magnetron development intended for use in power transmission, not for radar. A lack of appreciation of radar's potential and the rivalry that existed within the Japanese military caused Japan's development of radar to be retarded considerably. It was not until late 1941 that Japan placed into service its first full radar system [2–4]. When experimenting with the development of radar, Japanese engineers and researchers recognized microwave radiation as a potential heat source when they discovered that a sweet potato near the microwave generator was cooked.

High-frequency induction heating was commonly used before the discovery of microwaves; a patent for dielectric heating by means of high-frequency induction was issued in 1933 [5]. Accordingly, many specialists knew about the phenomenon of microwave heating. However, it is thanks to Spencer's contribution that this phenomenon could be developed further into microwave products, which, for example, was achieved in the 1970s when Japanese scientists converted a microwave generator into a domestic microwave oven using a simple, reliable, and inexpensive magnetron that allowed for the heating of simple foods and for materials processing [6].

Domestic and industrial microwave ovens generally operate at a frequency of 2.45 GHz, which corresponds to a wavelength of 12.24 cm and an energy of 1.02×10^{-5} eV, or at a frequency of about 915 MHz (wavelength, 32.8 cm) that can provide up to 1000 kW power in large process heating applications [7]. Since those early days, microwave chemistry has become a rapidly developing branch in numerous fields of fundamental and applied research, in both academia and industry.

Microwave radiation can be classified into two principal application technologies: communications and heat source. In order to carry information, communications—microwaves—typically use modulated waves in terms of frequency, phase, and amplitude. For microwave heating, output power and efficient irradiation

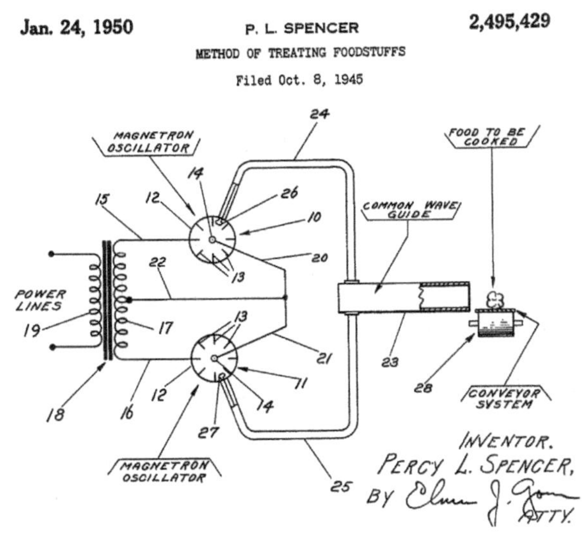

Fig. 1.1 Schematic image of a proposed method of heat-treating foodstuffs with microwave radiation [1]

apparatus are important factors. Therefore, the composition and fabrication of microwave devices may be significantly different, which calls for bringing together various technologies. Microwaves were also suggested by Nikola Tesla (Sect. 3.2) as a means to transmit electrical energy over long distances. This concept was adopted by the Solar Power Satellite (SPS) program [8] to supply electricity generated at the SPS to the surface of the Earth that required the fusion (coupling) of high-power handling techniques (in the field of heating) with highly precise microwave transmission techniques as are available in the telecommunication field.

As the United States and the rest of the world began to face the realities of living with a limited supply of oil and gas, together with the political uncertainties that accompany impending fossil fuel scarcity, the search for reliable and safe means of using the radiant energy of the Sun intensified. Solar radiation is already used in many parts of the world for direct space heating and water heating. It can also produce electricity through photovoltaic and thermoelectric conversion of sunlight, which on Earth is only available during daylight hours, as energy is required and consumed around the clock. In the absence of inexpensive energy storage capabilities, nighttime and cloud cover limit the potential of terrestrial solar technologies (with the exception of ocean thermal energy conversion) to supply the amounts of energy required for use in homes, businesses, and industries. By placing the solar collectors in space where sunlight is intense and constant, and then "beaming" that Sun's energy to Earth in the form of microwaves that is converted into direct current (DC) collectors on Earth, the SPS system sought to guarantee a nonpolluting supply of electricity for terrestrial consumers. Microwave beams have been considered toward this application. Several radically different versions of SPS have been proposed, some of which have been described and analyzed [8]. It was estimated that only six solar-powered satellites could supply all the electrical energy for the United States!

The first serious discussion of the SPS concept originated from a 1968 report by P.E. Glaser in the United States [8]; a schematic of the concept is illustrated in Fig. 1.2. During the following few years, several companies conducted preliminary analyses with some support from the Advanced Programs Office of the National Aeronautics and Space Administration (NASA) [9]. In May 1973, the Subcommittee on Space Science and Applications of the House Science and Astronautics Committee held the first congressional hearings on the concept. Following those hearings, NASA began a series of experiments in microwave power transmission at the Jet Propulsion Laboratory in California. In 1975, NASA created the SPS study office at the Johnson Space Center that performed several additional systems studies. A number of papers were published, culminating in an extensive report that established most of the basis for the Department of Energy's (DOE) system design, after which the United States lost interest in the SPS project. However, Japan's Aerospace Exploration Agency took up the challenge and continues to this day to examine the SPS concept [10] making extensive use of microwaves as energy transmission beams.

Fig. 1.2 Concept for a satellite system for the generation of solar power [8]

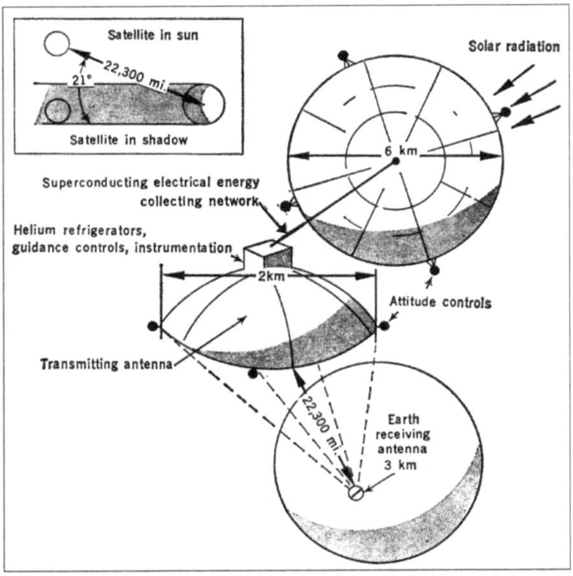

1.2 Some Applications of Microwave Heating

To the extent that microwaves are electromagnetic waves, they move at the speed of light in air and vacuum. Since microwaves directly heat a sample, the time of heat conduction is shortened considerably relative to conventional heating, regardless of however complex the reactor design might be, since reactor design has no influence on reaction times. Accordingly, microwaves have become an important heat source in a wide number of applications superseding the more traditional thermal methods. Figure 1.3 summarizes some of the applications where microwaves have become a common source of thermal energy in diverse fields such as in the attenuation of environmental pollution, medical applications, printing, paints, foodstuffs, the fabrication of thin films, agriculture, and the drying of wood, among several others [11].

1.3 Fields of Microwave Chemistry and Materials Processing

Even a researcher with no knowledge of microwave chemistry and materials processing can make use of microwaves. For instance, when using electron spin resonance (ESR) spectroscopy in chemistry, the researcher makes use of 9-GHz microwaves. When inductively coupled plasma mass spectrometry (ICP-MS) was introduced in the laboratory, it was also necessary to buy the microwave sample pretreatment unit at the same time. High-frequency microwaves are used as the

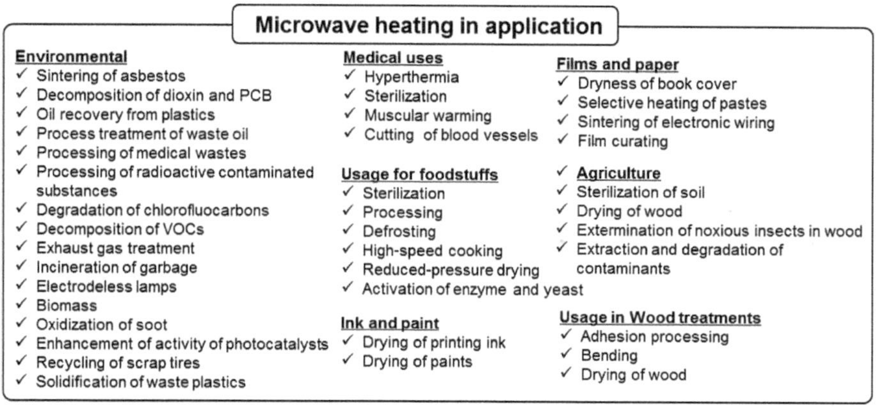

Fig. 1.3 Fields of applications of microwave irradiation [11]. Copyright 2013 by S. Horikoshi

plasma source for surface coating of materials and the surface treatment of semiconductors. Microwaves have also been used as a plasma source in the fabrication of industrial diamonds. Apart from these applications, microwave radiation has been used extensively in the fields of chemistry and material processing since the 1980s. The current wide scope of applications of microwaves in chemistry is summarized in Fig. 1.4, the major ones being organic chemistry, analytical chemistry, biochemistry, polymer chemistry, catalysis, photochemistry, and inorganic chemistry of materials.

Microwave heating can drive various chemical reactions that may not be otherwise possible with conventional heating (e.g., regioselectivity, regular radical reaction, molecule orientation, high crystallization, anisotropic crystal, specific solid-phase diffusion, and strong reducing reactions). Microwave heating leads to

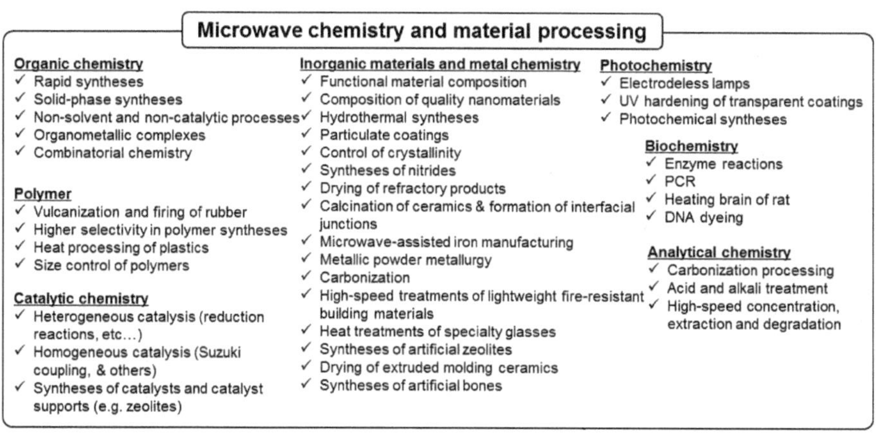

Fig. 1.4 Fields of applications with microwave heating [11]. Copyright 2013 by S. Horikoshi

the formation of temperature gradients at the microscale in catalyzed reactions and the sintering of solids. If these gradients can be managed, it is possible to open up new scientific opportunities in materials science.

Of late, microwave chemistry and materials processing have been the subjects of much excitement. Rapid and high-quality syntheses have been achieved using microwave heating techniques. Some of the fundamental research in microwave radiation has been developed by industry. The next stage will involve the improvement of efficient chemical syntheses as might be required for possible scaling up of industrial applications, *i.e.,* the preparation of significant quantities of high-value chemicals. For this to occur, however, it will necessitate a combination of microwave chemistry and highly coherent microwaves as established in the communications field. In this regard, coherence of electromagnetic waves is typically found in lasers. The amplification of a highly coherent wave by interference promises to enhance the efficiency of microwave chemistry that will be accompanied by significant energy savings. Unfortunately, it is often difficult for a chemist to understand, let alone master, the essence of microwave radiation. Issues that must always be controlled are the properties of the sample, heating the sample, and the proper application of the electromagnetic waves. Various factors impact microwave chemistry; their interconnections are summarized in Fig. 1.5 [12]. This book puts the

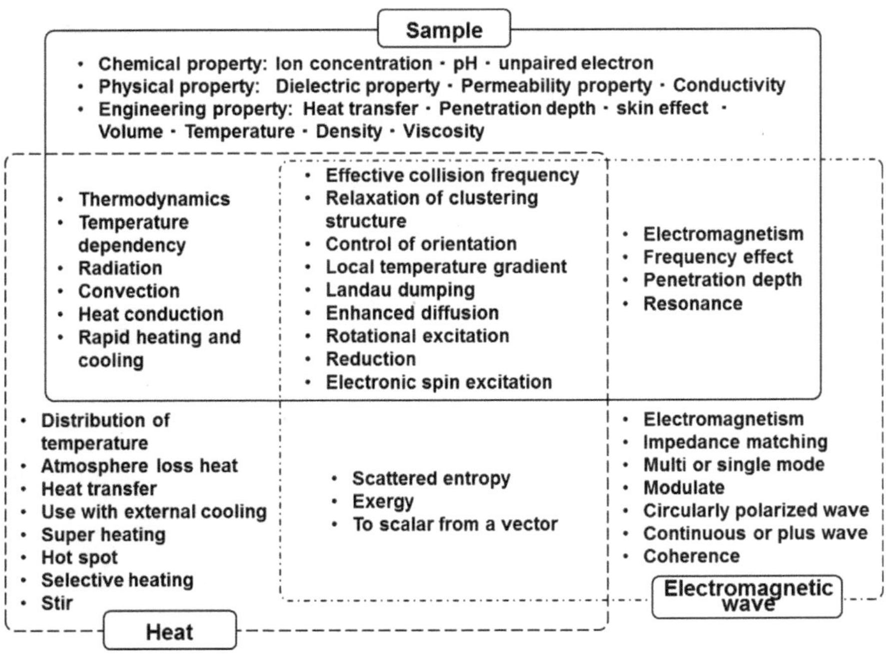

Fig. 1.5 Interconnections between factors that impact microwave chemistry, e.g., nature of the samples, heat and electromagnetic waves (microwaves) in the fields of chemical reactions and materials processing. Reproduced from Ref. [12]. Copyright 2013 by S. Horikoshi

emphasis in understanding these various important factors so that a novice researcher in the use of microwave radiation might be able to master this technology.

1.4 Overview of Microwave Chemistry

Before the turning into the twenty-first century, presentations that dealt with microwave chemistry at conferences were quite rare. This is no longer the case, however, as studies using microwave radiation and the relevant chemical equipment are becoming quite prominent in various laboratories. As a heat source, microwaves have become an additional useful tool for the chemist in carrying out chemical reactions because of rapid and efficient temperature increases with the microwaves penetrating throughout the samples, in contrast to a conventional heat source (e.g., steam, electric heater). The article by Adam titled *Microwave chemistry: Out of the kitchen* that appeared in 2003 in the journal *Nature* [13] has led to an ever increased use of microwaves, particularly in chemistry and the field of materials science. In this section, we introduce and briefly describe the use of microwaves and microwave equipment in each of these fields; a more elaborate discussion of each is presented in later chapters.

Microwaves in organic syntheses (Sect. 9.1): Two pioneering papers that made use of microwaves in organic synthesis were published in 1986 by the groups of Gedye and Giguere. Gedye et al. [14] carried out organic syntheses using a Teflon reactor and a commercial/domestic microwave oven, which resulted in a remarkable decrease of reaction time compared to conventional heating methods. On their part, Giguere et al. [15] pointed out that reactor, solvent, and temperature management were important factors in microwave-assisted organic syntheses. Research into such applications of microwaves increased gradually. Not surprisingly, the number of publications in microwave chemistry has increased dramatically since the year 2000 [16] due in large part to the availability of commercial microwave devices intended for organic synthesis. This led many chemists to convert from the domestic microwave oven to these newly fabricated devices. Safety, reproducibility, control of microwave output power, temperature, and pressure are some of the important and attractive features of such devices. In the first decade of this twenty-first century, the number of research papers in microwave organic chemistry has increased almost exponentially: more than eight times! Unfortunately, this boom in studies of microwave chemistry often lacked the high quality required in serious research. When researchers used domestic microwave ovens to carry out microwave-assisted organic syntheses, the studies often lacked correct temperature measurements, and not least microwave power control could not be achieved under the prevailing experimental conditions. Reproducibility in several of these studies was not a factor that was considered, and as a result, research papers using the microwave heating technique were not accepted by several journals such as *The Journal of Organic Chemistry* as clearly described in the *Guidelines for authors* [see e.g., 17]. Two reasons for such rejections by the journals were the nonstandardized experimental protocols in microwave chemistry and the different microwave equipments used from one laboratory to another.

Currently, by combining with robot technology, microwave organic synthesis equipment has become fully automated and consequently plays an active part in such fields as combinatorial chemistry. This has also brought research in microwave chemistry from the level of the academic laboratory to that of industry, where microwaves are used as a heating resource in organic syntheses. One may now ask: what is the attraction of microwaves in organic synthesis? The biggest attraction is the dramatic enhancement of reaction rates. An example is the Suzuki-Miyaura cross-coupling reaction that, when performed with microwaves, occurs in a reaction time of about 50 seconds, whereas when using the more traditional heating methods, the reaction time is several hours [18]. Similar findings have been reported for various other organic reactions. Several examples of solvent-free and non-catalytic synthesis reactions have also been reported. Not surprisingly, the use of microwaves in chemical syntheses is particularly attractive in the field of Green Chemistry. Many excellent books are available that describe some of the details of microwave-assisted organic syntheses—see for example the book by de la Hoz and Loupy [19].

Microwaves in polymer science (Sect. 9.2): In the field of polymer synthesis, drying and polymerization of an epoxy resin were reported at the end of the 1960s as an example of the use of microwaves [20]; resins used as denture bases are known examples. However, there are only a scant number of papers in the literature on the synthesis of macromolecules in comparison with the number of reports on organic syntheses. Vulcanization of rubber has a long history of applying microwave energy as part of the industrial equipment [21]. A 2010 patent application by Takahashi [22] describes the microwave-assisted vulcanization of rubber tires for application to pre-molded profiles in an autoclave, resulting in a significant reduction of vulcanization time when compared to conventional methods. This allowed for greater productivity and the elimination of some steps of the chemical transformation of rubber. In a technical note, Parodi [23] described the advantages of microwave vulcanization over other existing technologies. The principal advantages of the microwave technology are summarized in Table 1.1.

Rapid polymerization of amino acids by microwave heating was reported in 1990 by Yanagawa et al. [24] who described the efficient reaction of amino acid amides to produce polypeptides in response to microwave heating during repeated hydration–dehydration cycles. The polypeptides formed from a mixture of glycinamide, alaninamide, valinamide, and aspartic acid α-amide had molecular weights ranging from 1000 to 4000 Daltons. Except for glycine, the amino acids were incorporated into the polypeptides in proportion to the starting concentrations. The polypeptides had some definite secondary structure (e.g., an α-helix and β-sheets) in aqueous media. The reaction provided a convenient method for both abiotic peptide formation and chemical synthesis of peptides. In another study, Ito et al. [25] reported that polypeptides synthesized from a mixture of amino acid amides by microwave heating during repeated hydration–dehydration cycles showed hydrolase- and oxidoreductase-like catalytic activities. In this regard, the resulting

Table 1.1 Microwave vulcanization versus other rubber processing methods [23]

Microwave vulcanization	Other processing methods
Much faster and deeper heating of rubber items	Relative to all vulcanization methods (*advantage increasingly strong with the rising thickness of rubber articles to be processed*)
Much lower specific energy consumption (*Watts/kg of rubber processed*)	Relative to all vulcanization methods
Superior working safety	Over molten nitrate salt baths, molten metal baths, and overheated steam under pressure
Absence of any environmental impact risks and ecological issues	As compared to molten salt baths
Higher manufacturing throughput rates	Relative to all vulcanization methods
Minimized plant maintenance costs	As compared to molten salt baths and vulcanization by overheated steam under pressure
Absence of rubber pigment or dye discoloration phenomena	Relative to vulcanization by overheated steam and molten salt baths
Negligible process start-up times (*no time lags and minimal generation of rubber scraps to reach steady-state conditions*)	As compared to all vulcanization methods

polypeptides catalyzed the reduction of the ferricyanide ion [$Fe(CN)_6^{3-}$] with NADH. The polypeptides seemed to have a strong affinity for adenine nucleotides, because the reaction was inhibited by adenine derivatives such as NAD^+ and AppA. Ito and coworkers [25] also entertained the notion for the emergence of primitive protein enzymes.

The synthesis of polymers typically leads to an increase in the viscosity of the sample. Experiments carried out in a microwave oven frequently led to a thermal runaway of the sample. However, such problems have been mitigated with the availability of dedicated devices for microwave chemical synthesis. In recent years, research examples of various polymerizations have been reported such as, for example, the synthesis of polylactic acid, a simple process occurring in relatively short time when using microwaves [26].

Catalyzed reactions (Sect. 9.4): Catalyzed reactions driven by microwave heating are reaction systems that perform well with the characteristics of microwaves. Since the 1990s, several reactions have been carried out in the presence of solid catalysts for a potential application of microwave radiation in materials synthesis: for example, the synthesis of acetylene by oxidization of methane using particulate metal catalysts [27]. In contrast to microwave heating, when a conventional heating method was used, catalyst activity stopped when the reaction was only ca. 25% complete, whereas with microwave heating the reaction proceeded to about 90% completion. Research into the dechlorination of chlorinated substrates with heterogeneous metal catalysts was also examined in the initial stages (1990s) of utilization of microwave radiation. The influence of both the microwave

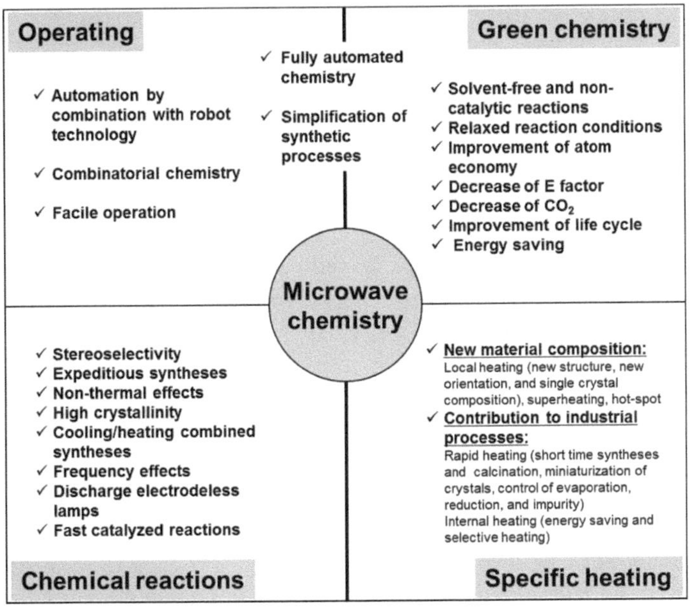

Fig. 1.6 Benefits of microwave radiation in microwave chemistry. Reproduced from Ref. [12]. Copyright 2013 by S. Horikoshi

frequency and pulsed microwave irradiation was considered; optimal reaction conditions were also determined in the presence of a solid catalyst [28]. Of particular interest are catalysts that absorb microwave radiation, including the concept of differential heating upon the quality of a reaction group involving the formation of hot spots on the catalyst surface. Accordingly, the next section examines selective microwave heating of a catalyst (metallic or organic catalyst) in heterogeneous media. Studies into various catalyzed reactions have been reported [29], and the influence of microwaves in photocatalytic reactions has also been shown to be somewhat unique [30].

Microwave-assisted chemical reactions in the liquid phase can be subdivided into four different areas: (i) Operating, (ii) Green Chemistry, (iii) Chemical Reactions, and (iv) Specific Heating (Fig. 1.6). The various features of microwave chemistry are by no means limited to those illustrated. Nonetheless, the following section examines some of the more relevant features of microwave chemistry.

1.5 Overview of Microwave Usage in Materials Processing

Ceramic sintering (Sect. 10.1): In the period spanning the 1980–1990s, in the field of materials processing, a study of microwave heating as an energy saving and low-cost process was applied to sintering of various ceramics. The number of

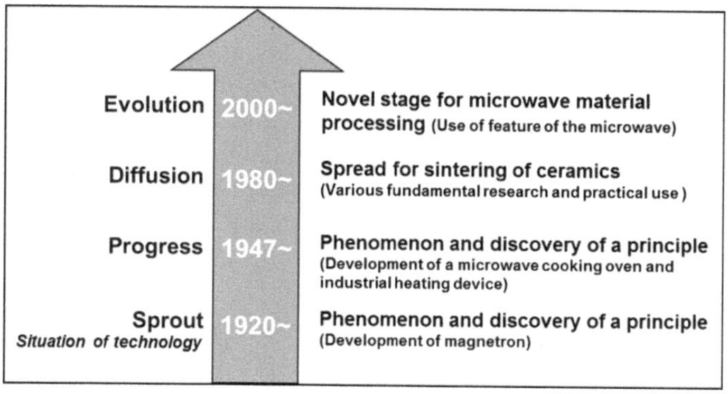

Fig. 1.7 Track of microwave materials processing. Reproduced from Ref. [12]. Copyright 2013 by S. Horikoshi

studies utilizing the various features of microwave heating has increased dramatically since the beginnings of the twenty-first century. Microwave heating has shifted to a new study stage (Fig. 1.7).

Calcination and sintering of ceramics by microwaves were reported as early as 1972 [31]. Fundamental and application studies soon followed around 1980. Conventional heating methods for such tasks apply heat conduction and radiation heating. When applying microwave radiation, the sample itself generates heat, which yields a sintered compact product at low temperature in short time. A symposium held as part of a Conference on Materials Science, sponsored by the Materials Research Society (MRS), revealed that materials processing by microwaves was a hot topic that drew a great deal of interest [32]. Indications from the Conference also showed that microwaves could not only be used to determine the quantity of water in ceramics but could also be used as a technique to detect and analyze the formation of fissures in ceramics and metals [32]. The basic interactions between the microwaves and materials were also widely discussed. Various kinds of ceramics have been treated by the microwave heating method [see e.g., 33, 34]. In recent years, calcination of special ceramics with a specific microwave method has also been reported [35].

Heating of metals (Sect. 10.1.3): Metals possess high conductivity and reflect electromagnetic waves. Nonetheless, sintering of metallic powders by microwave radiation was reported in a 1999 study by Roy et al. [36] who demonstrated that microwaves can penetrate into discrete metallic-powdered aggregates with the microwave energy being absorbed inside the bulk. Takayama [37] experimentally confirmed that microwaves penetrated and heated the inner core of a pelletized copper powder. Until the report by Roy et al. [36], metallic materials, especially sintering of metals, had been outside the scope of microwave heating. Accordingly, their study received much attention by microwave researchers.

Enthusiastic discussion is continuing in an attempt to understand the fundamental principles of the mechanism(s) of microwave heating of metals. Since the experiments of Roy et al. [36], the theory of microwave heating of metals has been considered in a number of studies. Skin effect, Eddy currents, and native oxide films are some of the keywords used to understand the microwave heating of metals. In this regard, Ignatenko and Tanaka [38] investigated the heating mechanism of compacted copper powders by a numerical analysis. Understanding the microwave heating mechanism(s) of metals has led to the development of several microwave material processes, such as metal sintering, nanoparticle synthesis, and the manufacturing of additives.

1.6 Overview of Microwave Usages in Other Sciences

Microwave extraction: Microwave extraction technology is in use in a variety of fields. Microwave extraction is roughly divided into the extraction of an active ingredient, pretreatment of an analytical sample, and extraction/removal of contaminants from polluted soils. As an example, in the early stages of the extraction of an active ingredient, we note the extraction of an essential oil from vegetable oil [39]. Moreover, extraction of useful ingredients from citrus fruits using microwaves requires only a few minutes relative to more conventional extraction methods that require as much as 18 days in some cases [40]. Microwaves have performed well as standard analytical methods in the pretreatment (decomposition) of a sample for use in atomic absorption spectrometry (AAS), inductively coupled plasma atomic emission spectroscopy (ICP-AES), and inductively coupled plasma mass spectrometry (ICP-MS). Equipment for microwave pretreatment is the most sold equipment in the microwave chemical field, more so than that designed for organic synthesis.

Microwave discharge electrodeless lamps (MDELs): Research on electrodeless and wireless lamps involving a microwave energy source is surprisingly over a century old. In the 1890s, Tesla [41] demonstrated in his lectures and articles that wired and wireless transfers of power to electrodeless fluorescent and incandescent lamps were possible. A research report by Inoue [42] in the late twentieth century examined the principles and the mechanism(s) of excitation of such lamps by microwave radiation. However, both the size of the microwave generator as the energy source of the electrodeless lamp and the corresponding cost were major barriers to a practical application. Recent years have witnessed increased practical applications of MDELs as a result of their high performance, miniaturization, and the lower costs of magnetrons and semiconductor microwave generators. Different types of microwave discharge electrodeless lamps (MDELs) are used in various industrial processes depending on the types of applications, such as in paint hardening (drying) equipment [43], a technology developed in the United States in the early 1970s to rapidly cast a polymer by means of UV hardening of a monomer material. The process yielded thin films of variable thicknesses compared to the more conventional drying process. Moreover, volatile organic compounds (VOCs) generated from the evaporation of organic solvents could be controlled in this

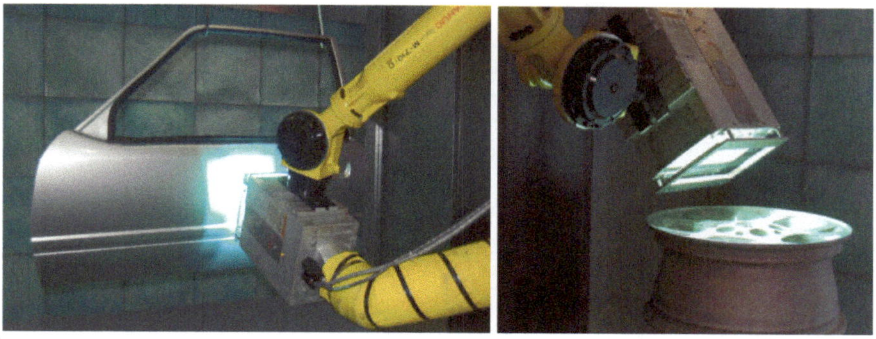

Fig. 1.8 Coating of photo-curing resin using a microwave discharged electrodeless lamp (photographs courtesy of Japan Fusion UV Systems)

system. The MDELs emit a stable light irradiance for long time periods so that the UV hardening method with an MDEL system fits well into an industrial process. At present, MDELs are widely used in the printing process, paints (Fig. 1.8), coatings of a DVD surface, and the plastic coating of optical fibers [43]. Microwave electrodeless lamps have also been proposed as environmental treatment lamps [44]. Since the lamps are electrodeless (i.e., wireless), they can be used to directly treat wastewaters, i.e., the contaminants present in wastewaters can be decomposed with the VUV/UV light emitted by these lamps. Particular types of MDELs are the high surface area systems provided by putting bead-like MDELs into a reactor [44]. MDELs have also been proposed as a light source to activate a photocatalyst [44].

1.7 Coffee Break 1: Raytheon Corporation

Beginning with the invention of the microwave oven, microwave energy has been applied to the heating of foods and foodstuffs since 1945. Today's microwave ovens are used all over the world. In 2013, it was estimated that there were approximately of 1 billion microwave ovens in consumers' homes [45]. Microwave energy is used in the processing of many food products such as meats, fish, vegetables, and more. Some of these applications will be described below. How was microwave heating discovered or invented? The story is long and complex, but some highlights are summarized in Table 1.2.

Help from across the Atlantic: the Raytheon Corporation pioneered the development of today's microwave ovens and was also instrumental in the development of radar during World War II, providing help from across the Atlantic. As the war progressed, British industry was unable to mass-produce the tens of thousands of magnetron tubes (the heart of radar's operation) needed for the total war effort. Across the Atlantic in the United States, a small and relatively unknown firm, Raytheon, had been experimenting with microwave tubes while producing

Table 1.2 History of the development of microwave technology and the microwave oven

Year	Item
1832	Michael Faraday speculates that electromagnetic phenomena are the same wave motion as sound and light
1846	James Clerk Maxwell provides a mathematical description of electromagnetic waves (EM)
1888	Heinrich Hertz propagates radio waves and demonstrates they will bounce off objects
1904	Christian Hulzmeyer awarded a patent on the principle of radar: reflection of EM waves by metallic objects making it possible to locate ships through fog
1900 to 1940	Several Swiss and German researchers develop early models of magnetrons
February 1940	Professors Boot and Randall at the University of Manchester (UK) build an electronic vacuum tube, the cavity magnetron, capable of generating 1000 times more power than previously possible. This was to be used in radar systems during World War II
1940	Percy Spencer at Raytheon discovers how to mass-produce these magnetrons
1945	Percy Spencer, working in the radar laboratory at Raytheon, discovers the principle of microwave heating of food, resulting in the first patent for a microwave oven [46]
1947	Raytheon introduces first commercial microwave oven
Late 1960s	New Japan Radio devises an inexpensive air-cooled magnetron that is the basis for today's microwave oven cooker magnetrons
1970s	The countertop microwave oven becomes a commercial success
Today	Microwave oven saturation, i.e., percentage of homes possessing microwave ovens is approximately 93–95% in the United States, 97% in Japan, and 90–95% in the UK

transmitting tubes [47]. At the suggestion of ***MIT's Radiation Laboratory***, a meeting was arranged between British scientists and Raytheon engineer Percy L. Spencer. Impressed by Spencer's innovations and through the MIT Radiation Laboratory, the UK awarded a contract to Raytheon to supply the magnetrons. Ultimately, Raytheon became the major Allied supplier of magnetrons during the war years. And for the future of allied defense, the events also brought together, for the first time on a common mission, Percy Spencer's Raytheon from the U.S. and Alfred Charles Cossor's A.C. Cossor Co. in the UK, pioneers in radar technology. As World War II was ending, the management of Raytheon, whose basic business was built around supplying equipment for military radar, recognized that there would be a serious decline in revenue and needed another major source of income (Fig. 1.9). Fortunately, Percy Spencer, head of new tube and engineering production, had an idea and suggested that Raytheon use its expertise in radar by making an oven that used microwave energy to heat foods. In the decades prior to 1945, researchers in Switzerland and Germany saw the capability of microwave energy to heat materials, but at that point, no one had translated that information into the concept of cooking with that energy. ***A little-known fact about Percy Spencer*** is

Fig. 1.9 Discussion led by Percy Spencer on a variety of magnetron programs (1949). From left to right, Raytheon chairman Laurence K. Marshall, General Omar Bradley, Raytheon consultant Dr. Edward Bowles, Col. Willi Mathews, and Percy Spencer. Reproduced with permission from Ref. [48]. Copyright 2008 by the Wicked Local

that he discovered a way to mass-produce magnetrons for use in radar, increasing production from 7 to 2600 per day, an important factor in the Allies winning World War II.

According to legend, while walking past an open radar waveguide, a chocolate bar in Percy's pocket became warm and melted, triggering his idea of heating food materials. So he tried microwaving foods such as corn to produce popcorn; while exposing an egg to the microwave energy from a magnetron, the egg exploded in the face of an onlooker. Spencer and his colleagues built a prototype oven by inserting a magnetron into a trashcan and, when that worked, went on to better test systems. Once satisfied that it was possible to heat and cook things such as vegetables very quickly by properly applying microwave energy, work continued at Raytheon to develop the first commercial microwave ovens, introduced in 1947 and intended for use in restaurants, not homes. These ovens weighed over 300 kg and stood about 1.5 m tall, and were powered by two 800-Watt water-cooled magnetrons operating at 240 V. Raytheon introduced its first consumer oven in 1967, but innovation continued at Raytheon and many other companies; with the adoption of rugged inexpensive air-cooled cooker magnetrons, this led to the design of countertop microwave ovens in kitchens around the world. Amazingly, Spencer was a self-taught tube expert who never finished grammar school, but had over 300 patents by the end of his life.

References

1. P.L. Spencer, Method of treating foodstuff, U.S. Patent 2,495,429, 24 Jan 1950
2. R.I. Wilkinson, Short survey of Japanese radar—Part I. Trans. AIEE **65**, 370 (1946)

3. S. Nakajima, The history of Japanese radar development to 1945, in R. Burns, *Radar Development to 1945*, Peter Peregrinus Ltd, 1988
4. S. Nakajima, Japanese radar development prior to 1945. IEEE Antennas Propag. Mag. **34**, 17–22 (1992)
5. T. McArthur, US Patent 1,900,573, 7 Mar 1933
6. R. Cherbański, E. Molga, Intensification of desorption processes by use of microwaves—an overview of possible applications and industrial perspectives. Chem. Eng. Process. **1**, 48–58 (2008)
7. C. Eskicioglu, N. Terzian, K.J. Kennedy, R.L. Droste, M. Hamoda, Athermal microwave effects for enhancing digestibility of waste activated sludge. Water Res. **41**, 2457–2466 (2007)
8. E. Glaser, Power from the sun: its future science **162**, 857–861 (1968)
9. E. Glaser, O.E. Maynard, J. Mockovciak, E.L. Ralph, Feasibility study of a satellite solar power station. Arthur D. Little Inc., NASA CR-2357 (contract No. NAS 3-16804), February 1974
10. http://www.ard.jaxa.jp/eng/research/ssps/hmi-index.html
11. S. Horikoshi, N. Serpone, in *Microwaves in Nanoparticle Synthesis: Fundamentals and Applications,* eds. S. Horikoshi, N. Serpone. (Wiley-VCH Verlag GmbH, Weinheim, Germany), 2013, Chapter 2
12. S. Horikoshi, N. Shinohara, H. Takizawa, J. Fukushima, *Microwave chemistry*, Sankyo Publishing Co. Ltd., 2013
13. D. Adam, Microwave chemistry: out of the kitchen. Nature **421**, 571–572 (2003)
14. R. Gedye, F. Smith, K. Westaway, H. Ali, L. Baldisera, L. Laberge, J.R. Rousell, The use of microwave ovens for rapid organic synthesis. Tetrahedron Lett. **27**, 279–282 (1986)
15. R.J. Giguere, T.L. Bray, S.M. Duncan, G. Majetich, Application of commercial microwave ovens to organic synthesis. Tetrahedron Lett. **27**, 4945–4948 (1986)
16. C.O. Kappe, A. Stadler, in *Microwaves in Organic and Medicinal Chemistry*, eds. by R. Mannhold, H. Kubinyi, G. Folkers (Wiley-VCH Verlag GmbH, Weinheim, Germany,2005) Chapter 1, pp. 4–5
17. http://pubs.acs.org/paragonplus/submission/joceah/joceah_authguide.pdf
18. V.V. Namboodiri, V.S. Varma, Microwave-accelerated Suzuki cross-coupling reaction in polyethylene glycol (PEG). Green Chem. **3**, 146–148 (2001)
19. A. de la Hoz, A. Loupy (eds.), *Microwaves in organic synthesis* (Wiley-VCH Verlag GmbH, Weinheim, Germany, 2012)
20. M. Nishi, Studies on the curing of denture base resins with microwave irradiation: with particular reference to heat-curing resins. J. Osaka Dep. Univ. **2**, 23–40 (1968)
21. R.J. Meredith, *Engineers' handbook of industrial microwave heating (Power & energy series)* (The Institute of Electrical Engineers, UK, 1998)
22. J. Takahashi, Vulcanization process of rubber tires with the use of microwaves, United States Patent Application Publication, US 2010/0090373 A1, 15 Apr 2010 (http://www.faqs.org/patents/app/20100090373)
23. F. Parodi, Microwave chemicals for the rubber industry: novel, specialty microwave heating susceptors for the fast UHF vulcanization of white & colored rubber compounds. (2001–2011), (http://users.libero.it/fgparodi/fap_6a2-en.html)
24. H. Yanagawa, K. Kojima, M. Ito, H. Handa, Synthesis of polypeptides by microwave heating I. Formation of polypeptides during repeated hydration-dehydration cycles and their characterization. J. Mol. Evol. **31**, 180–186 (1990)
25. M. Ito, N. Handa, H. Yanagawa, Synthesis of polypeptides by microwave heating I. Formation of polypeptides during repeated hydration-dehydration cycles and their characterization. J. Mol. Evol. **31**, 187–194 (1990)
26. T. Nakamura, R. Nagahata, K. Kunii, H. Soga, S. Sugimoto, K. Takeuchi, Large-scale polycondensation of lactic acid using microwave batch reactors. Org. Process Res. Dev. **14**, 781–786 (2010)

27. J.K.S. Wan, Microwaves and chemistry: the catalysis of an exciting marriage. Res. Chem. Intermed. **2**, 147–158 (1993)
28. J.K.S. Wan, M.S. Ioffe, surface heating and energy transfer in pulsed microwave catalytic systems: a microwave-induced acoustic study. Res. Chem. Intermed. **20**, 115–132 (1994)
29. S. Horikoshi, N. Serpone (eds.), *Microwaves in Catalysis: Methodology and Applications* (Wiley-VCH Verlag GmbH, Weinheim, Germany, 2015)
30. S. Horikoshi, N. Serpone, On the influence of the microwaves' thermal and non-thermal effects in Titania photoassisted reactions. Catal. Today **224**, 225–235 (2014)
31. K. Shimomura, T. Miyazaki, N. Taniguchi, A. Tutiya, Annual Report. RIKEN, Japan **48**, 11–19 (1972)
32. W.H. Sutton, Microwave processing of ceramic materials. Am. Ceram. Soc. Bull. **68**, 376–386 (1989)
33. Materials Research Advisory Board, *Microwave Processing of Materials, National Research Council, Publication NMAB-473* (National Academy Press, Washington, D.C., 1994)
34. D.C. Folz, J.H. Booske, D.E. Clark, J.F Gerling (eds.) Microwave and Radio Frequency Applications, in *Proceedings from the Third World Congress on Microwave and RF Processing*, (American Ceramic Society, Westerville, OH, 2003)
35. H. Takizawa, Institute Elect. Eng. Jpn. **132**, 17–19 (2012)
36. R. Roy, D. Agrawal, J. Cheng, S. Gedevanishvlli, Full sintering of powdered-metal bodies in a microwave field. Nature **399**, 668–670 (1999)
37. S. Takayama, Y. Saito, M. Sato, T. Nagasaka, T. Muroga, Y. Ninomiya, Sintering behavior of metal powders involving microwave-enhanced chemical reaction. Jpn. J. Appl. Phys. **45**, 1816–1822 (2006)
38. M. Ignatenko, M. Tanaka, Numerical analysis of the microwave heating of compacted copper powders in single-mode cavity. Jpn. J. Appl. Phys. **50**, 097302–097309 (2011)
39. J.R.J. Paré, M. Sigouin, J. Lapointe, U.S. Patent 5,002,784, 26 Mar 1991
40. M. Tsukayama, R. Ichikawa, K. Yamamoto, T. Sasaki, Y. Kawamura, Microwave-Assisted Rapid Extraction of Polymethoxyflavones from Dried Peels of Hort. ex Tanaka. Nippon Shokuhin Kagaku Kogaku Kaishi **56**, 359–362 (2009)
41. N. Tesla, *Lecture Delivered Before the American Institute of Electrical Engineers* (Columbia College, NewYork, NY, USA, 1891)
42. A. Inoue, Next generation new lamp. J. IEE Jpn. **177**, 155–158 (1997)
43. R. Suzuki, M. Saito, K. Watanabe, Light emitting phenomena in microwave discharge lamps. J. Illum. Engng. Inst. Jpn. **72**, 74–78, (1988)
44. S. Horikoshi, N. Serpone, Novel designs of microwave discharge electrodeless lamps (MDEL) in photochemical applications. Use in advanced oxidation processes. Photochem. Photobiol. Sci. **8**, 1087–1104 (2009)
45. J. Osepchuk, Microwave Ovens: An Overview, in *Proceedings of IMPI Symposium*, (Rhode Island, USA, 2013)
46. P. Spencer, Method of treating foodstuffs, US Patent 2,495,429, 1950
47. http://www.raytheon.co.uk/ourcompany/history
48. A.R. Earls, R.E. Edwards, *Raytheon Company: The First Sixty Years* (Images of America), Wicked Local, NY, USA, 2008)

Chapter 2
The Nature of Heat

Abstract When doing research on microwave-assisted chemistry and materials processing, what type of heat is used most often? This chapter discusses qualitatively issues such as heat, temperature, thermodynamics, and heat transfer without unnecessary use of equations. Understanding this chapter will provide a better understanding of later chapters. In addition, the coffee break talks about relationships between microwaves and foods, as well as the history of microwave cooking ovens.

Keywords Heat · Joule · Calorie · Maxwell · Chronological history
Temperature · Thermal energy · Heat transfer · Conduction heating
Convection heating · Radiation heating · Microwave cooking ovens

2.1 What is Heat?

The goal of this chapter is to describe various facets of heat and heat transfer. Research into microwave chemistry and materials processing through microwave heating can lead to some surprises and strange phenomena, not otherwise observed by conventional heating. The mechanisms of these phenomena have yet to be fully understood. First, however, we need to consider ***the nature of heat***.

2.2 Historical Aspects of Heat

The nature of heat was clearly described in the mid-eighteenth century by the Scottish scientist Joseph Black who distinguished between the quantity of heat in a substrate and its temperature. He realized that thermometers could be used to determine the quantity of heat if temperatures were measured over a period of time, while the body was either heated or cooled. His experiments used two similar glass flasks, followed by pouring the same quantity of water into both and placing them in a freezing mixture. In one flask, he had added a small quantity of alcohol to

S. Horikoshi et al., *Microwave Chemical and Materials Processing*,
https://doi.org/10.1007/978-981-10-6466-1_2

prevent freezing. Subsequent to the removal of the flasks from the bath, he noted that water was frozen in one flask, while in the other the contents remained in the liquid state, though both flasks were at the same temperature. The two flasks were then allowed to warm up gradually. The temperature of the flask containing water and alcohol warmed up several degrees, while ice in the other flask remained at its freezing point. As the flasks had absorbed heat at the same rate, Black argued that the heat absorbed by the ice after 10 h should have raised the temperature of the same quantity of water by 78 °C (140 °F), which he described as being the latent heat of fusion of water. Black then proceeded to extend the experiments to measuring the latent heat of vaporization of water [1]. Black's theory of latent heat marked the beginnings of thermodynamics. He also showed that different substances have different specific heats. The theory proved important not only in the development of abstract science but also most importantly in the development of the steam engine. The latent heat of water is large compared with many other liquids, which led James Watt to attempt improving the efficiency of the steam engine invented by Thomas Newcomen [1].

By the late 1700s, the experiments of Fahrenheit, Black, and others had established a systematic, quantitative way of measuring temperatures, heat flows, and heat capacities. However, these experiments threw no new light on exactly what was flowing, i.e., was heat just another invisible fluid? In 1787, Lavoisier thought so, calling it a *caloric fluid* (from the Greek word for heat). The existence of such a fluid was thought plausible at the time as heat flowed from a hot body to a cold body. In addition, the quantitative calorimetric experiments of Black and others had established heat as a *conserved quantity*. Because of this heat flow, it was believed that its particles repelled each other [2].

The first serious enquiry on the soundness of the caloric theory of heat took place in a cannon factory in Bavaria under the direction of Benjamin Thompson (also known as Count Rumford of the Holy Roman Empire). Ever the skeptical thinker, though Thompson was studying cannon boring, he was really thinking about whether or not Lavoisier's calorific fluid really had any validity. In the factory, the cannon cylinders were bored using an iron bit inside a brass cylinder which led to the generation of frictional heat, which, within the caloric theory, was accounted for by the pressure and movement squeezing out caloric fluid. Thompson subsequently measured how much heat was produced for an extended period by immersing the hot brass cylinder into water. To his astonishment, the water (2 gallons) began to boil without fire.

The relation between work and heat was clarified by the study of James Prescott Joule (England) who reported that when a motor was energized by a battery, the conducting wire was heated. This led Joule to write in his 1844 paper: *the mechanical power exerted in turning a magneto-electric machine is converted into the heat evolved by the passage of the currents of induction through its coils, whereas the motive power of the electromagnetic engine is obtained at the expense of the heat due to the chemical reactions of the battery by which it is worked* [3]. Joule further suggested that heat is the movement form of energy, equivalent to heat and dynamic work, and equivalent to heat and kinetic work. To establish the

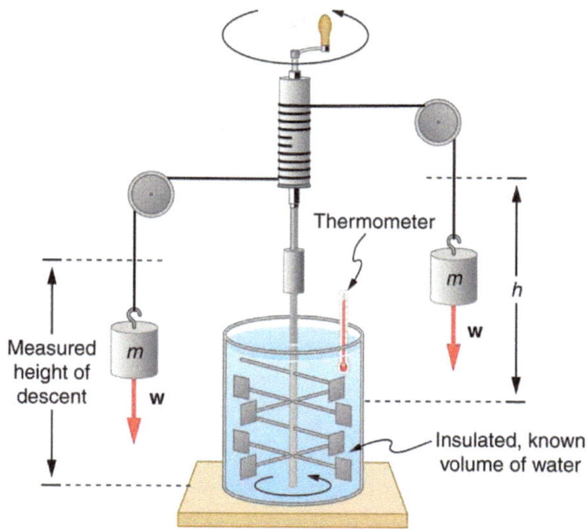

Fig. 2.1 Schematic depiction of Joule's experiment that established the equivalence of heat and work. Reproduced from Ref. [4]

equivalence between heat and work, he performed an experiment in which water was stirred by the power of two weights as illustrated in Fig. 2.1. At about the same time, the equivalence of dynamic work and heat was also inferred by Julius Robert von Mayer (Germany).

At the 1847 annual meeting of the British Association for the Advancement of Science, Joule attempted to discredit the caloric theory of heat and the theory of the heat engine articulated by Sadi Carnot and Émile Clapeyron; he argued for the mutual convertibility of heat and mechanical work and for their mechanical equivalence. William Thomson (also known as Lord Kelvin; England) was skeptical of Joule's viewpoints and though he felt that Joule's results necessitated a theoretical explanation, he remained deeply committed to the Carnot–Clapeyron theory. Lord Kelvin predicted that the melting point of ice must fall with pressure; otherwise, its expansion on freezing could be exploited in a *perpetuum mobile* [5]. Dissatisfied with the gas thermometer that provided only an operational definition of temperature, he extended the Carnot–Clapeyron theory further and proposed an absolute temperature scale in which …*a unit of heat descending from a body A at the temperature* T *of this scale, to a body B at the temperature (*T−1*), would give out the same mechanical effect* [work], *whatever be the number* T…; Kelvin further noted that such a scale would be …*quite independent of the physical properties of any specific substance*… [6]. Kelvin also speculated that a point would be reached at which no further heat (caloric) could be transferred—this is now referred to as the point of *absolute zero*, which Guillaume Amontons had mentioned in 1702 and that Carnot had published in 1824 in *Réflexions sur la Puissance Motrice du Feu (*Reflections on the Motive Power of Heat). At the time, −267 °C (zero degrees Kelvin) was estimated as being the absolute zero temperature [7]. In addition, Kelvin described the internal energy of a system as being the sum of the kinetic

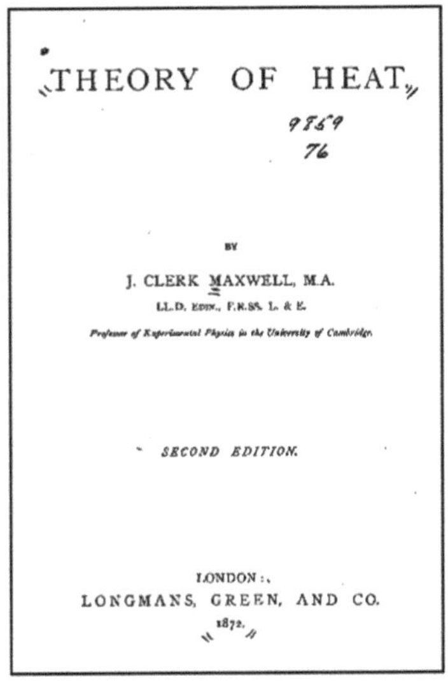
THEORY OF HEAT

BY

J. CLERK MAXWELL, M.A.

LL.D. EDIN., F.R.SS. L. & E.

Professor of Experimental Physics in the University of Cambridge

SECOND EDITION.

LONDON:

LONGMANS, GREEN, AND CO.

1872.

Fig. 2.2 Book cover of "Theory of Heat" by James Clerk Maxwell (1871); Reproduced from Ref. [8]

energy of an atom or molecule and the system's potential energy. Specifically, he stated that *the gap in the internal energy with the change of state of the system is equal to the sum of the work added from outside and the calorie*. Later, heat was summarized as a physical point by James Clerk Maxwell (England) in his 1871 book "Theory of Heat" (Fig. 2.2).

Table 2.1 summarizes the chronological history into the nature of heat and the contributions of the many individuals involved.

Carnot also made significant contributions to the determination of what heat is all about, when he attempted to answer two questions about the operation of heat engines [9]: (i) *Is the work available from a heat source potentially unbounded?* and (ii) *Can heat engines in principle be improved by replacing the steam with some other working fluid or gas?* One important conclusion that Carnot came to was that *the motive power of heat is independent of the agents employed to realize it; its quantity is fixed solely by the temperatures of the bodies between which is effected, finally, the transfer of caloric.* [10]. In other words, the efficiency of a reversible heat engine is independent of the agents employed to achieve it, its quantity depending solely on the difference in the temperatures of the bodies and the reversible transfer of entropy, ΔS.

2.3 Heat *Versus* Temperature

Clearly, heat is the form of energy transferred between two substances each of which is at a different temperature. At this stage, it is common knowledge that the direction of heat flow occurs from the substance at the higher temperature to the

Table 2.1 Chronological history of the nature of heat

Year	People (country)	Research
3000 BC	Egypt	The ancients viewed heat as that related to fire. Ancient Egyptians in 3000 BC viewed heat as having mythological origins
500 BC	Philosopher, Heraclitus (Greece)	The universe was postulated as a continuous state of flux or permanent condition of change as a result of transformations of fire. Heraclitus summarized his philosophy as "All things are an exchange for fire"
460 BC	Hippocrates (Greece)	Father of medicine. He postulated that Heat is a quantity that functions to animate and derives from an internal fire located in the left ventricle
1200s	A. A. Baydawi (Islamic philosophy)	Natural heat would be the heat of a fiery atom that is broken, and heat may occur through motion change, the proof of this being through experiment
1600s	F. Bacon (England)	Heat itself, its inherent nature and essence is motion and nothing else
1600s		Beginning of research on the basic physical quantity of pressure, temperature, and the capacity. Heat is a substance based upon the phlogiston theory, and the temperature was believed to be a material release in caloric theory
1643	E. Torricelli (Italy)	Torricellian vacuum was made with a mercurial column and existence of the atmospheric pressure was found
1660	R. Boyle (England)	The pressure and volume of a gas are in inverse proportion to each other at a fixed temperature (Boyle's law)
1669	J.J. Becher (Germany)	The basis of the caloric theory is proposed
1701	S.I. Newton (England)	A law of Newtonian cooling in heat transfer was announced
1724	D.G. Fahrenheit (Germany)	Supercooling phenomenon of water was found. The boiling point of water changes with pressure
1742	A. Celsius (Sweden)	The centigrade scale was proposed in which the temperature between the freezing point of water and its boiling point is divided into 100 equal parts
1763	J. Black (England)	Specific heat and the concept of latent heat were defined.
1774	A.-L. Lavoisier (France)	Burning is explained as a "combination with oxygen"
1798	C. Rumford (England)	It was shown that heat occurs by work, and the caloric theory was overturned
1800s		Practical use of a heat engine was developed at this time. It was suggested that heat is one form of energy
1822	J.B.J. Fourier (France)	The heat conduction phenomenon was studied by the "analytic theory of heat" and Fourier's law was announced
1824	N.L.S. Carnot (France)	Integrity of the reversible cycle was indicated from "consideration about the power of the heat" (Carnot cycle)

(continued)

Table 2.1 (continued)

Year	People (country)	Research
1843	J.P. Joule (England)	That the energy of 10 J is needed to lift a 1 kg of a substance by 1 m was clarified
1843	B.P.E. Clapeyron (France)	Thermal equilibrium and the leading role of thermodynamic theory were investigated
1850	R.J.E. Clausius (Germany)	The basis for the second law of thermodynamics was established
1859	G.R. Kirchhoff (Germany)	Kirchhoff's principle on heat radiation of a black body was reported
1859	J.C. Maxwell (England)	The basis for the Kinetic theory of gases was defined
1865	R.J.E. Clausius (Germany)	Entropy was defined
1875	J.W. Gibbs (USA)	The groundwork for chemical thermodynamics was laid from Gibbs' free energy
1875	J.C. Maxwell (England)	"Maxwell's devil" was imagined from a study of molecular theory of the specific heat, and irreversible progress of thermal phenomenon was defined
1877	L. Boltzmann (Austria)	A law of thermodynamics was dealt with statistically, and a relation between the entropy and state probability was found
1884	L. Boltzmann (Austria)	Stefan–Boltzmann's law of radiation about heat radiation was advocated

substance at the lower temperature. The units used to describe heat are **calories** or **Joules**. Heat transfer through conduction can occur by contact between the source and the destination body; it can also occur by radiation between remote bodies, conduction and radiation through a thick solid wall, way of an intermediate fluid body (as in convective circulation), or a combination of these [11]. Because heat refers to a quantity of energy transferred between two bodies, it is not a function of the state of either of the bodies, in contrast to temperature and internal energy. Instead, according to the first law of thermodynamics, heat exchanged during some process contributes to the change in the internal energy, and the amount of heat can be quantified by the equivalent amount of work that would bring about the same change [12].

Too often people erroneously use the word heat and temperature interchangeably to mean the same thing. Specifically, the word temperature refers to a measure of how hot or how cold a substance is. In other words, temperature refers to the average kinetic energy per molecule of a substance; it is measured in degrees Celsius (C), degrees Fahrenheit (F) or, in scientific articles, temperature is given in degrees Kelvin (K). Hence, simply stated, temperature is how hot or cold an object is, while heat is the energy that flows from the hotter substrate to the cooler substrate [13].

When energy is added to a substance by heat and/or by doing work on it, then the added energy is saved in the substance such that elements and atomic motions become active and *the internal energy* increases. The internal energy decreases when the energy goes out of a substance through heat loss and/or by doing work, whereas temperature is regarded as the index of changes in the quantity of energy of a substance. Therefore, the temperature of a substance reflects the linear measure that shows macroscopically the state of the thermal balance and microscopically shows the movement inside the substance and the quantity of state of the oscillation energy. In classical mechanics, the temperature at which material (atomic) motion stops completely is the absolute zero point (−273 °C), i.e., 0 K or absolute zero.

Temperature increases typically originate by applying heat to a substance. On the other hand, it is also possible to do work on a system and increase its temperature (e.g., through friction). In this sense, ***heat*** and ***work*** are often confused but can be distinguished by examining an example. When energy is added to a system to cause movement (work) of all the system's components, it suggests movement of the energy (Fig. 2.3a) so that it is possible to regard kinetic energy as the uniform motion of the internal energy. Also, when heat energy is added to a system, the constituents move randomly while undergoing vibrations and rotations. Thus, the energy will move from the high temperature side to the cold side (Fig. 2.3b). As such, heat can be regarded as movement of this random energy. In the case of *microwave heating*, heat occurs via a kinetic energy-like action—even though no heat energy, but microwave energy is added to the system. In the latter case, heat is created that spreads throughout the system.

Heat can be generated from various energy sources. The various interconnections amongst these sources are illustrated in Fig. 2.4 with the principal focus on heat energy (thermal energy). The most common heating method used in pre-historic times was fire generated by a sudden chemical reaction with oxygen, which results in a display of a flame and the generation of heat and light. However, the burning of important natural resources to generate heat energy and exhaust gases (e.g., the greenhouse gas CO_2) does not seem an economic source of heat. On the other hand, friction, which is a form of mechanical energy that generates heat, has been used for centuries.

To express quantitatively the amount of heat produced or applied to a system, the unit ***calorie*** was introduced to represent bringing 1.0 g of water from 14.5 to 15.5 °C; one calorie is equivalent to 4.186 J, thanks in part to James Prescott Joule (England) who had determined the amount of mechanical work needed to produce

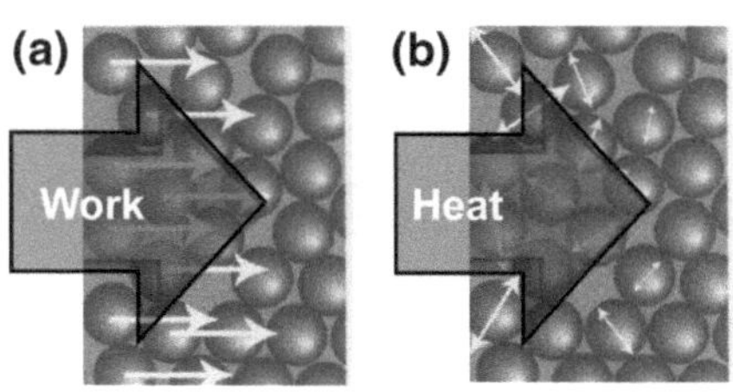

Fig. 2.3 Images of energy transfer to a substance by (**a**) doing work on the system and (**b**) applying heat to the system

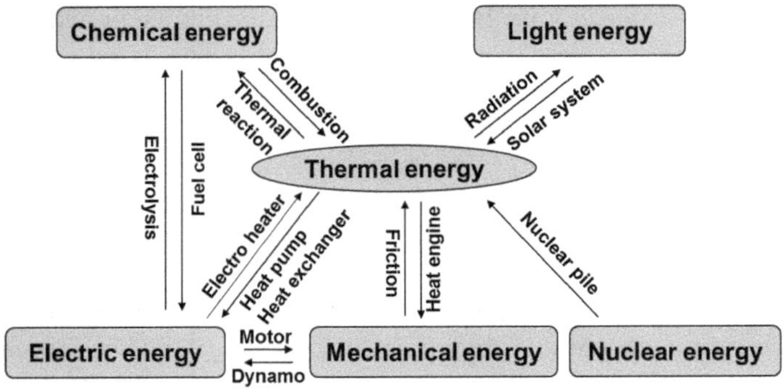

Fig. 2.4 Illustration of heat sources to connect with heat, i.e., thermal energy

an equivalent amount of heat, now referred to as *the mechanical equivalent of heat*. However, friction is not a suitable source of heat either because the absolute amount of energy is small. Even electrical energy generates heat. Heat by impedance, *i.e.,* electric current (Joule heat), was also proposed by Joule; this heat source is used in many daily activities. Heat can also be used to produce chemical energy. Finally, heat can be classified in different ways: as heat of vaporization, aggregation, fusion, solidification, and sublimation in various chemical reactions. These various forms of heat are often indicated in thermochemical reactions.

2.4 Thermodynamics

Thermodynamics is the branch of science that is concerned with *heat* and *temperature* and their relation to *energy* and *work*. The behavior of these four quantities is governed by the laws of thermodynamics (see Table 2.2) irrespective of the composition or specific properties of the system [14]. Thermodynamics applies to a wide variety of topics in science and engineering, especially in physical chemistry, chemical engineering, and mechanical engineering. Historically, thermodynamics developed out of a desire to increase the efficiency of the early steam engines

Table 2.2 The physical quantities and the contribution of heat to the laws of thermodynamics

Thermodynamic equilibrium	Physical quantity	Contribution to the heat
Zeroth law	Temperature (T)	Refers to two objects in thermal equilibrium
First law	Internal energy (U)	Explains the relationship between heat and work
Second law	Entropy (S)	This parameter could decide the direction of movement of a change of state

(particularly) through the work of the French physicist Nicolas Léonard Sadi Carnot (1824) who believed that engine efficiency was the key that could help France win the Napoleonic Wars [15]. Lord Kelvin was the first to formulate a concise definition of thermodynamics in 1854 when he stated that *thermodynamics is the subject of the relation of heat to forces acting between contiguous parts of bodies, and the relation of heat to electrical agency* [16].

A temperature gradient between two locations causes the flow of heat along a (thermally) conducting path between those two locations as long as the temperature difference is maintained. This flow of heat continues until the two objects reach the same temperature, i.e., the two objects are in thermal equilibrium at which point the flow of heat appears to no longer take place macroscopically. This principle is sometimes referred to as the Zeroth Law of thermodynamics because the First and Second Laws of thermodynamics had already been articulated. This Zeroth Law governs all objects—this is the tendency toward thermal equilibrium.

The First Law of thermodynamics relates the various forms of kinetic and potential energy in a system to the work that a system can perform and to the transfer of heat. This law is sometimes taken as the definition of internal energy and introduces an additional state variable: ***Enthalpy***, a measure of the total energy in a thermodynamic system. The First Law of thermodynamics allows for many possible states of a system to exist.

The Second Law of thermodynamics states that the total entropy (also taken a measure of the disorder in a macroscopic system) of an isolated system always increases over time or remains constant in ideal cases where the system is in a steady state or undergoing a reversible process. The increase in entropy accounts for the irreversibility of natural processes and the asymmetry between the future and the past. Historically, the Second Law was an empirical finding that was accepted as an axiom of thermodynamic theory and has been expressed in many ways. As noted above, its first formulation is credited to Sadi Carnot (France 1824) who showed that there is an upper limit to the efficiency of conversion of heat to work in a heat engine.

2.5 Heat Transfer

The rate of heat transfer is dependent on the temperatures of the systems and the properties of the intervening medium through which the heat is transferred. The three fundamental modes of heat transfer are (i) heat transfer via conduction, (ii) heat transfer via convection, and (iii) heat transfer via radiation (Fig. 2.5).

Heat transfer via conduction: Conduction is the transfer of heat between substances that are in direct contact with each other. The better the heat conductor is, the more rapidly heat will be transferred. Metals are good conductors of heat. Conduction occurs when a substance is heated, following which the atoms or molecules gain more energy and vibrate more. These atoms or molecules bump into nearby atoms or molecules and transfer some of their vibration energy to them. This continues and passes the energy from the hot end (or hot

Fig. 2.5 Cartoons illustrating the three fundamental modes of heat transfer. HT: high temperature, LT: low temperature

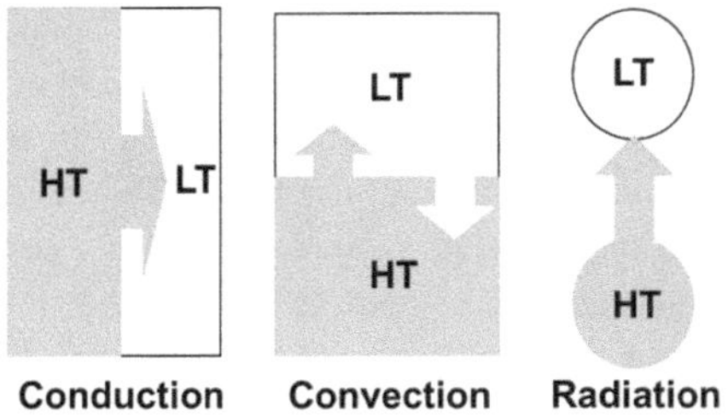

object) down to the cold end of the substance (or colder object). The temperatures of the two sides (or the two objects) become the same and soon reach thermal equilibrium.

Heat transfer via convection: Thermal energy is transferred from hot places to cold places by convection. Convection occurs when warmer areas of a liquid or gas rise to cooler areas in the liquid or gas. The cooler liquid or gas then takes the place of the warmer areas. This results in a continuous circulation pattern. Water boiling in a pan is a good example of this heat convection. Another good example of convection is in the atmosphere. The Sun warms the Earth's surface, following which the warm air moves upward and the cool air moves in. Convective heat transfer is typically taken as *free or natural convection heat transfer* that happens naturally, while *forced convection heat transfer* makes water flow by force.

Radiative heat transfer: Radiation is a method of heat transfer that does not rely upon any contact between the heat source and the heated object, as is the case with conduction and convection. Heat can be transmitted through empty space by thermal radiation, often called *Infrared radiation*, and is a type of electromagnetic radiation. No mass is exchanged and no medium is required for the radiative process. Examples of radiation are the heat transmitted from the sun through the vacuum of space, or the heat released from the filament of a light bulb. Some of the radiative energy that reaches a cold object is absorbed, some of it is reflected, and some is transmitted through the object. Such a phenomenon is analogous to microwave heating: the incident microwave energy absorbed by the substance is partially reflected and through penetration is partially transmitted (see Sect. 4.1).

2.6 Coffee Break 2: Background on the Relationship Between Microwaves and Foods

It was well known that radio waves could heat dielectric materials. The use of dielectric heating in industrial and medical contexts was common. The idea of heating food with radio waves was not new, however. Bell Laboratories, General Electric (GE), and Radio Corporation of America (RCA) had all been working on variations of the technology for some time. Indeed, at the 1933 World's Fair in

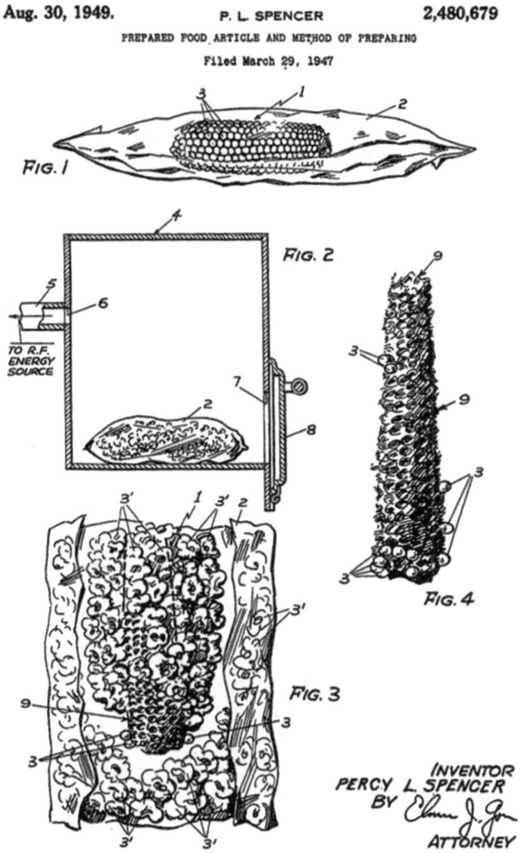

Illustration: United States Patent and Trademark Office

A 1947 patent application by Raytheon's Percy Spencer described the use of microwaves for heating foods, including popcorn.

Fig. 2.6 Illustration in the use of microwaves for heating foods, including popcorn. Reproduced with permission from Ref. [18]. Copyright 1947 by the US Patent Office

Chicago, Westinghouse demonstrated a 10-kilowatt shortwave radio transmitter that cooked steaks and potatoes between two metal plates [17]. But nothing came of these culinary adventures. Raytheon engineer Percy Spencer took it further. A 1958 article in Reader's Digest described Spencer's accidental discovery that microwaves could heat food quickly. He sent out for a package of corn grains. When he held it near a magnetron, popcorn exploded all over the lab. Next morning, he brought in a kettle, cut a hole on the side, and put an uncooked egg (in its shell) into the pot. Then, he moved a magnetron against the hole and turned on the power. A skeptical engineer peeked over the top of the pot just in time to catch a face full of cooked egg. The reason? The yolk cooked faster than the outside, causing the

Fig. 2.7 Pictures of the first commercial Raytheon microwave oven (Radarange): 340 kg, 1.5 m tall, cost approximately $ 5000. **(a)** Oven front, **(b)** inside the oven, **(c)** use at a restaurant. Reproduced with permission from Ref. [19] for **(a)** and Ref. [20] for **(b)** and **(c)**. Copyright 2013 and 2008 by the International Microwave Power Institute and Wicked Local

egg to burst. In 1946, Spencer filed for patents on the use of microwaves for cooking food (e.g., Fig. 2.6). One of his patents even illustrated the popping of popcorn, cob, and all [18].

Legends exist about a serendipitous discovery of microwave cooking by Percy Spencer but all remember the discovery as a gradual process involving both chance and the deliberate efforts by many individuals, *e.g.*, feelings of warmth near radiating tubes, experimenting with popcorn, etc. all of which led to the development of Raytheon's first microwave ovens. Still, it was Percy Spencer's discovery and participation that were key in convincing the company into exploiting this new cooking technology.

Despite the potential for near-instant snack foods, the first commercial Radarange microwave ovens unveiled by Raytheon in 1946 were intended for use in restaurants and for reheating meals on trains. Figure 2.7 shows the first Radarange models. They were massive and expensive appliances built around two 800-Watt magnetron tubes that had to be continuously water cooled. By 1955, Raytheon had begun licensing its microwave technology, and the first microwave oven designed for consumers went on sale at Tappan (USA). The Tappan RL-1 was wall mounted; it combined gas cooking with microwave assistance and cost US $1295 (almost $11,000 today), which put it outside the reach of most people. As microwave ovens became more common throughout the 1970s, concerns arose about the effects of microwave radiation on humans. As the New York Times described in 1974 [21]: If Marianne Leonard had to surrender just one of the appliances in her stainless steel and butcher block kitchen, which would it be? Her microwave oven? *That's the one thing I could do without,* the Westport, Conn., woman said. *Everything cooks so darn fast. You open the door and it's not done. You leave it in another 30 s and it's hard as a rock. I love it* Phyllis Brodsky said of the microwave oven in her sleek, all-white kitchen in Croton-on-Hudson, N. Y. *It's wonderful for heating funny things like a second cup of coffee. It's great for cooking a hot dog in 30 s.*

References

1. https://www.britannica.com/biography/Joseph-Black
2. http://galileo.phys.virginia.edu/classes/152.mf1i.spring02/What%20is%20Heat.htm
3. J.P. Joule, On the changes of temperature produced by the rarefaction and condensation of air. Proc. Royal Soc. London **5**, 517–518 (1844)
4. http://voer.edu.vn/m/heat/750462cf
5. https://en.wikipedia.org/wiki/William_Thomson,_1st_Baron_Kelvin
6. W. Thomson, On an absolute thermometric scale founded on carnot's theory of the motive power of heat, and calculated from regnault's observations. Math. Phys. Pap. **1**, 100–106 (1848)
7. https://en.wikipedia.org/wiki/William_Thomson,_1st_Baron_Kelvin
8. https://archive.org/details/theoryheat06maxwgoog
9. https://en.wikipedia.org/wiki/Nicolas_L%C3%A9onard_Sadi_Carnot#Reflections_on_the_Motive_Power_of_Fire
10. S. Carnot, *Reflections on the Motive Power of Heat*, ed. by R.H. Thurston (editor and translator) (J. Wiley & Sons, New York, 1890), p. 68
11. S. Chandrasekhar, *Hydrodynamic and Hydromagnetic Stability* (Oxford University Press, UK, 1961)
12. https://en.wikipedia.org/wiki/Heat#cite_note-2

13. http://study.com/academy/lesson/what-is-heat-definition-lesson-quiz.html
14. https://en.wikipedia.org/wiki/Thermodynamics
15. Rudolph Clausius, *On the Motive Power of Heat, and on the Laws Which can be Deduced from it for the Theory of Heat* (Poggendorff's Annalen der Physik, LXXIX (Dover Reprint), 1850), p. 1850
16. William Thomson, *Mathematical and Physical Papers*. ed. by C.J. Clay, M.A. & Son, vol. 1 (Cambridge University Press, UK, 1882), p. 232
17. http://spectrum.ieee.org/geek-life/history/a-brief-history-of-the-microwave-oven. (Accessed Mar 2017)
18. Spencer, P.L., Prepared food article and method of preparing, US patent 2,480,679 assignors to Raytheon Manufacturing Company, 1948
19. J. Osepchuk, Microwave ovens: an overview: 2013, Proceedings of IMPI Symposium, Rhode Island, USA
20. A.R. Earls, R.E. Edwards, *Raytheon Company: The First Sixty Years* (Images of America), Wicked Local, NY, USA, (2008)
21. http://www.nytimes.com/1974/03/19/archives/microwave-ovens-amid-some-doubts-their-popularity-grows.html?_r=1

Chapter 3
Electromagnetic Fields and Electromagnetic Waves

Abstract Microwaves are not a simple heat source. Substances are heated directly by electromagnetic waves at energies commensurate with microwaves. Therefore, an understanding of the nature of these electromagnetic waves is the dominating discussion on microwave heating. Moreover, the coffee break continues the discussion regarding relationships between microwaves and foods, and the history of microwave cooking ovens. The chapter explains the basics of electromagnetic waves and how these electromagnetic waves are being used. Furthermore, the characteristics of microwaves in electromagnetic waves are clarified. The coffee break describes the use of microwaves at various frequencies.

Keywords Electromagnetic waves · Electric field · Magnetic field
Radio waves · Maxwell's equations · Shortening of wavelength
Atomic species · Molecular species · Molecular assemblies
2.45-GHz microwaves · 915-MHz microwaves

3.1 The Nature of Electromagnetic Fields and Electromagnetic Waves

An electromagnetic field is a physical field produced by electrically charged objects [1] that affects the behavior of charged objects in the vicinity of the field. It extends indefinitely throughout space, describes the electromagnetic interaction, and is one of the four fundamental forces of nature [2]. The electromagnetic field comprises a combination of the electric field and the magnetic field, the former being produced by stationary charges, while the latter is generated by moving charges (currents).

By comparison, an electromagnetic wave is a wave of electromagnetic energy that is radiated by an accelerating or oscillating electric charge; it propagates through vacuum or a material medium as a periodic disturbance of the electromagnetic field at frequencies within the electromagnetic spectrum (Fig. 3.1).

Electromagnetic waves consist of both an electric and a magnetic component, and originate from vibrations of charged particles (e.g., electrons) as a result of various

S. Horikoshi et al., *Microwave Chemical and Materials Processing*,
https://doi.org/10.1007/978-981-10-6466-1_3

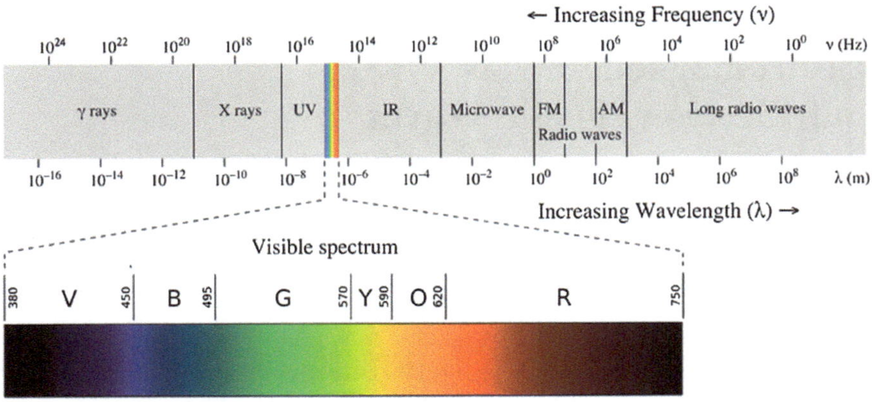

Fig. 3.1 Electromagnetic spectrum illustrating the eight components, which make up the spectrum together with their respective frequencies/wavelengths. Reproduced from Ref. [3]

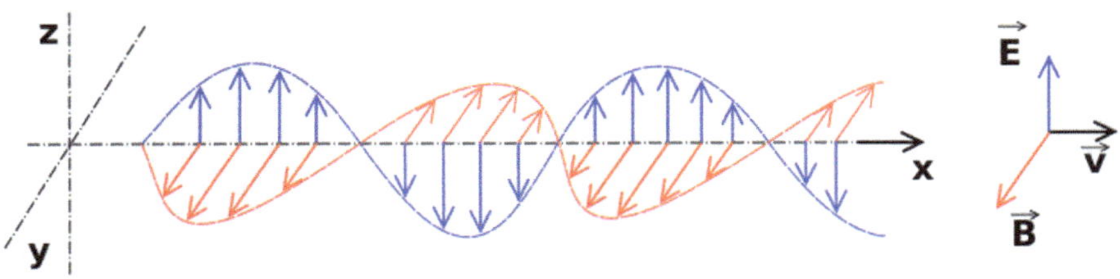

Fig. 3.2 Illustration of electromagnetic waves that compose electromagnetic radiation viewed as self-propagating transverse oscillating waves of electric and magnetic fields. The electric field (E) is in a vertical plane and the magnetic field (B) in a horizontal plane, i.e., at right angles to each other. Reproduced from Ref. [4]

forces that act upon them. Such vibrations of charged particles emit energy known as electromagnetic radiation. The oscillations of electromagnetic waves are perpendicular to the directions of the wave's travel (**x**) (Fig. 3.2); more precisely, the electric component (**E**) oscillates in one plane, while the magnetic component (**B**) oscillates in a different plane at 90 degrees (orthogonally) to the former. In vacuum, these waves travel at the speed of light, i.e., at 3×10^8 m s^{-1} in air or vacuum.

3.2 History of Electromagnetic Waves

Times before Christ (B.C.): Magnetism and electrical phenomena were known long before the birth of Jesus Christ. For example, around the sixth century B.C., it was known that a stone in a magnesium area (magnet) could attract a stone containing iron. At around the same time, **Thales of Miletus** {ca. 624–546 B.C.; a pre-Socratic Greek/Phoenician philosopher, mathematician, and astronomer from Miletus in

Asia Minor (currently aka Milet in Turkey) and one of the Seven Sages of Greece} reported that when fur was rubbed with amber, dust attached itself onto the amber surface, i.e., amber was electrified by friction; at the time this surprised many people. Electricity and magnetism are invisible forces, and although magnetism was recognized implicitly, electricity was not familiar or used in practice until after the 1600s.

Recent times: In the sixteenth century, the English Doctor William Gilbert discovered the science of magnetism and coined the word *electricity*; his findings on the Earth's magnetic properties explained how a compass, invented centuries earlier, actually worked. He is considered the father of both electricity and magnetism as a result of his publication *De Magnete, Magneticisique Corporibus* (year 1600) in which he explained his theories of magnetism. Gilbert proposed that electricity and magnetism were distinct effects, both capable of causing attraction and repulsion of objects. Mariners of the time were aware that lightning could disrupt a compass needle; yet the link between lightning and electricity was not confirmed until Benjamin Franklin's experiments of 1752. One of the first to discover and publish a link between man-made electric current and magnetism was Romagnosi, who in 1802 noted that connecting a wire across a voltaic pile deflected a nearby compass needle. This effect did not become widely known until 1820 when Ørsted performed a similar experiment [5], which ultimately led Ampère to develop a theory of electromagnetism that set it on a solid mathematical foundation.

The direct interconnection between electricity and magnetism was explained for the first time by Hans Christian Ørsted (Denmark) during a lecture in April 1820; he noticed a compass needle being deflected from magnetic north when an electric current from a battery was switched On and Off. His initial interpretation was that magnetic effects radiate from all sides of a wire carrying an electric current, as do light and heat radiating away from a source. Months later, Ørsted began more intensive investigations and soon thereafter published his findings in which he showed that an electric current produces a circular magnetic field as it flows through a wire (see Fig. 3.5a). This discovery was not due to mere chance, as Ørsted had been looking for a relation between electricity and magnetism for several years.

After more than 40 years since the lecture by Ørsted, a major advance was the development of a theory of electromagnetism (now known as classical electromagnetism) by various physicists over the course of the nineteenth century. It culminated in the work of James Clerk Maxwell (England; Fig. 3.3) who unified the preceding developments into a single theory and discovered the electromagnetic nature of light. Maxwell predicted that the wave motion of electricity and magnetism are propagated in a vacuum. Because the speed of propagation of an electromagnetic wave was equal to the speed of light, he suggested that light is also an electromagnetic wave. The existence of the electromagnetic wave was subsequently confirmed experimentally in 1888 by Heinrich Rudolf Hertz of Germany (Fig. 3.3). In classical electromagnetism, the behavior of the electromagnetic field is described by a set of equations now known as Maxwell's equations (see below).

Although the nature of light had been examined in the centuries before the coming of Christ, radio waves were only actively developed in the first decade of

Fig. 3.3 Photographs of James Clerk Maxwell (left) and Heinrich Rudolf Hertz (right). Reproduced from Ref. [6]

the twentieth century, a result of studies performed on wireless electrical transmission by Nikola Tesla (Austria) in 1904 (Fig. 3.4). After learning of Hertz's experiments with electromagnetic waves, Tesla's ideas for a World Wireless system originated from experiments that began in the early 1890s using induction coil transformers and spark gaps [7]. While repeating Hertz's experiments, Tesla improved Hertz's wireless transmitter by developing various alternator apparatuses and his own high-tension transformer, now known as the Tesla coil [8]. Tesla's primary interest in wireless phenomena was a power distribution system, by investigating early wireless lighting [9] as he demonstrated in the early 1890s in his various lectures in England and France that comprised *Experiments with Alternate Currents of High Potential and High Frequency*. Tesla's demonstration of wireless lighting [10] was presented in Philadelphia (Pennsylvania) and St. Louis (Missouri) in 1893, [11] and also included lighting Geissler tubes wirelessly [12]. Tesla filed numerous patents associated with transformer design, transmission methods, tuning circuits, and methods of signaling. In one of these [13], he described a plan to have some thirty Wardenclyffe-style telecommunications stations (referred to as Tesla Towers) positioned around the world, to be tied to existing telephone and telegraph systems (Fig. 3.4). Tesla theorized that if he injected an electric current into the Earth at just the right frequency, he could harness the Earth's own electrical charge and cause it to resonate at a frequency that would be amplified in *standing waves*, which could then be tapped anywhere on the planet to run devices or, through modulation, carry a signal [14]. His system was based more on nineteenth-century ideas of electrical conduction and telegraphy than on newer theories of air-borne electromagnetic waves, with the electrical charge being conducted through the ground and being returned through the air.

Fig. 3.4 The Wardenclyffe Power Plant prototype built in Shoreham, New York in 1901–1902, intended by Nikola Tesla to be a "World Wireless" telecommunications facility. Reproduced from Ref. [15]

3.3 The Nature of Microwaves

Microwaves and radio waves consist of electric and magnetic fields that propagate through space, changing in strength with time, and existing at right angles to each other (Fig. 3.2). A radio wave can be considered a physical object in a vacuum that possesses energy, momentum, and angular momentum. The common characteristics of electromagnetic waves (light microwaves and radio waves) are that (i) they have no mass, (ii) they are free to move at the speed of light, (iii) they cannot collide with each other, (iv) they possess significant energy, and (v) they are reflected from a metal surface. Electromagnetic waves are classified by their wavelength and frequency because they have the characteristics of wave motion (Table 1).

Microwave radiation is electromagnetic radiation spanning the frequency range 300 GHz–300 MHz (i.e., from a wavelength of 1 mm to 1 m). Overall, the types of electromagnetic radiation are broadly classified into the following: Gamma radiation, X-ray radiation, ultraviolet radiation, visible radiation, infrared radiation, terahertz radiation, microwave radiation, and radio waves (Fig. 3.1). Note that in the absence of a medium, a sound wave cannot be propagated, contrary to an electromagnetic wave that can be transmitted through vacuum.

A radio wave is a type of electromagnetic radiation that has wavelengths longer than infrared light. Radio waves have frequencies as high as 300 GHz to as low as 3 kHz; some definitions describe waves from 300 MHz to 300 GHz as microwaves. At 300 GHz, the corresponding wavelength is 1 mm, at 300 MHz it is 1 m, and at 3 kHz it is 100 km. Like all other electromagnetic waves, they travel at the speed of light in a vacuum. Naturally occurring radio waves are generated by lightning, or by astronomical objects. Moreover, an electromagnetic radio wave is a wave that is propagated by a change in spatial electric field and magnetic field. Accordingly, it is relevant to examine the nature of these fields.

Table 3.1 Types of electromagnetic radiation: names, frequencies, wavelengths, and energy content

Class			Frequency	Wavelength	Energy
Ionizing radiation	γ	Gamma rays	300 EHz	1 pm	1.24 meV
	HX	Hard X-rays	30 EHz	10 pm	124 keV
	SX	Soft X-rays	3 EHz	100 pm	12.4 keV
			300 PHz	1 nm	1.24 keV
	EUV	Extreme ultraviolet	30 PHz	10 nm	124 eV
Light	NUV	Near ultraviolet	3 PHz	100 nm	12.4 eV
	NIR	Near-infrared	300 THz	1 μm	1.24 eV
	MIR	Mid-infrared	30 THz	10 μm	124 meV
	FIR	Far infrared	3 THz	100 μm	12.4 meV
Microwaves and Radio waves	EHF	Extremely high frequency	300 GHz	1 mm	1.24 meV
	SHF	Super high frequency	30 GHz	1 cm	124 μeV
	UHF	Ultra high frequency	3 GHz	1 dm	12.4 μeV
	VHF	Very high frequency	300 MHz	1 m	1.24 μeV
	HF	High frequency	30 MHz	10 m	124 neV
	MF	Medium frequency	3 MHz	100 m	12.4 neV
	LF	Low frequency	300 kHz	1 km	1.24 neV
	VLF	Very low frequency	30 kHz	10 km	124 peV
	ULF	Ultra low frequency	3 kHz	100 km	12.4 peV
	SLF	Super low frequency https://en.wikipedia.org/wiki/Super_low_frequency	300 Hz	1 Mm	1.24 peV
			30 Hz	10 Mm	124 feV
	ELF	Extremely low frequency	3 Hz	100 Mm	12.4 feV

Reproduced from Ref. [16]

An electric field is a vector field that associates, to each point in space, the Coulomb force that would be experienced per unit of electric charge, i.e., by an infinitesimal test charge at that point [17]. Electric fields converge and diverge at electric charges and can be induced by time-varying magnetic fields. By comparison, a magnetic field is the magnetic effect of electric currents and magnetic materials; at any given point, it is specified by both a *direction* and a *magnitude* (or strength) and thus can be considered a vector field. As noted earlier, the electric field combines with the magnetic field to form the electromagnetic field. Electric and magnetic fields are orthogonal to each other.

3.4 Maxwell's Equations

Maxwell's equations (**I–IV**) describe how electric and magnetic fields are generated by charges, currents, and corresponding changes with each other. An important consequence of the equations is that fluctuating electric and magnetic fields are predicted to propagate at the speed of light. This electromagnetic radiation manifests itself in many ways from radio waves to light to X-rays to γ-rays—see the electromagnetic spectrum displayed in Fig. 3.1.

$\nabla \cdot D = \rho$ (the divergence of the electric displacement field (Maxwell's equation I)

$$\nabla \cdot B = 0 \qquad \text{(Maxwell's equation II)}$$

$$\nabla \times E = -\frac{\partial B}{\partial t} \qquad \text{(Maxwell's equation III)}$$

$$\nabla \times H = \frac{\partial D}{\partial t} + J \qquad \text{(Maxwell's equation IV)}$$

where $\boldsymbol{H}$ is the magnetic field (A m^{-1}); $\boldsymbol{E}$ denotes the electric field (V m^{-1}); $\boldsymbol{B}$ (= $\boldsymbol{\mu}$ $\boldsymbol{H}$) refers to the magnetic flux density (T); $\boldsymbol{D}$ (= $\boldsymbol{\varepsilon}$ $\boldsymbol{E}$) is the electric flux density (C m^{-2}); $\boldsymbol{J}$ (= $\boldsymbol{\sigma}$ $\boldsymbol{E}$) is the current density (A m^{-2}); $\boldsymbol{\mu}$ denotes the magnetic permeability (H m^{-1}); $\boldsymbol{\varepsilon}$ refers to the dielectric constant (F m^{-1}); $\boldsymbol{\sigma}$ denotes the conductivity (Ω^{-1} m^{-1}); and $\boldsymbol{\rho}$ is the electric charge density (C m^{-3}).

Equation I describes Gauss's law for electricity; it indicates how the electric flux out of any closed surface is proportional to the total charge enclosed within the surface—in applying Gauss's law to the electric field of a point charge, it can be shown that it is consistent with Coulomb's law.

Equation II represents Gauss's law for magnetism; it shows that the net magnetic flux out of any closed surface is zero, which amounts to a statement about the sources of the magnetic field. In the case of a magnetic dipole, in any closed surface, the magnetic flux directed inward toward the South Pole equals the flux outward from the North Pole, with the net flux being always zero for dipole sources.

Equation III is Faraday's electromagnetic induction law; it is a basic law of electromagnetism that predicts how a magnetic field will interact with an electric circuit to produce an electromotive force (EMF)—a phenomenon called electromagnetic induction (Fig. 3.5a), which is the fundamental operating principle that underlies transformers, inductors, and many types of electrical motors, generators, and solenoids.

Equation IV is Ampère's law, which indicates that the magnetic field in space around an electric current is proportional to the electric current that serves as its source, just as the electric field in space is proportional to the charge that serves as its source. Ampère's law also states that for any closed-loop path, the sum of the length elements times the magnetic field in the direction of the length element equals the permeability times the electric current enclosed in the loop. More simply

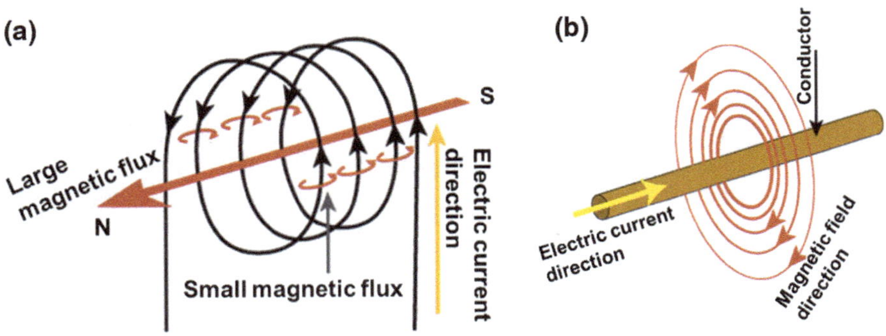

Fig. 3.5 (**a**) Faraday's electromagnetic induction law, (**b**) Enhanced Ampère's law

stated, in the case of a static electric field, the line integral of the magnetic field around a closed loop is proportional to the electric current flowing through the loop (Fig. 3.5b).

When a microwave is transmitted in air or vacuum, the traverse speed is equal to the speed of light; however, when passing through a dielectric substance, e.g., water, the speed is reduced and the wavelength of the microwave also changes; it decreases. For instance, the speed of light is ca. 3×10^8 m s^{-1} in air, but is slowed down to ca. 2.25×10^8 m s^{-1} in water. When microwaves (2.45 GHz = 12.24 cm wavelength) irradiate a water sample, the wavelength of the microwaves is reduced to 9.18 cm (Fig. 3.6). The speed of an electromagnetic wave ***c*** is defined by Maxwell's equations V, which demonstrates that the material permittivity $\boldsymbol{\varepsilon_o}$ and the magnetic permeability $\boldsymbol{\mu_o}$ both govern the speed of the electromagnetic wave. Note that the frequency of the microwaves does not change, only the wavelength changes as it travels through different materials; it returns to its free-space wavelength once the microwaves have passed through the substance. Note that the wavelength in air or vacuum is called the free-space wavelength (λ).

Fig. 3.6 Shortening of the wavelength (λ) of microwaves inside a dielectric substance (e.g., water)

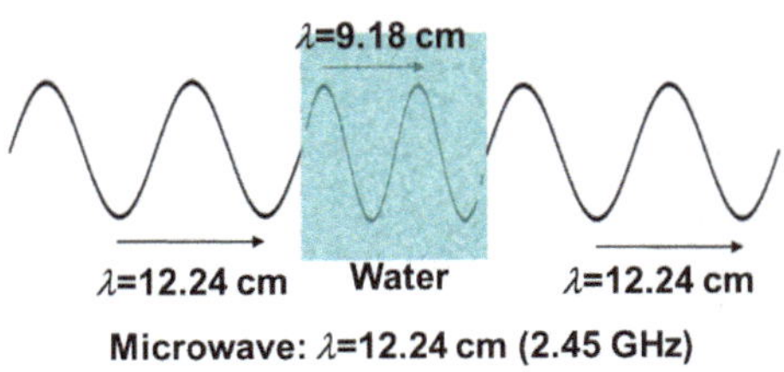

3.5 Microwaves as Electromagnetic Waves

3.5.1 *History of the Name "Microwave"*

A radio wave is a non-natural artificial electromagnetic wave occurring from low frequencies in the kHz scale (10^3 Hz; i.e., 1000 cycles/second or 1000 s^{-1}) to the MHz (10^6 Hz) and GHz scale (10^9 Hz). Upon further technological development, radio waves are now classified according to the frequency range: **UHF** (3 GHz–300 MHz), **VHF** (300–30 MHz), **HF** (30–3 MHz), **MF** (3 MHz to 300 kHz), **LF** (300–30 kHz), **VLF** (30–3 kHz), and **ULF** (3 kHz–300 Hz)—see Table 1. Subsequently, electromagnetic waves were named *microwaves* because the wavelength was very short, i.e., *micro*. There are technical applications of radio waves at the centimeter, millimeter, and micrometer wavelengths, with terms **SHF** and **EHF** referring to *centimeter* and *millimeter waves*, respectively. The *centimeter wave* is now more commonly referred to as a *microwave*.

3.5.2 *Differences in the Features of Communication by Light and Microwaves*

The main use of microwaves is in the communication field. The use of microwaves as a heat source in industry is said to be less than ca. 1% relative to the communication field. Both microwaves and optical means are used for communications; Table 2 summarizes the differences between optical communication and microwave communication. Because the wavelength of light is short compared with microwave wavelengths, much more information can be sent in a short time using the optical technology; however, this requires an optical fiber cable for the latter because of the poor material permeability of the light path. On the other hand, only a small volume of information can be sent via microwaves compared with light. Nonetheless, microwaves can be used for wireless communication such as **LAN** and cellular phones, among others, because of the high permeability of substances.

Table 2 Differences between optical communications and microwave communications

Subject	Optical communications	Microwave communications
Telecommunication medium	Optical fiber	Space
Practicable area	Area for construction of a cable	The Earth and the universe
Information transmission and amount and speed	High	Low
Communication partner	Specific	Unspecified

3.5.3 *Responses of Substances to Electromagnetic Waves*

The influence of electromagnetic waves on dielectric materials in the various ranges of the electromagnetic spectrum is summarized in Fig. 3.7 [18], which also shows the various analytical instruments and their usefulness in the ranges indicated.

Atomic species: If a dielectric is placed in an alternating electric field of electromagnetic waves, the positions of the atomic nucleus (protons) and the electrons that constitute the dielectric tend to follow the course of the electric field. The spatial relationship of a proton and an electron in an atom becomes distorted when subjected to the electric field (referred to as electronic polarization). Since the electric dipole that reflects the electric deviation (a proton and an electron) inside an atom is very small, a resonance phenomenon will occur with electromagnetic waves of short wavelengths such as, for example, X-rays and ultraviolet rays.

Molecular species: The distortion of the electric charge by electromagnetic waves also occurs in a molecule with an electrical dipole moment: for example, in a salt solution. In a solution of NaCl salt, composed of Na^+ and Cl^- ions, distortion of the structure arises when subjected to an alternating electric field (referred to as ionic polarization). Compared with an atom, the electric dipole moment of a molecule is large, so that resonance occurs in the infrared domain, i.e., at longer wavelengths. The inter-ionic distance and the types of ions involved dictate the magnitude of ionic polarization. Moreover, changes in bond lengths in an organic molecule occur through stretching and bending vibrations. These phenomena occur

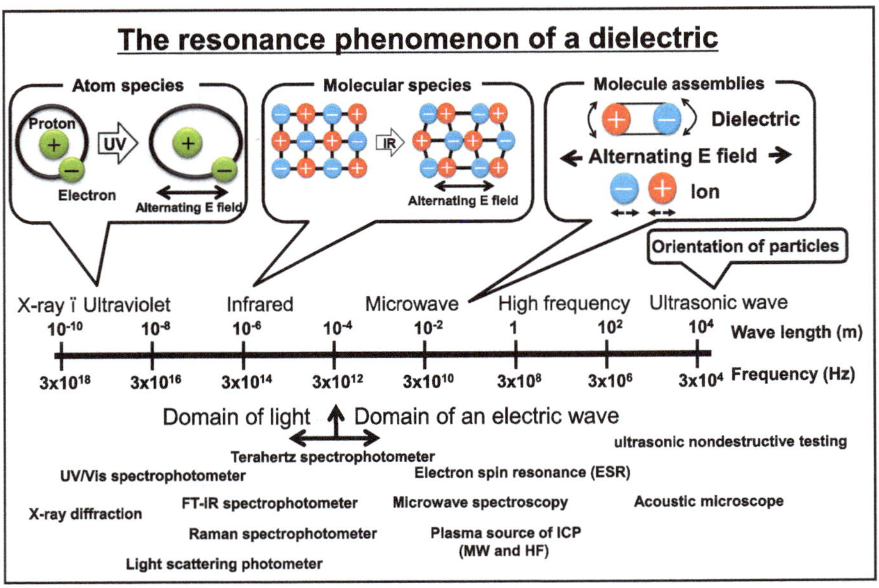

Fig. 3.7 Resonance of a dielectric to electromagnetic waves, and positioning of the analytical equipment. Reproduced from Ref. [18] Copyright 2013 by Wiley-VCH Verlag GmbH

upon absorbing the electromagnetic radiation in various wavelength regions and depend on the chemical structure of the species; they can be observed by means of an infrared absorption spectrophotometer.

Molecular assemblies: Orientation polarization of a dielectric possessing an electric dipole takes place upon interaction with the electromagnetic waves in the microwave range. Moreover, in the case of ions in a solution, Joule heating takes place by space charge polarization. For example, when microwave heating an electrolyte/water solution, dielectric heating and Joule heating occur simultaneously compared with microwaving pure water, and thus the exothermic efficiency becomes remarkably high. Accordingly, dielectric heating by orientation polarization of water and resistance heating by the Joule process are enhanced in electrolyte/water media.

3.6 Coffee Break 3: Frequencies Used for Food Heating

Microwaves are a form of electromagnetic energy, similar to gamma rays, X-rays, ultraviolet/visible light, and infrared radiation. Each is defined by specific wavelengths and frequencies and all are parts of the electromagnetic spectrum (Fig. 3.8). The heating of foods with microwaves uses only two frequencies 2450 MHz (= 2.45 GHz; wavelength = 12.2 cm) and 915 MHz (wavelength = 32.8 cm).

Why was the 2450 MHz frequency chosen for microwave oven magnetrons and who did it? In order to explain this, it is first necessary to define ISM (Industrial Scientific and Medical) bands or frequencies, which have been set aside for the exclusive use of scientific and medicinal purposes. When microwave cooking was

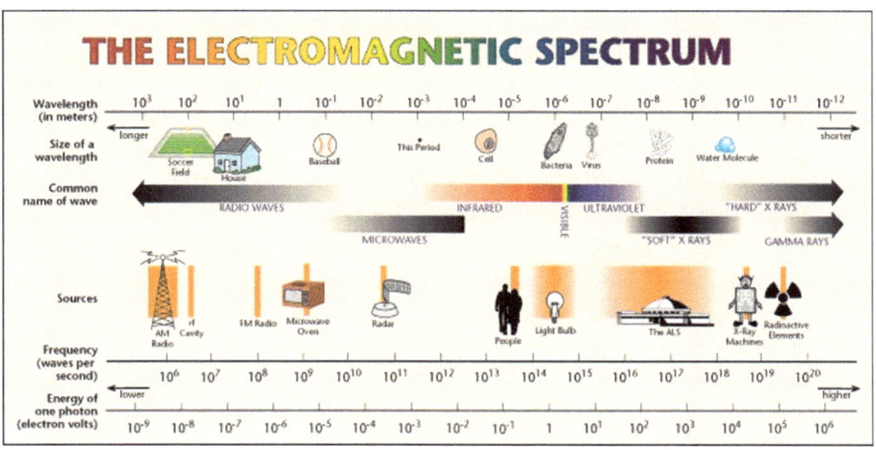

Fig. 3.8 The electromagnetic spectrum. Note that microwaves have very low photon energy and are "non-ionizing". Reproduced with permission from Ref. [19]. Copyright 2013 and 2008 by CISCO

being developed, two USA companies (Raytheon and General Electric) petitioned the USA's Federal Communications Commission (FCC) for the two respective microwave frequencies, 2450 and 915 MHz. Raytheon was developing a microwave cooking device to cook smaller foods such as hamburgers and frankfurters, whereas GE intended to incorporate microwave energy into its standing ranges to more quickly cook larger foods such as roasts and turkeys. Best cooking results were achieved when the wavelength of the microwaves was in the order of the size of the food: 2450 MHz for hamburgers and 915 MHz for turkeys. However, another factor that governs which frequency is best relates to the dimensions of the oven cavity (the metal box in which the food is placed), and for best results that should be 2–4 times the wavelength. Accordingly, the 915 MHz frequency was and is impractical for a countertop microwave oven—the oven would be too large!

The 915 MHz frequency is used aggressively in the United States for what reason? This frequency has been used for analog mobile phones in Japan since 1979 (downstream 860–885 MHz and upstream 915–940 MHz). In recent years, this frequency is usually used for cellular and smartphones; its use for heating is prohibited in Japan and in many other countries. In the United States, while 915 MHz is impractical for microwave ovens, it is used widely for industrial processing of foods and many other materials. The major advantage of this frequency is that its long wavelength allows much larger depth of penetration of microwave energy into materials, i.e., the microwave energy penetrates more deeply inside materials, making it possible to microwave-heat larger volume food and nonfood materials. Note that there is no exact correlation of wavelength to penetration depth since the material's dielectric properties also play an important role (dielectric properties are intrinsic properties of all materials that govern how well those materials heat at a particular frequency). This greater penetration depth, combined with the greater efficiency, higher power, and lower equipment cost at this frequency, makes the 915-MHz frequency preferable for industrial processing. In the United States, this frequency is commonly used for such food processing applications as the microwave cooking of bacon that may require 500–1000 kW of microwave power. Interestingly, this is not an ISM band in most of the rest of the world, but can be used in those places by providing suitable shielding to prevent microwave radiation leakage, including at the harmonic frequencies, to non-detectable limits. In the UK, there is an ISM band at 894 MHz, and 922 MHz in Australia used for industrial microwave processing. Worldwide there is also an ISM band at 5800 MHz (5.8 GHz) for which Panasonic manufactures the required magnetrons. Thus far, however, they have been used for research and there are no process applications to date.

If microwaves are a form of energy, not heat, then how hot do materials become inside microwave ovens? Since microwaves cause materials to heat, the longer the material is exposed to microwave energy, the hotter it will become. In effect, the only limit to the temperature is the destruction temperature of that material, making it possible to reach temperatures of 1000 °C and higher in a microwave oven!

Do all materials heat when exposed to microwaves? Yes, but some materials heat more easily than others, and some, such as borosilicate glass, heat very little. The reasons are complex, but essentially boil down to two major factors: (i) ***The Dielectric Properties*** of the material: intrinsic properties of every material that govern how well that material will absorb microwave energy and convert it into heat; there are three major material properties defined as the ***Dielectric Constant*** ε', the ***Dielectric Loss Factor*** ε'', and the ***Dissipation Factor*** tan δ; (ii) ***The Specific Heat Capacity*** of the material—essentially, how much energy is required to raise the temperature of the material a specific amount. For example, if material A has a specific heat capacity half as great as material B, then A may become twice as hot as material B when heated for the same amount of time in the same microwave oven.

References

1. Richard Feynman (1970). The Feynman Lectures on Physics, Vol 2, Addison Wesley Longman. ISBN 978-0-201-02115-8 (http://feynmanlectures.caltech.edu/II_toc.html)
2. https://en.wikipedia.org/wiki/Electromagnetic_field#cite_note-1
3. https://commons.wikimedia.org/wiki/File:EM_spectrumrevised.png
4. https://en.wikipedia.org/wiki/Electromagnetic_radiation
5. https://en.wikipedia.org/wiki/Electromagnetism#cite_note-Oersted-3
6. http://juanramonruizcategorias.blogspot.jp
7. J.A. O'Neill, *Prodigal Genius: The Life of Nikola Tesla* (Cosimo Inc., New York, NY, 2006), p. 86; M.J. Seifer, *Wizard: The Life and Times of Nikola Tesla* (Citadel Press, New York, NY, 1998), p. 1721
8. N. Tesla, *Alternating Electric Current Generator*, U.S. Patent 447,921A, 10 Mar 1891
9. Brian Regal, *Radio—The life story of a technology* (Greenwood Press, Westport Connecticut USA, 2005). Chapter 2
10. W. B. Carlson, *Tesla: Inventor of the Electrical Age*, (Princeton University Press, 2013), p. 132
11. N. Tesla, *On Light and Other High Frequency Phenomena*, Lecture delivered before the Franklin Institute, Philadelphia, February 1893, and before the National Electric Light Association, St. Louis, March 1893
12. https://en.wikipedia.org/wiki/Geissler_tube
13. N. Tesla, *Apparatus for transmitting electrical energy*, United States Patent 1119732 A, December 1, 2014; see also https://www.google.com/patents/US1119732
14. W. B. Carlson, *Tesla: Inventor of the Electrical Age* (Princeton University Press, 2013), pp. 210–211
15. https://en.wikipedia.org/wiki/World_Wireless_System
16. https://en.wikipedia.org/wiki/Electromagnetic_spectrum
17. http://feynmanlectures.caltech.edu/II_toc.html
18. S. Horikoshi, N. Serpone, in *Microwaves in Nanoparticle Synthesis: Fundamentals and Applications*, ed. by S. Horikoshi, N. Serpone, (Wiley-VCH Verlag GmbH, Weinheim, Germany, 2013), Chapter 3
19. https://learningnetwork.cisco.com/docs/DOC-31289

Chapter 4
Microwave Heating

Abstract Microwave heating is a characteristic heating phenomenon that cannot be imitated by other heating methods; it can lead to internal heating, selective heating, formation of hotspots, local heating, superheating, and nonuniform heating —the penetration depth of microwaves is discussed. As well, a comparison is made in the use of microwaves with regard to heating efficiency, temperature control, environmental impact, low heat conduction sample, specimen geometry, and location of heat with other heat sources (electric furnace, gas, steam, high-frequency, and infrared). If microwaves are used effectively, innovative heating can be executed by using microwaves. The chapter describes microwave heating mechanisms and microwave behavior toward liquids and solids, together with the characteristics of microwave heating against other heat sources and microwave proprietary heating phenomena based on specific cases. The coffee break talks about the engineering and style of microwave ovens.

Keywords Dipolar rotation · Dipolar polarization · Internal heating
Selective heating · Hotspots · Local heating · Superheating · Penetration depth
Nonuniform heating · Microwave leakages · Domestic microwave ovens

4.1 Types of Microwave Heating

4.1.1 Overview of Microwave Heating

People are exposed to various types of energies: for example, sunlight photons, wind power, and various types of noises (vibrational energy), among many others. Our attempts to use these energy sources led to the development of chemistry, engineering, hydrodynamics, thermodynamics, photochemistry, and others. It is relevant to ask at this point: what kind of energy are microwaves? Microwaves are a form of electromagnetic radiation that, as noted earlier, is electromagnetic radiation with wavelengths ranging from one meter to one millimeter that correspond to frequencies from 300 MHz to 300 GHz. This also includes both **UHF** and **EHF**

S. Horikoshi et al., *Microwave Chemical and Materials Processing*,
https://doi.org/10.1007/978-981-10-6466-1_4

(millimeter waves). Moreover, microwaves include the entire **SHF** band (3–30 GHz, or λ = 10–1 cm) at minimum, with **RF** engineering often restricting the range between 1 and 100 GHz (λ between 30 and 0.3 cm) [1]. Microwaves do not contain sufficient energy to chemically ionize or change substances, and so are examples of *non-ionizing radiation*. Note that the term *radiation* refers to energy radiating from a source and is not to be confused with radioactivity. Although low levels of microwave radiation have not been conclusively shown to have significant adverse biological effects, some studies have suggested that long-term exposure may have carcinogenic effects [2]. It is important to note, however, that exposure to any high-intensity radiation source can cause heating and burns, a property that cannot be exclusively attributed to microwaves only. Also, microwave exposure is not cumulative, i.e., a first exposure does not add to a second exposure, and so on, i.e., each exposure is separate, as opposed to *X*-rays that are cumulative.

Microwaves possess less energy than visible light, yet microwave energy can be converted into heat when substances absorb microwave radiation. In other words, just as a conventional heat source can drive chemical reactions, absorption of microwave energy can do the same because the reactants convert this energy into heat as a result of the presence of an electric field and a magnetic field. Although the principles underlying how this occurs are still not completely understood, microwaves are being used in various applications, e.g., as a heat source and as a means to transmit information—cell phones are an example. The utility of thermal energy (heat) was recognized long before people developed a theory to explain this type of energy. For example, the steam engine developed by Robert Stephenson in 1829 was used in the world's first passenger rail system (Fig. 4.1). The field of thermodynamics came well after this application of heat (in 1954) as a result of the work of Lord Kelvin. Next, microwave heating will be described and discussed.

Fig. 4.1 Image of an application of the steam engine by Stephenson [3]

4.1.2 Microwave Heating of Substrates in Solutions

Microwave heating is often referred to as dielectric heating, and to understand this designation we need to understand the interaction(s) between microwave radiation and liquid or solid substances. The principal components of microwaves are the electric and magnetic fields (see Fig. 3.2 in Chap. 3). The interaction between microwaves and substances subjected to microwave irradiation typically involves the electric field and the electric dipole moment of the constituents of the liquid or solid substances [4]. For example, water is a polar molecule with an electric dipole moment originating from the separated electric charges on the atoms that make up the water molecule (polarization)—see Fig. 4.2. However, if the liquid or solid consists of more than one molecular substrate, all those molecules are in constant continuous motion so that the electric dipole moment of one molecule may partially cancel the dipole of the other molecule.

Water exists as a dynamic cluster of molecules whose polarization is significantly diminished [Fig. 4.2]. However, when a microwave electric field (***E***) is applied to such a cluster, the structure of the water cluster becomes compromised and broken up allowing the electric dipole moments of the water molecules to align themselves parallel to the microwaves' electric field. Thus, a torque (*T*) is generated causing the molecules to rotate in a manner equivalent to a torque applied to a rotary shaft (Fig. 4.3). This movement, called orientation polarization, is one of the interactions between the microwaves and the materials and causes an alteration of the electric field.

The typical microwave frequency most often used in microwave chemistry is 2.45 GHz; that is, 2.45×10^9 Hz (or cycles per second), so that movement in the direction of the electric field is performed 2.45×10^9 times per second; as the field alternates, the motion occurs in the opposite direction in $1/(2.45 \times 10^9)$ seconds (i.e., at 4.1×10^{-10} s or 0.41 ns) later (Fig. 4.3). In general, molecular rotation occurs in materials containing polar molecules having an electrical dipole moment,

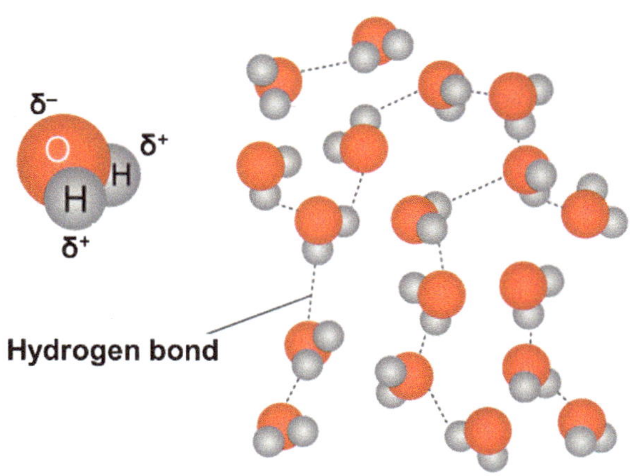

Fig. 4.2 Image of a water cluster illustrating the various interactions between hydrogen atoms with a partial positive electrical charge (δ^+) and the oxygen atoms with a partial negative electrical charge (δ^-)

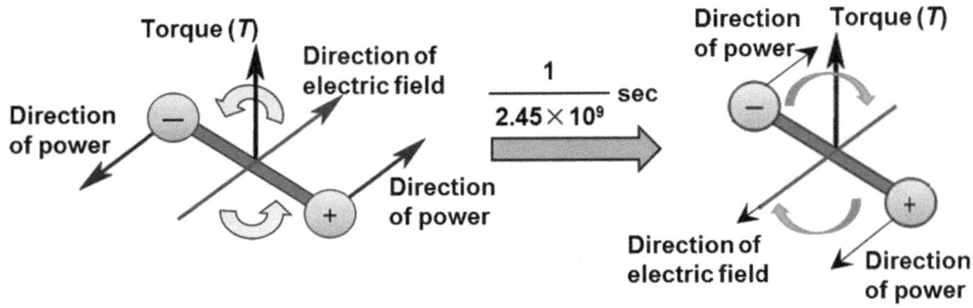

Fig. 4.3 Illustration of the rotary movement of a polar molecule caused by the electric field of 2.45-GHz microwaves

with the consequence that they will attempt to align themselves in an electromagnetic field. If the field is oscillating, as it is in an electromagnetic wave or in a rapidly oscillating electric field, these molecules rotate continuously by attempting alignment with the reversing field. This is known as dipolar rotation or dipolar polarization. As the field alternates, the molecules reverse direction. Rotating molecules push, pull, and collide with other molecules (through electrical forces), and distribute the resulting random kinetic energy to adjacent molecules and atoms in the material; ultimately, this appears as heat—referred to as dielectric heating. In actual fact, however, molecular rotation of the water molecule does not perfectly follow the oscillating frequency of the microwaves. In the presence of a highly polar molecule or material, tuning the microwave electric field is difficult and the calorific heat is also large, i.e., changing the molecules' rotational frequencies. Note that the heating efficiency cannot be understood solely by the actual polarity resulting from the elements that constitute the material. There is also something that changes in a cluster structure and/or nano-domain structure in liquid substances—for example, in alcohols [5].

4.1.3 Microwave Heating of a Solid Substance

A solid can be classified as either a conductor or as an insulator depending on the interaction between the oscillating electric field and the solid. When an electric field is applied to a conductor, such as a metal, movement of free electrons occurs and causes a surface current to form. Metals reflect microwaves, so no microwave heating ensues. Such phenomenon also occurs with highly electrically conductive liquid substances. An insulator does not transmit electricity because the electronic chain is held together by strong covalent bonds. Nonetheless, when the elements or unit cell of a solid substance (the domain) possess polar properties, polarization then forms the microwaves' electric field. This is called a dielectric property and the insulator is then called a dielectric.

A solid is different from a liquid substance as there is almost no fluid motion of the constituent atoms or constituent molecules in the former. When the temporal change of the electric field cannot be absorbed because of a gap between the locations of atoms (lattice points) and/or because of a structural distortion, the material is then said to be exothermic. Therefore, a dielectric substance possessing very low polarization does not interact well with the microwave radiation, but microwaves do nonetheless penetrate such dielectrics. In different forms of the same substance (e.g., liquid water, steam, ice), microwave absorption properties of such a dielectric are governed by the molecular density. For example, water can be heated by microwaves, whereas water vapor cannot be heated because the molecular density of water vapor is far smaller than liquid water. For this reason, molecules in water vapor cannot follow the microwaves' electric field. Approximately 70% of water molecules in liquid water are hydrogen bonded to each other into nano-clusters and, on absorption of microwaves, these water molecule clusters cannot follow the microwaves' alternating electric field unless these clusters are disrupted by the radiation, while the remaining molecules do accumulate the microwave energy and convert it into heat. By contrast, ice possesses a strongly hydrogen bonded regular structure and so each water molecule in ice will not be affected by the microwaves' oscillating electric field as the molecules are held rigidly in the ice structure. A much higher electromagnetic frequency is required to affect this tightly held structure: for instance, infrared radiation. For solids that also include magnetic substrates and ionic substrates, absorption of microwaves causes these substrates to be heated by Joules heating and/or by the microwaves' magnetic field (see Sect. 8.4.3).

The interactions between various properties of a substance and the microwaves are summarized in Fig. 4.4. A substance that absorbs microwave energy is heated by dielectric loss, magnetic loss, and Joule loss. On the other hand, microwaves that

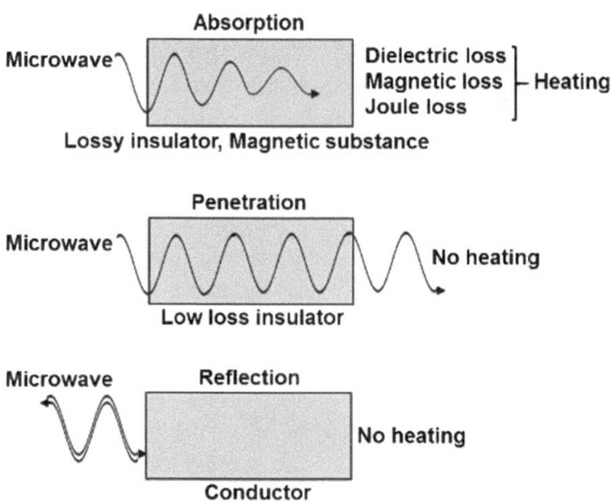

Fig. 4.4 Illustrations of the interactions between microwaves and the various materials (absorption, penetration and reflection)

penetrate an insulator are not absorbed and there is no generation of heat. A conductor also does not generate heat since the microwaves are reflected by such materials.

4.1.4 Difference(s) Between Microwave Heating and Conventional Heating

What is the difference between microwave heating and conventional heating? Conventional heating depends on the conduction of heat from the surface to the interior, i.e., conductive heat transfer, often a slow process. The efficiency of microwave heating of a sample is governed by the absorption and loss of microwave energy to the interior of the substance. Microwave heating occurs by direct transfer of energy, thereby heating at the molecular scale and at the nano-cluster scale without depending on heat transfer. Chemical and/or materials processing by conventional heating cannot be compared directly with microwave heating. The characteristic merits of microwave heating are noted below.

Internal heating: The rate of heat conduction by conventional heat transfer (e.g., with an oil bath) into dielectrics (insulators) is slow, so that a relatively long time is needed to heat the inner core of dielectrics. Microwave heating is a more effective and efficient internal heating method, contrary to conventional heating which also heats the walls of the reactor through convection or conduction heating (see Sect. 4.2).

Selective heating: Microwave heating efficiency is governed by the absorption and loss of microwave energy by a substance, i.e., as microwave energy heats a material, this energy is lost from the material causing it to heat. When microwave energy is applied to a heterogeneous sample with different properties, the temperature of each substance in the sample will be different. It is also possible to heat only a part of the sample (see Sect. 4.3).

Energy saving: As an internal heat source, microwave absorption can heat target compounds without heating the surroundings, thereby saving time and energy. It can also heat thin objects throughout their volume instead of near its outer surface, leading to more uniform heating. This also prevents case hardening or overheating of the surface, which often occurs with conventional heating.

Low environmental impact: Microwave heating alone does not use fossil fuels, although they may be used to generate the microwave energy. However, microwave heating may be combined with conventional heating methods, especially in industry, Nonetheless, it is necessary to understand some of the characteristics of microwave heating so as to be able to control microwave heating and avoid leaving an environmental footprint.

Discharge phenomenon: Microwave irradiation of metals or highly conducting substances often leads to the generation of an electrical discharge at the sample surface, producing arcs or sparks, or burning of the surface.

Penetration depth of microwaves: Material heating occurs by absorption of microwave energy through the surface of the material, and loss, i.e., use of that microwave energy by the material. The penetration depth of microwaves into the sample limits the extent of microwave heating.

Nonuniform heating: It is not possible to heat a sample uniformly through internal heating and selective heating. To assess the distribution of temperature in a sample, it is important that precise temperature measurements be done, a task often difficult to achieve, and which may require special temperature measurement equipment such as fiber optic thermometers.

Microwave leakages: It is necessary to put a sample in a metallic applicator because microwaves are contained within metals by a radio-type action.

4.1.5 Features of Microwave Heating Relative to Other Heating Methods

Some of the important factors that industry considers in the operation of microwave equipment as a heat source are the operating costs, energy conservation, and operating personnel requirements. Table 4.1 compares microwave heating with other heating sources such as the electric furnace, steam, and high-frequency and infrared radiation generally used by the chemical and materials industries. Microwave heating compares favorably well with other heating sources. For example, when heating a sample using an electric furnace, the refractory material in the electric furnace is heated after which the sample is heated by heat transfer. Therefore, the initial temperature of the sample is lower than the temperature of the refractory material in the electric furnace. The quantity of energy used for heating the sample is only about 20–30% of the total power consumption (i.e., ca. 70–80% of the power is lost). On the other hand, heating a sample by the microwave heating method consumes approximately 70% of the microwave power. There are examples in which, by using microwaves, the manufacturing time of a product is reduced by 80%. Also, equipment size may be reduced by 90%. On the other hand, microwave heating may present some disadvantages when heating metallic substrates and highly conductive samples. The high cost of microwave equipment can also be a drawback when using microwaves.

4.2 Direct Heating of Materials

4.2.1 Internal Heating and External Heating

When a dielectric substance is microwave-irradiated, the microwaves permeate into its inner core, where they are absorbed and converted into heat, after which the heat

Table 4.1 Comparison of properties of various heat sources: electric furnace, steam, and high-frequency and infrared radiation with regard to heating efficiency and various other factors

Properties	Electric furnace	Gas	Steam	Microwaves	High frequency	Infrared
Heating efficiency	Low	Low	Low	High	High	Medium
Temperature control	Low	High	Low	High	High	Medium
Environmental impact	Low	Medium	High (a fossil fuel is used mainly)	Low	Low	Low
Low heat conduction sample	Inappropriate	Inappropriate	Inappropriate	Appropriate	Appropriate	Appropriate
Specimen geometry	Optional	Optional	Optional	Optional	Limited	Optional
Location of heat	Surface	Surface	Surface	Whole	Whole	Surface

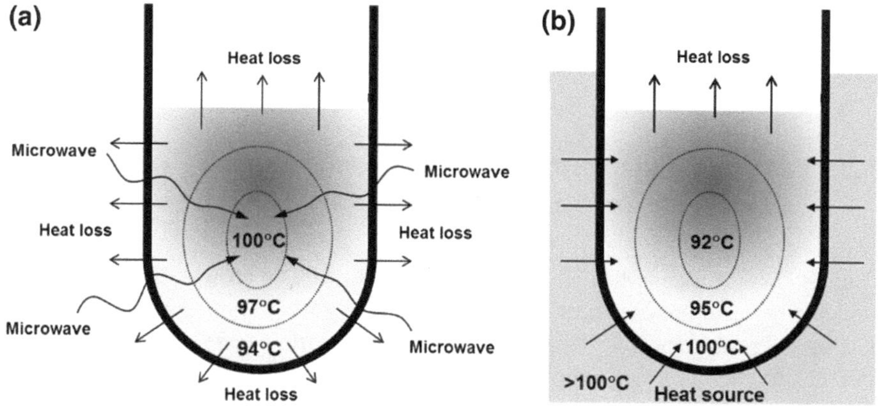

Fig. 4.5 Illustrations comparing the distribution of temperature in a dielectric sample upon **(a)** microwave heating and **(b)** conventional heating (heat transfer)

is lost to the dielectric's surroundings. If the sample is smaller than the penetration distance of the microwaves, the dielectric sample is heated relatively uniformly. Containers or reactors that do not absorb microwave radiation are useful in improving the efficiency of direct heating by microwaves. For example, quartz, Teflon, polyether ether ketone (PEEK), and certain ceramics (e.g., alumina) are often used to contain samples during microwave heating. However, when such containers/reactors are used, the heat of the dielectric substance is ultimately lost to the containers' walls and then to the surrounding atmosphere. For example, when water is microwave-heated in a quartz test tube, the heat is lost at the water–quartz wall interface and creates a temperature gradient between the inner core of the water sample and the interface (Fig. 4.5).

On the other hand, when a sample is heated by heat transfer from a conventional heat source, a large temperature gradient forms far from the location of the heat source, completely opposite to microwave heating. Conventional conductive heat transfer is classified as *external heating*, whereas microwave heating is classified as *internal heating* (see Fig. 4.5). In Fig. 4.6, various heat sources have been classified

Fig. 4.6 Classification of several heat sources as internal heating or external heating

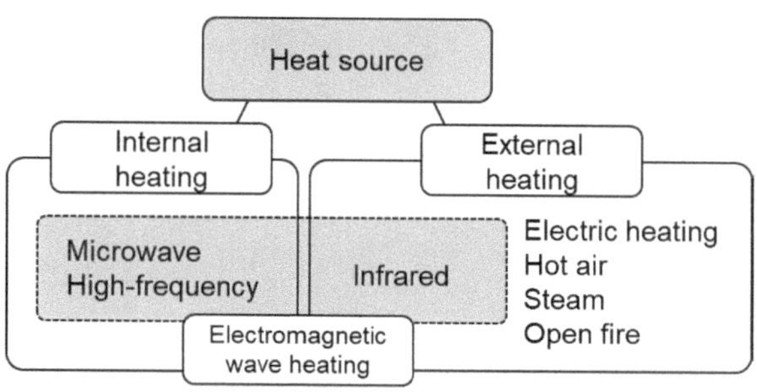

as either internal heating or external heating. Note that infrared radiation, a form of electromagnetic wave heating, is classified as external heating.

4.2.2 Precise Temperature Measurement by Using Microwave Internal Heating

Though little known, when microwave irradiation of a microwave absorber ends, the absorber sample loses heat fairly rapidly to the surrounding atmosphere. When pulsing microwave power ON and OFF, while monitoring sample temperature, it is possible to measure the sample temperature with great precision. For example, Fig. 4.7 compares the temperature changes of a water sample heated by microwaves to that when heating it on an electric cooking plate. Under microwave heating, the water temperature rose to 84 °C after a 50-second heating period, but when the microwave irradiation terminated, the temperature of the sample fell to 25 °C within ca. 40 s (about a 60° drop). By comparison, when the same sample was heated using an electric cooking plate, not only was the temperature rise less (ca. 45 °C), but also 40 s after turning off the heater the temperature of the water fell by only 6 °C owing to the influence of the residual heat of the container and the heater; this residual heat does not occur with microwave heating, hence the much larger temperature drop. Accordingly, even if the ON/OFF cycle were repeated for a heater, it would not be possible to control the temperature precisely.

4.2.3 Applications of Internal Heating

Exposing wood substrates to the Sun's radiation, which comprises UV/visible light and infrared radiation, typically dries wood substrates with high moisture content.

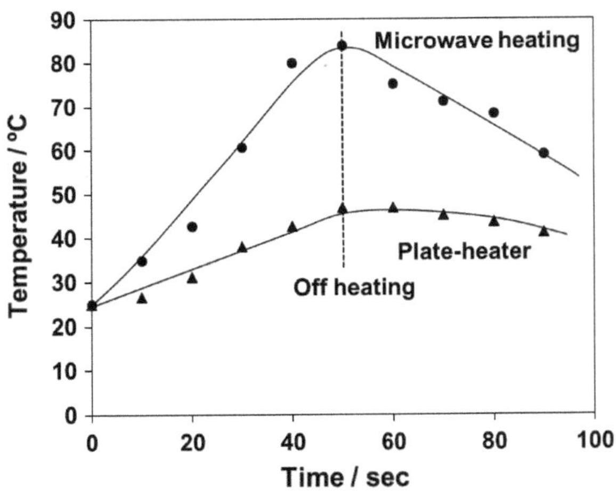

Fig. 4.7 Temperature changes of a water sample subjected to microwave heating and conventional heating using an electric cooking plate

Sometimes, the drying time is reduced by external heating using the heat from steam. However, to achieve a temperature of 100 °C to dry, the inner core requires an external temperature significantly above 100 °C. The resulting large temperature difference between the surface and the center of the wood may cause a reduction in the quality of the wood substrate since the ensuing expansion causes cracks and warps the wood (Fig. 4.8a). Internal microwave heating may also be applied to drying wood. Moreover, when using internal heating, the temperature at the wood surface may be lower than that at the center of the wood substrate because of loss of heat to the atmosphere. There may also be an internal pressure gradient that causes water moisture to move automatically from the inner core of the wood sample to the surface; this is known as *water pumping*. As a result, wood dries uniformly from the center (Fig. 4.8b) by internal microwave heating and, because of the short times afforded by microwave irradiation, cracks and warping are significantly diminished. Compared to the steam drying method, microwave internal heating can also save about 48% energy.

Microwave radiation can also be used to sterilize foodstuffs by internal heating; however, sterilization of the food's surface cannot be done by this method. In such cases, a combination of sterilization through internal microwave heating combined with surface sterilization by ultraviolet light, and near- or far-infrared radiation has proven useful and practical. The UV light is suitable for surface sterilization of food because the penetration depth of UV light into the substance (food) is shallow. Hybrid sterilization with UV light and microwaves is a germicidal method that makes use of the properties and advantages of both sterilization methods (Fig. 4.9).

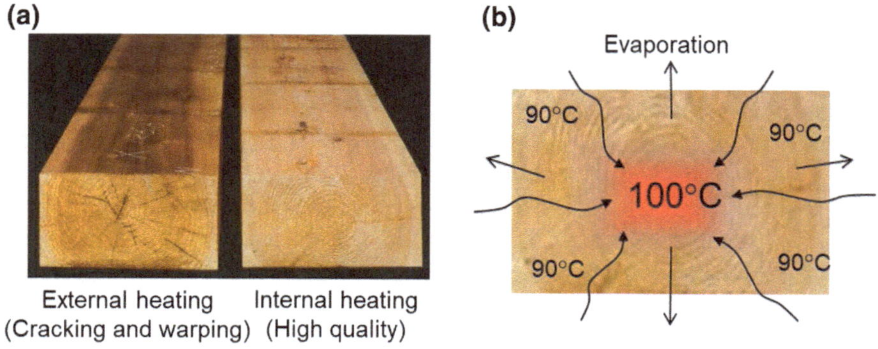

Fig. 4.8 **(a)** Difference between external heating and internal heating in wood; high-frequency heating was used in this process. **(b)** Image of the temperature distribution in wood and flow of water moisture from its inner core. Photograph provided by courtesy of Yamamoto Vinita Co. Ltd.

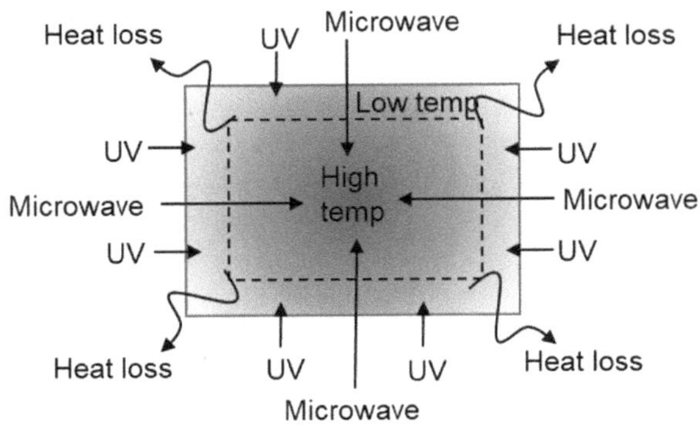

Fig. 4.9 Illustration of the microwave/UV light (MW/UV) hybrid sterilization method as may be applied to sterilize food. Reproduced from ref. [6]. Copyright 2013 by S. Horikoshi

4.3 Selective Heating

4.3.1 Fundamental Selective Heating

Heating substances with microwaves are most efficient when they are strong microwave absorbers and when the penetration depth of microwaves is suitable; such materials are highly polar. Conversely, the microwave heating efficiency of nonpolar substances is low and the penetration depth is large. Therefore, when microwaves are applied to a mixture composed of different polar and nonpolar substances, the more polar substances will be heated selectively. Such a phenomenon is well known for heating food in a domestic microwave oven. For example, when food contained in a plastic lunch box is heated in a microwave oven, the nonpolar plastic box is not heated, which means that the polar food is selectively heated; the plastic box may be heated by conduction from the hot food. It is imperative that users have some knowledge of how to use a microwave oven and what materials (e.g., plastics) can be safely used. For instance, if the food tray is made of polystyrene (PS; melting temperature 99–121 °C), the microwave-irradiated tray would be easily heated and may melt in the oven. This indicates that the choice of the wrong plastic material can cause an undesirable incident. Instead, such food trays are normally made of polypropylene (melting temperature, 130–171 °C), a material that is safe to use when microwave cooking or reheating of food. For example, when spaghetti with Bolognese ragu sauce is heated in a microwave oven, the noodles are heated very little by the microwaves, while the sauce is rapidly heated because of the salinity content in the sauce. As a result of the effect that salt has upon the dielectric properties of the sauce, the penetration depth of the microwaves is shallow and the microwave absorption ratio is very high near the surface, so that the surface of the sauce will be heated selectively. In recent years, olive oil, garlic, and salt are used to make *pasta con aglio e olio*; in this case, microwaves are well suited to cook this pasta. When a layered Bolognese ragu sauce with a low salt concentration in the upper layer and a larger concentration of salt in the bottom layer is placed in a microwave oven, microwaves will heat the sauce layers to different temperatures, selectively and unevenly—some areas are much hotter than others. Hence, the consumer who uses a microwave oven must understand the features of microwave heating and choose the appropriate conditions; otherwise, microwave heating results may be disappointing.

4.3.2 Application of Microwave Selective Heating

Microwave equipment used to dry starch adhesive on a delivery paper slips used for home delivery uses selective heating of the starch adhesive, not heating the paper (Fig. 4.10), whereas if steam or infrared heating were used for drying the starch adhesive, the paper and the ink would all be heated and might deteriorate. In earlier times, drying was carried out under natural atmospheric conditions. Today, however, using the equipment illustrated in Fig. 4.10, the delivery slips move at high speed on a conveyor belt through the microwave applicator where microwave radiation is momentarily applied to the delivery slips to effect drying only the starch adhesive.

Organic syntheses have been performed using some of the features of microwave heating. The efficiency of microwave heating changes according to the dielectric loss of a substance. Organic syntheses that use this characteristic have been reported by Raner et al. [7] to produce 4′-ethoxyphenylvinyl ketone. Differential microwave heating is particularly advantageous in carrying out Hoffmann elimination reactions (Fig. 4.11). The synthetic protocol involves a mixture of *N*-[2-(4′-ethoxybenzoyl)ethyl]-*N,N,N*-trimethylammonium iodide in water/chloroform solvents that is subsequently microwave-heated with stirring, for 1 min at 110 °C (temperature of the aqueous phase). After commencement of microwave heating, the two-phase water/chloroform system (1:1 by volume; 100 mL) for 40 s, the temperatures of the aqueous and organic solvent phases are 105 and 48 °C, respectively, owing to differences in the dielectric properties of the two solvents. A sizable differential can be maintained for several minutes before cooling. Comparable conditions are difficult to obtain by traditional heating

Fig. 4.10 Photograph illustrating equipment used to microwave-dry a starch adhesive placed on paper slips for a home delivery service by selective heating. Photograph provided by courtesy of Fuji Electronic Industrial Co. Ltd.

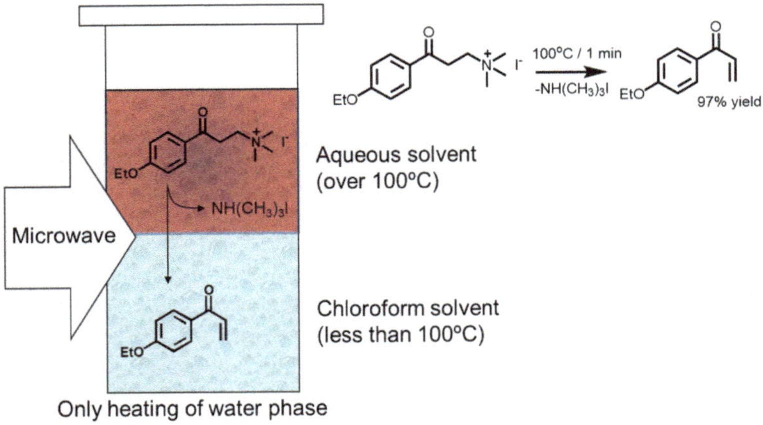

Fig. 4.11 Hoffmann elimination reaction taking place by selective heating. Reproduced from ref. [7]. Copyright 1995 by the American Chemical Society

methods. During the reaction, the product 4′-ethoxyphenylvinyl ketone is extracted and diluted into the poorly microwave absorbing and cooler organic phase. The mixture is then subjected to rapid post-reaction cooling; the ketonic monomer is obtained at a yield of 97% [7]. Selective microwave heating can also occur in heterogeneous catalyzed reactions.

Material synthesis of composite grains with the mesoscopic structures of SnO_2 and TiO_2 has been reported using microwave heating [8]. SnO_2/TiO_2 nanoscale mesoscopic structures were formed by microwave selective heating (SnO_2: high microwave heating efficiency; TiO_2: low microwave heating efficiency). The sample produces a supersaturated solid solution under microwave irradiation, after which the nanoscale mesoscopic structures are generated in the microphase (see Sect. 10.1.6 for greater details).

A catalyst system consisting of activated carbon as the support for a metallic catalyst has been used in environmental cleanup and organic syntheses. If the activated carbon support/metal catalyst system were distributed in a nonpolar solvent, only the catalyst would be heated selectively by the microwaves (Fig. 4.12). The temperature at the catalyst is remarkably high relative to the nonpolar solvent. By contrast, under conventional heating conditions, the temperature of the catalyst would be no higher than the temperature of the solvent. When the solution reaches the reaction temperature, the temperature on the surface of the catalyst can be relatively high [9]. On the other hand, if the catalyst system were to be irradiated with microwaves with excessive output power, hotspots would be formed on the catalyst surface [10], an event that could be deleterious and could deactivate the catalyst. Section 9.4 should be consulted for greater details regarding effects of microwaves on catalytic systems and related reactions.

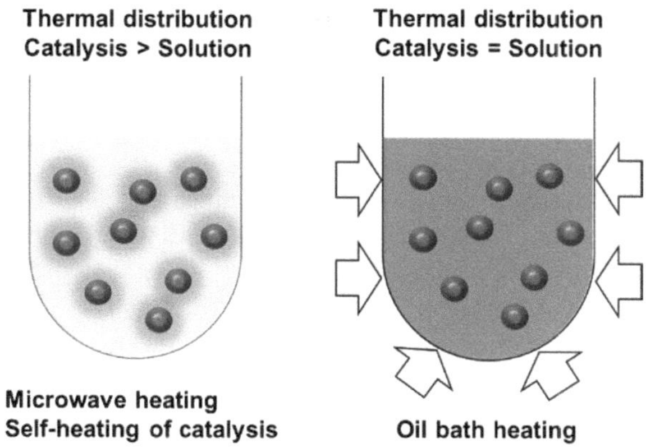

Fig. 4.12 Illustrations displaying the difference(s) between microwave selective heating and traditional heating of a heterogeneous catalyzed reaction

4.4 Hotspots or Local Heating

4.4.1 What Is a Hotspot?

The word ***hotspot*** has been used in the microwave heating field for several years, although no clear explanation as to its origin has been forthcoming. The term hotspot in microwave chemistry and materials processing refers to an inhomogeneous dissipation of microwave energy through selective heating of different parts of a material that causes the generation of temperature gradients within various domains or locations within the material or sample [see e.g., 11]. Where high-temperature micro-domains (hotspots) are formed when the material is microwave-irradiated, these hotspots must undergo fast heat transfer processes for the material to be brought to thermal equilibrium. However, where conditions are such that the rate of heat transfer is slow, the formation of steady-state hotspots occurs that, in principle, can enhance the rate of a chemical process occurring within or near the hotspots. However, the formation of hotspots can also prove deleterious to the smooth progress of chemical processes.

4.4.2 Hotspot Formation in Catalyzed Reactions—Background

Chemical reactions driven by microwave selective heating in the presence of heterogeneous catalysts have been examined extensively for nearly three decades (see Sect. 9.4). Results have shown that selective heating of solid catalysts is the most effective means in microwave chemistry. One should keep in mind, however, that hotspots can form at the surface of a solid catalyst subjected to microwave irradiation, a rather common occurrence that can cause damage to a microwave reactor. Because of such a possibility, commercial microwave chemical equipment manufacturers recommend not to use such equipment to carry out chemical reactions with heterogeneous catalysts. Accordingly, whenever chemical syntheses involve microwave-assisted heterogeneous catalyzed reactions, it is imperative to control hotspot formation. Better still, it would be wise to choose conditions that avoid such formation altogether.

Several researchers have attempted to elucidate the existence of hotspots in chemicals subjected to microwave radiation. Indeed, heating a solid catalyst by microwave radiation frequently generates hotspots because of differential heating. An initial report by Mingos and coworkers [12] showed significant apparent shifts in the equilibrium constant in the microwave-assisted gas phase decomposition of H_2S catalyzed by metal sulfides on a γ-Al_2O_3 support (solid/gas system). These authors attributed the shifts to the development of hotspots with dimensions of 90–1000 μm in the catalytic beds. In solid/liquid systems, hotspots can reach temperatures by ca. 100–200 K above the temperature of the remaining bulk that can cause a possible reorganization of the catalyst under microwave irradiation conditions. In this regard, Tsukahara et al. [13] observed the occurrence of nonequilibrium local heating of DMSO molecules in the proximity of Co and Fe particles as a result of a faster input of microwave power than heat loss induced by temperature gradients established between the high-temperature micro-domains and the surrounding low-temperature domains. Such nonequilibrium local heating led to enhanced dechlorination of 2-chloroethylbenzene and 4-phenylbutylchloride. In that instance, the reaction temperature of the microwave system was ca. 55 K greater (528 K) than the bulk solution temperature (473 K) under microwave irradiation.

The occurrence and the existence of hotspots induced by microwave radiation on materials have often been described by indirect suppositions, but have seldom been observed visually. For instance, Gutmann et al. [14] showed that formation of an organomagnesium Grignard reagent could be activated and deactivated by the microwaves' electric field (***E***-field); they attributed the activation event as originating from either a *cleansing effect* of the magnesium surface or the formation of more reactive spherical magnesium particles owing to mild electrostatic discharges (arcing, hotspots) between individual magnesium turnings. However, neither the cause of the generation of the hotspots nor their features were examined [14]. Microwave heating can be used to promote heterogeneous reactions with carbon materials primarily because of the heating mechanism of microwave radiation and the phenomenon known as microplasmas [15].

It is relevant to examine an example where hotspots form and relate the mechanism of their generation with describing how their formation might be minimized or suppressed. Events occurring during the synthesis of 4-methylbiphenyl by the Suzuki–Miyaura coupling reaction (reaction 1) under microwave irradiation have been investigated by Horikoshi et al. [16] using a high-speed camera to provide the basis to explain the occurrence of hotspots. Figure 4.13 displays photographs recorded in real time, which show that the Pd/AC

Fig. 4.13 High-speed camera photographs of the electrical arc discharge occurring on the activated carbon in toluene solvent under the microwaves' *E*-field irradiation: **(a)** after 3 s of MW irradiation, **(b)** after 16 s of irradiation, **(c)** under continuous irradiation, and **(d)** emitted light reaching maximal intensity. Reproduced from Horikoshi et al. [16]. Copyright 2013 by Elsevier B.V

catalyst particulates immersed in toluene solvent were selectively heated by the microwaves' ***E***-field radiation.

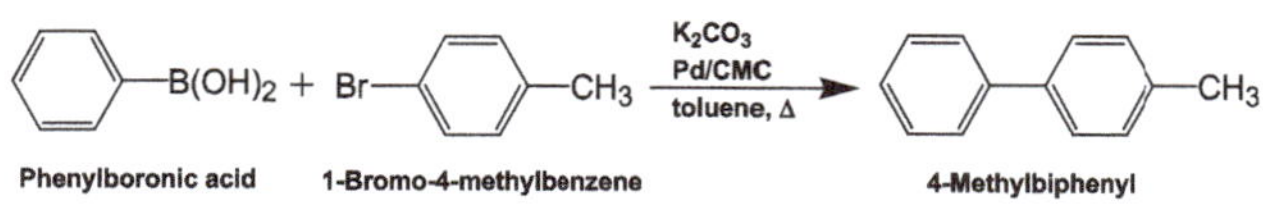

(reaction 1)

In the above case, as soon as microwave irradiation was initiated, bubbles could be seen on the activated carbon surface after only 3 s (Fig. 4.13a) owing to differential heating of the activated carbon particles. Also observed was the aggregation of the activated carbon particulates as if they had been magnetized by the microwave radiation. After 16 s, an orange-colored light was observed from the aggregated activated carbon particles (see the lower left side in Fig. 4.13b). Subsequently, the light emanating from the activated carbon surface increased, as did its intensity, together with a greater number of bubbles formed around the emitted light (Fig. 4.13c). The emitted light became even more intense on the activated carbon surface exposed to continued microwave radiation. Figure 4.13d shows the light originating from hotspots that reached maximum intensity, after which the luminescence from the activated carbon surface disappeared and the aggregated activated carbon particles were re-dispersed.

Chemical yields of 4-methylbiphenyl produced in toluene solvent from the Suzuki–Miyaura coupling reaction (reaction 1) under microwave ***E***-field and ***H***-field irradiation, together with those from the more conventional oil bath heating method, are reported in Fig. 4.14. Under ***E***-field conditions, the yield was about 6% after a 60-minute irradiation period at an input microwave power of 70 W, whereas the corresponding yield under ***H***-field irradiation conditions was 34% (also after 60 min). With the oil bath heating method employed for comparison with the microwave method, the chemical yield of 4-methylbiphenyl was 21% for a heating time of 60 min; when hotspots formed, the efficiency of the synthesis of 4-methylbiphenyl tended to decrease [10, 16]. On the other hand, when no hotspots formed, as occurred under ***H***-field irradiation, the synthesis efficiency of 4-methylbiphenyl was significantly greater (nearly sixfold greater) as depicted in Fig. 4.14. Accordingly, these authors found it relevant to investigate the influence that hotspots may have on the possible occurrence of secondary reactions, other than the principal Suzuki–Miyaura cross-coupling reaction. With the microwaves output power fixed at 70 W, generation of hotspots occurred with ***E***-field irradiation, but none under ***H***-field irradiation. This also led to a closer examination of the Suzuki–Miyaura cross-coupling reaction and side reactions that might include the formation of biphenyl through a Suzuki–Miyaura homo-coupling process and formation of 4,4′-dimethylbiphenyl through an Ullmann-type coupling (reaction 2) [17] under hotspot and no hotspot formation conditions.

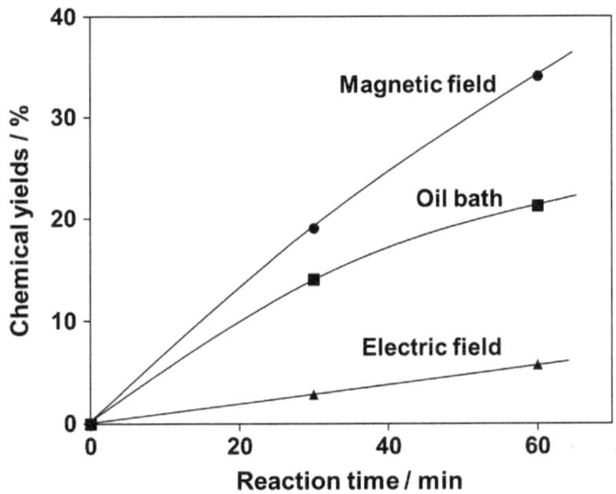

Fig. 4.14 Product yields of 4-methylbiphenyl in toluene solvent under irradiation with the microwaves' magnetic field and electric field components, and oil bath heating. Reproduced from Horikoshi et al. [16]. Copyright 2013 by Elsevier B.V.

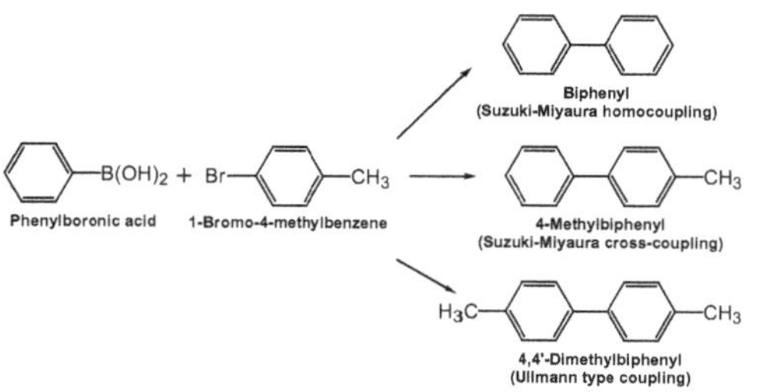

(reaction 2)

When hotspots do occur, the cause for the decrease of yields in a chemical synthesis may be due to a decline in catalyst activity. In the present case, the formation of hotspots on the catalyst surface caused the yield of 4-methylbiphenyl to decrease, while the amount of by-product(s) generated increased. To understand this assertion, Fig. 4.15 illustrates scanning electron microscopic (SEM) images of the Pd/AC surface before and after the formation of hotspots [10]. The Pd catalyst particles are dispersed rather uniformly on the AC surface prior to microwave irradiation (Fig. 4.15a). However, large Pd aggregates are formed after 30 min of

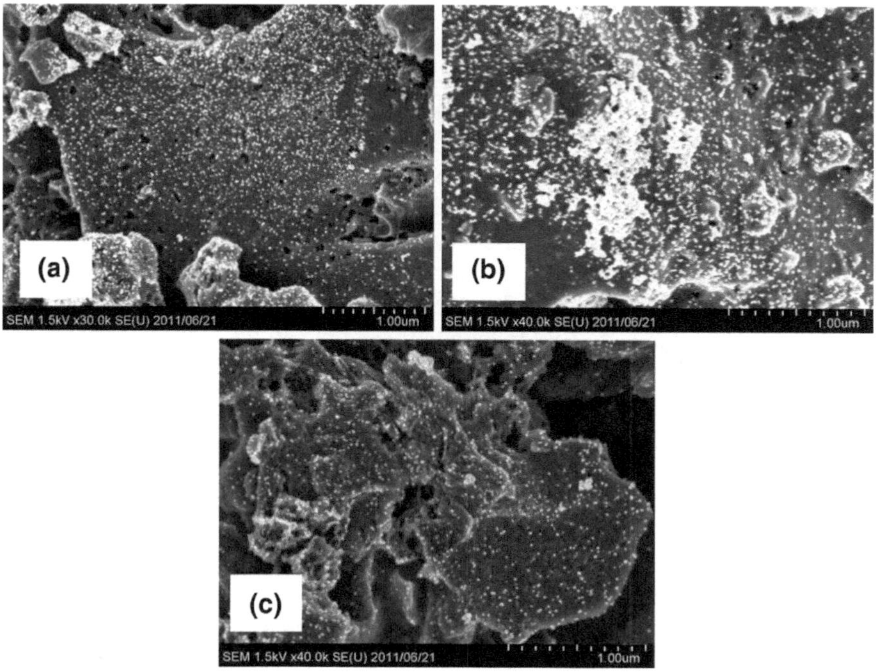

Fig. 4.15 SEM images of Pd/AC catalyst surface: (**a**) 0-min irradiation, (**b**) after 30 min of microwave irradiation and formation of hotspots, and (**c**) under conditions where no hotspots formed. Reproduced from Horikoshi et al. [10]. Copyright 2011 by the American Chemical Society

microwave irradiation of the Pd/AC catalyst under ***E***-field heating conditions (Fig. 4.15b). By contrast, no aggregation of Pd particles occurred, and no hotspots formed, under conditions when heating was carried out by the microwaves' magnetic ***H***-field (Fig. 4.15c). Evidently, aggregation of Pd particles and generation of hotspots (electric discharge) on the catalyst surface are events that occurred only under ***E***-field heating conditions. Consequently, the lower chemical yields of 4-methylbiphenyl (12 vs. 22% after 2 h of microwave irradiation) were attributed, in part, to the formation of less reactive larger Pd aggregates as one of several related factors that affect yields, i.e., such factors as (i) a lower number of reactive sites on the Pd particles through annihilation by the Joule heating effect, (ii) possible obstruction of reactive sites on the Pd particles by a small layer of carbonaceous residue that might have formed by dielectric heating of the AC support under the microwaves' ***E***-field, and finally (iii) to an excessive number of hotspots formed on the Pd/AC catalyst surface. It was deduced that the decrease in efficiency was due to aggregation of the Pd catalyst because of the hotspots on the activated carbon surface [10].

4.4.3 Mechanism(s) of Formation of Hotspots

To further understand the phenomenon of hotspots, the distribution and the relationship of the microwaves' ***E***-field intensity (2.45 GHz) within the spatial volume between the activated carbon particles were examined by a simulation technique using an RF module and the COMSOL Multiphysics software Version 4.3a [16]. The distances between the activated carbon particles (size, 0.65-mm diameter) in toluene were set up at intervals of 0, 0.001, and 0.1 mm. Figure 4.16a-i shows that at a distance of 0 mm (particles in contact with each other), the maximal electric field distribution was seen at both ends of the particles and decayed at longer distances from the particles; the electric field intensity between the two particles was 813 V m^{-1} (Fig. 4.16a-ii). On the other hand, when the distance between the activated carbon particles was set at 0.001 mm (Fig. 4.16b-i), the field was concentrated entirely within the gap that separated the activated carbon particles; the maximum *E*-field intensity was 11838 V m^{-1} (Fig. 4.16b-ii). Compared with the activated carbon particles that were glued to each other (0 mm), the change in the electric field intensity was nearly 15 times greater if the particles were ever so slightly separated by a mere 0.001 mm, not to mention the location of the electric field observed in the volume between the particles. Extending the distance between two activated carbon particles to 0.1 mm (Fig. 4.16c-i) caused the maximal ***E***-field intensity to decrease to 1253 V m^{-1} (Fig. 4.16c-ii).

Clearly, hotspots are generated whenever the particles are separated by very small distances; however, when the distance was 0 mm, no hotspots formed, at least none were observed. Such a phenomenon recalls the *Surface Enhanced Plasmon Resonance* (SEPR) phenomenon that occurs between metal nanoparticles, wherein the electric field of light is concentrated between nanoparticles within a narrow gap

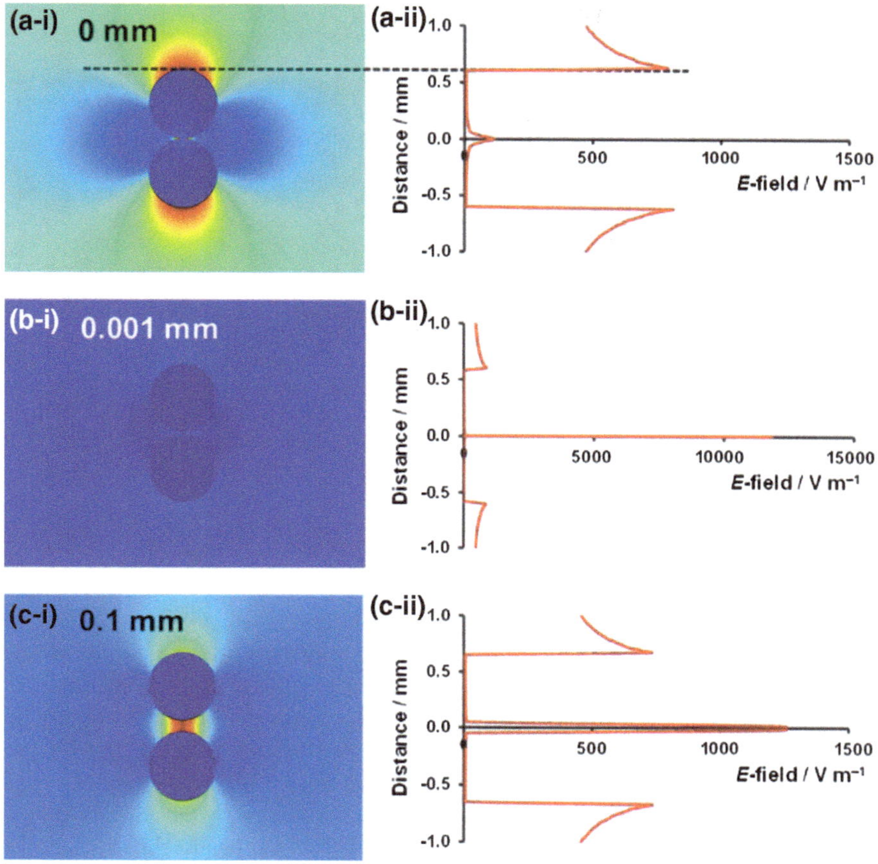

Fig. 4.16 Distribution of the microwaves' *E*-field (2.45 GHz) around two activated carbon particles in toluene solvent simulated with the COMSOL Multiphysics software version 4.3a. Colored images **(a-i)**, **(b-i)**, and **(c-i)** display the *E*-field intensity between two activated carbon particles in toluene together with the magnitude of the gap between the particles, whereas **(a-ii)**, **(b-ii)**, and **(c-ii)** graphics show the magnitude of the E-field (V m^{-1}) within this gap. Reproduced from Horikoshi et al. [16]. Copyright 2013 by Elsevier B.V.

[18]. To experimentally confirm the phenomenon predicted from the simulations with respect to the effect of distances between particles and to the generation of hotspots, activated carbon particles were pasted on a glass plate at various distances using an adhesive made of polyvinyl acetate resin. Hotspots were easily generated (Fig. 4.17) at distances between the activated carbon particles comparable to those used in the simulation. The distance between particles is therefore an important parameter in the formation of hotspots. It is interesting to note that generation of hotspots can be suppressed in a reaction taking place in a fixed bed reactor under flow-through conditions.

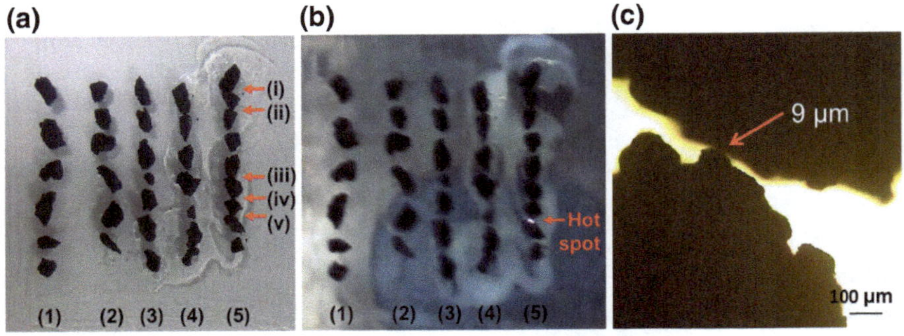

Fig. 4.17 Observation of hotspots generated in activated carbon particles fixed at various intervals on a glass plate: (**a**) photograph of the initial sample; (**b**) formation of hotspots between the AC particles under microwave irradiation; (**c**) observed distance between the particles in the fourth column with an optical microscope. Reproduced from Horikoshi et al. [16]. Copyright 2013 by Elsevier B.V.

4.4.4 Control of the Occurrence of Hotspots

As inferred above, the occurrence of hotspots on a solid catalyst surface can also be deleterious to microwave-assisted catalyzed reactions. Accordingly, studies have been undertaken to further clarify their generation and their influence on chemical reactions, and to find means on how they can be controlled. When considering the industrialization of a chemical process involving microwave radiation, this hotspot phenomenon peculiar to microwaves needs to be minimized, if not entirely suppressed, which can be achieved by using only the microwaves' magnetic field [10].

The generation of hotspots can also be avoided by replacing the AC catalyst support by carbon micro-coil (CMC) supports (Fig. 4.18) [19] that proved to be effective catalyst supports in microwave organic chemistry, even though the available surface area was significantly smaller than that of activated carbon particles. The CMCs also proved to be better microwave absorbers than the ACs and thus optimal for the selective heating of metal catalyst deposits.

Chemical reactions can also be carried out with microwave dielectric heating accompanied by the more conventional heating method, with the benefit that the applied microwave power can be reduced. For instance, using an internal/external hybrid dielectric heating methodology, in which indirect microwave irradiation is achieved with the use of microwave absorbers (external heating) and direct heating of the Pd/AC catalyst particles with the microwaves (internal heating), led to a significant enhancement of the chemical yields of 4-methylbiphenyl from the Suzuki–Miyaura coupling reaction in toluene solvent [20]. An important result of this methodology is that the generation of hotspots can be controlled by the distribution of the microwave energy and low-temperature gradient. Another advantage of the methodology is that at a high temperature it can be maintained with significant energy saving.

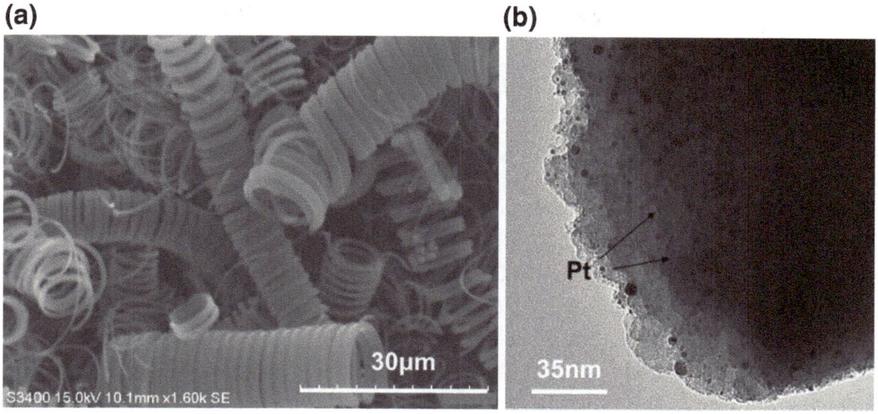

Fig. 4.18 (**a**) Low-resolution SEM image of Pt/CMCs (Hitachi scanning electron microscope S-3400 N) and (**b**) TEM image of Pt deposits on CMCs at greater resolution. Reproduced from Horikoshi et al. [19]. Copyright 2012 by Elsevier B.V.

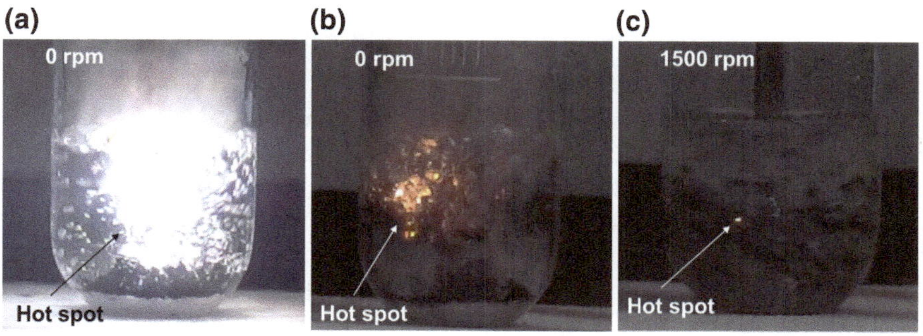

Fig. 4.19 High-speed camera photographs of the electrical arc discharge occurring on the Pd/AC catalyst surface during the reaction in the 25-mm-wide tube reactor under 550-Watt microwave irradiation and at a stirring rate of (**a**) 0 rpm, (**b**) 0 rpm, and (**c**) 1500 rpm. Reproduced from Horikoshi et al. [21]. Copyright 2014 by the American Chemical Society

The occurrence of hotspots in a heterogeneous catalyzed process can also be controlled largely by the experimental conditions used [21] such as (i) by the size of the reactor, (ii) the extent to which the solid catalyst particles (e.g., Pd/AC) are dispersed within the reaction volume through changes in stirring rates, i.e., their dispersiveness, (iii) by irradiation with pulsed *versus* continuous microwaves, (iv) by the microwaves' output power levels, and (v) by irradiating with microwave radiation in the presence or absence of a standing wave, i.e., a non-standing wave condition. Increasing the rate of stirring of the reactants tends to minimize the occurrence of hotspots as demonstrated in the photographs of Fig. 4.19 that displays the generation of hotspots in the heterogeneous sample at 0 revolutions per minute (rpm; no mixing) and under conditions of significant mixing (stirring at

1500 rpm) [21]. The hotspots that formed under non-stirring conditions showed up as intense white sparkles (Fig. 4.19a). With the turbulence created by the microwave heating, many orange-like colored hotspots were observed over a wide area (Fig. 4.19b). By contrast, no hotspots displaying the whitish intense sparkles were observed at 500 rpm stirring; only the orange-colored hotspots were seen (not shown in the figure). The number and frequency of formation of the orange-colored hotspots decreased considerably on increasing the stirring rate to 1500 rpm (Fig. 4.19c); in fact, there was a hint of only one hotspot.

In addition, to the extent that toluene solvent in the process examined could be heated with 5.8-GHz microwaves (a frequency change from the usual 2.45-GHz microwaves), the temperature gradient between the catalyst and the solution bulk decreased such that there was a remarkable drop in the generation of hotspots [22]. It is also possible to minimize hotspot formation by carrying out the heterogeneous catalyzed reaction in a fixed bed reactor [23].

4.5 Hotspots in Microwave Sintering

4.5.1 Background for Hotspots in Solid Processing

The term *hotspot* refers to a discharge between materials that occur under a highly intense microwave electric field. In solid processing with microwave heating, this term has two meanings: the first is higher temperature by a discharge on a microscale on a solid surface (see above); the second is the temperature distribution (selective heating) on a large-scale sample.

Commonly, 1000 °C or greater temperatures are required to synthesize or sinter solid materials. However, if a discharge occurs, then microwaves are concentrated on a portion of the sample such that bulk temperature cannot increase: sintering and syntheses do not proceed under these conditions. On the other hand, sometimes a sample cannot be heated uniformly and only a portion of the sample is heated (selective heating). The visible image of temperature variation is shown in Fig. 4.20; even if each of the powdered specimens has the same composition, partial heating of such samples leads to the formation of temperature gradients at the microscale owing either to the electromagnetic field distribution in the microwave applicators or otherwise to differences in the shape of the samples.

For hotspots used for selective heating, microwave energy may be concentrated at a high-temperature portion of the specimen at which a substantive increase of the local temperature occurs. A runaway phenomenon subsequently occurs [24]. Generally, the dielectric loss factor tends to be larger as temperature increases. Thus, for some reason, local overheating develops when partial overheating occurs and microwave absorption is accelerated at that location of the sample. This runaway phenomenon is often a problem in microwave heating because such local temperature gradients will cause differences in thermal expansion between the

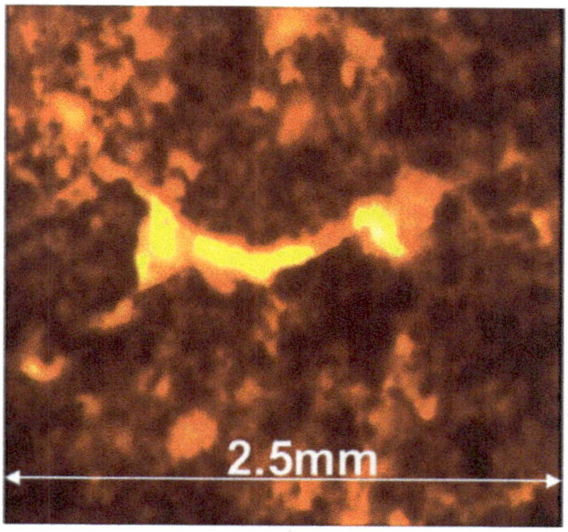

Fig. 4.20 Photograph of hotspots under microwave irradiation of TiO_{2-x} materials

heated and less-heated portions of the sample. Such differences lead to stress deformation and cracks in the sample, or in the crucible that contained the sample. Figure 4.21 shows a heat insulator and an alumina insulation crucible after the experiment. After microwave irradiation, some portion of the sample overheated causing the crucible to crack because of thermal stress. In addition, after hotspot formation, some of the sintered sample along with the crucible and heat-insulating material can melt. Figure 4.22 displays a portion of the sample that melted under microwave irradiation. When it is not possible to control hotspot formation, the center of the sample can melt while the peripheral portion remains untouched as shown in Fig. 4.22.

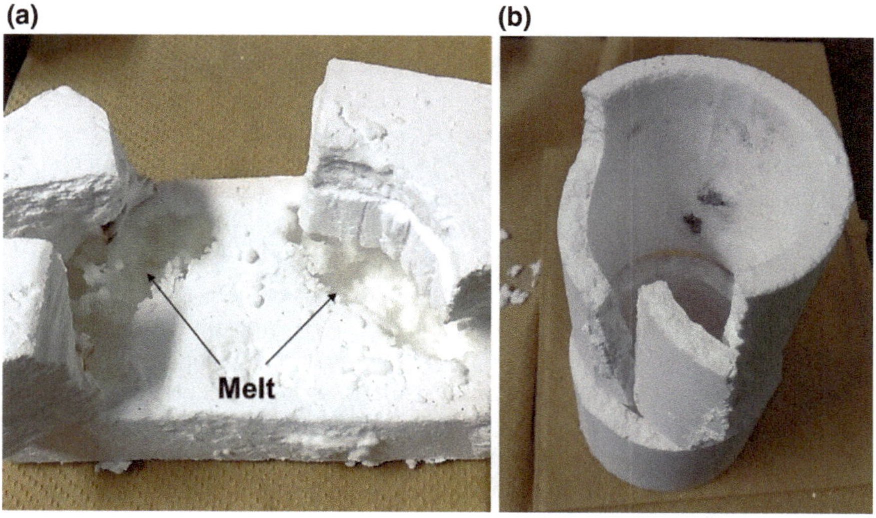

Fig. 4.21 (**a**) Heat insulator after an experiment; (**b**) Alumina insulation crucible after an experiment

Fig. 4.22 Display of a sample of Fe_3O_4 that melted at the center after being microwave-irradiated

4.5.2 *Principles and Control of Occurrence of Hotspots in Samples*

One of the causes of the occurrence of hotspots is the rapid change of dielectric constant with temperature increase. Chromia (Cr_2O_3) is a typical example that displays a rapid rise of the loss factor ε'' with increase in the temperature. Selective microwave heating of a specific material can also result in the generation of hotspots. For example, since magnetite (Fe_3O_4) has a large loss factor ε'' at ambient temperature, it is easy to reach high temperatures in short time. Unlike Fe_3O_4, however, hematite (Fe_2O_3) has a small dielectric loss factor ε'' at room temperature so that when mixtures of magnetite and hematite are irradiated with microwaves, only magnetite is heated that could result in the formation of hotspots. This is related to the temperature dependence of the dielectric constant and the conductivity of metal oxides. In other words, along with the lattice vibrations, which are the basis of the dielectric loss becoming larger, ionic conductivity and electronic conductivity also increase, and microwave absorption capacity is higher. For example, in alumina, microwave absorption increases dramatically at about 600 °C. The extent of heat generated rapidly increases above this temperature, which leads to the formation of hotspots. Other causes of the occurrence of hotspots are local arcing between materials.

Before one can control the generation of hotspots, one needs to understand the heating behavior of the material itself. In particular, it is possible to predict when this runaway phenomenon will occur if one can measure the dielectric loss up to a high temperature and obtain a plot of dielectric loss *versus* temperature. The temperature dependence of the dielectric loss factor, ε'', is reduced as the frequency of the microwaves is increased. A curved line of dissipation loss from normal temperature to high temperature is made to estimate the restraint condition, i.e., control condition, of the hotspot from the condition of the sample. The restraint condition of the hotspot from the viewpoint of the sample can be anticipated by the curved

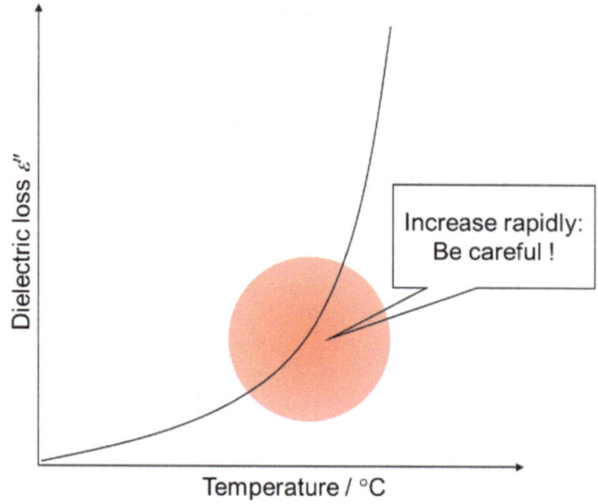

Fig. 4.23 Prediction of hotspot formation by relation between dielectric loss factor ε'' and temperature

line of dielectric loss factor ε'' to the temperature change as shown in the image displayed in Fig. 4.23. Temperature gradients tend to predict the location where dielectric loss factor ε'' changes suddenly.

On the other hand, in microwave heating of a solid powder, a difference in heating behavior may occur because of differences in the shape of the powder and the size of the particles. Because a solid sample has no fluid character, temperatures may be nonuniform and as a result a hotspot may form inside the sample. A temperature gradient within powders may be large in high-temperature ranges, and there are some concerns that hotspots may be generated. Temperature distribution in the solid sample can be equalized by traditional heat transfer, by controlling the microwave output power, thereby controlling the formation of hotspots. An adiabatic method with heat insulation can also restrain the formation of hotspots (see Sect. 9.1).

4.5.3 Principles and Control of Hotspots and Electromagnetic Waves

In case of microwave irradiation using a multi-mode applicator, the microwaves that are not absorbed by a sample are reflected off the walls inside the applicator. A reflected wave will overlap with a different reflected wave resulting in a standing wave (see Sect. 6.4.1). Microwave heating is location-specific, i.e., it varies from location to location within an applicator because the electromagnetic field strength of a standing wave is very high, whereas in an adjacent location it will have very low electromagnetic field strength. In order to have a sufficient number of standing waves so that the distribution of electromagnetic field intensity disappears, the size of the applicator should be more than 100 times the microwave wavelength (see Sect. 6.4.3). Therefore, the ideal size of the multi-mode applicator used in 2.45 GHz (wavelength: ca. 12.24 cm) microwave will be beyond 12 m, which is impractical for use in microwave chemistry or materials processing. Instead, the width of most domestic microwave domestic ovens rarely exceeds 50 cm. Thus, uneven temperature distribution and hotspots are often a problem in these common-size applicators. Higher frequencies have shorter wavelengths and may generate many more standing waves, and thus provide the potential for more uniform heating [25]. Using high-frequency microwaves (e.g., 28 GHz) is a good approach to prevent many of these nonuniformity problems [26, 27]. Another approach is to control the size of the sample such that it is small enough as not to disturb the field distribution but can be placed in a location of high electromagnetic field density. It is also possible to use a single-mode applicator instead of the common multi-mode applicator, such as seen in domestic microwave ovens. However, care must be taken that the sample is not so large that in the single-mode applicator its outer parts are in areas of lower field intensity; otherwise, it will heat nonuniformly. In summary, the problems concerning hotspots can be approached by controlling the material's characteristics and microwave heating behavior, as well as the microwave electromagnetic field distribution.

4.6 Superheating of Liquids

4.6.1 What Is Superheating?

Superheating of solvents is a condition that often occurs during microwave heating. For instance, if a micro-spatula were put into methanol that had been microwave-irradiated, it might result in a violent eruption as the methanol changed explosively from a liquid into a gas as illustrated in Fig. 4.24. The condition of such a solution is referred to as a metastable state. Unless the superheated solvent is disturbed by some vibrational action, the solution remains in a metastable state. Such phenomena can cause serious accidents that typically occur when using a domestic microwave oven [28]. The U.S. Food and Drug Administration (FDA) has received reports of serious skin burns or scalding injuries around people's hands and faces as a result of hot water erupting out of a cup after it had been heated in a microwave oven. Internal overheating of water in a cup results in superheated water (temperature greater than its boiling temperature) without appearing to boil. Superheating is unlikely if foreign materials such as instant coffee or sugar are added before microwaving. If superheating has occurred, any slight disturbance or movement such as picking up the cup, or pouring in a spoon full of instant coffee, may result in a violent eruption with the boiling water exploding out of the cup. ***What Can Consumers Do to Avoid Superheated Water?*** The following suggestions are worth noting: (i) follow the precautions and recommendations

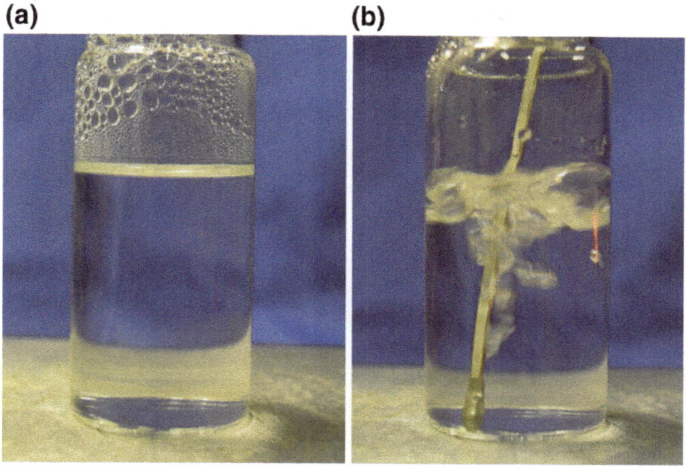

Fig. 4.24 Pictures of (**a**) methanol under superheating conditions, and (**b**) violent eruption of methanol on putting a micro-spatula into the superheated methanol

found in the microwave oven instruction manuals, specifically the heating time; (ii) do not use excessive amounts of time when heating water or liquids in the microwave oven; (iii) determine the best time setting to heat the water just to the desired temperature and use that time setting regularly; and (iv) add instant coffee and other solid ingredients to the water prior to microwaving.

Meyer [29] was the first to report the superheating phenomenon, well before the development of microwave ovens. Meyer reported that water could exist at normal atmospheric conditions and at temperatures greater than 100 °C, without any sensing of boiling. Optic fiber thermometers were not widely used in the early stages of development of microwave chemistry, so that temperatures were determined under reflux conditions of the solvents or solutions. Several curious reports appeared in the past: for example, one can often get greater yields of products under microwave heating compared to oil bath heating at the same temperature. It is possible that this may be because the temperature exceeded the boiling point, indicating superheating. When temperatures are not measured accurately, it can lead to incorrect conclusions. A liquid under superheating conditions is very unstable; an accident can easily occur through the solvent spout from a reflux condenser during an experiment. At present, thermometers and stirring peripherals are attached to the microwave chemical equipment in a manner intended to prevent superheating [30].

It is especially important to recognize that bumping of low-boiling-point solvents can occur on superheating, a disadvantage when performing chemical reactions. If we assume that water reaches a temperature of 101 °C through microwave superheating, and taking into account the latent heat of vaporization of water (2.26 MJ Kg^{-1} at 100 °C and normal atmospheric pressure) and the specific heat capacity of water (4.2 Kj Kg^{-1}; heat to raise the temperature of 1 kg of water by 1 °C), when water begins to boil it will quickly cool to 100 °C and the heat liberated turns water into steam. Cooling 1 kg of water by 1 °C gives off 4.2 kJ of heat, enough to evaporate ca. 1.9 mL (4.2 kJ/2.26 MJ Kg^{-1}) of water (note that 1 kg of water is ≈ 1 L) yielding ca. 3.2 L of steam ((1.9/18)×8.314 × (374)/101325 ≈ 0.0032 m^3) that is ejected violently from the reactor. Note that one of this book's authors (Schiffmann) has measured temperatures as high as 116 °C in a microwave-heated water sample without any visible boiling.

4.6.2 Mechanistic Stages of Superheating

Heating a solution by conventional heat transfer leads to the formation of a temperature gradient between the walls of the container and the bulk of the liquid. As the temperature of the liquid rises, the equilibrium vapor pressure also increases and when it reaches the outside air pressure (1 atm) it causes nucleating gas bubbles to form in micro-defects in the walls of the container, i.e., inside the water. When these gas bubbles are able to overcome the interfacial surface tension holding them to the walls, they are released, creating the boiling phenomenon. The boiling state indicates that the added heat energy from the heat source and the heat escaping to the atmosphere are identical and the temperature no longer changes (Fig. 4.25) [31].

Contrary to conventional heating, microwave heating efficiently transfers thermal energy into solution. The heating rate by the microwave heating method is greater than the slower rate of heat lost to the atmosphere, such that the liquid will be in a nonequilibrium state beyond the boiling point [31]. For example, Fig. 4.26 illustrates a pure methanol solution being heated by microwave radiation at three output power levels: 50, 100, and 150 W. The heating of methanol by the 50-Watt microwaves leads to superheating by +6 °C above the boiling point of methanol (64 °C), whereas superheating with the 100-Watt microwaves the methanol reaches a temperature of ca. 86 °C and with the 150-Watt microwaves the methanol is superheated to a maximal temperature of 92 °C. Through bumping and subsequent intense eruptive boiling, the temperature of methanol returns to its normal boiling point. Furthermore, it is difficult to cause conventional-type boiling activity because the container wall is cold compared to the internal solution temperature. The rate of

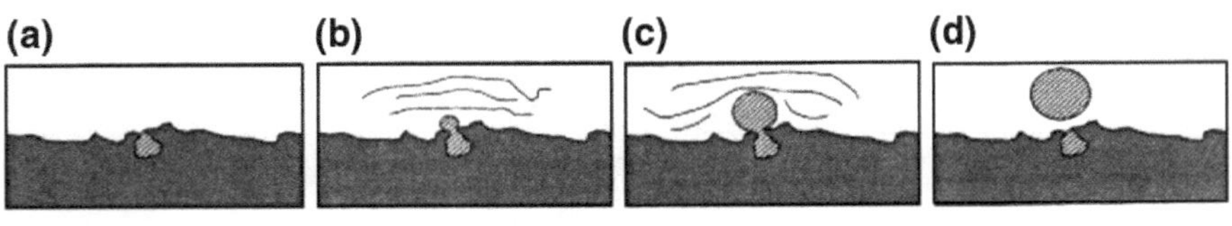

Fig. 4.25 Developmental process of a boiling nucleating seed. (**a**) embryo trapped by solvent, (**b**) hot zone forms around embryo, embryo growth, (**c**) bubble growth approaches conical size, (**d**) bubble detaches, boiling commences, Reproduced from [31]. Copyright 1992 by the Royal Society of Chemistry

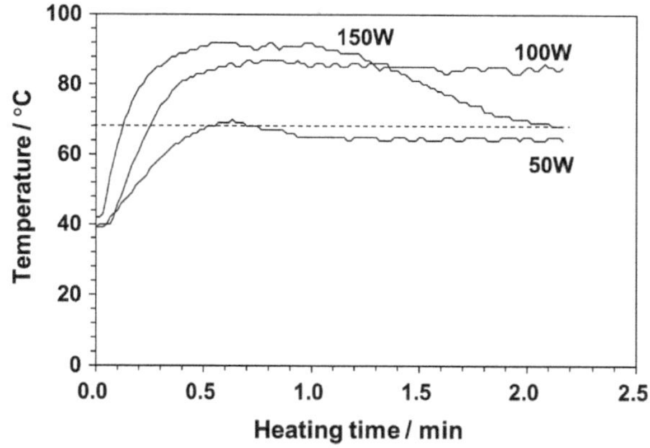

Fig. 4.26 Temperature profile of microwave heating of methanol at reflux under microwave heating at output power levels of 50, 100, and 150 W. The horizontal-dashed line represents the normal boiling point of methanol

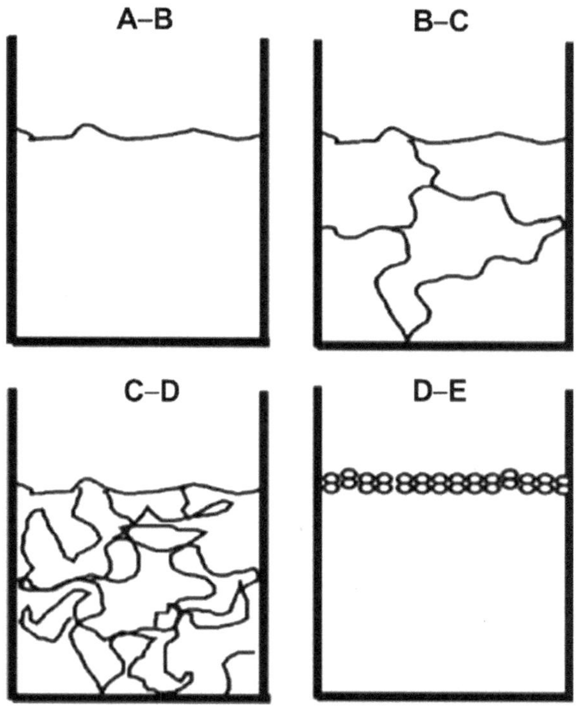

Fig. 4.27 Visual observation of microwave superheating phenomenon. Reproduced from [32]. Copyright 2001 by Wiley-VCH Verlag GmbH

occurrence of superheating is enhanced when there are fewer defects on the inner surface of a new container (reactor) when performing microwave chemistry.

Chemat and Esveld performed a very careful examination of the superheating of a solution and described some mechanistic stages on how superheating occurs [32]. A visual observation of ethyl alcohol revealed an ensemble of several stages leading up to the superheated boiling phenomenon (Fig. 4.27). In the initial stage, the pure liquid is calm and undergoes no boiling (stage AB). When the liquid reaches an overheated temperature higher than the boiling point, fine scores (convection currents) appear. The liquid is divided into a number of areas by these scores (stage BC). At point C, the liquid is infinitely divided and seems to be in an unstable state, and an intense convection movement appears inside the liquid (stage C). The liquid boils instantaneously and sometimes bumping occurs. The temperature decreases and the liquid becomes more stable (stage CD), after which microwave boiling starts at the free liquid surface (stage DE).

4.6.3 Applications of Superheating to Chemical Reactions

The chemist can use the superheating phenomenon to advantage if it is controlled such that it is not a physical danger. The boiling point of the solvent used in a chemical reaction is one of several important factors in organic syntheses. Reaction efficiencies increase with increase in reaction temperature in organic syntheses. Therefore, in a closed container, a reaction depends upon temperature conditions beyond the boiling point. Even if a closed container was not used under microwave heating, it would still be possible for the reaction to proceed at temperatures beyond the boiling point via the superheating phenomenon.

Superheating data for some 23 common solvents subjected to 5.8 and 2.45-GHz microwave radiation have been reported by Horikoshi et al. [33]; they are summarized in Table 4.1. Except for ethylene glycol and pyridine, in all other cases superheating occurred upon heating with the 2.45-GHz microwaves. With a 5.8-GHz single-mode microwave applicator, superheating occurred in almost all polar organic solvents (note that changing the equipment and the reactor can lead to values different from those of Table 4.2). For DMF, acetic acid, 1-propanol, acetone, ethyl acetate, THF, and dichloromethane, the increase in temperature is 20 °C or more above the boiling point. For nonpolar solvents, since heating hardly occurs with the 2.45-GHz microwaves, unless it is under optimal heating conditions, no superheating was observed. However, if the frequency is changed to 5.8 GHz, superheating of nonpolar solvents then becomes easily observable (see Sect. 5.7 for details about frequency effects).

As an example of microwave superheating occurring in polar solvents, Klán et al. examined a temperature-dependent photochemical reaction involving the Norrish type II reaction of valerophenones [34]. Equimolecular mixtures of excited alkylphenyl ketones with hydrogen on γ-carbon react on their alkyl chains according to the Norrish type II reaction via the triplet state to produce triplet 1,4-biradicals subsequent to irradiation at wavelengths $\geq$280 nm in various solvents under otherwise identical experimental photochemical conditions. The fragmentation/cyclization ratios varied from 5 to 8 and were characteristic for given reaction conditions. The photochemical efficiency (fragmentation/cyclization ratio) was temperature-dependent and the magnitude was most likely related to solvent alkalinity. Superheating occurring under microwave irradiation was most likely responsible for the observed modification of selectivity.

Table 4.2 Boiling points and difference between superheating temperature and boiling point for some 23 common solvents under non-stirring conditions (Sample volume: 30 mL, Microwave output power: 30 W). Reproduced from [33]. Copyright 2008 by the American Chemical Society

Solvent	Nature of solvents[a]	Boiling point/°C	Difference between superheating temperature and boiling point/°C (minutes to reach superheating temperature)	
			2.45 GHz	5.80 GHz
Ethylene glycol	Protic/polar	198	−(39)	2 (39)
DMSO	Aprotic/polar	189	2 (30)	13 (27)
DMF	Aprotic/polar	153	10 (25)	24 (25)
Acetic anhydride	Aprotic/polar	140	2 (25)	18 (15)
Pyridine	Aprotic/polar	115	−(30)	7 (7)
Acetic acid	Protic/polar	101	25 (20)	29 (9)
Water	Protic/polar	100	2 (17)	−(30)
1-propanol	Protic/polar	97	20 (10)	20 (10)
2-propanol	Protic/polar	82	15 (6)	10 (6)
Ethanol	Protic/polar	78	20 (5)	19 (8)
Methanol	Protic/polar	65	24 (8)	15 (6)
Acetone	Aprotic/polar	57	15 (5)	22 (5)
Xylene	Nonpolar	139	−(66)	1 (66)
Toluene	Nonpolar	111	−(30)	3 (30)
Triethylamine	Nonpolar	90	−(30)	8 (9)
Cyclohexane	Nonpolar	81	−(30)	5 (27)
Benzene	Nonpolar	80	−(30)	8 (28)
Ethyl acetate	Nonpolar	77	−(30)	26 (7.8)
Hexane	Nonpolar	69	−(30)	1 (28)
THF	Nonpolar	66	2 (20)	22 (5)
Dichloromethane	Nonpolar	40	−(30)	22 (2)
n-pentane	Nonpolar	36	−(30)	11 (10)
Diethyl ether	Nonpolar	35	5 (14)	5 (5)

[a]In general terms, any solvent that contains labile H^+ is called a protic solvent. The molecules of such solvents readily donate protons (H^+) to reagents

A simple microwave irradiation method described by Pol et al. [35] is for the large-scale synthesis of sub-micrometer-sized TiO_2 rods performed at normal atmospheric pressure and boiling temperature of the solvent. These authors emphasized that only 1–3 min of microwave irradiation was adequate to react tetra-isopropyl orthotitanate with ethylene glycol to produce rods of titanium glycolate with diameters of ~0.4 μm and lengths up to 5 μm. The relatively close boiling points of the solvent (ethylene glycol 198 °C) and the solute (232 °C) apparently enhanced the coordination tendency of the solute and the solvent, thereby influencing the superheating phenomenon that accelerated the formation of 1D sub-micrometer structures of a titanium glycolate complex.

4.7 Coffee Break 4: What Is a Microwave Oven?

An early consumer microwave oven on display at a House Secrets Exhibition was a 1959 Swedish Husqvarna, Electronic 2001 "Cupol" microwave oven (Fig. 4.28) [36]. Various oven designs were proposed as the market for microwave domestic ovens grew, leading to today's general-purpose design.

The components of a typical microwave oven are ***a magnetron, a cavity, a power supply, a timer, power selection, and other devices*** (Fig. 4.29a).

1. ***The Magnetron***: an air-cooled vacuum tube that generates microwave energy at 2450 ± 50 MHz; this energy is then launched (inserted) into the microwave oven cavity.
2. ***The Cavity***: a carefully-designed metal box made of either stainless steel or cold-rolled steel (in which case it is painted) that contains the microwave energy (microwaves cannot penetrate metals). This energy effectively bounces off the walls of the cavity and heats the food. The oven cavity usually contains some device to improve the dispersion of the microwave energy to provide more uniform heating, i.e., overcoming the *hot* and *cold* spots: areas of intense and

Fig. 4.28 The early Swedish "Cupol" microwave oven (1959). Reproduced with permission from ref. [36]

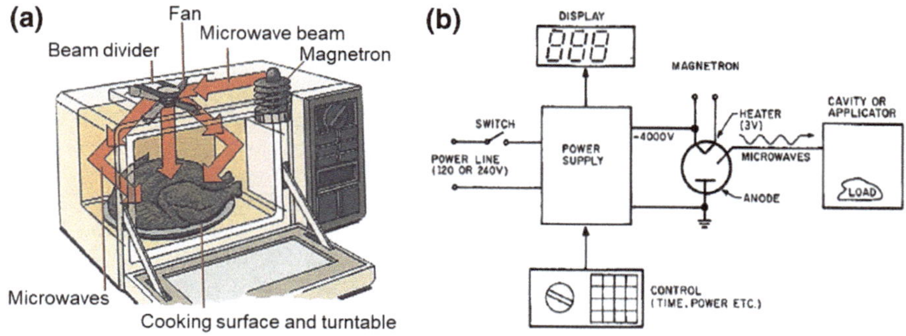

Fig. 4.29 (**a**) Cutout image of a typical microwave oven. The "fan" is a mode stirrer. Reproduced with permission from The International Microwave Power Institute. (**b**) A block diagram illustrating the main components. Reproduced with permission from ref. [37]. Copyright 1993 by Van Nostrand Reinhold

little microwave energy. The two major types are "mode stirrers" (located inside the ceiling of the cavity; they resemble slowly rotating fans but with each fan blade pitched at a different angle from the others) and turntables usually made of glass but could also be metallic if there is hot air in the oven, which rotates the food throughout the oven cavity. The access door is also specially designed to prevent leakage of the microwave radiation.

3. ***The Power supply***: it consists of a transformer, a diode, and a condenser; its purpose is to convert the electrical mains AC voltage (100, 120, 240 v at 50 or 60 Hz, depending on geography) to 4000 v DC, which then excites the magnetron to produce microwaves at 2450 MHz that are then launched into the oven as shown in Fig. 4.29b.
4. ***The Timer***: digital or analog timer is used to select the cooking time.
5. ***The Power selection***: it is done through a microprocessor; it allows the consumer to select the percentage of microwave power for the cooking cycle, e.g., 30% defrosting, 50% medium power, etc. Nearly, all microwave ovens control power by pulsing the full power On and Off during the cooking cycle: for example, for medium (50%) power the microwave energy may be ON for 10 s then OFF for 10 s, then On for 10 s and Off for 10 s, etc. These times may vary, so it might be 1 s On and 1 s Off, or 30 s On and 30 s Off, etc. There is only one microwave oven (Panasonic) currently on the market that uses solid-state means to truly vary the power, similar to the dimmer light switch that dims a light in the room.

Table 4.3 Some critical factors of domestic microwave ovens

Item	Definition
Power	The output power of the magnetron, i.e., the power that is used to heat the food is rated in watts and this may vary, oven to oven, from 400 to 1200 W. In general, the higher the wattage, the faster the food will heat. However, the higher the wattage, the greater will be the current required to operate the oven, so this may limit combining a grill element with the microwaves: for example, because the common kitchen electricity is 120 v and 15 amperes in the USA, such combination ovens are not common
Size	The cavity size may vary widely from as little as 17 L to as much as 57 L and greater. This difference is useful as kitchen sizes and available countertop space vary widely in the world
Magnetron launch	Older microwave ovens had microwave energy inserted into the cavity from either the top or top and bottom of the cavity. In the last 10 or so years, this changed so that microwave energy is now commonly launched into the oven through an opening in the right side wall (looking at the front of the oven)
Types of ovens (Fig. 4.30)	The most common types of microwave ovens are those that are found on kitchen countertops; however, there are other designs: (i) over the cooking range, (ii) under the countertop (drawer ovens), (iii) built-in microwave ovens, i.e., usually larger wall ovens
Others	The presence or absence of auxiliary heating and microprocessor control

6. ***The other devices***: grill elements, steam, and hot air for auxiliary heating are used in some ovens to supplement the microwave energy. This is particularly important for browning, crisping, and otherwise cooking food, rather than simply reheating it. Microprocessors and computer chips may do more to take the guesswork out of microwave cooking.

With over a billion (thousand millions) microwave ovens in the world, there are many differences other than cosmetic; these are summarized in Table 4.3. Types of commercially available microwave ovens are illustrated in Fig. 4.30.

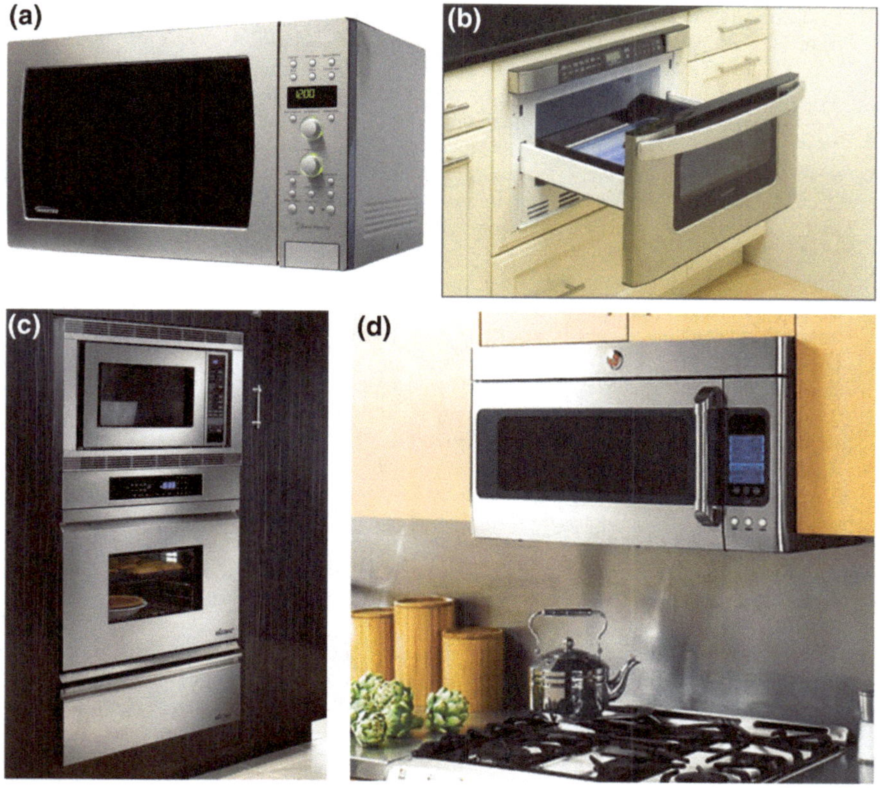

Fig. 4.30 Different types of domestic microwave ovens: (**a**) countertop, (**b**) under the counter, (**c**) built-in wall oven, (**d**) over the range

References

1. https://en.wikipedia.org/wiki/Microwave
2. J.R. Goldsmith, Epidemiologic evidence relevant to radar (microwave) effects. Environ. Health Perspect. **105**(Suppl. 6), 1579–1587 (1997)
3. S. Smiles, *The Life of George Stephenson and of his son Robert Stephenson* (Harper & Brothers, New York, 1868), p. 321. (https://commons.wikimedia.org/w/index.php?curid=31589133)
4. Agilent, *Basics of Measuring the Dielectric Properties of Materials*, Application Note (Agilent Technologies, Inc. 2006). (http://academy.cba.mit.edu/classes/input_devices/meas.pdf)
5. T. Sumi, R. Dillert, S. Horikoshi, Novel microwave thermodynamic model for alcohol with clustering structure in nonpolar solution. J. Phys. Chem. B **119**, 14479–14485 (2015)
6. S. Horikoshi, N. Shinohara, H. Takizawa, J. Fukushima, *Microwave Chemistry* (Sankyo Publishing Co. Ltd., 2013)
7. K.D. Raner, C.R. Strauss, R.W. Trainor, J.S. Thorn, A new microwave reactor for batchwise organic synthesis. J. Org. Chem. **60**, 2456–2460 (1995)

8. H. Takizawa, A. Hagiya, T. Aoyagi, Y. Hayashi, Preparation of mesoscopic TiO_2–SnO_2 composite grains by spinodal decomposition under 28 GHz microwave irradiation. Chem. Lett. **37**, 714–715 (2008)
9. Y. Suttisawat, H. Sakai, M. Abe, P. Rangsunvigit, S. Horikoshi, Microwave effect in the dehydrogenation of tetralin and decalin with a fixed-bed reactor. Intern. J. Hydrogen Energy **37**, 3242–3250 (2012)
10. S. Horikoshi, A. Osawa, M. Abe, N. Serpone, On the generation of hotspots by microwave electric and magnetic fields and their impact on a microwave-assisted heterogeneous reaction in the presence of metallic Pd nanoparticles on an activated carbon support. J. Phys. Chem. C **115**, 23030–23035 (2011)
11. http://en.wikipedia.org/wiki/Microwave_chemistry
12. S. Zhang, D.O. Hayward, D.M.P. Mingos, Apparent equilibrium shifts and Hotspot formation for catalytic reactions induced by microwave dielectric heating. Chem. Commun. 975–976 (1999)
13. Y. Tsukahara, A. Higashi, T. Yamauchi, T. Nakamura, N. Yasuda, A. Baba, Y. Wada, *In situ* observation of nonequilibrium local heating as an origin of special effect of microwave on chemistry. J. Phys. Chem. C **114**, 8965–8970 (2010)
14. B. Gutmann, A.M. Schwan, B. Reichart, C. Gspan, F. Hofer, C.O. Kappe, Activation and deactivation of a chemical transformation by an electromagnetic field: evidence for specific microwave effects in the formation of grignard reagents. Angew. Chem. Int. Ed. **50**, 4636–4640 (2011)
15. J.A. Menéndez, E.J. Juárez-Pérez, E. Ruisánchez, J.M. Bermúdez, A. Arenillas, Ball lightning plasma and plasma arc formation during the microwave heating of carbons. Carbon **49**, 346–349 (2011)
16. S. Horikoshi, A. Osawa, S. Sakamoto, N. Serpone, Control of microwave-generated hotspots. Part V. Mechanisms of hotspot generation and aggregation of catalyst in a microwave-assisted reaction in toluene catalyzed by Pd-loaded AC particulates, Appl. Catal. A: Gen. **460–461**, 52–60 (2013)
17. G. Cravotto, M. Beggiato, A. Penoni, G. Palmisano, S. Tollari, J.-M. Lévêques, W. Bonrath, High-intensity ultrasound and microwave, alone or combined, promote Pd/C-catalyzed aryl–aryl couplings. Tetrahedron Lett. **6**, 2267–2271 (2005)
18. J.M. McMahon, S. Li, L.K. Ausman, G.C. Schatz, Modeling the effect of small gaps in surface-enhanced Raman spectroscopy. J. Phys. Chem. C **116**, 1627–1637 (2012)
19. S. Horikoshi, Y. Suttisawat, A. Osawa, C. Takayama, X. Chen, S. Yang, H. Sakai, M. Abe, N. Serpone, Organic syntheses by microwave selective heating of novel metal/CMC catalysts—the Suzuki–Miyaura coupling reaction in toluene and the dehydrogenation of tetralin in solvent- free media. J. Catal. **289**, 266–271 (2012)
20. S. Horikoshi, A. Osawa, S. Sakamoto, and N. Serpone Control of microwave-generated Hotspots. Part IV. Control of Hotspots on a heterogeneous microwave-absorber catalyst surface by a hybrid internal/external heating method, Chem. Eng. Process. **69**, 52–56 (2013)
21. S. Horikoshi, M. Kamata, T. Mitani, N. Serpone, Control of microwave-generated hotspots. 6. Generation of hotspots in dispersed catalyst particulates and factors that affect catalyzed organic syntheses in heterogeneous media. Ind. Eng. Chem. Res. **53**, 14941–14947 (2014)
22. S. Horikoshi, N. Serpone, in *Microwaves in Organic Synthesis*, ed by A. de la Hoz, A. Loupy, 3rd ed. (Wiley, Weinheim, 2012), pp. 377–423
23. S. Horikoshi, M. Kamata, T. Sumi, N. Serpone, Selective heating of Pd/AC catalyst in heterogeneous systems for the microwave-assisted continuous hydrogen evolution from organic hydrides: temperature distribution in the fixed-bed reactor. Inter J. Hydrogen Energ. **41**, 12029–12037 (2016)
24. A.C. Metaxas, R.J. Meredith, *Industrial of Microwave Heating* (Peter Peregrines, London, 1983)
25. E.T. Thostenson, T.-W. Chou, Microwave processing: fundamentals and applications. Composites A **30**, 1055–1071 (1999)

26. B. Meng, B.D.B. Klein, J.H. Booske, R.F. Cooper, Microwave absorption in insulating dielectric ionic crystals including the role of point defects. Phys Rev. B **53**, 12777–12785 (1996)
27. J.E. Eldridge, P.R. Staal, Far-infrared dispersive-reflection measurements on NaCl, compared with calculations based on cubic and quartic anharmonicity. I. Room temperature. Phys. Rev. B **16**, 4608–4618 (1977)
28. http://www.fda.gov/RadiationEmittingProducts/RadiationEmittingProductsandProcedures/HomeBusinessandEntertainment/ucm142506.htm
29. J. Meyer, Zur Kenntnis des negative Druckes in Flüssigkeiten. Abhandl. Dent. Bunsen Ges. **III**(1) and 6, (1911)
30. A. Ferrari, J. Hunt, A. Stiegman, G.B. Dudley, Microwave-assisted superheating and/or microwave-specific superboiling (nucleation-limited boiling) of liquids occurs under certain conditions but is mitigated by stirring. Molecules **20**, 21672–21680 (2015)
31. D.R. Baghurs, D. Michael, P. Mingos, Superheating effects associated with microwave dielectric heating. Chem. Commun. 674–677 (1992)
32. F. Chemat, E. Esveld, Microwave super-heated boiling of organic liquids: Origin, effect and application. Chem. Eng. Technol. **24**, 735–744 (2001)
33. S. Horikoshi, S. Iida, M. Kajitani, S. Sato, N. Serpone, Chemical reactions with a novel 5.8-GHz microwave apparatus. 1. Characterization of properties of common solvents and application in a Diels-Alder organic synthesis. Org. Process Res. Dev. **12**, 257–263 (2008)
34. P. Klán, J. Literák, S. Relich, Molecular photochemical thermometers: investigation of microwave superheating effects by temperature dependent photochemical processes. J. Photochem. Photobiol. A **143**, 49–57 (2001)
35. V.G. Pol, Y. Langzam, A. Zaban, Application of microwave superheating for the synthesis of TiO_2 rods. Langmuir **23**, 11211–11216 (2007)
36. https://museumvictoria.com.au/about/mv-blog/jun-2011/five-things-about-microwaves/
37. C.R. Buffler, *Microwave Cooking and Processing* (Van Nostrand Reinhold, New York, 1993)

Chapter 5
Physics of Microwave Heating

Abstract Microwave heating generates heat by absorption and loss of energy. Accordingly, how does a substance turn microwave into heat? How does one measure the efficiency of microwave heating? These topics require to be understood. As such, this chapter provides some commentaries on these topics. Furthermore, a method for measuring the heating efficiency of microwaves, impedance matching, is also explained. There are books out there that presume that microwave heating refers to heating in which dielectric heating proceeds. This is not entirely correct, as there are several types of microwave heating phenomena. This chapter also describes the type of microwave heating through illustrations and some equations. Furthermore, actual examples and chemical reactions are given; such topics as the penetration depth of microwaves, the skin effect, the frequency effect, simulation of electromagnetic waves, and electrothermal and transmission modes are discussed. The coffee break shows the proper usage of microwave ovens.

Keywords Dielectric properties · Dielectric constant · Dielectric loss · Tan δ
Permeability · Transmission/reflection line method · Open-ended coaxial probe method · Free-space method · Resonant method · Impedance · Dipole rotation
Conduction loss heating · Dielectric loss heating · Magnetic loss heating
Impedance · Dipole rotation · Conduction loss heating · Penetration depth
Skin depth · Industrial · Scientific and medical (ISM) bands · 915-MHz frequency
2.45-GHz frequency · 5.80-GHz frequency · Frequency effect
Diels–Alder reaction · Synthesis of ionic liquids · Nanoparticle synthesis
Electromagnetic simulations · Thermodynamics simulations · Transmission modes

5.1 Dielectric Properties

Microwaves are electromagnetic waves that mimic an alternating current (electric and magnetic fields). Interactions between microwaves and materials depend upon the physical properties of the materials; that is, how the alternating electric and magnetic fields interact with materials. When considering microwave heating, it is

S. Horikoshi et al., *Microwave Chemical and Materials Processing*,
https://doi.org/10.1007/978-981-10-6466-1_5

necessary to understand the electrical properties of the material since the alternating electric field interacts with the electrons of the material. Therefore, the electrical nature of the substance is related to its interaction with microwaves. Classification of classes of materials on the basis of their electrical conductivity is reported in Table 5.1. Materials that possess high electrical conductivity are conductors and examples are *metallic in nature*. Materials that display lesser electrical conductivity are the *semiconductors* (e.g., silicon). Materials that do not exhibit electrical conductivity are referred to as *dielectric* or *insulating materials*; an example is quartz, one of many insulating materials.

A material is classified as *dielectric* if it has the ability to store energy upon application of an external electric field. If a direct current (DC) voltage source is placed across a parallel plate capacitor (Fig. 5.1), more charge is stored when a dielectric material is located between the electrodes. The dielectric material increases the storage capacity of the capacitor by neutralizing the charges at the electrodes, which ordinarily would contribute to the external field. The capacitance with the dielectric material is related to its *dielectric constant*.

In Fig. 5.1, C and C_0 are, respectively, the capacitance with and without the dielectric; $\kappa' = \varepsilon_r'$ is the real dielectric constant or permittivity; A and t denote, respectively, the area of the capacitor plates and the distance between them (Fig. 5.1). The capacitance of the dielectric material is related to the dielectric constant as indicated in the equations reported in Fig. 5.1. If an AC sinusoidal voltage source V is placed across the same capacitor (Fig. 5.2), the resulting current will be made up of a charging current I_c and a loss current I_l that is related to the dielectric constant. The losses in the material can be represented as a conductance (G) in parallel with a capacitor (C).

Table 5.1 Electrical resistivity of materials (the reciprocal of conductivity) and carrier density at room temperature

Materials	Electrical resistivity ρ/Ω cm	Carrier density/cm^{-3}
Metal	10^{-6}	$>10^{22}$
Semiconductor	10^{-6}–10^{9}	10^{13}–10^{17}
Dielectric (insulating)	10^{14}	$<10^{13}$

$$C_0 = \frac{A}{t}$$

$$C = C_0 K'$$

$$K' = \varepsilon_r' = \frac{C}{C_0}$$

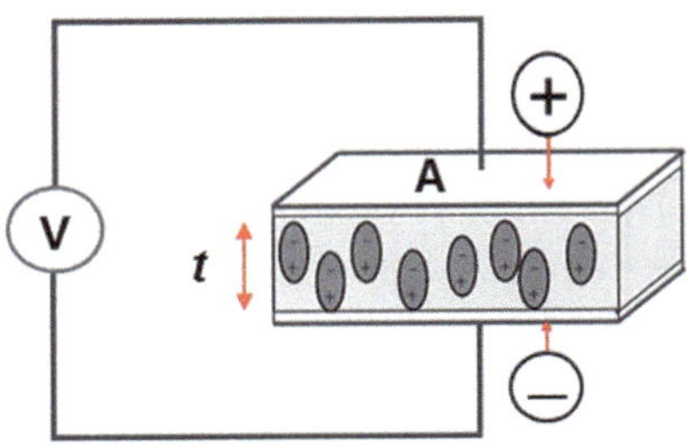

Fig. 5.1 Parallel plate capacitor, DC case. Reproduced from Ref. [1]

When an electric field interacts with an insulator, there is no current flow in the material. Electrons and ions in the material move to opposite sides under the electric field; this displacement of electrons and ions generates an electric dipole, a phenomenon referred to as *dielectric polarization*. The *electric flux density* inside the material, induced as a result of the dielectric polarization, is expressed by the product of the external electric field and the dielectric constant.

The *complex dielectric constant* κ consists of a real part κ' that represents the storage and an imaginary part κ'' that represents the loss. The following notations are often used interchangeably to denote the complex dielectric constant (Eq. 5.1):

$$\kappa = \kappa^* = \varepsilon_\mathrm{r} = \varepsilon_\mathrm{r}^* \tag{5.1}$$

From the point of view of electromagnetic theory, the definition of electric displacement (electric flux density: D_f) is given by Eq. 5.2:

$$D_\mathrm{f} = \varepsilon E \tag{5.2}$$

From Eq. 5.2, the dielectric constant is defined as the ratio between the electric flux density generated in the material and the electric field when the electric field interacts with the material.

Under microwave irradiation, the external electric field in Eq. 5.2 changes with time. Although an electromagnetic wave is represented by trigonometric functions, for mathematical convenience it can be expressed by an exponential function and complex numbers (Euler's formula: $\mathrm{e}^{j\theta} = \cos\,\theta + j\,\sin\,\theta$). As described above, under an alternating electromagnetic field, the dielectric constant becomes a complex number (Eq. 5.3):

$$\varepsilon = \frac{\tilde{D}}{\breve{E}} = \frac{D}{E}\mathrm{e}^{j\theta} = \frac{D}{E}(\cos\theta + j\sin\theta) = \varepsilon' - j\varepsilon'' \tag{5.3}$$

In Eq. 5.3, ε' is the dielectric constant (the real part of the dielectric constant when an alternating electromagnetic field is applied to the material) and ε'' is the *dielectric loss* (the imaginary part of the dielectric constant when an alternating electromagnetic field is applied to the material). The dielectric constant, ε', represents the degree of polarization in the material under an applied electric field: the effect of electrical energy. If polarization is zero, the value of ε' is zero. The

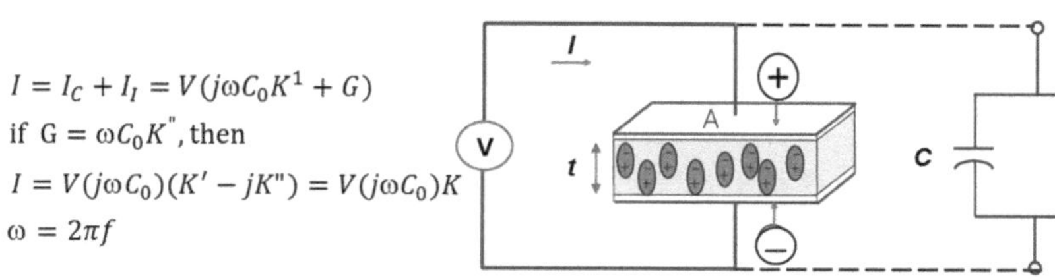

Fig. 5.2 Parallel plate capacitor, AC case. Reproduced from Ref. [1]

dielectric loss, ε'', refers to the degree of energy loss in the material under an applied electric field, the loss being converted to heat. If the time difference is zero, the value of ε'' is also zero. The dielectric constant (κ) is equivalent to the relative *permittivity* (ε_r) or the absolute permittivity (ε) relative to the permittivity of free space (ε_o: $\varepsilon_o \approx 1/36\pi \times 10^{-9}$ s). The real part of permittivity (ε_r') is a measure of how much energy from an external electric field is stored in a material. The imaginary part of permittivity (ε_r'') is called the loss factor and is a measure of how dissipative or *lossy* a material is to an external electric field; that is, how much energy is lost (converted to heat) in the material. The imaginary part of permittivity, ε_r'', is always greater than zero and is usually much smaller than ε_r'. The loss factor includes the effects of both dielectric loss and conductivity.

The dielectric polarization is time-dependent and, if accurately measured, it changes slightly with time. This time lag can be represented geometrically. If the degree and the time lag of the dielectric polarization are represented by a real component and an imaginary component, respectively, the vector diagram is then given as illustrated in Fig. 5.3. The phase of the real and imaginary components is shifted 90°. The vector sum forms an angle δ to the real axis ε'. The relative loss of materials can be represented by tan δ, which is defined by the ratio of the dielectric loss and the dielectric constant. When complex permittivity is drawn as a simple vector diagram, the real and imaginary components are 90° out of phase. The vector sum forms an angle δ (the "loss tangent") with the real axis ε_r'. The relative *lossiness* of a material is the ratio of the energy lost to the energy stored.

The *loss tangent*, or tan δ, represents the ratio of the imaginary part of the dielectric constant to the real part. In the equation of Fig. 5.3, *D* denotes the *dissipation factor* and *Q the quality factor*. The loss tangent is called *tan delta* (i.e., tan δ), or loss tangent, or dissipation factor. Sometimes, the terms *quality factor* or *Q-factor* are used with respect to an electronic microwave material; this factor is the reciprocal of the loss tangent. For very low-loss materials, since tan $\delta \approx \delta$, the loss tangent can be expressed in angle units as milliradians or microradians. This is represented as the dielectric loss tangent. We can estimate the extent to which materials are heated by microwave radiation by considering the dielectric loss and the relative loss as indicators. Since the dielectric constant takes a different value depending on the material, we use a value of the ratio against the permittivity of vacuum ($\varepsilon_o = 8.85 \times 10^{12}$ F m^{-1}); this value is referred to as the *relative dielectric constant* (ε_r'). Similarly, for dielectric loss, the ratio of the permittivity of a material

$$\tan\delta = \frac{\varepsilon_r''}{\varepsilon_r'} = D = \frac{1}{Q}$$

$$= \frac{\textit{Energy lost per cycle}}{\textit{Energy stored per cycle}}$$

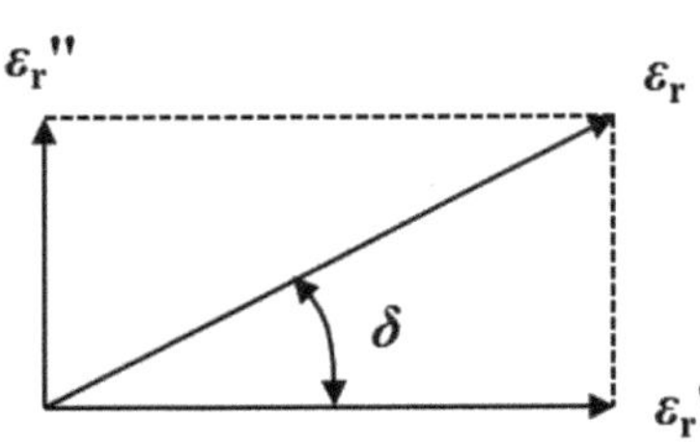

Fig. 5.3 Loss tangent vector diagram. Reproduced from Ref. [1]

to that of a vacuum is represented as the relative dielectric loss: ε_r''. Hence, the loss tangent is given by the ratios expressed in Eq. 5.4:

$$\tan \delta = \frac{\varepsilon''}{\varepsilon'} = \frac{\varepsilon_r''}{\varepsilon_r'} \tag{5.4}$$

Figure 5.4 shows the relationship between the relative dielectric constant and the dielectric loss of powdered materials and several liquids. The dielectric constants of the powders were measured at a microwave frequency of 6 GHz using the perturbation method (Fig. 5.4a) [2]. Carbon has both a large relative dielectric constant and dielectric loss factor. In an oxide consisting of molecules of the same type, the dielectric loss factor is largely different depending on the valence. For example, in iron oxides, the dielectric loss factor of Fe_2O_3 is less than 0.01 at the same level as SiO_2, but the dielectric loss factors of FeO and Fe_3O_4 are greater than 1, approximately two orders of magnitude larger. Figure 5.4b shows the dielectric constant and dielectric loss of liquids [3]. Polar solvents possess greater dielectric loss factors than nonpolar solvents. In particular, water and methanol have a dielectric loss greater than 10. On the other hand, organic solvents and gasoline have smaller dielectric loss factors, and thus microwave heating of such solvents tends to be difficult.

Note that the dielectric polarization consists of (i) electronic polarization, (ii) atomic polarization, (iii) orientation polarization, and (iv) ionic polarization; each of these is determined by the frequency.

In the microwave frequency band, the peak of the complex dielectric constant is in the region of orientation polarization. Further, as inferred from Fig. 3.7 of Chap. 3, in the microwave band, atomic polarization and electron polarization can respond quickly to changes in the microwave's electric field. A material may have several dielectric mechanisms for polarization effects that contribute to its overall permittivity (Fig. 3.7 of Chap. 3). Electric charge carriers in a dielectric material are

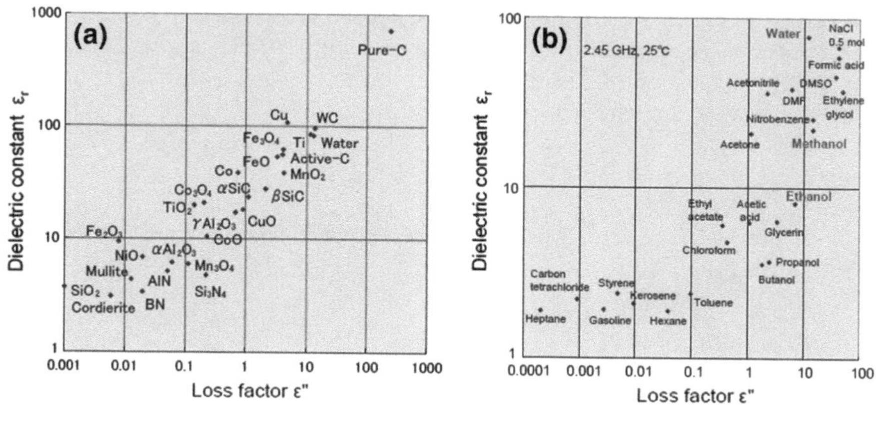

Fig. 5.4 Complex permittivity and loss factor of various powers and liquids in the microwave region. (**a**) powder materials (measured at 6 GHz) and (**b**) liquids (reference at 2.45 GHz). Reproduced from Ref. [2]. Copyright by the GCMEA Committee

so arranged that they can be displaced by an electric field. The charges become polarized to compensate for the electric field, so that the positive and negative charges move in opposite directions. At the microscopic level, several dielectric mechanisms can contribute to the dielectric behavior. Dipole orientation and ionic conduction interact strongly at microwave frequencies. Water molecules, for example, possess permanent dipoles, which rotate to follow an alternating electric field, making water quite lossy—this is why water-based food heats well in a microwave oven. Atomic and electronic mechanisms are relatively weak and usually constant over the microwave region. Each dielectric mechanism has a characteristic *cutoff frequency*. As frequency increases, the slow mechanisms drop out, in turn, leaving the faster ones to contribute to ε'. The loss factor (ε_r'') will peak correspondingly at each critical frequency. The magnitude and *cutoff frequencies* of each mechanism are unique to each material. Water has a strong dipolar effect at low frequencies—but its dielectric constant rolls off dramatically around 22 GHz. On the other hand, PTFE has no dipolar mechanisms and its permittivity is remarkably constant well into the millimeter wave region.

A resonant effect is usually associated with electronic or atomic polarization, whereas a relaxation effect is usually associated with orientation polarization, which depends on frequency because it changes with the nature of the microwave radiation. Therefore, the electric flux density of an insulator at a given time t is separated into a term $D_p(t)$ that can follow and a term $D_d(t)$ that can delay a change of the microwave electric field (Eqs. 5.5 and 5.6):

$$D_{\mathrm{p}}(t) = \varepsilon_0\varepsilon_{\mathrm{r}}(\infty)E(t) \tag{5.5}$$

$$D_{\mathrm{d}}(t) = \int_0^t E(u)f(t-u)\mathrm{d}u \tag{5.6}$$

where $f(t-u)$ is a function of the dielectric constant. This means that the microwave electric field at an earlier time u affects the dielectric constant at current time t (that function is referred to as the *after-effect function*). Thus, the electric flux density $D(t)$ is the sum of a term $D_{\mathrm{p}}(t)$ that can follow and a term $D_{\mathrm{d}}(t)$ that can delay a change of the microwave electric field (Eq. 5.7),

$$D(t) = D_{\mathrm{p}}(t) + D_{\mathrm{d}}(t) \tag{5.7}$$

If we now let ε be $\varepsilon_0\varepsilon_{\mathrm{r}}^*$, where $\varepsilon_{\mathrm{r}}^*$ is the complex relative dielectric constant (Eqs. 5.8–5.11),

$$\varepsilon_0\varepsilon_{\mathrm{r}}^*(\omega)E(t) = \varepsilon_0\varepsilon_{\mathrm{r}}(\infty)E(t) + \int_0^t E(u)f(t-u)\mathrm{d}u \tag{5.8}$$

$$\varepsilon_0\{\varepsilon_r^*(\omega) - \varepsilon_r(\infty)\}E(t) = \int_0^t E(u)f(t-u)\mathrm{d}u \tag{5.9}$$

$$\varepsilon_0\{\varepsilon_r^*(\omega) - \varepsilon_r(\infty)\} = \int_0^t \mathrm{e}^{j\omega(u-t)}f(t-u)\mathrm{d}u \tag{5.10}$$

$$\varepsilon_0\{\varepsilon_r^*(\omega) - \varepsilon_r(\infty)\} = -\int_t^0 \mathrm{e}^{-j\omega x}f(x)\mathrm{d}x = \int_0^t \mathrm{e}^{-j\omega x}f(x)\mathrm{d}x \tag{5.11}$$

and if we let the current time be *far ahead* (i.e., infinity) and consider the integral range $x = t - u = \infty$ (Eq. 5.12),

$$\varepsilon_0\{\varepsilon_r^*(\omega) - \varepsilon_r(\infty)\} == \int_0^\infty \mathrm{e}^{-j\omega x}f(x)\mathrm{d}x \tag{5.12}$$

and if the *after-effect function* is expressed by Eq. 5.13,

$$f(x) = \varepsilon_0\{\varepsilon_r(0) - \varepsilon_r(\infty)\}\frac{\mathrm{e}^{-x/\tau}}{\tau} \tag{5.13}$$

then we obtain a complex relative dielectric constant given by Eqs. 5.14 and 5.15:

$$\begin{aligned}
\varepsilon_0\{\varepsilon_r^*(\omega) - \varepsilon_r(\infty)\} &= \int_0^\infty \mathrm{e}^{-j\omega x}f(x)\mathrm{d}x \\
&= \int_0^\infty \mathrm{e}^{-j\omega x}\left[\varepsilon_0\{\varepsilon_r(0) - \varepsilon_r(\infty)\}\frac{\mathrm{e}^{-x/\tau}}{\tau}\right] \\
&= \int_0^\infty \varepsilon_0\{\varepsilon_r(0) - \varepsilon_r(\infty)\}\frac{\mathrm{e}^{-x(1+j\omega\tau)/\tau}}{\tau}\mathrm{d}x \\
&= \varepsilon_0\{\varepsilon_r(0) - \varepsilon_r(\infty)\}\frac{1}{\tau}\left[-\frac{\tau}{1+j\omega\tau}\mathrm{e}^{-x(1+j\omega\tau)/\tau}\right]_\tau^\infty \\
&= \frac{\varepsilon_0\{\varepsilon_r(0) - \varepsilon_r(\infty)\}}{1+j\omega\tau}[(-0) - (-1)] \\
&= \frac{\varepsilon_0\{\varepsilon_r(0) - \varepsilon_r(\infty)\}}{1+j\omega\tau}
\end{aligned} \tag{5.14}$$

$$\varepsilon_r^*(\omega) = \varepsilon_r(\infty) + \frac{\varepsilon_r(0) - \varepsilon_r(\infty)}{1+j\omega\tau} \tag{5.15}$$

Equation 5.15 shows that the complex relative dielectric constant is a function of the frequency—it is known as Debye's relaxation equation. The relationship between microwave heating and the complex dielectric constant is described next in Sect. 5.2. In addition, the dependence of the dielectric constant on temperature will be described later (see below, Sect. 5.3.4).

5.2 Permeability

Permeability (μ) describes the interaction of a material with a magnetic field. A similar analysis can be performed for permeability using an inductor with resistance to represent core losses in a magnetic material (Fig. 5.5). If a DC current source is placed across an inductor, the inductance with the core material can be related to the permeability. In the equations of Fig. 5.5, L is the inductance with the material, L_o is the free-space inductance of the coil, and μ' is the real permeability. If an AC sinusoidal current source is placed across the same inductor, the resulting voltage will be made up of an induced voltage and a loss voltage that is related to permeability. The core loss can be represented by a resistance (R) in series with an inductor (L). The complex permeability (μ^* or μ) consists of a real part (μ') representing the energy storage term, and an imaginary part (μ'') representing the energy loss term. The relative permittivity μ_r is the permittivity relative to free space (Eq. 5.16):

$$\mu_r = \frac{m}{m_0} = \mu_r - j\mu_r'' \tag{5.16}$$

Note that $\mu_o = 4\pi \times 10^{-7}$ H m^{-1} is the free-space permeability.

Some materials, such as iron (ferrites), cobalt, nickel, and their alloys have appreciable magnetic properties; however, many materials are non-magnetic making the permeability very close to the permeability of free space ($\mu_r = 1$). All materials, on the other hand, have dielectric properties, so the focus of this discussion is mostly on permittivity measurements.

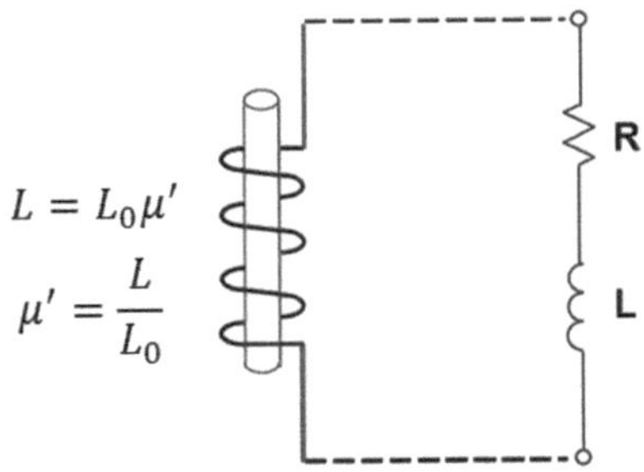

Fig. 5.5 Image of an inductor. Reproduced from Ref. [1]

5.3 Measurement of Electric and Magnetic Permeabilities

The heating efficiency of a sample exposed to microwave radiation depends upon the electric and magnetic permeabilities of the sample. Many researchers have reported the electric permeability and magnetic permeability of pure materials; however, when using a mixture in a chemical reaction it is best to measure these in the actual sample. It would not be possible to assess the microwave heating efficiency without first measuring the permeability. For example, impurities (e.g., small quantities of some ionic species) present in a polar solvent would significantly improve the microwave heating efficiency.

There exist many methods for measuring the complex permittivity and permeability of samples, each method being limited to, among others, specific microwave frequencies, materials, and applications. Accordingly, this section introduces and describes the four methods used to measure the permeability. Technical books may also be consulted for more details; the research paper by Venkatesh and Raghavan [4] is also worth consulting as it provides an overview of the various techniques. Heat radiated to the atmosphere from a microwave-heated sample may be large which will affect the microwave heating efficiency of the sample. Even if the heating efficiency were estimated by calculating the electric and magnetic permeabilities, a judicious researcher should always determine the difference between the microwave heating efficiency and the radiated heat to the atmosphere of an actual sample.

Table 5.2 identifies examples of classes of materials and the dielectric properties determined using various measurement methods. These methods are best used at some specified frequency, and at some given degree of microwave absorption depending upon the nature of the sample (liquid, bulk, or powder) and on electrical characteristics.

The **Transmission/Reflection line method** is a popular broadband measurement technique in which only the fundamental waveguide modes (TEM mode in coaxial line and TE mode in waveguides) are assumed to propagate. The **Open-ended coaxial probe method** is a nondestructive technique that assumes only the TEM or

Table 5.2 Comparison between the measurement methods

Measurement techniques	Features	Dielectric properties
Transmission/reflection line method	Low-loss samples, solids	ε_r, μ_r
Open-ended coaxial probe method	Liquids, biological specimens, semi-solids, high loss samples	ε_r
Free-space method	High-temperature samples, large flat solids, gases, hot liquids	ε_r, μ_r
Resonant method (Cavity)	Rod-shaped solid materials, liquids, low-loss samples, small samples	ε_r, μ_r

TE modes propagate. The **Resonant method** provides high accuracies and assumes TE or TM propagation modes. The **Free-space method** is for broadband applications and assumes that only TEM modes propagate.

5.3.1 The Transmission/Reflection Line Method

The transmission/reflection line method involves placing a sample in a section of the waveguide or coaxial line followed by measuring the two complex parts of the scattering parameters using a vector network analyzer [5]. This vector network analyzer is an instrument that measures the network parameters of electrical networks. At present, network analyzers commonly measure scattering parameters (that is, the elements of a scattering matrix, or S-matrix) because reflection and transmission of electrical networks are easy to measure at high frequencies. Network analyzers are used mostly at high frequencies; operating frequencies may range from 5 Hz to 1.05 THz. Scattering parameters (i.e., S-parameters) describe the electrical behavior of linear electrical networks when undergoing various steady-state stimuli by electrical signals. This method is used for measurements of high loss materials. It can be classified into small groups of the transmission line method, an S-parameter method, and a free-space method.

The relevant scattering parameters relate closely to the complex permittivity and permeability of the material. The conversion of S-parameters to complex dielectric parameters is calculated using a computer program for solving the relevant equations. In many cases, the method requires initial preparation of the sample (e.g., machining) so that the sample fits tightly into the waveguide or the coaxial line. Calibration in transmission line measurements uses various terminations that produce different resonant behaviors in the transmission line (see Fig. 5.6) [6]. For good dielectric measurements, the maximum electric field is required, which can be achieved by open circuit or other capacitive termination, whereas calibration in coaxial line measurements can be made using a short circuit, an open circuit, or a matched load termination. Whatever the method used, it provides for the measurement of the permittivity and permeability of dielectric materials.

5.3.2 The Open-Ended Coaxial Probe Method

The open-ended coaxial probe method has been used for years as a nondestructive testing procedure and is the most often used procedure in microwave chemistry. With this equipment and a probe, dielectric factors of the solution or of the sample can be measured, regardless of whether or not a researcher has the appropriate technical knowledge of the system. Measurements are not unlike measuring the pH of a solution with a pH meter. The coaxial probe is inexpensive and the measurement method is relatively simple. It is possible to measure the permittivity of a

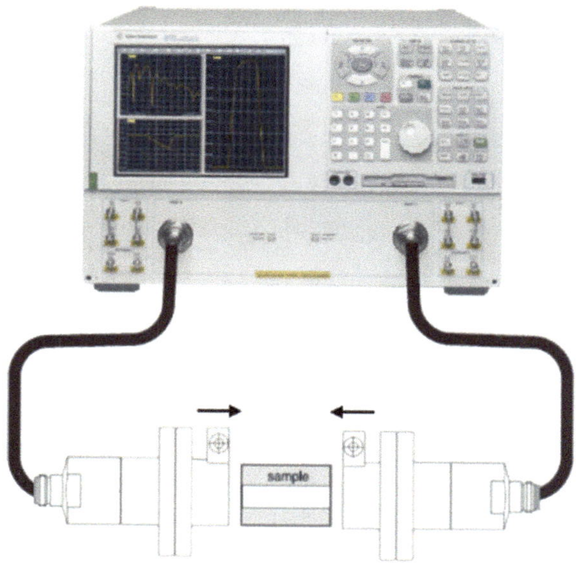

Fig. 5.6 Image illustrating the measurement of permeability using the transmission line method. Reproduced from Ref. [6]. Copyright 2013 by Wiley-VCH

sample and its dissipation loss with temperature changes, because sample temperatures from −40 to +200 °C can be measured with the probe for low- and high-temperature systems. For example, the three measuring probes are (i) the high-temperature probe (Fig. 5.7a), (ii) the slim form probe (Fig. 5.7b), and (iii) the performance probe (Fig. 5.7c) [1].

In this open-ended coaxial probe method, the probe is pressed against a solid specimen or else is immersed into a liquid sample, followed by measuring the reflection coefficient subsequently used to determine the permittivity. Note that for some measurements, it may not be possible to cut or otherwise modify the sample

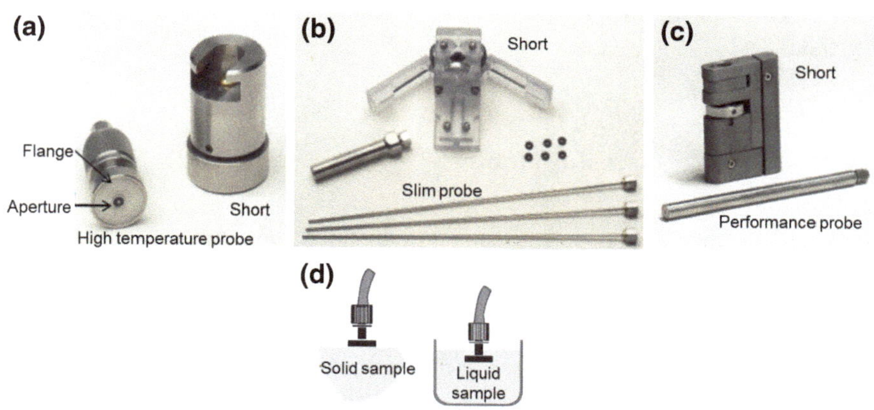

Fig. 5.7 Types of probes (**a**) high-temperature probe, (**b**) slim probe, and (**c**) performance probe; (**d**) images of measurement of samples using an open coaxial probe. Reproduced from Ref. [1]

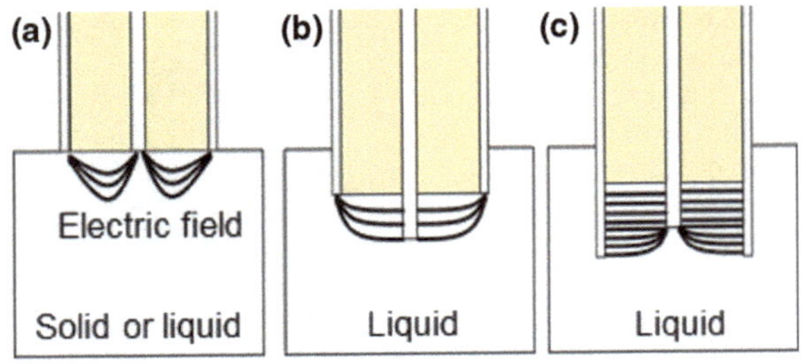

Fig. 5.8 Images of types of coaxial probes

of a material for measurement. This is especially important in the case of performing *in vivo* measurements on biological specimens because the material characteristics may change. In this case, the probe may be placed in close contact with the sample without causing any changes in the material's characteristics.

The reflection coefficient is measured using a vector network analyzer with a probe system that is first calibrated so that reflection coefficient measurements are referenced to the probe's aperture plane. This can be done in two ways, one of which uses reference liquids for direct calibration at the open end of the probe—a very direct and simple procedure. Uncertainties in the measurement are due to uncertainties in the characterization of the reference liquids, and to the selection of reference liquids as calibration standards. The reference liquid is used as a calibration standard and must be a liquid with ***known*** dielectric properties: for example, water, a saline solution, or methanol are usually selected as reference liquids.

To perform an actual measurement, the end of the coaxial probe is immersed in the liquid or in the case of solids is pressed tightly against the surface of the solid sample. Dissipation loss is measured for high loss solids and liquid samples. The three kinds of *maser* (microwave amplification by stimulated emission of radiation) coaxial probes are illustrated in Fig. 5.8. Measurement and cleaning after the measurement of a solid sample are simple for the type of maser displayed in Fig. 5.8a. The types shown in Fig. 5.8b and 5.8c are used for liquid samples for which measurement sensitivity is much greater than for the type depicted in Fig. 5.8a.

5.3.3 The Free-Space Method

The free-space method uses antennas to focus the microwave energy at or through a slab or sheet of the material (Fig. 5.9) [1]. This method is a noncontact method that can be applied to materials to be tested under high temperatures and in hostile environments. The measurement requires the materials to be large and flat. As a non-contact method, the free-space measuring technique does not require the sample to be inserted in the waveguide and resonator. *Time-Domain Gating* can cancel the influence of secondary reflection, which is a problem of the transmission/reflection line method. The precision of the free-space method depends upon the

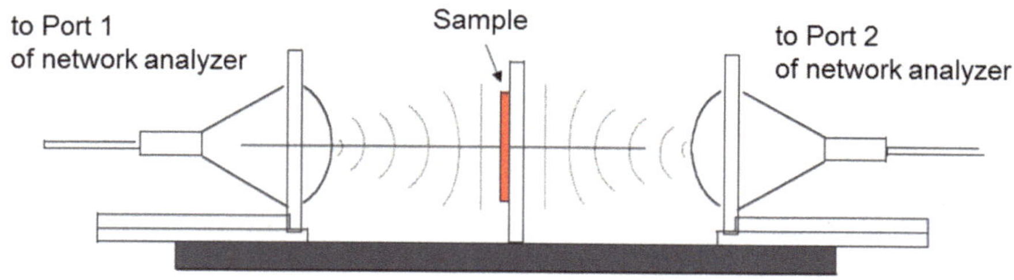

Fig. 5.9 Setup of measurement of a sample using the free-space method. Reproduced from Ref. [1]

environment. For example, when the room temperature changes during the measurement, the measured value is changed; when the cable connected to a network analyzer is moved, the measured value is changed. Hence, control of these and other factors are required to eliminate measurement errors.

5.3.4 The Resonant Method

Resonant measurements are the most accurate procedures for obtaining the permittivity and permeability of samples. This technique measures a small sample using a microwave resonator. The dielectric loss and the magnetic permeability can be measured by the changes in the resonant frequency and the quality factor (Q-factor) of the cavity resonator shown in Fig. 5.10. (Sect. 5.4.2). It is particularly well suited for measurements of low dissipation loss materials. However, it is not possible to measure large samples because of changes they may cause in the resonant conditions in the cavity. There are limitations on the frequencies and loss characteristics of the materials that can be measured with this method. Accordingly, there are many types of resonant methods available such as re-entrant cavities, split cylinder resonators, cavity resonators, and Fabry–Perot resonators among others. Two types of resonant measurements are commonly used: (1) perturbation methods are suitable for all permittivity measurements, magnetic materials, and medium-to-high loss materials; and (2) the low-loss measurement method is applicable for low-loss materials using larger samples. However, the perturbation method is more popular, especially using the TM cavity geometry as those shown in Fig. 5.10a, b. A cavity resonator method is the method with which the electric permeability can be determined most precisely for a solid sample, more so than a liquid. The formula of the material constant is easily obtained when the perturbation method is used for a minute sample inserted in a cavity resonator. Thus, a cavity resonator method can separate the magnetic permeability from the permittivity using a field distribution in the cavity resonator and changing the location of the sample insert.

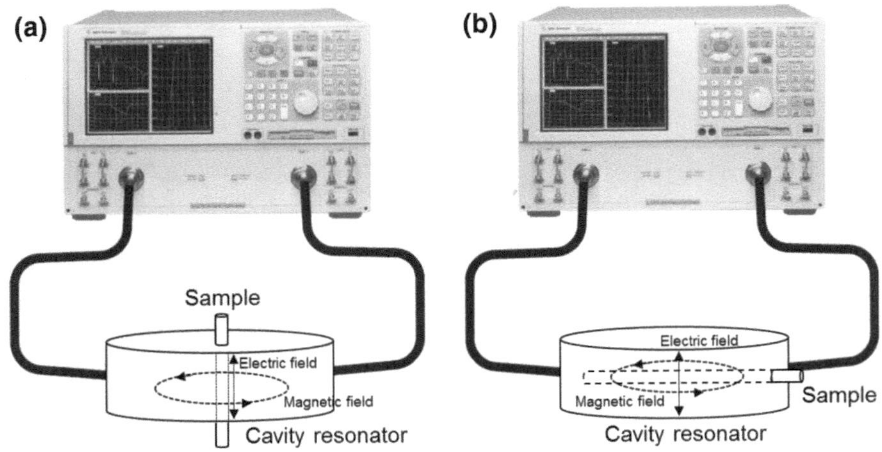

Fig. 5.10 Measurement of slim sample using cavity resonator for (**a**) electric permeability setup and (**b**) magnetic permeability

5.4 Adjustment of the Impedance

5.4.1 What Is Meant by Impedance in the Present Context?

It is necessary to understand what impedance is in relation to the microwave heating efficiency of a sample located in an applicator. Impedance is easily imagined when it is considered to be the resistance to an alternating current (AC). Simply stated, the characteristic impedance (Z) by microwave heating is the ratio of the alternating electric field to the alternating magnetic field (Eq. 5.17) as shown by the electric permeability and magnetic permittivity (or conductivity) in Maxwell's equations:

$$Z = \frac{E}{H} = \sqrt{\frac{\mu}{\varepsilon}} \tag{5.17}$$

Hence, impedance is the electrical characteristic that can be determined from the magnetic permeability and the permittivity; it can be defined for gas and vacuum. The impedance in vacuum is referred to as the "*impedance offree space*", and is ca. 120 $\pi\Omega$ (i.e., ca. 377 Ω). When microwaves are transmitted through a substance different in impedance (including vacuum and gas) a fraction of the microwaves is reflected at the boundary between each substance in the applicator, while the other fraction is absorbed by the substance (i.e., penetrates into the substance). The behavior for reflection, absorption, or penetration of microwaves depends on the microwave irradiation angle relative to the surface of the material and its separation from the source (e.g., sample and reactor); the characteristic impedance also changes as the material also changes when heated (see Fig. 5.11).

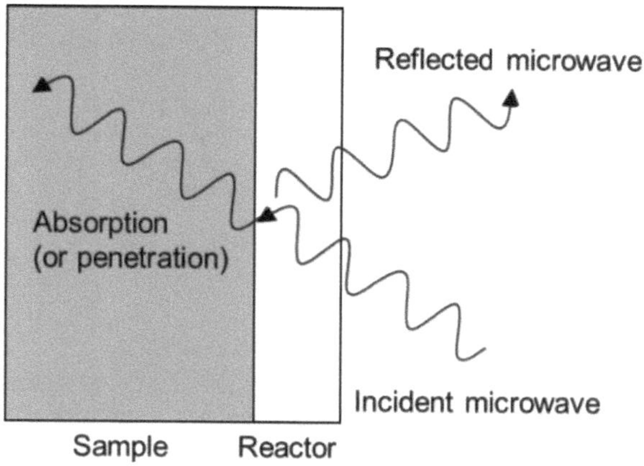

Fig. 5.11 Image of reflection and absorption (or penetration) of microwaves at the phase boundary of the material different in impedance

5.4.2 Impedance in Equipment and Sample

If the microwave radiation emitted from a generator does not easily enter into an applicator, the applied microwave energy cannot be used for heating. Such phenomenon between the sample and the air can occur in microwave heating situations. Accordingly, if the *impedance does not match*, the heating efficiency is small because a large fraction of the microwaves is reflected. Clearly, it is necessary for the characteristic impedance of the system to match the impedance of the sample so as to improve the microwave heating efficiency (Fig. 5.12). For example, in a microwave-assisted organic synthesis, the formation of the desired product causes the permittivity of the initial sample to be changed; even if impedance matching occurred at first, it may become invalid during the process. In addition, even though the electric permeability and magnetic permittivity may be the same for the substances, sometimes the impedance may be different. For instance, when applying

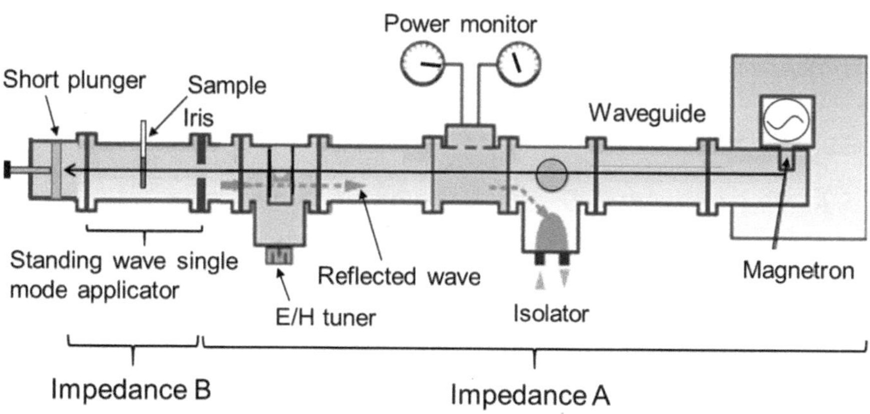

Fig. 5.12 Image illustrating the possible adjustments of impedance A and impedance B

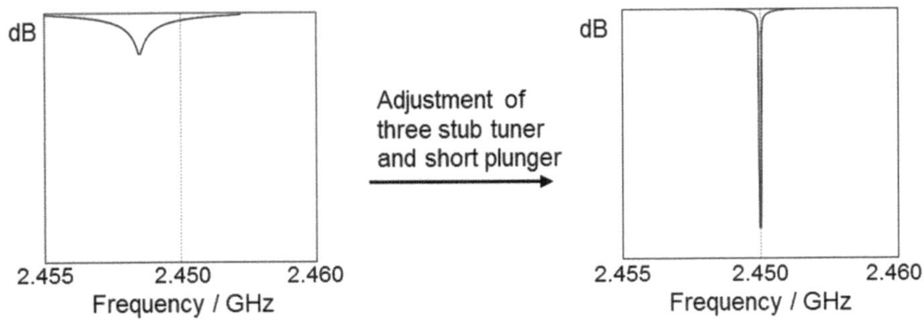

Fig. 5.13 Image illustrating the non-adjustment (left) and adjustment (right) of the voltage standing wave ratio (VSWR) after impedance adjustment in a microwave equipment

microwaves using a waveguide, the size and shape of the waveguide influence the distribution of the inner electric and magnetic fields inside the waveguide, which causes the impedance to also change significantly according to Eq. 5.17. To resolve this problem, a tuner is installed to adjust the impedance between the generator and the applicator (see tuner in Sect. 6.3.5). It is also possible to change the impedance by changing the shape of the applicator (see short plunger in Sect. 6.3.5).

The impedance can also be adjusted by connecting a tuner, an iris, and a short plunger to the equipment (Sect. 6.3.6). Generally, the adjustment of the impedance in the microwave equipment is done after installing the sample in the microwave equipment (Fig. 5.13). A vector network analyzer is used to provide the microwave radiation instead of a microwave generator. First, the three-stub tuner and the short plunger are opened, which is then followed by adjusting the tuner and the plunger while monitoring the resonance condition. The voltage standing wave ratio (VSWR) is determined by a vector network analyzer, and the frequency is scanned so that a peak may become sharp at a given microwave frequency. The standing wave ratio (SWR) expresses the ratio of the standing wave between the incident microwave and the reflected microwave, with the VSWR being the standard voltage.

5.5 Microwave Heating Mechanism

5.5.1 Phenomena of Dipole Rotation on Application of Microwaves

A molecule is formed when atoms combine by sharing one or more of their electrons. This rearrangement of electrons may cause an imbalance in charge distribution, thereby creating a permanent electric dipole moment. These moments are oriented in a random manner in the absence of an electric field so that no polarization exists. When microwave energy is applied to this state, the electric field

Fig. 5.14 Image of dipole rotation for a polar molecule in an electric field (***E***-field)

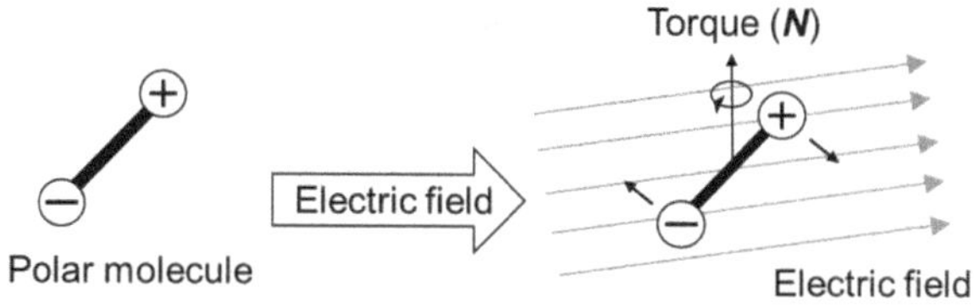

influences the electric dipole (Fig. 5.14). The microwaves' electric field (***E***-field) creates a torque (*N*) on the electric dipole, and the dipole will consequently rotate to align itself with the electric field, thus causing orientation polarization to occur [7]. As the field changes direction, the torque also changes direction, so that the orientation polarization changes with the changing direction of the microwave electric field. There is a time delay between the frequency of the microwave alternating electric field and the electric dipoles continual rotational motion, as it tries to keep up with the field's changing polarity. At the general frequency of 2.45-GHz, the microwaves vibrate 2,450,000,000 times per second, such that a molecular cluster cannot follow this vibration through the power chain with the surrounding molecules. This delay changes to heat energy (random kinetic energy) as loss of the electromagnetic wave energy. The friction accompanying the orientation of the dipole will contribute to dielectric losses. The dipole rotation causes a variation in both ε_r' (the relative dielectric constant) and ε_r'' (the relative dielectric loss) at the relaxation frequency, which usually occurs in the microwave region. Water is an example of a substance that exhibits a strong orientation polarization.

Electronic polarization occurs in neutral atoms when an electric field displaces the nucleus with respect to the electrons that surround it. Atomic polarization occurs when adjacent positive and negative ions *stretch* under an applied electric field. For many dry solids, these are the dominant polarization mechanisms at microwave frequencies, although the actual resonance occurs at a much higher frequency. In the infrared and visible light regions, the inertia of the orbiting electrons must be taken into account. Atoms can be modeled as oscillators with a damping effect similar to a mechanical spring and mass system. The amplitude of the oscillations will be small for any frequency other than the resonant frequency. Far below resonance, the electronic and atomic mechanisms contribute only a small constant amount to ε_r' and are almost lossless. The resonant frequency is identified by a resonant response in ε_r' and a peak of maximum absorption in ε_r''. Above resonance, the contribution from these mechanisms disappears.

A solid with a partial electric charge (dielectric) is also an insulator. The slight distortion of the atomic positions (lattice points) in a lattice, and the strain of the lattice of a crystalline solid cannot follow the time changes of the microwave electric field. Microwave heating of a solid occurs via these phenomena [8]. On the other hand, heating of solid substances possessing a magnetic dipole moment occurs by the microwaves' magnetic field component. The heating process is the same phenomenon as by the microwaves' electric field [9]. Generation of heat by magnetic loss heating is expected only in magnetic (solid) materials. Joule heating progresses by the interaction of an electric or a magnetic field with solid particles of activated carbon and with those that possess conductivity-like metallic properties [10].

5.5.2 Relationship of Microwave Heating Behavior with the Materials' Physical Properties

The microwave heating behavior depends on the nature of the material and the microwave irradiation conditions. The electrical conductivity (σ), permittivity (ε), and permeability (μ) are three physical properties of materials with respect to microwaves and depend upon the physical properties of the materials (dipole moment, mass, etc.) as well as the state of matter (intermolecular force, boundary conditions, etc.). The intensity of the electric and magnetic fields and the frequency are examples of microwave conditions.

Corresponding to the above three physical properties, there are three types of heating phenomena caused by microwave irradiation: (i) conduction loss heating, (ii) dielectric heating, and (iii) magnetic loss heating. Accordingly, the thermal energy P produced per unit volume originating from microwave radiation is given by Eq. 5.18 [11]:

$$P = \frac{1}{2}\sigma|E|^2 + \pi f \varepsilon_0 \varepsilon_r'' |E|^2 + \pi f \mu_0 \mu_r'' |H|^2 \quad (5.18)$$

where $|E|$ and $|H|$ denote the strength of the microwaves' electric and magnetic fields, respectively; f is the frequency of the microwaves; ε_o is the dielectric constant in vacuum; ε_r'' is the relative dielectric loss factor; μ_o is the magnetic permeability of vacuum; and μ_r'' is the relative magnetic loss. The first term in Eq. 5.18 expresses ***conduction loss heating***; the second term denotes ***dielectric loss heating***, whereas ***magnetic loss heating*** is given by the third term.

5.5.3 Conduction Loss Heating (Eddy Current Loss and Joule Loss)

If a current is generated in the materials, Joule loss occurs. In this case, the relationship between the electric field E, the current density i, and the conductivity σ is given by Eq. 5.19:

$$i = \sigma E = \sigma E_0 \sin \omega t \quad (5.19)$$

Defining the microwave frequency as f, the energy loss w in the material during one cycle of microwave ($1/f = 2\pi/\omega$) is expressed by Eq. 5.20:

$$\begin{aligned} w &= \int_0^{2\pi/\omega} iE\mathrm{d}t = \int_0^{2\pi/\omega} \sigma E^2 \mathrm{d}t = \int_0^{2\pi/\omega} \sigma E_0^2 \sin^2 \omega t \mathrm{d}t \\ &= \sigma E_0^2 \int_0^{2\pi/\omega} \frac{1-\cos 2\omega t}{2} \mathrm{d}t = \sigma E_0^2 \left[\frac{1}{2} t - \frac{1}{4\omega} \sin 2\omega t \right]_0^{2\pi/\omega} = \frac{\pi}{\omega} \sigma E_0^2 \end{aligned} \quad (5.20)$$

Since the energy loss is repeated f times (=$\omega/2\pi$) per second, the energy W that the material receives from the electric field per unit time and unit volume is given by Eq. 5.21:

$$W = \frac{\omega}{2\pi} \frac{\pi}{\omega} \sigma E_0^2 = \frac{1}{2} \sigma E_0^2 \quad (5.21)$$

Clearly, Eq. 5.21 does not include ω, which relates to the microwave frequency.

The hydroxide ion is a typical ionic species with both ionic and dipolar characteristics. The conductive loss effect can become larger than dipolar relaxation in solutions that contain large quantities of ionic salts. Conduction losses tend to be slight at ambient temperatures in the case of solids but can change substantially with increase in temperature. A typical example is alumina (Al_2O_3) whose dielectric loss is negligibly small ($\sim 10^{-3}$) at ambient temperature but can reach fusion levels, i.e., high dielectric loss, in very short times (minutes) in a microwave cavity. This effect originates from a strong increase of conduction losses associated with the thermal activation of electrons, which migrate from the oxygen's $2p$ valence band to the aluminum's $3s3p$ conduction band. Moreover, conduction losses in solids are usually enhanced by defects in materials. Defects sharply reduce the energy needed to generate electrons and holes in the conduction and valence bands, respectively. In the particular case of carbon black powder, conduction losses are rather high, so much so that this material is often used as an additive to induce losses, i.e., heat, within solids whose dielectric losses are too small.

Microwave heating of a metallic sample concerns the second and third terms of Eq. 5.18. More information and examples of metal heating are described in Sect. 9.3.

5.5.4 *Dielectric Heating and Magnetic Loss Heating—An Introduction*

When high-frequency electric and magnetic fields are applied to a material, the charge distribution within the dipole moment changes (liquid sample), and the material may then shift (solid sample) and/or deflection of spin may occur as portrayed in Fig. 5.15.

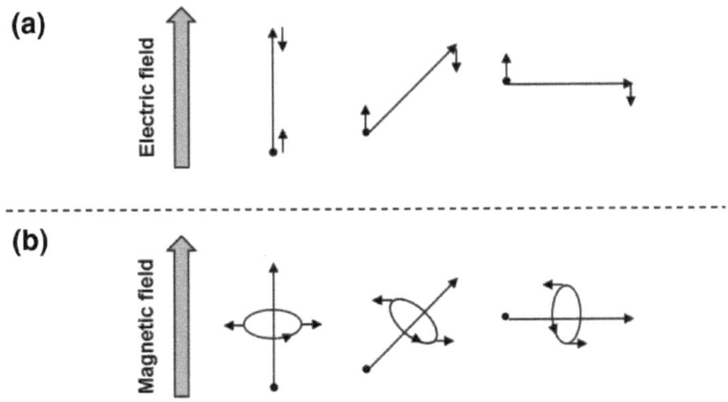

Fig. 5.15 Images of (**a**) dipole moment changes and (**b**) deflection of spin

Loss behavior depends upon the change of the alternating microwave electromagnetic field that is very fast compared to direct current, in which case the change of charge distribution (the dipole moment and/or the deflection of the spin) can (i) follow, or (ii) follow with delays, or (iii) cannot follow at all. For the first two cases, the response coefficient is just the dielectric constant and the magnetic permeability. If the charge distribution and the deflection of the spin follow without delay in an alternating electromagnetic field, the materials receive the energy from the electromagnetic field and store it, i.e., they gain energy. When the electromagnetic field is no longer applied, the energy is released, i.e., it propagates through the material; propagation of electromagnetic waves is illustrated in Fig. 5.16a). On the other hand, if the charge distribution and the deflection of spin were delayed, i.e., too slow, in an alternating electromagnetic field, the energy relaxes in the substance and becomes a loss (Fig. 5.16b). This is because under applied microwaves the dielectric constant has a real part (gain) and an imaginary part (loss). The next section describes these images by mathematical formulas.

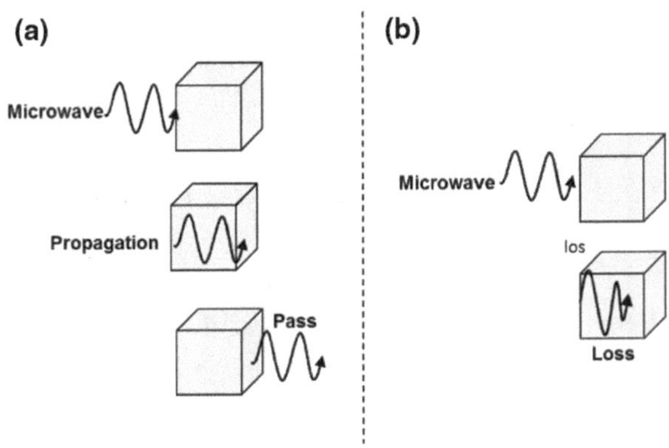

Fig. 5.16 Images of (**a**) no absorption of microwaves and transmission of microwaves through the sample; (**b**) absorption of microwaves and generation of heat

5.5.5 *Dielectric Heating—Energy Loss in a Microwave Field*

The *electric flux density* in materials changes according to the change of the external field. The change of the electric flux density is divided into two parts: (a) follows with no delay and (b) follows with delay; the change is described by Eq. 5.22:

$$D(t) = D_p(t) + D_d(t) = \varepsilon_0 \varepsilon_r(\infty) E(t) + \int_0^t E(u) f(t-u) \mathrm{d}u \tag{5.22}$$

When the electric field vibration at time t is given by $E_0\ e^{j\omega t}$, the electric flux density that follows without delay (a) is expressed by $D_p\ e^{j\omega t}$, while the electric flux density (b) that follows the phase delay δ is expressed as $D_d\ e^{j(\omega t - \delta)}$ lead from Eqs. 5.23 through 5.25.

$$\varepsilon_0 \varepsilon_r(\omega) E_0 e^{j\omega t} = D_p e^{j\omega t} + D_d e^{j(\omega t - \delta)} \tag{5.23}$$

$$\varepsilon_0 \varepsilon_r(\omega) E_0 = D_p + D_d e^{-j\delta} \tag{5.24}$$

$$\varepsilon_0 \left\{ \varepsilon_r'(\omega) - i\varepsilon_r''(\omega) \right\} E_0 = D_p + D_d \cos\delta - jD_p \sin\delta \tag{5.25}$$

If the electric flux density in the dielectric substance changes only by dD when the energy change dU per unit volume, the energy loss w in the dielectric materials during one cycle ($1/f = 2\pi/\omega$) can then be described by Eq. 5.26:

$$w = \int_0^{2\pi/\omega} \mathrm{d}U = \int_0^{2\pi/\omega} E\mathrm{d}D = \int_0^{2\pi/\omega} E \frac{\mathrm{d}D}{\mathrm{d}t} \mathrm{d}t \tag{5.26}$$

When E is expressed as a cosine function, D then consists of the parts of the follow-up with delay and without delay (Eqs. 5.27 and 5.28):

$$E = E_0 \cos\omega t \tag{5.27}$$

$$D = D_p \cos\omega t + D_d \cos(\omega t - \delta) \tag{5.28}$$

Rearranging D using the addition theorem {cos($x - y$) = cos × cos + sin × sin} followed by differentiation with respect to t yields Eq. 5.29:

$$\begin{aligned} \frac{\partial D}{\partial t} &= -\omega\left(D_p \cos\omega t + D_d \cos\delta\right) \sin\omega t + \omega D_d \sin\delta \cos\omega t \\ &= -\omega\varepsilon_0\varepsilon_r'(\omega) E_0 \sin\omega t + \omega\varepsilon_0\varepsilon_r''(\omega) E_0 \cos\omega t \end{aligned} \tag{5.29}$$

and the energy loss w then becomes as given by Eq. 5.30:

$$\begin{aligned} w &= \int_0^{2\pi/\omega} E \frac{\mathrm{d}D}{\mathrm{d}t} \mathrm{d}t \\ &= \int_0^{2\pi/\omega} E_0 \cos \omega t \left(-\omega \varepsilon_0 \varepsilon_r'(\omega) E_0 \sin \omega t + \omega \varepsilon_0 \varepsilon_r''(\omega) E_0 \cos \omega t \right) \mathrm{d}t \\ &= \varepsilon_0 E_0^2 \left[-\varepsilon_r'(\omega) \frac{\sin^2 \omega t}{2} + \varepsilon_r''(\omega) \frac{\sin 2\omega t}{4} \right]_0^{2\pi/\omega} + \varepsilon_0 \varepsilon_r'' E_0^2 \left[\frac{\omega t}{2} \right]_0^{2\pi/\omega} \\ &= \pi \varepsilon_0 \varepsilon_r'' E_0^2 \end{aligned} \tag{5.30}$$

Since the energy loss is repeated f times (=$\omega/2\pi$) per second, the energy loss P that the material receives from the electric field per unit time and unit volume is expressed by Eq. 5.31:

$$P = \pi f \varepsilon_0 \varepsilon_r'' E_0^2 = \frac{1}{2} \omega \varepsilon_0 \varepsilon_r'' E_0^2 \tag{5.31}$$

The dipolar polarization mechanism only concerns polar compounds, e.g., water, methanol, and ethanol, which possesses a permanent dipole moment [12]. With regard to the frequency of the applied electric field, the dipoles can behave in three ways: (i) at low frequencies, the dipoles keep in phase with the electric field and the temperature of the irradiated medium hardly changes; (ii) at high frequencies, the response time of the dipoles is too long to follow very rapid oscillations of the electric field and no energy is absorbed; and (iii) at the microwave frequencies, the dipoles' rotation lags behind the changes of the electric field and results in the heating of the irradiated material.

The relaxation time τ is a measure of the mobility of the molecules (dipoles) that exist in a material. It is the time required for a displaced system aligned in an electric field to return to 1/e of its random equilibrium value (or the time required for dipoles to become oriented in an electric field). Liquid and solid materials have molecules that are in a condensed state with limited freedom to move when an electric field is applied. Constant collisions cause internal friction so that the molecules turn slowly and exponentially approach the final state of orientation polarization with a relaxation time constant τ (Eq. 5.32):

$$\tau = \frac{1}{\omega_c} = \frac{1}{2\pi f_c} \tag{5.32}$$

When the field is switched off, the sequence is reversed and random distribution is restored with the same time constant. The relaxation frequency f_c is inversely related to the relaxation time.

At frequencies below relaxation, the alternating electric field is slow enough that the dipoles can keep pace with the field variations. Because the polarization is able to develop fully, the loss (ε_r'') is directly proportional to the frequency (Fig. 5.17). As the frequency increases, ε_r'' continues to increase but the storage (ε_r') begins to decrease due to the phase lag between the dipole alignment and the electric field. Above the relaxation frequency, both ε_r'' and ε_r' drop off as the electric field is too fast to influence the dipole rotation, and the orientation polarization disappears.

Debye relation: Materials that exhibit a single relaxation time constant can be modeled by the Debye relation, which appears as a characteristic response in permittivity as a function of frequency (Fig. 5.17). The parameter ε_r' is constant above and below the relaxation with the transition occurring near the relaxation frequency (22 GHz). Additionally, ε_r'' is small above and below relaxation and peaks in the transition region at the relaxation frequency. In calculating the above curves, the static (DC) value of the dielectric constant is $\varepsilon_s = 76.47$, the optical (infinite frequency) value of the dielectric constant is $\varepsilon_\infty = 4.9$, and the relaxation time $\tau = 7.2$ ps.

Cole–Cole diagram: The complex permittivity may also be shown on a Cole–Cole diagram by plotting the imaginary part (ε_r'') on the vertical axis and the real part (ε_r') on the horizontal axis with the frequency as the independent parameter (Fig. 5.18).

A Cole–Cole diagram is, to some extent, similar to the Smith chart. A material that has a single relaxation frequency, as exhibited by the Debye relation, will appear as a semicircle with its center lying on the horizontal $\varepsilon_r'' = 0$ axis and the peak of the loss factor occurring at $1/\tau$. A material with multiple relaxation frequencies will be a semicircle (symmetric distribution) or an arc (nonsymmetrical distribution) with its center lying below the horizontal $\varepsilon_r'' = 0$. The curve in Fig. 5.10 is a half-circle with its center on the *x*-axis; the radius of the curve is given by Eq. 5.33:

$$\frac{\varepsilon_S - \varepsilon_\infty}{2} \tag{5.33}$$

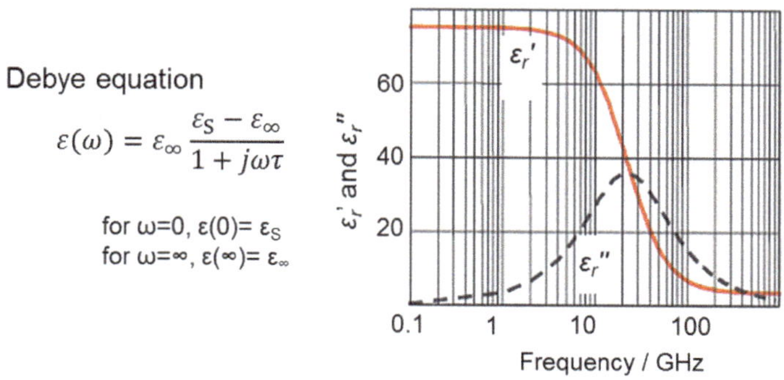

Fig. 5.17 Debye relaxation of water at 30 °C. Reproduced from Ref. [1]

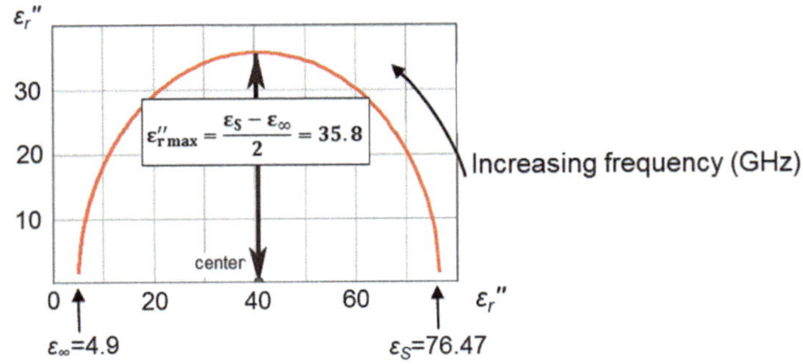

Fig. 5.18 Cole–Cole diagram of Fig. 5.10. Reproduced from Ref. [1]

The maximum imaginary part of the dielectric constant ε'_{rmax} will be equal to the radius. Note that the frequency moves counterclockwise on the curve of Fig. 5.10.

5.5.6 Magnetic Loss

Magnetic heating is the phase delay of the vibration of the magnetic dipole moment. Thus, a magnetic loss is obtained by substituting *H* for *E* and *m* for *e* (Sect. 8.4.3). The magnetic loss per unit time and volume, P_μ, is given by Eq. 5.34:

$$P_\mu = \frac{1}{2}\omega\mu_0\mu_r''H_0^2 \tag{5.34}$$

In metal oxides such as ferrites and other magnetic materials, magnetic losses occur in the microwave region. These losses are different from hysteresis or Eddy current losses because they are induced by the domain wall and electron spin resonance. Such materials are typically placed at positions of magnetic field maxima for optimal absorption of the microwave energy. Transition metal oxides (e.g., iron, nickel, and cobalt oxides) possess high magnetic losses and thus can be used as added loss impurities or additives to induce losses within those solids for which dielectric loss is too small. For instance, induction heating by the microwaves' magnetic field is expected to occur in catalyzed reactions involving solids, an example being the rapid induction heating of magnetite (Fe_3O_4) but not hematite (Fe_2O_3) as the latter is not a magnetic material [13–15]. More information and examples of heating of solid-state magnetic materials are provided in Sect. 8.4.3.

5.6 Penetration Depth and Skin Depth

5.6.1 *What Are They?*

The next topic we need to consider was a hot topic of discussion at International Conferences in the latter part of the first decade of the twenty-first century because experiments of microwave-assisted syntheses gave different results in various laboratories owing to the use of different equipments from different manufacturers. This likely occurred because of different penetration depths of the microwaves as the reactor size and shape in commercial microwave chemical equipment may have been different depending on the manufacturers, which may have caused variations in the heat distribution generated for a similar sample size. Even though the dielectric loss of a solvent may be high, the heating efficiency may sometimes be low because of the shallow penetration depth of the microwaves. Therefore, if a large size reactor is used, the microwaves may not heat the center of the sample as the microwave electric field (E) is attenuated as it traverses the medium and therefore the power is dissipated; this is a function of E^2 so that the power is reduced to an even greater extent (Fig. 5.19).

The penetration depth or skin depth is a measure of how deeply light or any other electromagnetic radiation can penetrate a material. It is defined as the depth at which the intensity of the radiation inside the material falls from its original value at (or more properly, just beneath) the surface by a specific amount. When performing microwave chemistry, the penetration depth is an important factor. Unfortunately, many researchers do not appear to appreciate or understand this important factor. The reason may be due to a lack of a suitable theory and a correspondingly appropriate database. Accordingly, we now describe and explain the relevance/significance of the penetration depth and the skin depth.

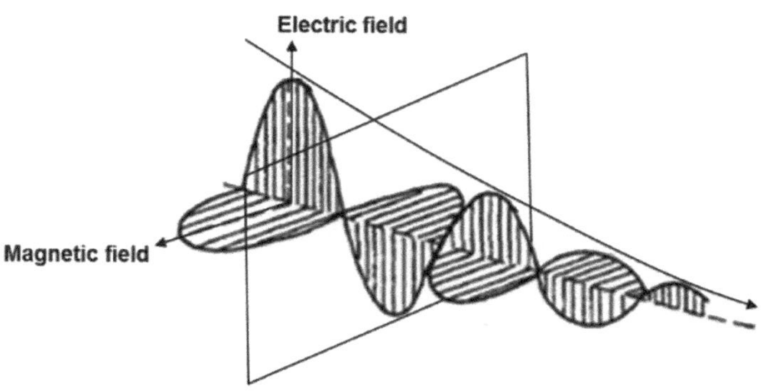

Fig. 5.19 Propagation of a plane wave into a lossy medium (lossy medium means a medium that absorbs microwave energy well and heats). Reproduced from Ref. [8]. Copyright 1983 by Peter Peregrinus Ltd.

5.6.2 *Penetration Depth*

The penetration depth D_p (in cgs units) is the depth at which microwaves penetrate into a material and the power flux has fallen to 1/e (=36.8%) of its surface value; That is, it denotes the depth at which the power density of the microwaves is reduced to 1/e of its initial value. D_p can be estimated from Eq. 5.35 [8],

$$D_p = \frac{\lambda}{4\pi}\left[\frac{2}{\varepsilon'\left(\sqrt{1+\left(\frac{\varepsilon''}{\varepsilon'}\right)}-1\right)}\right]^{\frac{1}{2}} \tag{5.35}$$

where λ is the wavelength of the radiation, $\lambda_{(2.45\ \mathrm{GHz})} = 12.24$ cm. Since the dielectric constant and the dielectric loss are components of this equation, we expect the penetration depth to change with increase of temperature. For example, even if we used a 10-cm-diameter reactor, the microwaves would not reach the center of the sample as the penetration depth at 25 °C is only ca. 1.8 cm for water. However, at 50 °C, the penetration depth is 3.1 cm while at 90 °C it is 5.4 cm, so that at the latter temperature the microwaves will reach the center of the water sample. On the other hand, if ions were added to water, the penetration depth then would become remarkably shallow. For instance, when NaCl (0.25 M) is added to pure water, the penetration depth changes to 0.5 cm at room temperature, while the penetration depth in pure water is 1.8 cm [10]. Therefore, in the case of solutions containing ions, microwave heating may generate hotspots at the surface of the reactor and it would then be necessary to carry out vigorous stirring to avoid these hotspots. Also, because microwave energy is constantly being lost as the microwaves enter the water or other material, the greatest intensity of energy is at the surface and constantly diminishes as it enters the material, so that mixing/stirring will always be a necessary component.

It is important to examine the heating efficiency and the penetration depth of microwaves into a substance so as to achieve optimal microwave heating. One must consider the reactor size with the penetration depth of the microwaves into the sample for uniform microwave heating to occur. For example, when the temperature at the solution center reached 75 °C with the microwave energy directed to only one side of the container of ethylene glycol in a quartz container of diameter 3.0 cm, the temperature of the solution at the container walls was 62 °C due to heat loss; the resulting temperature gradient was 13 °C (see Fig. 5.20). However, when the size of the reactor was changed to 1.5 cm and the solution was stirred (magnetic stirrer), the temperature gradient between the solution center and the container walls dropped to about 2 °C.

The penetration depths of microwaves into some common organic solvents and water at 25 °C and at 50 °C are reported in Fig. 5.21 [16]. The penetration depths D_p of microwaves into polar solvents at 25 °C are generally within 10 cm, for instance, the penetration depths of 2.45-GHz microwaves in ethylene glycol,

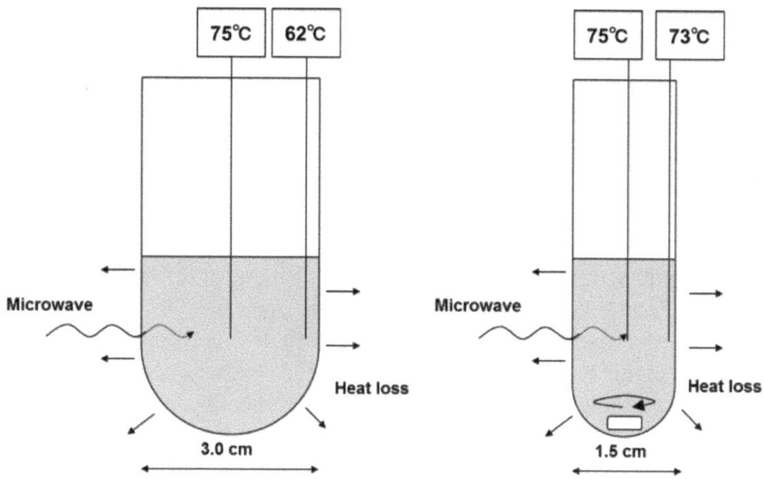

Fig. 5.20 Temperature gradient of ethylene glycol in a quartz reactor (3 and 1.5 cm diameter) with stirring

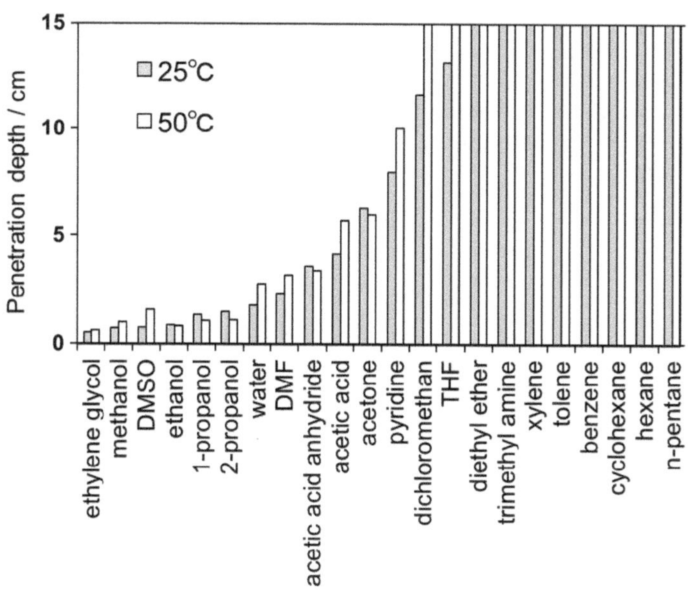

Fig. 5.21 Penetration depths of the 2.45-GHz microwaves for some common solvents. Figure drawn from the data of Ref. [16]. Copyright 2012 by Wiley-VCH Verlag GmbH

methanol, DMSO, and ethanol are less than 1 cm, so that if these solvents were contained in a large size reactor and subjected to microwave irradiation, the temperature at the center of the samples would be lower than at the reactor walls. As the temperature of a solvent rises, however, the microwaves can penetrate deeper into the sample. Therefore, in order to reach the appropriate reaction temperatures for a given reactor design, it is necessary to calculate the penetration depth of the microwaves over a range of temperatures. By contrast, the penetration depths of microwaves into nonpolar solvents are remarkably greater, as for example, the penetration depth of microwaves into xylene is 28 cm at 25 °C, while at a temperature of 50 °C the penetration depth is several meters.

In the case of microwave heating of solid samples, the penetration depth is a very important factor because no stirring is possible. Therefore, the sample size will be dictated by the balance between the microwave heating efficiency, the loss of heat to the atmosphere, and the penetration depth. When scaling up is considered in microwave chemistry, the reproducibility obtained at the laboratory scale will no longer be possible because of the restriction imposed by penetration depth when going to larger reactor size. In such a case, a flow-type reactor is often used to carry out microwave chemistry, preferably using a small reactor so that the penetration depth is no longer a limiting factor that might affect efficiency. On the other hand, when synthesizing a polymeric substance, the viscosity of the sample becomes a limiting factor for the flow-type reactor. A strong stirrer is usually installed in the batch reactor for polymer syntheses.

The distance for absorption of microwaves to reach 50% of the initial microwave power is given by the half power density depth D (Fig. 5.22) that can be estimated from Eq. 5.36:

$$D = \frac{3.32 \times 10^7}{f\sqrt{\varepsilon''}} \tag{5.36}$$

where f is the frequency of the microwaves and ε'' is the dielectric loss component of the substance.

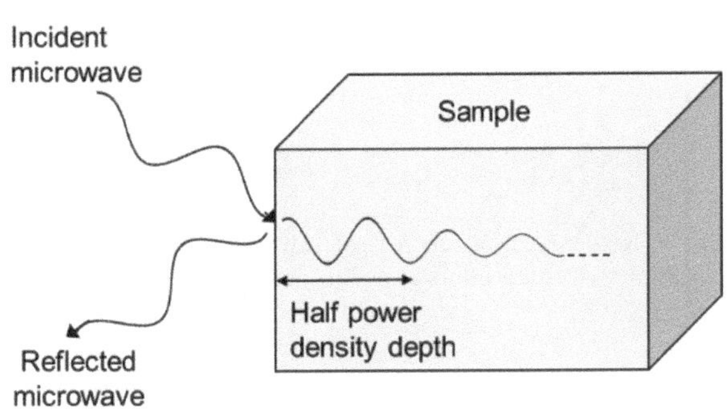

Fig. 5.22 Image of the half power density depth of the microwave

5.6.3 Penetration Depths of 915-MHz, 2.45-GHz, and 5.80-GHz Microwaves

Figure 5.23 illustrates the penetration depths of microwaves at different frequencies into some selected solvents. Penetration of the 5.8-GHz microwaves was shallower for all solvents compared to the depth of penetration by the 2.45-GHz microwaves. For instance, in the case of xylene, the 5.80-GHz microwaves can penetrate to a depth of ca. 1.24 cm, whereas the depth of penetration is about 28.32 cm for the 2.45-GHz microwaves. In particular, the largest difference in penetration depth between the 5.80-GHz and 2.45-GHz microwaves was observed for the nonpolar solvents. Note that, although the 2.45-GHz microwaves penetrate deeper into a solvent/solution, they have a comparatively low energy conversion, whereas the 5.80-GHz microwaves with their shallower penetration depth display a much greater energy conversion within the shallower depth. Accordingly, the energy efficiency for heating thin portions of the solvents/solutions is greater when using 5.80-GHz microwaves. Hence, the penetration depth of the microwaves becomes one of the principal factors that lead to fast-heating and superheating. In the case of 5.80-GHz irradiation, hexane displayed the deepest penetration depth (2.25 cm). On the other hand, in the case of 2.45-GHz microwaves, the penetration depth into ethyl acetate was the shallowest of the nonpolar solvents (11.05 cm).

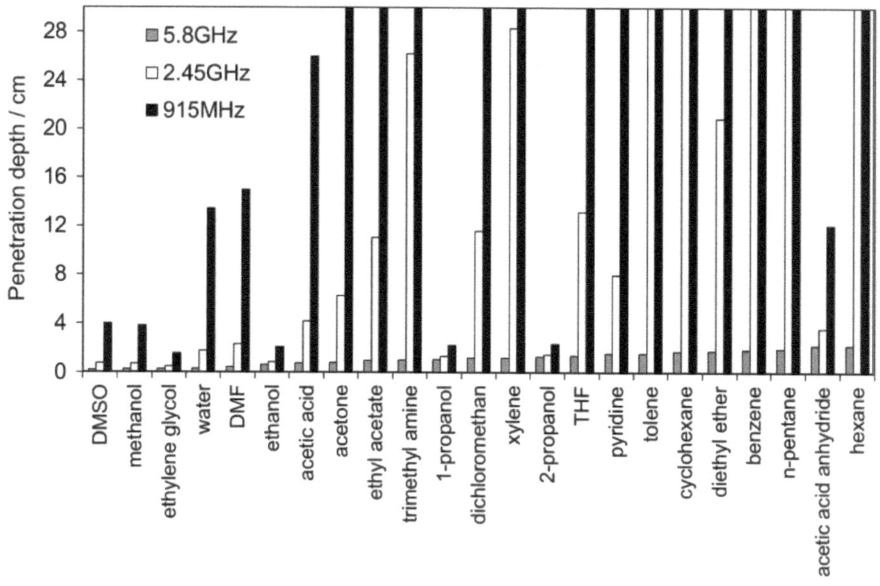

Fig. 5.23 Penetration depths of microwaves at a frequency of 5.80 GHz, 2.45 GHz, and 915 MHz in water and in 22 pure organic solvents (100 mL) at ambient temperature. Figure was drawn from data reported in Ref. [16]. Copyright 2012 by Wiley-VCH Verlag GmbH

The heating efficiency of a solvent with a shallow penetration depth such as DMSO (0.25 cm), methanol (0.28 cm), ethylene glycol (0.29 cm), DMF (0.46 cm), and water (3.4 mm) is not high under 5.80-GHz conditions. Accordingly, when 5.80-GHz microwaves are used with a large reactor, the microwave heating efficiency is not high. Thus, a change in penetration depth of the sample involved in a reaction should be reflected in the design of the microwave reactor.

At the 915-MHz frequency, the characteristic dielectric loss factors (ε_r'') when compared to 2.45-GHz were greater for ethylene glycol, ethanol, 1-propanol, 2-propanol, and acetic acid anhydride. Thus, the optimal frequency for the heating of alcohol (but not methanol) is 915 MHz. The penetration depth at the 915-MHz frequency for all solvents, other than alcohols, is very deep. The largest cylindrical reactor that can be accommodated in a 915-MHz structure is ca. 24.8 cm, so that it is suitable for heating polar solvents.

5.6.4 Skin Depth

Everyone understands that light, an electromagnetic wave, is reflected at a metal surface. Microwaves, which are also electromagnetic waves, are mostly reflected when used to irradiate highly conducting substances such as metals that possess limited impedance. The skin effect refers to the tendency of an alternating electric current to become distributed within a conductor such that the current density is largest near the surface of the conductor but decreases at greater depths in the conductor [17]. The electric current flows mainly at the *skin* of the conductor, between the outer surface and a level called the *skin depth*. The skin effect causes the effective resistance of the conductor to increase at higher frequencies where the skin depth is smaller, thus reducing the effective cross section of the conductor. The skin effect is due to opposing Eddy currents induced by the changing magnetic field resulting from the alternating current [17].

The skin depth, ξ, can be calculated from Eq. 5.37 in which ρ is the bulk resistivity of the sample ($\Omega \times 10^{-8}$ m), μ_r is the relative magnetic permeability, μ_0 is the magnetic permeability constant ($4\pi \times 10^{-7}$), and f is the frequency of the microwaves. The skin depths of various metals are listed in Table 5.3.

$$\xi = \sqrt{\frac{\rho}{\pi \cdot f \cdot \mu_r \cdot \mu_0}} \tag{5.37}$$

In the case of bulk metal, even if the surface were exothermic when heated with microwaves (most are reflected), the resulting heat would be scattered and lost by the heat conduction mechanism. Therefore, the temperature of the metal sample does not rise. However, it is possible to heat metallic microparticles using microwaves. For example, irradiating a sample consisting of 1-μm copper powder particles (skin depth, ca. 1.3 μm) with 2.45-GHz microwaves causes microwave

Table 5.3 Skin depths of various metals

Material	Chemical formula	Bulk resistivity at 20 °C/Ω × 10^{-8} m	Relative permeability/$\mu\ \mu_0^{-1}$	Skin depth/μm
Aluminum	Al	2.65	1	1.66
Beryllium	Be	3.3	1	1.85
Brass	70%Cu + 30%Zn	7	1	2.69
Bronze	89%Cu + 11%Sn	15	1	3.94
Chromium	Cr	13.2	1	3.69
Copper	Cu	1.69	1	1.32
Gold	Au	2.2	1	1.51
Graphite	C	783.7	1	28.5
Inconel alloy 600	72%Ni + 16% Cr + 8%Fe	103	1	10.3
Lead	Pb	20.6	1	4.61
Magnesium	Mg	4.2	1	2.08
Mercury	Hg	95.9	1	9.96
Molybdenum	Mo	5.7	1	2.43
Palladium	Pd	10.8	1	3.34
Platinum	Pt	10.58	1	3.31
Rhodium	Rh	4.7	1	2.20
Silver	Ag	1.63	1	1.30
Tin (pure)	Sn	12.6	1	3.61
Titanium	Ti	54	1	7.47
Tungsten	W	5.4	1	2.36

heating of the copper sample, a phenomenon reported by Roy and coworkers and who developed this microwave metallic heating method [13] (see Sect. 10.1.3).

5.7 Frequency Effect

5.7.1 Is It Possible to Use Various Frequencies?

As reported elsewhere in this book, microwaves are electromagnetic waves that span the frequency range between 0.3 and 300 GHz. Accordingly, we do not need to adhere to the commonly used microwaves at the 2.45-GHz frequency. However, when examining frequency effects, two barriers have obstructed the researcher's curiosity. First, microwave devices that can generate microwave frequencies other than 0.915, 2.45, and 5.80 GHz are very expensive compared to those devices operating at the common frequencies. Currently, chemical reaction equipment that

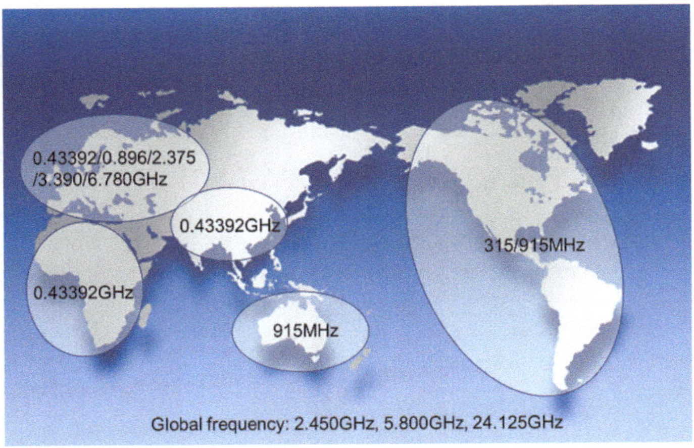

Fig. 5.24 Global distribution map of frequency bands (GHz) used for industrial, medical, and scientific (ISM) purposes [16]. Copyright 2012 by Wiley-VCH Verlag GmbH

can generate different frequencies is available at costs similar to those available at the 2.45-GHz frequency. The second barrier has been with the industrial, scientific, and medical (ISM) bands that have been reserved internationally for the use of radio frequency energies employed for industrial, scientific, and medical purposes other than communications [18]. Examples of applications in these bands include radio frequency process heating, microwave ovens, and medical diathermy machines. The powerful emissions of these devices can create electromagnetic interference and disrupt radio communications that use that frequency, so that these devices have been limited only to certain frequency bands. The ISM bands are defined by the ITU-R (International Telecommunication Union—Radio Communications Sector). The globally used frequencies are 2.45, 5.80, and 24 GHz; individual countries have established other frequencies. The global distribution map of frequencies used is reported in Fig. 5.24 [16].

5.7.2 Historical Overview of Microwave Frequency Effects in Chemical Reactions and Sintering

The microwave (MW) frequency response and its impact on chemical reactions have been examined in only a few cases; that is, research reports regarding frequency effects in microwave-assisted organic syntheses have tended to be rather scarce. In the field of inorganic chemistry, Möller and Linn [19] used 5.80-GHz microwaves to calcine ceramics, while Takizawa and coworkers [20] utilized 28-GHz microwaves for the rapid synthesis of $(In_{0.67}Fe_{0.33})_2O_3$, whereas Malinger et al. [21] used 5.5-GHz microwaves in the rapid syntheses of cryptomelane-type manganese oxides for catalysis. In the field of organic chemistry, Gedye and Wei

[22] used a variable-frequency microwave oven to effect the Knoevenagel reaction of acetophenone with ethyl cyanoacetate in the presence of piperidine (no solvent; microwave frequency, 8.1 GHz); with anisole as the solvent, the reaction was carried out at 12.2 GHz. Séguin et al. [23] reported a new IR reactor cell built with a microwave cavity operating at a frequency of 5.80 GHz; this equipment allowed efficient and time-resolved heating of the catalyst. The latter authors also showed that the 5.80-GHz microwave frequency is the optimal frequency for catalyst heating. Microwave-induced oligomerization of methane with nickel, iron, and activated carbon catalysts has been examined relative to microwave frequency and other parameters by Conde and Suib [24]; an increase in the microwave frequency enhanced the activity of the catalysts.

5.7.3 *Microwave Chemical Equipment for 0.915, 2.45, and 5.80 GHz Frequencies*

Figure 5.25 illustrates three different microwave chemical reaction systems that use single-mode microwave irradiation applicators at 2.45-GHz, 915-MHz (i.e., 0.915 GHz), and 5.80-GHz frequencies. Each microwave system is constructed with a waveguide as the resonant cavity, a short plunger, a three-stub tuner (and/or an iris), a power monitor, and an isolator. Impedance matching is accomplished by measuring the output power and the reflected power using the power monitor. The microwave source for each system is a semiconductor generator.

Differences in the sizes of the waveguide are responsible for variations in the output power density of the microwave radiation. In this regard, Fig. 5.26 illustrates the various waveguides seen in the photographs of Fig. 5.25. The cross-sectional area of the waveguide at the common microwave frequency of 2.45 GHz is 59.645 cm^2 (10.92 cm × 5.46 cm). On the other hand, the cross-sectional areas of the waveguides of the 915-MHz and 5.80-GHz apparatuses are, respectively, 306.64 cm^2 (24.77 cm × 12.38 cm) and 8.0 cm^2 (4.0 cm × 2.0 cm). For cross-sectional area comparison, the ratios of the power density (Watt cm^{-1}) in the waveguides for the 2.45-GHz microwaves relative to the 915-MHz microwaves and the 2.45-GHz microwaves relative to the 5.80-GHz microwaves are, respectively, 0.19 and 7.46. The corresponding wavelengths of each of the frequencies are 32.79 cm (915-MHz) > 12.24 cm (2.45-GHz) > 5.17 cm (5.80-GHz). The power density is remarkably lower in the 915-MHz apparatus, which can accommodate a reactor of a large size. On the other hand, superheating of nonpolar solvents can be achieved with 5.80-GHz microwaves (see Sect. 4.6.3) because the higher microwave frequency in that cross-sectional area of the waveguide is some threefold smaller than that of the 2.45-GHz microwaves. Moreover, the overall structure of the device comprises a much smaller waveguide.

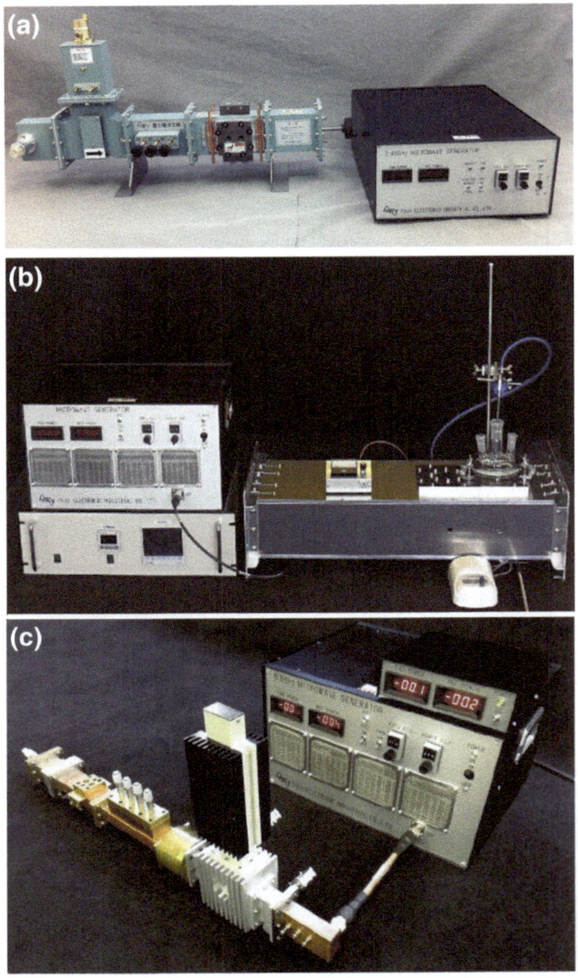

Fig. 5.25 Commercial single-mode semiconductor microwave generator systems operating at frequencies of (**a**) 2.45 GHz (FSU-201VP-01), (**b**) 915 MHz (FSU-301VP-01), and (**c**) 5.80 GHz (FSU-131VP-01). Photographs provided by courtesy of Fuji Electronic Industrial Co. Ltd. [16]. Copyright 2012 by Wiley-VCH Verlag GmbH

5.7.4 Frequency Effects for the Common Solvents

Frequency effects are seen even for a simple chemical structure such as that of water. Changes in the dielectric constant and dielectric loss factors of water as a function of the microwave frequency in the frequency range 0.1–1000 GHz are illustrated in Fig. 5.27 [25]. The dielectric loss increases at first with an increase in frequency reaching a maximum at ca. 10 GHz, and then decreases with further increase in frequency. By contrast, the dielectric constant is relatively constant at first and then decreases with increase in the microwave frequency.

The most important characteristics of the 23 solvents examined at two microwave frequencies and ambient temperature are the penetration depth of the microwaves into the bulk of the solution and the dielectric parameters of these

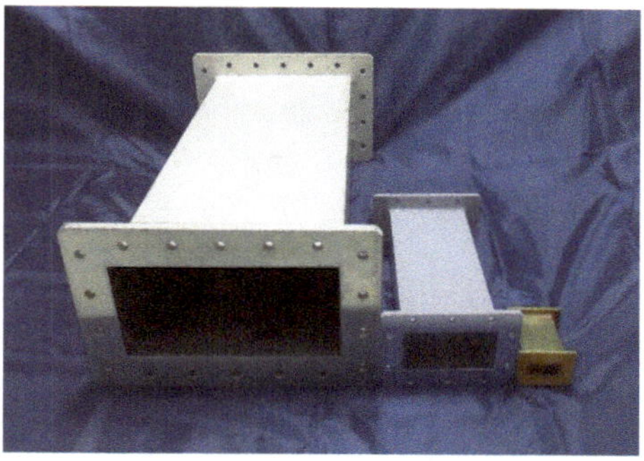

Fig. 5.26 Photograph displaying the different sizes and sectional areas of the waveguides used for 915-MHz microwaves (left), 2.45-GHz microwaves (center) and 5.80-GHz microwaves (right). Reproduced from [16]. Copyright 2012 by Wiley-VCH Verlag GmbH

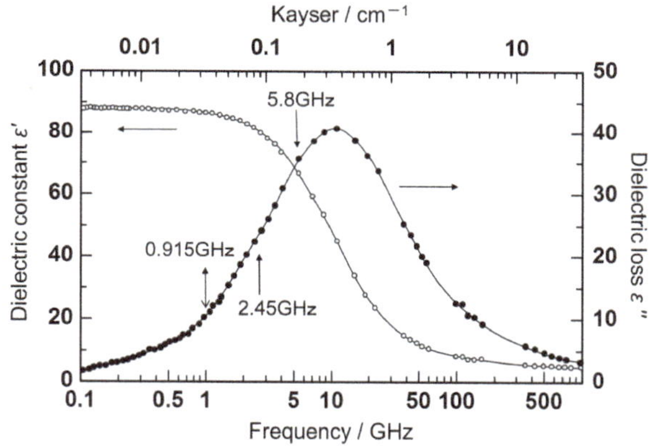

Fig. 5.27 Changes in the dielectric constant (ε′) and dielectric loss (ε″) of water at 25 °C as a function of frequency. Reproduced with permission from [25]

solvents: the dielectric constant (ε_r'), the dissipation factors [sometimes also referred to as the dissipation loss tangent, tan δ ($=\varepsilon_r''/\varepsilon_r'$)], and the dielectric loss factors (ε_r'') [16]. Note that the dielectric loss factors include the leakage currents. Results show that the dielectric constants (ε_r') of all the solvents at the 2.45-GHz microwave frequency are greater than the analogous constants at 5.80 GHz (Fig. 5.28a). In particular, such solvents as DMSO, methanol, ethylene glycol, and ethanol display significant variations in the dielectric constants. Hence, insofar as dielectric constants are concerned, there are but slight advantages in using the 5.80-GHz microwaves. The dielectric loss (ε_r'') at 5.80 GHz exceeds those at 2.45 GHz for most solvents except for the alcohols (Fig. 5.28b). Water, DMSO, methanol, DMF, and ethylene glycol display the largest ε_r'' values at 5.80 GHz, whereas at the 2.45-GHz microwave frequency the dielectric loss factors for methanol, ethylene glycol, ethanol, 1-propanol, 2-propanol, and acetic acid anhydride are greater than those at 5.80 GHz. Dissipation factors (tan δ) for the 5.8-GHz microwave radiation are greater than those for 2.45-GHz microwaves for almost all solvents except ethylene glycol, ethanol, 1-propanol, 2-propanol, and acetic acid anhydride for which tan δ are exceptionally greater at the latter microwave frequency (Fig. 5.28c). In heating solvents other than these, the expectation from the dissipation factors (tan δ) is that for each of the solvents the 5.80 GHz would be the optimal frequency.

Recall that the penetration depth of microwaves is the depth at which microwaves penetrate into the material and at which the power flux has fallen to 1/e (=0.368%) of its surface value. Penetration of the 5.80-GHz microwaves in all solvents was shallower compared to the depth of penetration by the 2.45-GHz microwaves (Fig. 5.28d). For instance, in the case of xylene, the 5.80-GHz microwaves could penetrate to a depth of ca. 1.24 cm, whereas the depth of penetration is about 28.32 cm for the 2.45-GHz microwaves. The largest difference in penetration depth between the 5.80-GHz and 2.45-GHz microwaves was observed for the nonpolar solvents. Note that although the 2.45-GHz microwaves penetrate deeper into a solvent/solution, they have a comparatively low energy conversion, whereas the 5.80-GHz microwaves with their shallower penetration depth display a much greater energy conversion within that shallow depth. Accordingly, the energy efficiency for heating thin portions of the solvents/solutions is greater when using 5.80-GHz microwaves. Hence, the penetration depth of the microwaves is one of the principal factors that lead to rapid heating and superheating (see Sect. 4.6.3). For the 915-MHz frequency, the characteristic dielectric loss factors (ε_r'') against the 2.45-GHz frequency are greater for ethylene glycol, ethanol, 1-propanol, 2-propanol, and acetic acid anhydride. Thus, the optimal frequency for the heating of alcohols (except for methanol) is 915 MHz. The penetration depth of microwaves at the 915-MHz frequency in all solvents other than alcohols is very

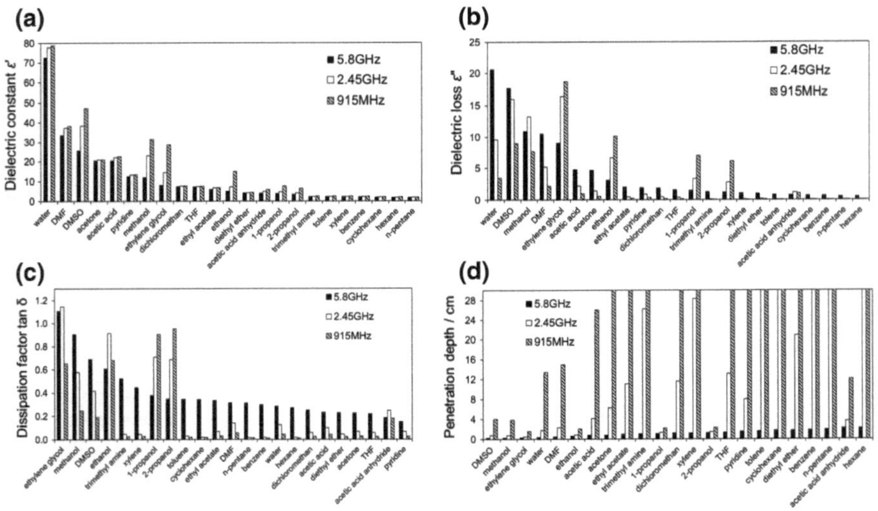

Fig. 5.28 Dielectric parameters at 5.80 GHz, 2.45 GHz, and 915 MHz for water and some 22 pure organic solvents at 25 ± 1 °C: (**a**) dielectric constant, ε_r'; (**b**) dielectric loss factor, ε_r''; (**c**) dissipation factor, tan δ; (**d**) penetration depth of the microwaves (cm). Volume of the sample was 100 mL. Reproduced from [26]. Copyright 2012 by Elsevier B.V

deep. The largest cylindrical reactor that can be accommodated into a 915-MHz waveguide is ca. 24.8 cm, so that it is suitable for heating polar solvents.

Changes in the dissipation factors (tan δ) that accompany microwave heating of various polar solvents at three different frequencies are displayed in Fig. 5.21 [26]; the temperature was measured with an optical fiber thermometer. An initial increase in the dissipation factor with increasing temperature was observed at 5.80 GHz for ethylene glycol, ethanol, 1-propanol, 2-propanol, and acetic anhydride, whereas a decrease was observed for the other solvents (Fig. 5.29a); when using 2.45-GHz microwaves, tan δ increased initially with temperature only for 1-propanol and 2-propanol (Fig. 5.29b). On application of the 915-MHz frequency, a decrease of the dissipation factor was observed with increasing temperature for all solvents (Fig. 5.29c).

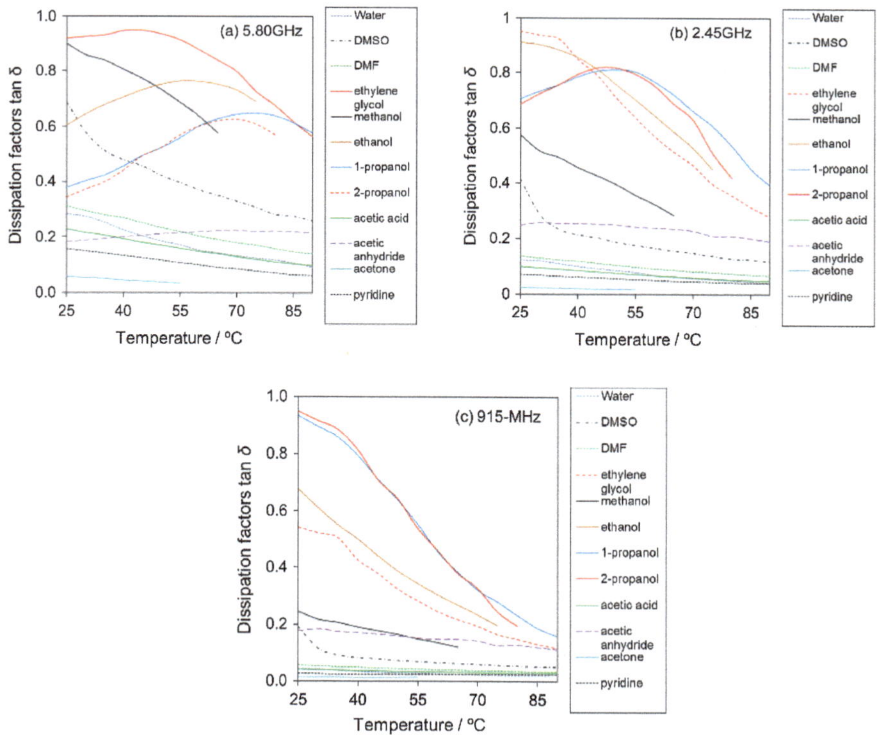

Fig. 5.29 Dissipation factors (tan δ) at a microwave frequency of (**a**) 5.80-GHz, (**b**) 2.45-GHz, and (**c**) 915-MHz for pure solvents at various temperatures: Volume of the sample, 100 mL. Reproduced from [26]. Copyright 2012 by Elsevier B.V

5.7.5 Rates of Temperature Increase of Common Solvents

Heating rates of common solvents determined in a quartz reactor irradiated continuously with 5-Watt microwaves from a semiconductor generator and a single-mode resonator at the 5.80 GHz, 2.45 GHz, and 915 MHz are reported in Fig. 5.30a–c. Samples of the solvents were located at the maximal position of the electric field. The heating rate was not affected by the penetration depth when using a 4-mm reactor. Ratios in heating rates between the 5.80-GHz and 2.45-GHz frequencies are shown in Fig. 5.30d based upon the heating rates from Fig. 5.30a, b. In some instances, differences were well over tenfold: *n*-pentane (18-fold), toluene (15-fold), and xylene (12-fold). Clearly, the largest differences between heating with 5.80-GHz versus 2.45-GHz microwaves were mostly displayed by the nonpolar solvents. The solvents with the highest heating rate under 2.45-GHz microwave heating were ethylene glycol, ethanol, 2-propanol, and 1-propanol (Fig. 5.30b).

The heating rates at the 915-MHz frequency were greatest for the alcohols followed by DMF, DMSO, and water (Fig. 5.30c). On the other hand, most nonpolar solvents were not heated by the 915-MHz radiation. In this experiment, a quartz test tube (diameter, 20 mm) containing a 10-mL sample was set at the maximal position of the electric field inside the 915-MHz waveguide. To the extent that the sample size was small relative to the cross-sectional area of the waveguide, heating was not very efficient when using the 4-mm reactor; the microwave applied power was 70 W at this frequency. The ratios of the heating rates at the 915-MHz frequency to the 2.45-GHz frequency are reported in Fig. 5.30e.

5.8 Frequency Effects in Organic Synthesis

5.8.1 Application to a Diels–Alder Reaction

This section examines reactions using nonpolar substances (reacting substances and solvents) and shows that some of the issues of microwave chemistry can be resolved by taking advantage of the frequency effect. The usefulness of 5.80-GHz microwaves in performing organic syntheses is demonstrated by the synthesis of 3,6-diphenyl-4-*n*-butylpyridazine (DBP) through a Diels–Alder process using two representative solvents of different polarities: ethyl acetate and xylene [27].

The Diels–Alder synthesis of 3,6-diphenyl-4-*n*-butylpyridazine (DBP) from diphenyl- tetrazine and 1-hexyne reagents in ethyl acetate and xylene solvents is illustrated in reaction 1; the resulting chemical yields under various experimental heating conditions after silica gel treatment of the resulting product(s) are listed in Table 5.4.

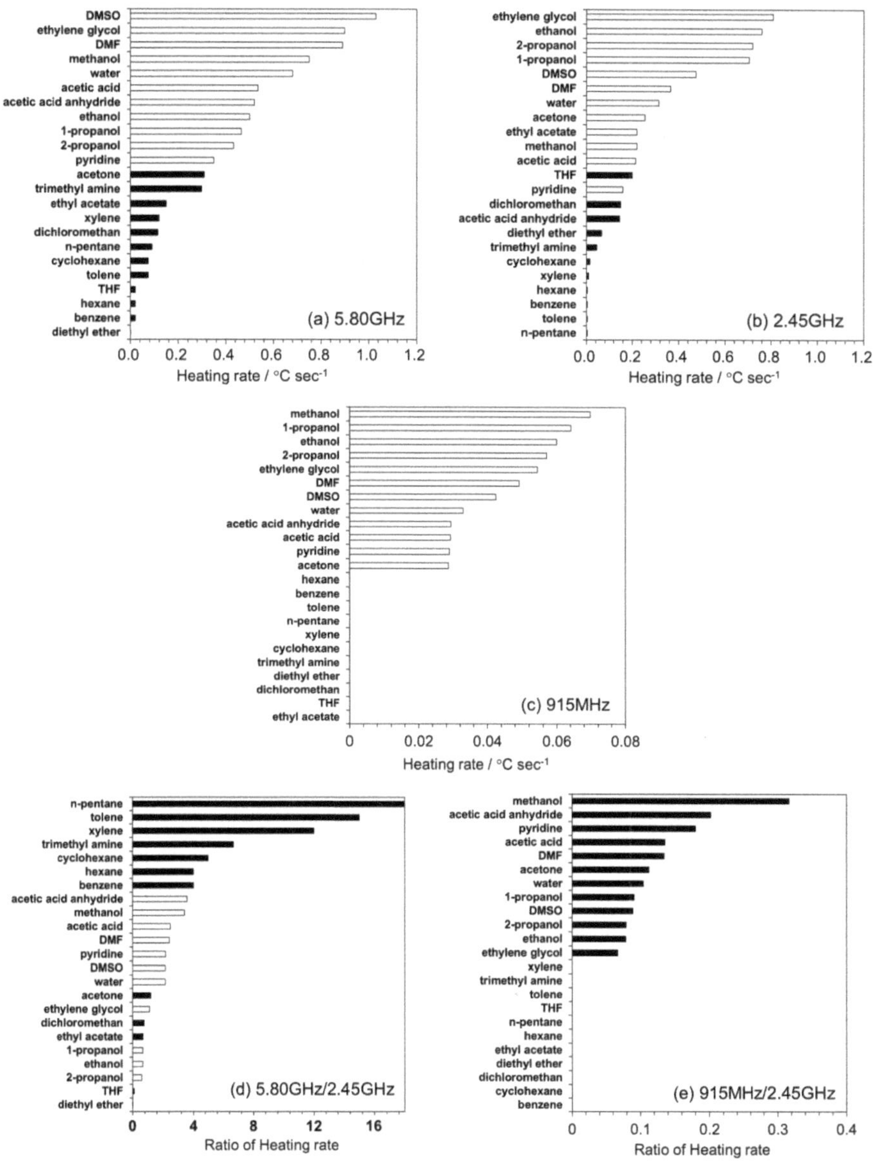

Fig. 5.30 Rates of increase of temperature for some 23 common solvents under non-stirring conditions under irradiation with (**a**) 5.80-GHz and (**b**) 2.45-GHz microwaves at a power level of 5 Watts using single-mode resonant systems with a semiconductor generator at 5-s intervals up to 30 s; (**c**) 915-MHz microwave irradiation (70 W; continuous irradiation) using a single-mode apparatus with semiconductor generator at 5 s intervals; (**d**) ratios of heating rates for the 5.80-GHz and 2.45-GHz frequencies; (**e**) ratios of heating rates for the 915-MHz and 2.45-GHz frequencies; black bars: polar solvents; empty bars: nonpolar solvents. Volumes of samples, 1 mL. Reproduced from [26]. Copyright 2012 by Elsevier B.V

Table 5.4 Yields of 3,6-diphenyl-4-n-butylpyridazine (DBP) under various experimental conditions

Heating method	Solvent	Temperature/°C	DBP yield/% after	
			90 min	160 min
5.8-GHz	Ethyl acetate	100	12	29
	Xylene	122	12	38
2.45-GHz	Ethyl acetate	43	0	0
	Xylene	33	0	0
Oil bath	Ethyl acetate	76	<1	2
	Xylene	135	5	7

Reproduced from [27]

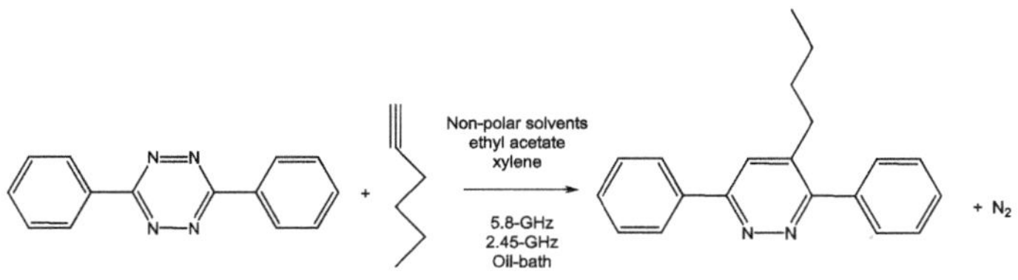

(reaction 1)

The yield of the DBP product under continuous 5.80-GHz microwave irradiation ranged between 12 and 29% for ethyl acetate; for xylene, the chemical yield was 12–38%. This was not unexpected for the synthesis of this pyridazine since irradiation of the ethyl acetate and xylene solvents by the 2.45-GHz microwaves was comparatively inefficient. For comparison, heating the xylene solvent with an oil bath led only to a 5–7% yield of DBP; note that the reactor was introduced into the oil bath which had been preheated at 80 ºC because the initial rate of temperature rise was rather slow with this method. Although the yields of the pyridazine product under 5.80-GHz microwave radiation were not spectacular, they were nonetheless significantly greater than conventional heating of the reacting solution, and far better than under 2.45-GHz irradiation.

Temperature-time profiles of the ethyl acetate and xylene solvents used are illustrated in Fig. 5.31a, b, which show that 5.80-GHz microwave heating was constantly higher than with the 2.45-GHz microwaves. Irradiation with 2.45-GHz microwaves for longer times had no effect upon the reaction. The oil bath heating of ethyl acetate was limited by its boiling point of 77 °C. However, heating of ethyl acetate with 5.80-GHz microwaves exceeded the boiling point (100 °C after 6 min). Such superheating cannot be achieved by existing conventional methods. The high chemical yield was caused by the temperature rise under the continuous 5.80-GHz microwave irradiation. The temperature-time profiles for the reacting substances and the solvents used in the open reactor system are shown in Fig. 5.23c. The temperature of 3,6-diphenyl-1,2,4,5-tetrazine (tetrazine) and of 1-hexyne increased rapidly under continuous irradiation with the 5.80-GHz

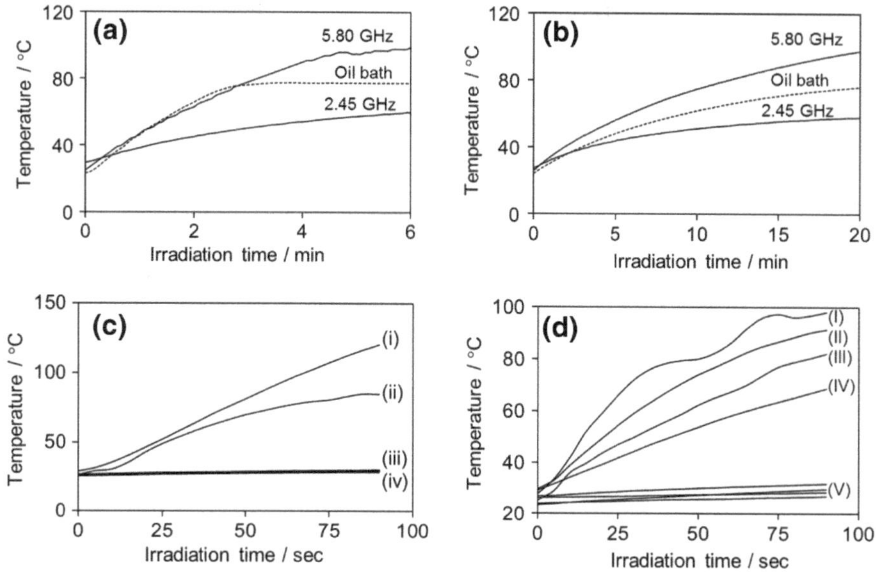

Fig. 5.31 Temperature-time profiles of (**a**) ethyl acetate, and (**b**) xylene under 5.80-GHz and 2.45-GHz microwave heating, and heating in an oil bath; (**c**) Temperature-time profiles of naked 3,6-diphenyl-1,2,4,5-tetrazine (i and iii) and 1-hexyne (ii and iv) when subjected to 5.80-GHz (i and ii) and 2.45-GHz (iii and iv) microwave continuously irradiation (in all cases the solvents were stirred); (**d**) Temperature-time profiles of pure solvent and solvent with reacting substances (sample solution) under 5.80-GHz (I: sample solution with ethyl acetate; II: ethyl acetate; III: sample solution with xylene; IV: xylene) and 2.45-GHz (V: each of the four samples). Reproduced from [27]. Copyright 2008 by the American Chemical Society

microwaves. Note that the tetrazine is a solid substance at ambient temperature. The heating of the reacting substances under otherwise identical conditions progressed effectively with the 5.80-GHz microwaves as shown in Fig. 5.23d. Temperature increases of the solvents were somewhat attenuated with the use of 2.45-GHz microwaves. In the case of the 5.80-GHz microwave system, the rates of increase of temperature of the sample solution were greater than those of the pure solvents used. That is, the rate of heating of the reacting agents also aided the microwave-assisted organic synthesis at the 5.80-GHz frequency.

5.8.2 Synthesis of a Room-Temperature Ionic Liquid (RTIL)

Both 5.80-GHz and 2.45-GHz microwaves were used to carry out the novel one-pot solvent-free synthesis of the room-temperature ionic liquid (RTIL) 1-butyl-3-methylimidazolium tetrafluoroborate ([bmim]BF_4) in a reactor containing the three required reagents, namely 1-methylimidazole, sodium tetrafluoroborate, and

1-chlorobutane (reaction 2) [28]. The synthesis yields of [bmim]BF_4 under various experimental conditions are listed in Table 5.4. Results clearly demonstrated that the one-step solvent-free synthesis of this RTIL was possible with microwave heating at 5.80-GHz and 2.45-GHz frequencies and by conventional heating. The synthetic efficiencies were significantly different as evidenced by the data in Table 5.5. In the batch mode, the chemical yields of [bmim]BF_4 (87% after a 30-min period) from the 5.80-GHz microwave heating were some three-to-fourfold greater than the yields obtained with either the 2.45-GHz microwave heating (28%) or the oil bath heating (21%), even though the temperature reached with 2.45-GHz microwaves was greater than 200 °C, whereas the corresponding temperature reached with the 5.80-GHz microwaves was less than 170 °C. The oil bath heating temperatures reached to synthesize this RTIL were similar to those of the 5.80-GHz microwaves.

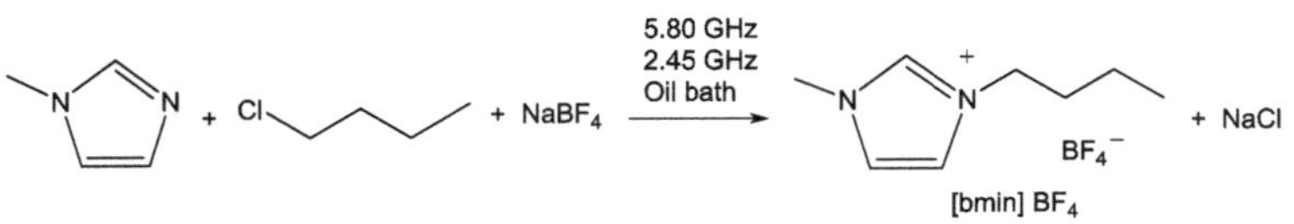

(reaction 2)

The temperature-time profiles of the mixture of reacting substrates in the synthesis of [bmim]BF_4 under irradiation with 5.80-GHz and 2.45-GHz microwaves were quite dissimilar. After 10 min into the process, upon heating with the 5.80-GHz microwaves the measured temperatures were constantly lower than with the 2.45-GHz microwaves. Temperature increase was not monotonic at the latter frequency. In fact, for the synthesis of [bmim]BF_4 by the batch process, unusual temperature increases were observed with the 2.45-GHz radiation after 3, 10, and 20 min. This irregular increase was likely due to the formation of the [bmim]BF_4 ionic liquid. That is, the heating efficiency with the 2.45-GHz microwaves changed because the composition of the reacting mixture changed with the formation of [bmim]BF_4. Contrary to the latter (2.45 GHz), the temperature profile observed for the 5.80-GHz microwave radiation was somewhat noisy beyond 20 min of

Table 5.5 Chemical yields of [bmim]BF_4 from the one-step solvent-free synthesis under irradiation at various microwave frequencies and experimental conditions. Microwave power level was constant at 30 W

Heating source	Reaction time/min	Temperature/°C	Yield/%
5.80-GHz	30	155	87
	60	169	84
2.45-GHz	30	204	28
	60	219	29
Oil bath	30	168	21
	60	176	25

Reproduced from [28]

irradiation, but otherwise of no consequence with regard to the smooth formation of the product. After a 10-min irradiation period, the differences in the temperature-time profiles of the individual reacting substrates 1-chlorobutane (CB), 1-methylimidazole (MI), and the solid $NaBF_4$ in the batch mode showed significant differences on irradiation with 2.45-GHz microwaves (CB: 84 °C, MI: 147 °C, and $NaBF_4$: 56 °C). Differences in the temperature profiles of the reacting materials subjected to the 5.80-GHz microwave radiation were smaller (CB: 105 °C, MI: 124 °C and $NaBF_4$: 143 °C, after 10 min) than with the 2.45-GHz microwaves. Closer examination of the results also revealed that the temperature rise in the synthesis of [bmim]BF_4 with the 2.45-GHz microwaves was more effective than with the 5.80-GHz microwaves.

The temperature distributions for each of the reacting substances (heterogeneous CB/MI/NB) were determined with a thermographic system which measures the temperature at the surface of the reactor; temperatures were measured through the slit in the corresponding 2.45-GHz and 5.80-GHz microwaves' waveguides [16]. Comparatively, uniform heating of the CB and MI layers was observed from continuous irradiation with 5.80-GHz microwaves for a 60-s period. Clearly, best microwave heating with the 2.45-GHz microwaves occurred in the 60-s period at the 1-methylimidazole layer that contained some dissolved $NaBF_4$. In the case of 2.45-GHz microwave heating, measurements of the temperature distributions by thermographic imaging clearly demonstrated that the 1-methylimidazole is a heat conductor that converts microwave energy into thermal energy. Note that the uniform heating distribution of the sample solution was confirmed by thermographic measurements on heating pure water under 2.45-GHz microwave irradiation. The temperatures of the CB, MI, and NB layers in Fig. 5.32, estimated from the thermographic images, were 73.3 °C (CB), 77.0 °C (MI), and 60.9 °C (NB) for continuous 5.80-GHz microwave irradiation and 61.6 °C (CB), 73.1 °C (MI), and 53.5 °C (NB) for continuous 2.45-GHz microwave irradiation. The differences in temperatures of the layers correlated with the result of differences in the dielectric loss factors of the pure substances. Note that the surface of the reactor was heated by thermal conductivity from the sample solution. Further microwave heating led to boiling of the sample mixture (boiling points of pure CB and MI liquid are 78 and 198 °C, respectively). Subsequently, the macroscopic temperature distribution of the solution became uniform as a result of convection.

5.8.3 Synthesis of Gemini Surfactants Under 915-MHz Microwave Irradiation

Dimeric surfactants represent a new class of surfactants and thus the name *Gemini surfactants*. Common Gemini cationic surfactants with different chain lengths C_{12}–C_2–C_{12} have been synthesized in ethanol solvent (reaction 3) using 915-MHz microwave irradiation (80 Watts) [29]. The synthesis of such surfactants using a 2.45-GHz multimode applicator was also performed for comparison.

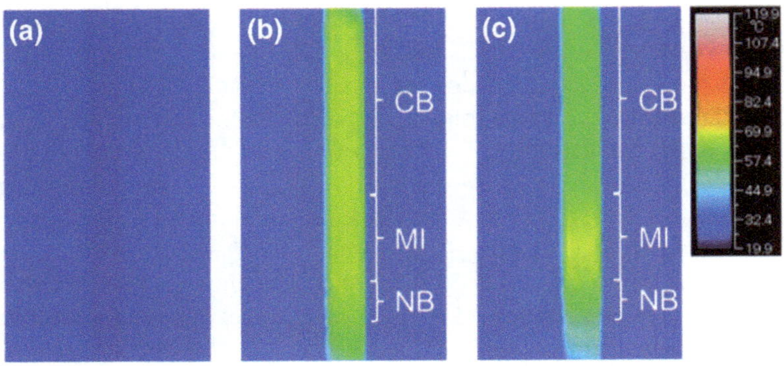

Fig. 5.32 Thermographic images indicating the temperature distributions in the profiles of the 1-chlorobutane (CB) layer, the 1-methylimidazole layer containing a small quantity of dissolved $NaBF_4$ (MI) layer, and the undissolved $NaBF_4$ salt (NB) layer under microwave heating (MW power, 10 W) and under non-stirring conditions. Times of continuous microwave irradiation were (**a**) 0 s, (**b**) 60 s at 5.80 GHz, (**c**) 60 s at 2.45 GHz. Reproduced from [16]

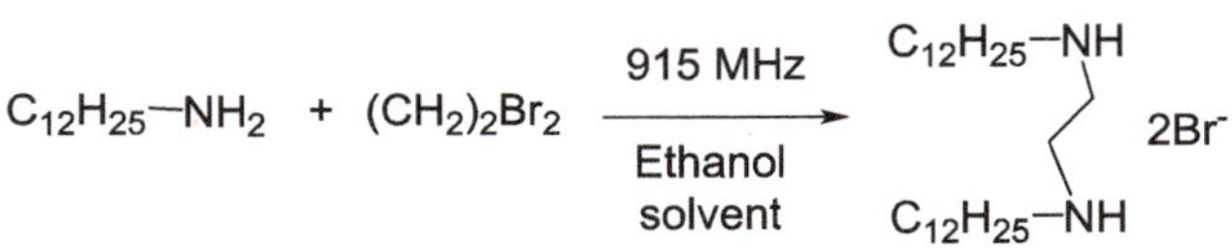

(reaction 3)

Product yields of C_{12}–C_2–C_{12} Gemini surfactants synthesized with 915-MHz and 2.45-GHz microwaves after a 60-min period were 34% (915-MHz) and 28% (2.45-GHz) (Table 5.6). The maximum quantity of the sample solution in a 2.45-GHz single-mode irradiation system was about 50 mL. However, single-mode irradiation can be performed effectively at the 915-MHz frequency even with a 500-mL scale quantity. Note that using the common 2.45-GHz microwaves to heat large-scale samples necessitates equipment that is equipped with a circulating reactor whether using a single-mode applicator or a multimode applicator.

Table 5.6 Chemical yields of a C_{12}–C_2–C_{12} Gemini surfactant in ethanol solvent under 915-MHz and 2.45-GHz irradiation conditions for a 60-min period

	Reflux time/min	Output power/W	C_{12}–C_2–C_{12} Gemini surfactant/%
915-MHz	60	70	34
2.45-GHz	60	150	28
Oil bath	3-days	–	41

Reproduced from [29]

5.8.4 Frequency Effects in Nanoparticle Synthesis

The synthesis of inorganic nanoparticles (e.g., metallic, metal oxides and metal chalcogenides, among others) has made effective use of dielectric heating provided by microwave radiation [30]. The major role of microwaves in nanoparticle synthesis is rapid and uniform heating. Uniform nano- and bulk particle sizes are typically obtained by uniform nuclear growth through microwave specific heating. In this regard, synthetic methods that utilize conventional convective heating, because of the need for high-temperature initiated nucleation followed by controlled precursor addition to the reaction media, rely upon a conduction pathway to drive the synthetic process in which the reactor acts as the intermediary to transfer the thermal energy from the external heat source to the solvent and ultimately to the reactants [31]. In addition, uniform nanoparticle synthesis by microwave heating supports the possibility of scale-up [32]. On the other hand, no synthesis of nanoparticles takes place when the reacting substrates do not absorb strongly the microwaves. Moreover, if a microwave absorber (e.g., activated carbon) unrelated to the reaction system were added it would become an impurity within the reaction. Accordingly, there are limitations as to the reacting substrates that can be used.

Let us now examine the synthesis of gold nanoparticles as model nanoparticles in nonpolar and polar media [33]. In particular, we examine the influence of the microwave frequency on the synthesis of gold nanoparticles in the presence of oleylamine in a nonpolar solvent of low dielectric constant (ε') that does not absorb 2.45-GHz microwave radiation and so is not dielectrically heated by these microwaves (reaction 4). Let us also examine the effect of microwave frequency on the shape and size of gold nanoparticles that are produced in a highly polar medium such as ethylene glycol, under otherwise identical temperature conditions to those used in nonpolar media, so as to probe other possible factors that could influence the nanoparticle synthesis.

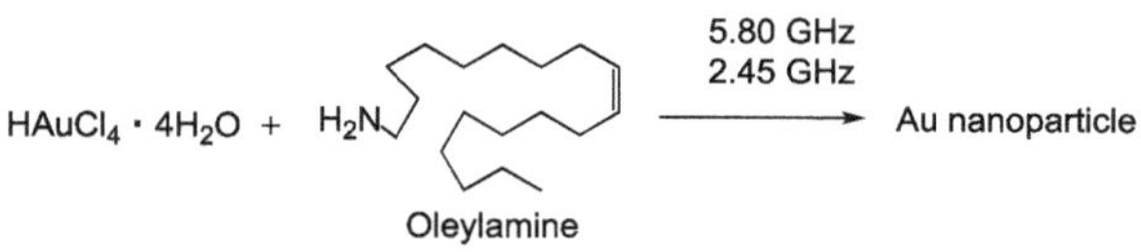

(reaction 4)

The temperature-time profiles of a mixture of oleylamine and aurochloric acid were first determined at the two microwave frequencies used (5.80 and 2.45 GHz). The initial rates of temperature increase for a 6-min irradiation period were 16.5 °C min^{-1} and 2.79 °C min^{-1} for the 5.80-GHz and 2.45-GHz microwaves, respectively, under a continuously applied microwave power of 45 Watts. The rate of increase of temperature was nearly fivefold faster for the 5.80-GHz microwaves than for the 2.45-GHz microwaves, with the latter increasing gradually with irradiation time.

The surface plasmon resonance absorption spectra of the resulting gold nanoparticles are displayed in Fig. 5.33 for irradiation times of 7 and 15 min for the 5.80-GHz microwaves and 15 and 30 min for the 2.45-GHz microwaves. A surface plasmon resonance (SPR) band was seen at 528 nm on irradiation with the 5.80-GHz microwaves for an irradiation period of 7 min and 15 min in accord with the absorption spectra of gold nanoparticles that had been reported earlier by Mohamed and coworkers [34] and by Yu and Andriola [35]. By contrast, no absorption bands were seen after a 15-min and 30-min heating time with the 2.45-GHz microwaves, which indicated that no gold nanocolloids formed under these conditions. Evidently, the Au^{3+} ions in $HAuCl_4$ are reduced by pure oleylamine acting as a reducing agent under the 5.80-GHz microwaves. That the product displayed only one Plasmon band at 528 nm was taken as an indication of the formation of spherical gold nanoparticles; note that oleylamine also acted as a capping agent [34].

Transmission electron microscopy (TEM) images of the resulting Au nanoparticles produced subsequent to the 5.80-GHz microwave heating for an irradiation time of 15 min are illustrated in Fig. 5.34, which show a fairly uniform particle size distribution of the spherical nanocolloids. A random sampling of 35 particles gave a size distribution of 9 $\pm$ 2 nm after the 15-min heating period. By contrast, the TEM images (not shown) from the 2.45-GHz microwave heating of the oleylamine/aurochloric acid mixture resulted in no formation of Au nanoparticles even after an irradiation period of 30 min, which confirm the results observed from the surface Plasmon resonance spectra of Fig. 5.33. These observations demonstrate that the lack of formation of gold nanoparticles under the 2.45-GHz microwaves was the result of not reaching the appropriate reactive temperature. Generally, it is assumed that microwave-assisted syntheses do not occur in a nonpolar solvent, as the latter is not a microwave absorber. However, using microwaves of higher frequency (e.g., 5.80 GHz) it is possible to synthesize gold nanoparticles, even in nonpolar solvents for the same microwave power and solution composition.

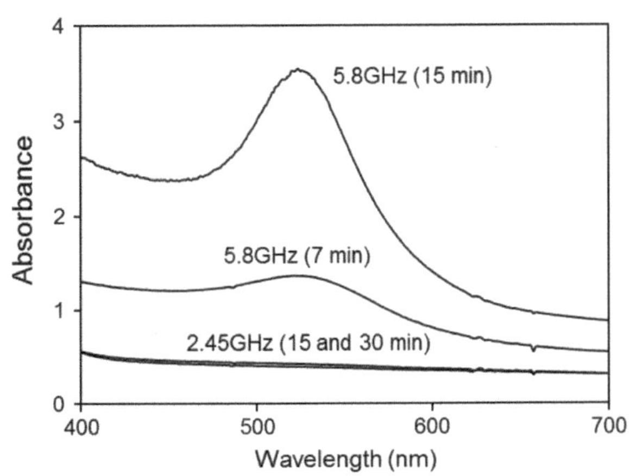

Fig. 5.33 UV–visible absorption spectra of the product from irradiating the oleylamine/aurochloric acid solution with 5.80-GHz and 2.45-GHz microwaves (power, 45 W) at the times indicated. Reproduced from [33]. Copyright 2011 by the Royal Society of Chemistry

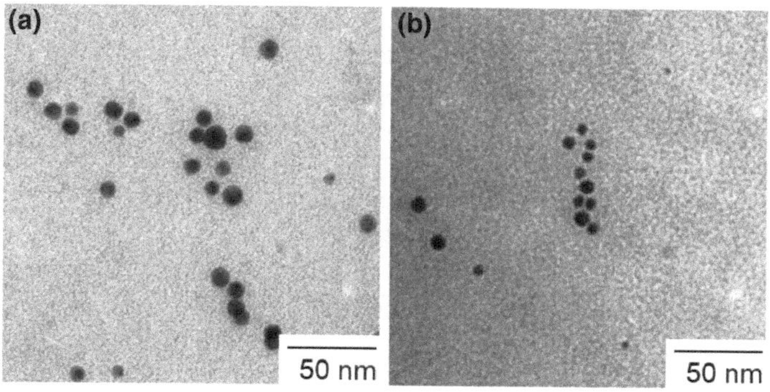

Fig. 5.34 TEM images of the gold nanoparticles produced after the (**a**) 7 min and (**b**) the 15 min of dielectric heating by the 5.8-GHz microwave radiation. Note the larger gold nanoparticles produced at the longer irradiation time

Gold nanoparticle synthesis by the polyol method has also been investigated in a polar medium such as in ethylene glycol solvent [36]. Only a small difference in heating efficiency was observed between the 5.80-GHz and 2.45-GHz microwaves. The corresponding measured dielectric loss factors were $\varepsilon_r'' = 9.8$ (5.80 GHz) and 14.7 (2.45 GHz), whereas the depth of penetration of the microwaves was estimated to be $D_p = 2$ mm (5.80 GHz) and 5 mm (2.45 GHz). Any temperature fluctuation caused by the frequency effect was canceled by the continuous microwave irradiation at 40-Watts power for both the 5.80-GHz and 2.45-GHz microwaves. The corresponding initial rates of temperature increase in the first 1 min of irradiation were both 86 °C min^{-1}. Interestingly, no differences were seen in the shape and quantity of Au nanoparticles produced when the reacting substrates were heated with the same heating profile in a polar solvent and the same two microwave frequencies were used [36]. Evidently, the frequency effect had no consequence in the formation of gold nanoparticles in polar media.

5.8.5 Summary Remarks on the Frequency Effect

The frequency effect is attractive in microwave-assisted chemistry and materials processing. When this effect can be managed, it can be applied to various applications. For instance, using a single-mode applicator, 915-MHz microwaves can irradiate a reactor of large size because of the inherent large size of the waveguide when compared to the waveguide when using 2.45-GHz microwaves. Furthermore, heating large quantities of alcohols is remarkably effective at the lower 915-MHz frequency compared with the 2.45-GHz and 5.80-GHz frequencies. At the industrial scale, it is well known that the extent of energy conversion (more than 80%) and the operating lifetime (more than 10,000 h) of a 915-MHz magnetron are two attractive features of this frequency compared to those of the 2.45-GHz magnetron (70% and

8,000 h). The conditions for which 915-MHz microwaves are more suitable for making large-sized equipment are well known. On the other hand, heating nonpolar solvents with 5.80-GHz microwaves is particularly effective owing to the higher densities of the electric and magnetic fields. Choosing the optimal frequency of the microwaves for each situation can further promote existing microwave-assisted chemistry. Moreover, the use of several microwave frequencies could also prove to be of great help to the chemical industry in microwave organic chemistry. However, in order to fully understand the microwave frequency effect(s), it is necessary to perform additional studies in this area so as to identify, and otherwise clarify, some of the factors that play an important role in microwave chemistry carried out at different frequencies.

5.9 Electromagnetic and Thermodynamics Simulations

A chemist who is interested in experimenting with microwave-assisted chemistry will likely inquire how microwave radiation is applied to a sample and how the resulting heat is distributed. Typically, a sensor is used to visualize microwaves (Sect. 6.7); however, it is difficult to imagine the distribution of microwaves in space because the observation of microwaves with a sensor reveals the field only at the location of the sensor. A similar problem arises when attempting to measure the temperature distribution with an optical fiber thermometer. Consequently, a simulation is often used to visualize the spatial distribution of microwaves and the distribution of temperature. Accordingly, this section will attempt to explain the significance of the use of an electromagnetic field simulation and a heat transfer simulation.

An electromagnetic field simulation models the target space (and/or sample) by dissecting the space (or sample) into what will look like a mesh, after which an equation is used to solve every mesh component. It is then possible to stack the results for the whole mesh to visualize the distribution of temperature in a sample. Two methods can be used to solve the relevant integral equations: (a) by the boundary element method (BEM) and (b) by the finite-difference time-domain method (FDTD). The electromagnetic field simulation is made possible given appropriate initial conditions and boundary conditions as required by Maxwell's equations to calculate the distribution of microwaves in space-time. The initial condition is the location of the microwave radiation, while the boundary condition is the location of the sample in space. There exist appropriate software packages for computers to solve the equations. A flow diagram of the calculation methodology will involve the following considerations:

1. It is necessary to restrict and simplify the model to avoid long calculation times typically seen for long complicated models. In addition, one must also consider the accuracy of the calculations.

2. The input data will be the values of several physical parameters. For instance, for dielectric factors one uses the actual measurements of the factors ***with changes in temperature***.
3. The boundary condition will typically be the location of the designated microwave radiation source.
4. Dissection of the space or sample into a mesh is subsequently made, and to reduce errors the mesh size is typically made less than one-tenth (1/10) of the wavelength of the microwave radiation.
5. Finally, the calculations are performed.

In the present instance, to illustrate some simulations carried out using the FDTD method, liquid samples (benzene, ethylene glycol, and water) with different microwave heating efficiencies were placed either in the multimode applicator or in the single-mode applicator, following which the changes in the electric field were simulated. The results of the electric field distribution and heat distribution in the multimode applicator are reported in Fig. 5.35.

Benzene, which has the lowest microwave heating efficiency of the three samples in the multimode applicator, showed no large changes in the distribution of the electric field, and so the temperature of the benzene hardly increased. For the ethylene glycol solution, the electric field does not reach the center because the penetration depth of the microwaves in ethylene glycol is shallow, which results in significant uneven heating. On the other hand, the microwaves permeate the water sample to the center, and as such heating occurs progressively and uniformly.

Results of the simulated electric field inside the resonator of a single-mode applicator for all three samples (benzene, ethylene glycol, and water) within the waveguide are shown in Fig. 5.36. In the case of benzene, no standing wave is disordered because no microwaves are absorbed by this nonpolar sample. On the other hand, the standing wave in the single-mode applicator is significantly disordered in the case of ethylene glycol; internal heating of the ethylene glycol sample occurs. The standing wave also changes significantly in the case of water; internal and surface heating also occur.

5.10 Transmission Modes

It is important to remember that air is also present when placing a sample of a pure material in an applicator; also, a boundary is generated in the reactor between the components in microwave heating, causing impedance changes. The distribution of an electromagnetic field in space changes as per its boundary conditions and/or with the shape of the waveguide. The expression used to classify this state is referred to as Transmission Modes. As electromagnetic waves, microwaves are transmitted into free space by the crossed waves of the electric and magnetic fields normal to the direction of propagation; these are known as a TEM (transverse electric magnetic wave) mode. In this section, the microwaves inside the waveguide are

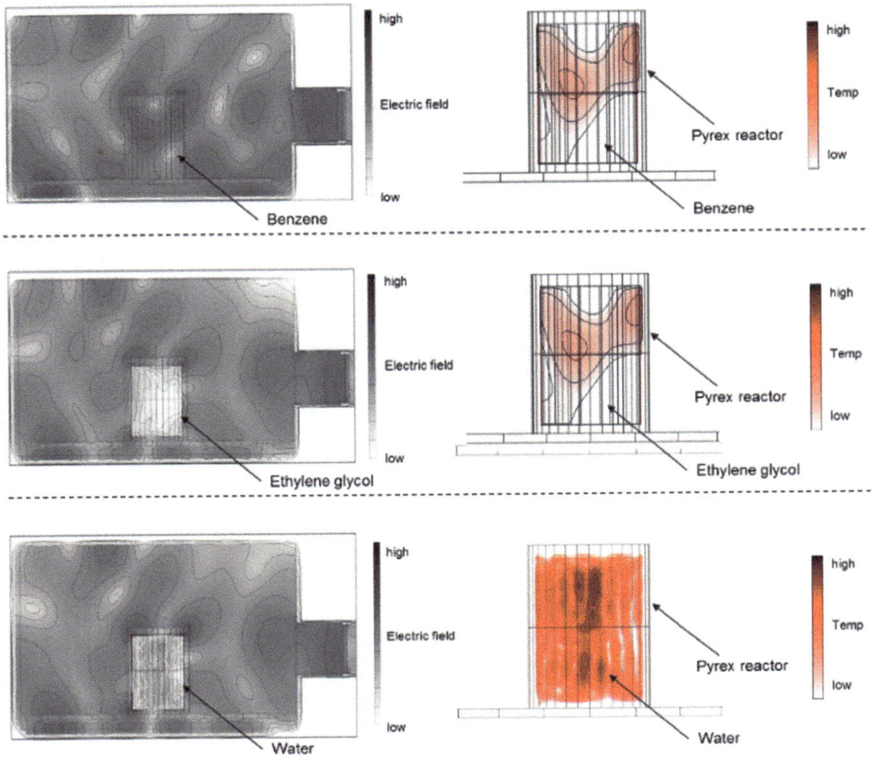

Fig. 5.35 Simulations of electric field and temperature distributions in three samples of different microwave heating efficiencies located in a multimode applicator. Equations were solved using the FDTD methodology. Samples analyzed and shown are benzene, ethylene glycol, and water

depicted as a sine wave. An actual electromagnetic field may be illustrated by the distribution shown in Fig. 5.37.

When an electromagnetic wave propagates down a waveguide, only one field—either electric or magnetic—will actually be transverse to the wave's direction of travel. The other field will *loop* longitudinally to the direction of travel, but still be perpendicular to the other field. Whichever field remains transverse to the direction of travel determines whether the wave propagates in the TE mode (Transverse Electric) or in the TM mode (Transverse Magnetic). An electric field of the TE mode exists in parallel to the short side of the waveguide, and its strength is distributed over the sine wave distribution along the long side, so that no electric field component exists along the tube's axial direction. On the other hand, a magnetic field is circular and exists along the long side direction of the waveguide and the tube's axial direction. Thus, only a magnetic field ingredient exists in the tube's axial direction, and the mode, which has no electric field components, is shown in the TE mode. By contrast, the electric and magnetic fields show a reverse mode with a TM mode.

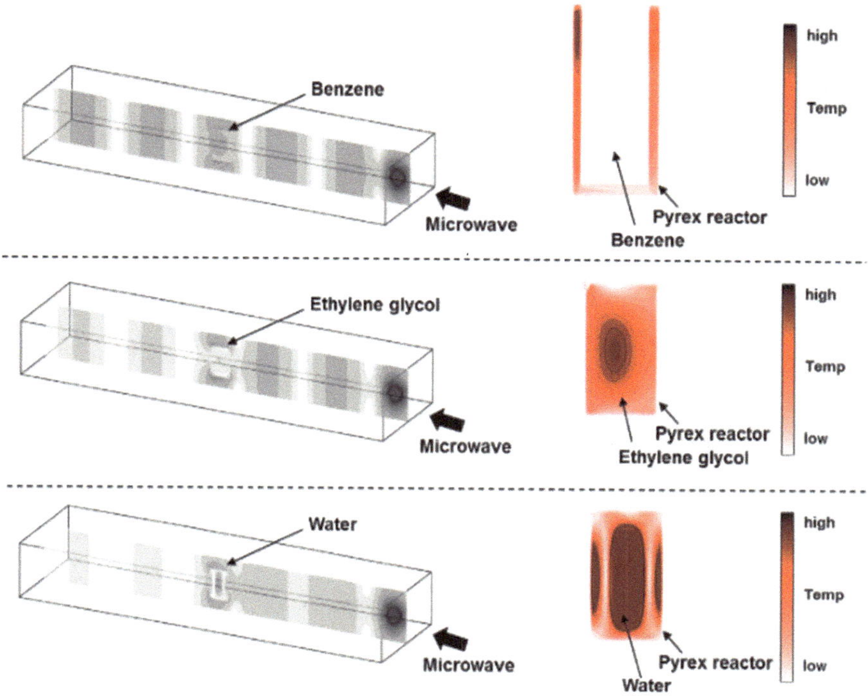

Fig. 5.36 Simulations of electric field and temperature distributions in three samples of different microwave heating efficiencies located in a single-mode applicator. Equations were solved using the FDTD methodology. Samples considered were benzene, ethylene glycol, and water

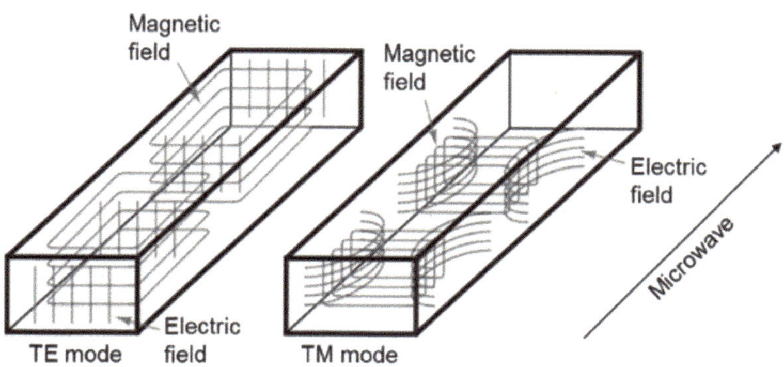

Fig. 5.37 Microwave distribution image of a waveguide illustrating the presence of a transverse electric mode (TE) and a transverse magnetic mode (TM)

The distribution of the electric field and the magnetic field in the waveguide is either in the TE or TM mode; generally, it is the TE mode. These modes change with a change in the cross-sectional area of the waveguide. The number of electric

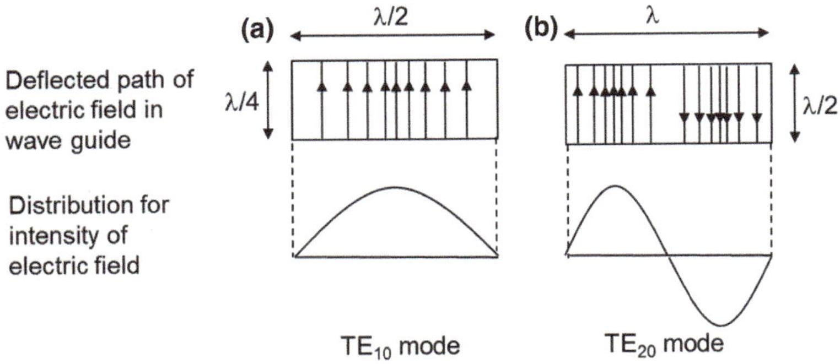

Fig. 5.38 Images of (**a**) a TE_{10} mode and (**b**) a TE_{20} mode in a waveguide

field intensity distribution in the long side of the waveguide and the number of the electric field intensity distribution in the short side are added behind the TE mode. In the TE_{10} mode, the first image on the left side of Fig. 5.38 displays the number of half wavelengths ($\lambda/2$) across the broad dimension of the electric field. The second image indicates that the length of the narrow dimension is less than $\lambda/2$ of the magnetic field. Also, TE_{10} means that during propagation through the waveguide, the electric field is perpendicular to the direction of propagation. It is equal to $\lambda/2$ across the broad dimension, while the magnetic field is less than $\lambda/2$. In the TE_{20} mode, the first image shows the number of full wavelengths λ across the broad dimension of the electric field, while the second image indicates that the length of the narrow dimension is less than the λ of the magnetic field. The TE_{20} means that the electric field is perpendicular to the direction of propagation.

When showing a standing wave of the resonance type in a single-mode applicator, other numbers may be added to the designation TE_{10} or TE_{20}. For instance, when there are two numbers at the peak top of the electric field inside the single-mode applicator, it is expressed as TE_{102} (see e.g., Fig. 5.39). When the

Fig. 5.39 Simulations of TE_{102} modes in a waveguide

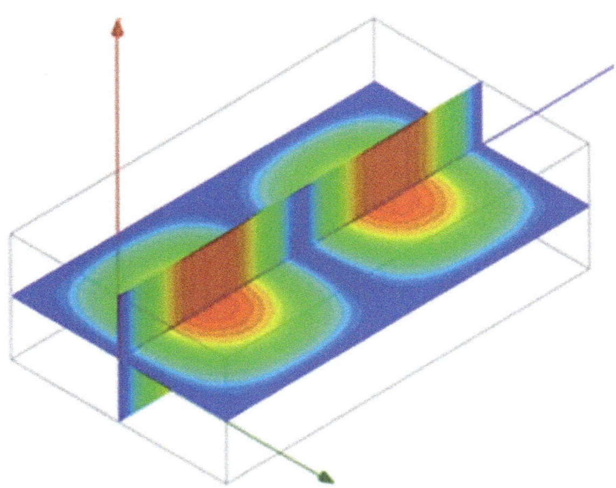

regularity of the mode is recognized, then it becomes possible to visualize the electric and magnetic fields inside the waveguide. For a single-mode applicator, only one transmission mode exists, whereas for a multimode applicator, more than one mode exists.

5.11 Coffee Break 5: Using the Microwave Oven

What makes heating foods with microwave energy different from heating them with conventional heating? Microwave heating can best be described as *bulk heating*. Conventional non-microwave ovens heat the surface of foods, after which that heat slowly migrates by conduction to the interior of the foodstuffs. However, in a microwave oven, the microwave energy penetrates the food, heating both the surface and the interior at the same time, so it becomes hot much faster than by conventional heat conduction. Nonetheless, shape, density, and volume of the food, along with its dielectric properties and specific heat capacity may create unusual heat patterns inside the foods. Microwave energy penetrates only a few centimeters into most foods before it is dissipated, causing some unusual temperature profiles. For instance,

1. microwaving a 10-cm-diameter beef roast is likely to result in the microwave energy not penetrating deeply enough to reach the center, which remains uncooked. Cooking can be completed by heat conduction outside the oven after first wrapping the roast with aluminum foil, and allowing it to stand for 10 or 15 min during which time the heat from the hot-microwaved areas is conducted inward to complete cooking. Note that there is no more microwave energy in the meat—that energy disappears as soon as the microwave oven is turned Off.
2. when microwaving a potato, the opposite may occur, i.e., the microwave energy concentrates inside the potato and if heated too long may lead to smoke and even fire. In this case, it is both the smaller size of the potato, permitting deeper penetration, and the dielectric properties that lead to concentrated central heating.
3. a peculiar phenomenon is that the air inside a microwave oven is cold unless some other auxiliary heating method is employed. This means that the surface of foods is never the hottest, but it will be someplace inside the food. This kind of center heating can create dangerous situations; for example, the interior of baby food in small glass jars may become hot enough to injure the infant by scalding its throat. These high interior temperatures may be hidden from the consumer because the cold air in the oven cools the food and glass surfaces, hiding the very hot interior. This cold surface phenomenon is also what makes it so difficult to brown and crisp foods in a microwave oven, unless some other auxiliary heat form, such as dry steam or hot air, is employed. Some browning and crisping can also be achieved by means of specially designed food ingredients, or with special packaging and cookware as will be explained below.

What are some of the problems related to formulating or designing foods for microwave ovens? Designing microwavable food products is especially difficult because of the large variety of microwave ovens in consumers' homes. Accordingly, the food products should be compatible with as many varieties of ovens as possible despite the variations in wattage, size, and otherwise. Some of the challenges that need to be faced are the following:

(a) ***Frozen foods*** will heat differently than refrigerated or room-temperature foods. This is because ice and frozen foods are very poor absorbers of microwave energy and heat very poorly, whereas nonfrozen foods are generally much better microwave absorbers and heat well and faster. Thus, as some portion of the food thaws, becomes unfrozen, it absorbs microwaves very well and heats quickly while the area next to it may remain frozen; frozen foods usually heat nonuniformly.
(b) ***Salt*** is an ingredient that may have a profound effect upon how well and uniformly a food product heats up. Salt reduces the ability of microwaves to penetrate into foods making it harder to heat the interior. Also, salt has a peculiar property, a phenomenon called *runaway heating*, i.e., the hotter a salt containing food becomes the faster it heats, making its heating very difficult to control (Fig. 5.40). Most prepared foods contain 1–2% salt, but, where possible, it is best to reduce this amount to as little salt as possible.
(c) ***Temperature***: water and high-water-containing foods such as vegetables, meats, and fish tend to heat more slowly the hotter they become (the opposite of the runaway salt effect noted above); this adds an element of control to prevent overheating of foods. It is like stepping on the brake in an automobile to slow it down (Fig. 5.41).

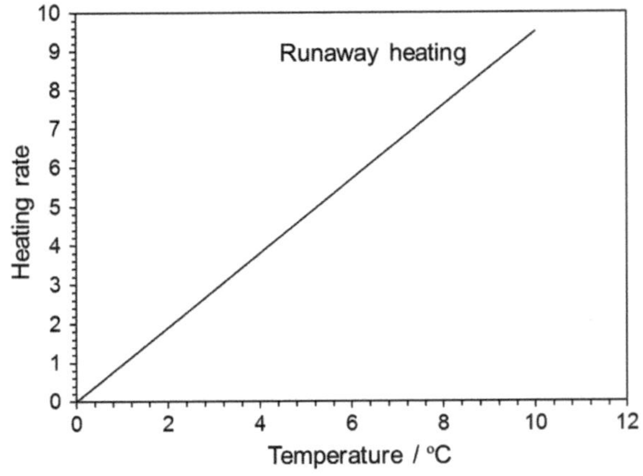

Fig. 5.40 Illustration of runaway heating, also called thermal runaway. The hotter the material becomes, the faster it heats! Copyright 2017 by R.F. Schiffmann

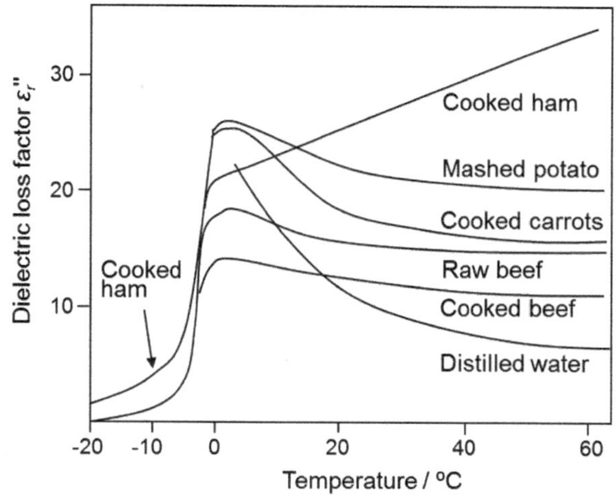

Fig. 5.41 Dielectric loss factor versus temperature. Note that as water and high-water-containing foods become hotter their loss factor decreases, meaning that they heat more slowly as they become hotter. Note also that Ham, which has a high salt content, heats faster as it becomes hotter. Reproduced with permission from [37]. Copyright 1971 by the International Microwave Power Institute

References

1. Keysight Application Note, Basics of Measuring the Dielectric Properties of Materials (http://literature.cdn.keysight.com/litweb/pdf/5989-2589EN.pdf#search=%27Keysight+Basics+of+Measuring+the+Dielectric+Properties+of+Materials%27)
2. H. Fukushima, Microwave processing and its applications to the future automobile, in *Proceedings of 1st Global Congress on Microwave Energy Applications*, Japan, 2008
3. C. Gabriel, S. Gabriel, E.H. Grant, E.H. Grant, B.S.J. Halstead, D.M.P. Mingos, Dielectric parameters relevant to microwave dielectric heating. Chem. Soc. Rev. **27**, 213–223 (1998)
4. M.S. Venkatesh, G.S.V. Raghavan, An overview of dielectric properties measuring techniques. Can. Biosyst. Eng. **47**, 7.15–7.30 (2005). http://www.csbe-scgab.ca/docs/journal/47/c0231.pdf#search=%27An+overview+of+dielectric+properties+measuring+techniques%27
5. J. Baker-Jarvis, M.D. Janezic, J.H. Grosvenor, R.G. Geyer, in *Transmission/reflection and short-circuit line methods for measuring permittivity and permeability*, NIST Technical Note 1355-R, U.S. Government printing office Washington, (1993)
6. N. Wagner, T. Bore, J.-C. Robinet, D. Coelho, F. Taillade, S. Delepine-Lesoille, Dielectric relaxation behavior of Callovo-Oxfordian clay rock: A hydraulic-mechanical-electromagnetic coupling approach. J. Geophys. Res. Solid Earth **118**, 4729–4744 (2013)
7. Agilent basics of measuring the dielectric properties of materials, Application Note, June 26, 2006 (http://www3.imperial.ac.uk/pls/portallive/docs/1/11949698.PDF)
8. A.C. Metaxas, R.J. Meredith, *Industrial Microwave Heating* (Peter Peregrinus, London, UK, 1983)
9. Z. Peng, J.-Y. Hwang, M. Andriese, Magnetic loss in microwave heating. Appl. Phys. Express **5**, 027304-1–027304-3 (2012)
10. S. Horikoshi, T. Sumi, N. Serpone, Unusual effect of the magnetic field component of the microwave radiation on aqueous electrolyte solutions. J. Microwave Power Electromagn. Energy **46**, 215–228 (2012)
11. S. Horikoshi, N. Serpone, Microwave frequency effect(s) in organic chemistry. Mini-Rev. Org. Chem. **8**, 299–305 (2011)
12. R. Cherbański, E. Molga, Intensification of desorption processes by use of microwaves—an overview of possible applications and industrial perspectives. Chem. Eng. Process. **1**, 48–58 (2008)

13. R. Roy, D. Agrawal, J. Cheng, S. Gedevanishvil, Full sintering of powdered-metal bodies in a microwave field. Nature **399**, 668 (1999)
14. J. Cheng, R. Roy, D. Agrawal, Experimental proof of major role of magnetic field losses in microwave heating of metal and metallic composites. J. Mater. Sci. Lett. **20**, 1561–1563 (2001)
15. J. Cheng, R. Roy, D. Agrawal, Radically different effects on materials by separated microwave electric and magnetic fields. Mater. Res. Innov. **5**, 170–177 (2002)
16. S. Horikoshi, N. Serpone, in *Microwaves in Organic Synthesis*, 3rd edn., ed. by A. de la Hoz, A. Loupy (Wiley-VCH Verlag GmbH, Weinheim, Germany, 2012)
17. https://en.wikipedia.org/wiki/Skin_effect
18. Technical report by A.C. Metaxas (http://www.pueschner.com/downloads/MicrowaveHeating.pdf#search='40680ISMband3390')
19. M. Möller, H. Linn, New microwave frequency 5.8 GHz for industrial applications. Key Eng. Mater. **264–268**, 735–739 (2004)
20. H. Takizawa, K. Uheda, T. Endo, Rapid formation and growth of bixbyite-type $(In_{0.67}Fe_{0.33})_2O_3$ by 28 GHz microwave irradiation. J. Am. Ceram. Soc. **83**, 2321–2323 (2000)
21. A.K. Malinger, Y.-S. Ding, S. Sithambaram, L. Espinal, S. Gomez, S.L. Suib, Microwave frequency effects on synthesis of cryptomelane-type manganese oxide and catalytic activity of cryptomelane precursor. J. Catal. **239**, 290–298 (2006)
22. R.N. Gedye, J.B. Wei, Rate enhancement of organic reactions by microwaves at atmospheric pressure. Can. J. Chem. **76**, 525–532 (1998)
23. E. Séguin, S. Thomas, P. Bazin, G. Bond, C. Henriques, F. Thibault-Starzyk, Infrared and microwaves at 5.8 GHz in a catalytic reactor. Phys. Chem. Chem. Phys. **11**, 1697–1701 (2009)
24. L.D. Conde, S.L. Suib, Catalyst nature and frequency effects on the oligomerization of methane via microwave heating. J. Phys. Chem. B **107**, 3663–3670 (2003)
25. J.B. Hasted, *Aqueous Dielectrics* (Chapman and Hall, UK, 1974)
26. S. Horikoshi, S. Matsuzaki, T. Mitani, N. Serpone, Microwave frequency effects on dielectric properties of some common solvents and on microwave-assisted syntheses: 2-Allylphenol and the C_{12}–C_2–C_{12} Gemini surfactant. Rad. Phys. Chem. **81**, 1885–1895 (2012)
27. S. Horikoshi, S. Iida, M. Kajitani, S. Sato, N. Serpone, Chemical reactions with a novel 5.8-GHz microwave apparatus. 1. Characterization of properties of common solvents and application in a Diels–Alder organic synthesis. Org. Process Res. Dev. **12**, 257–263 (2008)
28. S. Horikoshi, T. Hamamura, M. Kajitani, M. Yoshizawa-Fujita, N. Serpone, Green chemistry with a novel 5.8-GHz microwave apparatus. Prompt one-pot solvent-free synthesis of a major ionic liquid: The 1-butyl-3-methyl- imidazolium tetrafluoroborate system. Org. Process Res. Dev. **12**, 1089–1093 (2008)
29. S. Horikoshi, T. Sato, N. Serpone, Rapid synthesis of Gemini surfactants using a novel 915-MHz microwave apparatus. J. Oleo Sci. **62**, 39–44 (2013)
30. I. Bilecka, M. Niederberger, Microwave chemistry for inorganic nanomaterials synthesis. Nanoscale **2**, 1358–1374 (2010)
31. J.A. Gerbec, D. Magana, A. Washington, G.F. Strouse, Microwave-enhanced reaction rates for nanoparticle synthesis. J. Am. Chem. Soc. **127**, 15791–15800 (2005)
32. S. Horikoshi, H. Abe, K. Torigoe, M. Abe, N. Serpone, Access to small size distributions of nanoparticles by microwave-assisted synthesis. Formation of Ag nanoparticles in aqueous carboxymethylcellulose solutions in batch and continuous-flow reactors. Nanoscale **2**, 1441–1447 (2010)
33. S. Horikoshi, H. Abe, T. Sumi, K. Torigoe, H. Sakai, N. Serpone, M. Abe, Microwave frequency effect in the formation of Au nanocolloids in polar and non-polar solvents. Nanoscale **3**, 1697–1702 (2011)

34. M.B. Mohamed, K.M. Abouzeid, V. Abdelsayed, A.A. Aljarash, M.S. El-Shall, Growth mechanism of anisotropic gold nanocrystals via microwave synthesis: formation of dioleamide by gold nanocatalysis. ACS Nano **4**, 2766–2772 (2010)
35. L. Yu, A. Andriola, Quantitative gold nanoparticle analysis methods: a review. Talanta **82**, 869–875 (2010)
36. M. Tsuji, N. Miyamae, M. Nishio, S. Hikino, N. Ishigami, Shape selective oxidative etching and growth of single-twin plate-like and multiple-twin decahedral and icosahedral gold nanocrystals in the presence of au seeds under microwave heating. Bull. Chem. Soc. Jpn **80**, 2024–2038 (2007)
37. N. Bengtsson, P.O. Risman, Dielectric properties of foods at 3 GHz as determined by a cavity perturbation technique. J. Microwave Power Electromagn. Energy **6**, 107–123 (1971)

Chapter 6
Engineering of Microwave Heating

Abstract Researchers conducting microwave-assisted chemistry and materials processing use commercially available equipment. However, knowing the mechanisms of microwave devices and mastering them broadens the possibility of using new microwaves. This chapter illustrates and describes microwave generators, waveguides, coaxial cables, isolators, power monitors, microwave tuners, iris, and short plungers. Furthermore, the chapter clarifies the difference between single-mode and multimode microwave irradiation applicators. In addition, introduced are temperature measurement methods, leakage prevention methods against thermometer insertion, and leakage measuring instruments among and others. It also describes how to drill a hole in an applicator that can prevent leakage. In addition, the coffee break introduces various tools to cook food deliciously in microwave ovens.

Keywords Magnetron · Klystron · Traveling wave tube · Solid-state generator
LDMOS · GaN · Waveguide · Coaxial cable · Isolator · Microwave tuner
Iris · Short plunger · Single-mode applicator · Multimode applicator
Temperature measurement · Choke trap

6.1 Components in Microwave Heating Equipment

Microwave chemical equipment is typically composed of several integrated components, made in such a way that most researchers are unable to appreciate their inner structure. When researchers buy microwave chemical reaction equipment, all they have to do is choose the conditions: sample type (solid, liquid and the capacity), the desired temperature or temperatures (i.e., reaction temperature, reflux apparatus, cooling unit and thermometer) and reactor type (bath, flow, crucible) to perform the microwave-assisted chemical processes. However, if they understood the need for such components, they likely would advance and optimize conditions for greater yields of products in synthesis projects. This section is focused on explaining the equipment necessary to achieve proper microwave irradiation of

S. Horikoshi et al., *Microwave Chemical and Materials Processing*,
https://doi.org/10.1007/978-981-10-6466-1_6

reaction samples. When researchers need a deeper understanding on the use of microwave irradiation, they should consult an engineering book that deals with microwave heating (see e.g., [1]).

The composition of microwave application equipment is typically subdivided into the *microwave generation part* and the *applicator part* required for microwave irradiation of a sample (Fig. 6.1). The microwave generation part consists of an independent *microwave generator compartment* and a *microwave transmission compartment.* The microwave generator compartment consists of a *microwave source*, a *power supply*, and a *cooling unit*, whereas the microwave transmission compartment consists of a *waveguide or coaxial line*, an *isolator*, a *power monitor or power sensor*, and a *tuner*. The microwave energy is then launched into the *applicator* that contains the sample to be irradiated.

The components in the microwave generation compartment have a significant influence on the performance of cost of a larger scaled-up process: the application system. The selection of an appropriate microwave source and a power supply system is important. For example, a magnetron microwave generating source consumes much of the power of the whole microwave equipment (Fig. 6.2); the microwave change efficiency of the magnetron is approximately 60–85%. Consequently, only an energy conservation of ca. 25% can be achieved by a combination of the magnetron's performance and the performance of the magnetron's power supply system.

The role of the microwave transmission compartment to transmit microwaves efficiently to an applicator when the microwave generation compartment is used with a waveguide and/or coaxial line that transmit or carry the waves to irradiate the targeted sample. The efficiency can be further improved by tuners (e.g., E/H tuner and three-stub tuner). The microwaves that are not absorbed by a sample will be absorbed by the isolator. The power monitor or power sensor combination is used to ascertain the extent of microwaves' incident and reflective power. The iris and the short plunger are used to make the microwaves resonate with the tuner. Note that even though not all components may be functioning at their ultimate efficiency, microwave heating can still be performed. Low-cost microwave domestic ovens

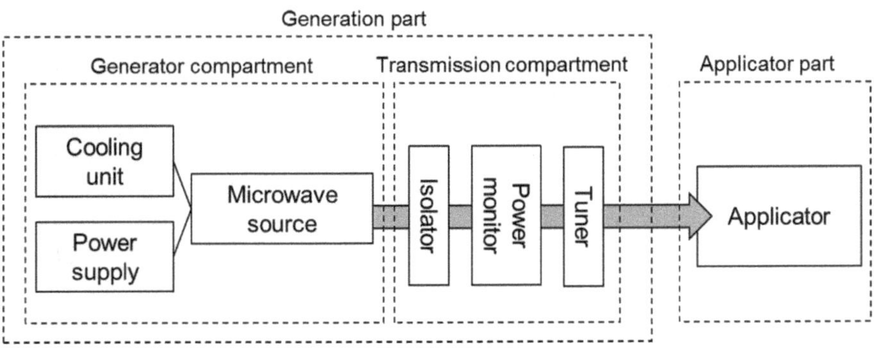

Fig. 6.1 Construction components of the microwave generating device

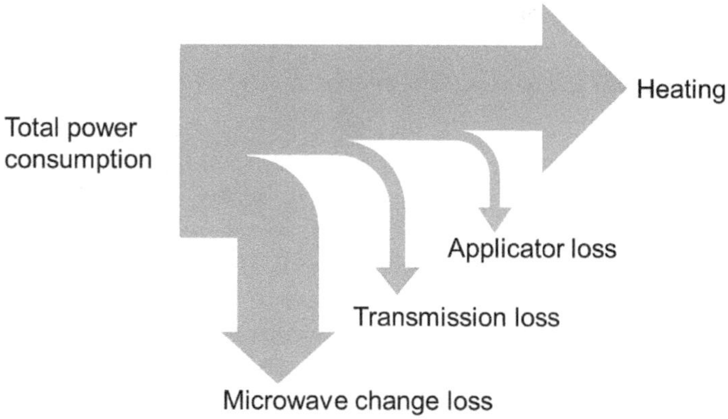

Fig. 6.2 Image illustrating the possible sources of loss of power consumption in microwave equipment

that are not equipped with an isolator, a tuner, and a power monitor are not suitable for microwave-assisted chemistry experiments. The irradiation efficiency of the microwaves is much improved by the structure and size of the applicator. The roles of the components in the microwave generation portion of the equipment are summarized in Table 6.1.

Table 6.1 Components of microwave chemistry equipment and their purposes

Compartment	Components	Outline of equipment
Generator compartment	Microwave generator	The equipment generating microwaves consists of a microwave generation tube (typically a magnetron, klystron, or TWT), or a semiconductor generator, together with a power supply and cooling unit
Transmission compartment	Waveguide or coaxial line	This is the metallic pipe or cable through which the microwaves are propagated to a sample
	Isolator	Protectively absorbs excess microwave energy not absorbed by the sample
	Power monitor or power sensor	Monitoring the incident and reflective microwave power
	Tuner	There are E/H and three-stub tuners whose adjustment improves the microwave irradiation efficiency. An iris, a short plunger and a dummy load are used in a single-mode applicator

6.2 Microwave Generators

6.2.1 Vacuum Tubes as Microwave Sources

The major microwave generation source in commercial microwave ovens, as well as in commercial microwave chemical equipment, is almost invariably a magnetron. While semiconductor generators can be used, they generate low power levels at high cost and act as radio waves in WiFi technology and in smart cellular phones. To achieve high power and the high frequencies required for microwave heating, most microwave sources employ vacuum tubes that have been used for microwave heating—they include magnetrons, klystrons, and traveling wave tubes (TWTs) and are summarized in Table 6.2. A klystron is often used in microwave materials processing.

Generation of radio waves evolved from the invention of the vacuum tube, which has the same structure as the triode (and transistor). A triode consists of a hot cathode electrode heated by a filament that releases electrons (thermionic emission), together with a flat metal plate electrode that attracts the electrons, and a grid consisting of a screen of wires between the cathode and the metal plate to control the current. These components are sealed inside a high vacuum (ca. 10^{-9} atm) glass container; the operating lifetime of a triode is approximately 2,000 h for small tubes, and 10,000 h for power tubes. Vacuum tubes different from the triode for the generation of microwaves were invented one after another during World Wars I and

Table 6.2 Kinds, performances, and uses of microwave tube generators

Generator	Power/kW	Frequency/GHz	Lifetime	Use
Magnetron	0.4–100	0.9–2.45	Several thousand hours	Microwave ovens, industrial microwave heating equipment and plasmas
	0.4–100 1–5000	0.9–2.45 1–17		Radars for ship, airlines, and meteorology
	1–5000	1–17		Particle accelerator
Klystron	10–1300	0.35–35	Several tens of thousands of hours	TV and communications
			50,000 h	Particle accelerator and dielectric heat
	1–100,000	0.32–35	Several tens of thousands of hours	Radar for shipping, airlines and weather
				Particle accelerator and dielectric heat
Traveling wave tube (TWT)	0.001–400	0.4–45	70,000 h	Satellite and space communications
				Microwave line

II. These are the magnetron, the klystron, and the traveling wave tubes (TWTs). The history of the magnetron began with a unit constructed by Albert Hull (General Electric's Research Laboratories, Schenectady, New York, USA, 1916) in which the anode and the cathode formed a coaxial structure; this particular unit could oscillate only at low frequency and thus could not produce microwaves. Nonetheless, Hull's work contributed to the initial basic structure of the magnetron. The split anode magnetron was developed by Kinjirou Okabe (Tohoku University, Japan) in 1927 and published in 1928. The water-cooled high-power magnetron was designed by Randall and Booth in 1940 (University of Birmingham, UK). For an overview of early magnetron designs see the article by Willshaw et al. [2].

During World War II, Randall and Booth's design was a source of high-power microwaves used in military radar equipment. Since then it has been further developed to cover a very wide frequency range (0.5–100 GHz) in a wide range of power outputs for radar, for communications, for the military and for microwave heating. The greatest single application of magnetrons, currently numbering about one billion, is for the domestic microwave cooking oven. These small magnetrons operate at a frequency of 2450 MHz (2.45 GHz) with power output ratings in the range of 600–1500 W. They are produced in large quantities by mass production methods that yield exceptionally low-cost products. Some industrial installations use large numbers of these domestic magnetrons because they are inexpensive and their power supplies are very simple. However, in such systems care must be taken to prevent the many magnetrons from interacting with each other. Early on it was recognized that microwaves could heat water in a dramatic fashion; domestic and commercial appliances began to appear soon thereafter in the United States (in the 1950s) for heating and cooking foodstuffs. Effective Japanese technology transfer and global marketing have led to the widespread domestic use of microwave ovens in the 1970s and 1980s. The successful development of small and cheap magnetrons by the New Japan Radio Co. Ltd. contributed to the large sale and use of microwave ovens.

6.2.2 Magnetron Generation of Microwaves

A magnetron is the microwave device wherein the microwaves are formed because of the presence of magnetic and electric fields. As in a vacuum tube, the application of an electric field in a magnetron makes electrons flow from a hot filament. Although, a magnetron consists of a very simple structure (Fig. 6.3), it required much technical knowledge to produce this device. The magnetron is a vacuum tube oscillator that generates high-power electromagnetic signals in the microwave frequency range. Its operation is based on the combined action of a magnetic field applied externally and an electric field between its electrodes. The tube consists of a diode (i.e., a cathode and an anode) surrounded by an external magnet. In the absence of the external magnetic field, the tube would simply function as a simple diode with the electrons flowing directly from the cathode to the anode. The

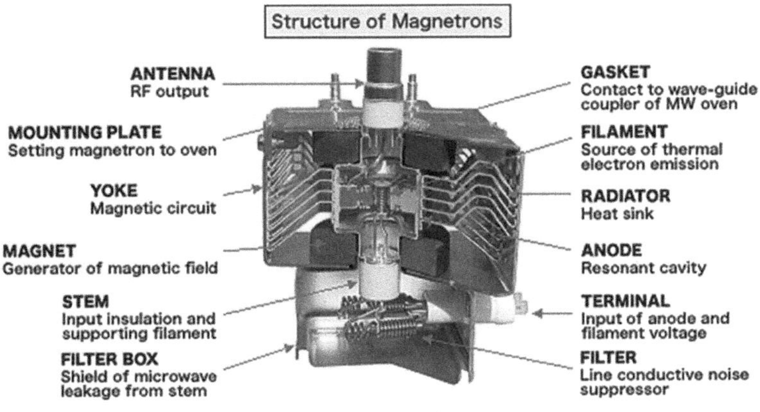

Fig. 6.3 Cut-off picture of a magnetron; reproduced from Ref. [3]

magnetic field forces the cathode-emitted electrons to assume a curved path and thus creates a rotating electron cloud about the tube axis; this cloud excites a microwave field as it passes tuned cavities in the anode block that then is antenna-launched out of the magnetron. Magnetrons are relatively highly efficient (about 60–85%) in converting electrical power input to microwave power output, and can generate microwave energies ranging from less than 1 kW to 1 MW or more. They are used extensively in radar systems, for industrial heating and in microwave ovens. The magnetrons used in domestic microwave ovens are inexpensive and relatively highly energy efficient in producing microwaves.

Magnetrons used in laboratory-scale microwave chemistry and in materials processing equipment usually have output powers of 600 W to 3 kW. Most industrial microwave systems demand power greater than 10 kW, often extending into the range of 100 kW to 1 MW frequently by means of combinations of lower power magnetrons. Laboratory-scale microwave equipment is commercially sold with a magnetron and is ready to operate. However, when using microwave equipment for an industrial production plant [4], it is advisable to have some knowledge of magnetron characteristics together with the following essential requirements: (i) high conversion efficiency from incoming power to usable microwave power output, especially important as the power rating rises; (ii) operation must always be within the prescribed frequency band; (iii) low capital cost per kilowatt of output power, important because the generator often represents more than half the cost of the installation; (iv) robustness in surviving conditions typical of industrial operation (e.g., hostile environments, electrical power surges and transients, arcing in the applicator and feed system, incorrect adjustment of operating conditions, and vibrations among others); (v) simple to operate with the absolute minimum of user-adjustable controls; (vi) simple to maintain, faultfind and remedy; (vii) low running cost, not only in electrical power consumed but also in replacement of consumables, notably the microwave power tube itself—this implies a long magnetron life; and (viii) comprehensive accurate documentation to enable installation, service, and operating teams to understand the equipment fully within their designated roles.

The lifetime of magnetrons has increased through the many years of their development [4]. The most usual end of useful life condition for a magnetron is cathode failure. In large-power magnetrons with directly heated tungsten filaments, this is a *sudden death* event without warning as the filament *open circuits* through melting at a hotspot. Fully rated power output is obtainable until this happens, typically for 8000–12,000 h at a frequency of 915 MHz, and 4000–6000 h at a frequency of 2450 MHz. Operating conditions also play a role on the longevity of a magnetron, and may significantly shorten the magnetron's life if the system is not well matched, or the magnetron not protected. Lower power magnetrons at 2450 MHz normally fail through loss of cathode emission. The initial burn-in period is typically 50 h of cathode/filament life, with the end point at 2000–5000 h depending on manufacturer and magnetron type.

6.2.3 Klystron and TWT Generators

The klystron was developed as an alternative method to generate microwaves other than the magnetron. The name *klystron* was suggested by Hermann Fränkel, a professor in the Classics Department at Stanford University during the period the klystron was under development [5]. The name originates from the root forms of the Greek verb κλυσ-(klys) that refers to the action of waves breaking against a shore, and from the suffix-τρον ("tron"), which means the place where the action takes place.

Essentially, the klystron is a linear beam vacuum tube that was invented in 1937 by the American electrical engineers Russell and Sigurd Varian at Stanford University (Palo Alto, CA, USA) [6]; it is used as an amplifier for high radio frequencies, from UHF up to the microwave range. Klystrons amplify RF signals by converting the kinetic energy in a DC electron beam into radio frequency power. Low-power klystrons are used as oscillators in terrestrial microwave relay communications links, while high-power klystrons are used as output tubes in UHF television transmitters, satellite communication, and radar transmitters, and to generate the driving power for modern particle accelerators. In a klystron, the electron beam interacts with radio waves as it passes through resonant cavities, i.e., metal boxes along the length of a tube. The electron beam first passes through a cavity to which the input signal is applied. The energy of the electron beam amplifies the signal, and the amplified signal is taken from a cavity at the other end of the tube. The output signal can be coupled back into the input cavity to make an electronic oscillator to generate microwaves [7]. Klystrons can be found at work in radar, satellite, and wideband high-power communication (very common in television broadcasting and EHF satellite terminals), in medicine (radiation oncology), and high-energy physics (particle accelerators and experimental reactors). Compared with a magnetron, the klystron can also be used for high-power irradiation by a high-frequency band; its efficiency relative to the magnetron is slightly lower at 50–70%. Klystrons are usually too expensive for industrial processing use.

The traveling wave tube (TWT) is a vacuum tube used in electronics to amplify radio frequency (RF) signals in the microwave range [7]. The TWT belongs to a category of *linear beam* tubes, such as the klystron, in which radio waves are amplified by absorbing power from a beam of electrons as it passes down the tube [7]. Although there are various types of TWTs, two major categories are (i) the Helix TWT in which the radio waves interact with the electron beam while traveling down a helical wire that surrounds the beam: these have wide bandwidth, though the output power is limited to a few hundred watts [8], and (ii) the coupled cavity TWT in which the radio waves interact with the beam in a series of cavity resonators through which the beam passes; these function as narrow-band power amplifiers. A major advantage of the TWTs over other microwave tubes is their ability to amplify a wide range of frequencies, i.e., over a wide bandwidth. However, relative to the magnetron, the efficiency of TWTs in converting electric power into microwaves is only 30–60%. TWTs cannot produce high-power microwaves, but can produce microwaves at various frequencies.

The original design and prototype of the TWT was provided by Andrei Haeff around 1931 while working as a doctoral student at the Kellogg Radiation Laboratory at Caltech (California Institute of Technology, Pasadena, CA, USA); his original patent, *Device for and Method of Controlling High Frequency Currents*, was filed in 1933 and granted in 1936 [9]. However, the invention of the TWT is often attributed to Rudolf Kompfner in 1942/1943, and not least to Nils Lindenblad at RCA (Radio Corporation of America) in the USA who also filed a patent for a device in May 1940 [10] that was remarkably similar to Kompfner's TWT. Both devices improved on Haeff's original design as they both used the then newly invented precision electron gun as the source of the electron beam, and they both directed the beam down the center of the helix instead of outside of it [11].

6.2.4 Semiconductor Generation of Microwaves

The performance of microwave generators was remarkably improved by the shift from a magnetron vacuum tube to a transistor semiconductor developed especially as an alternative method for the generation of microwaves. The development of the field-effect transistor (FET) and the high-electron-mobility transistor (HEMT) semiconductor devices occurred rapidly for the application of microwaves in the communications field. Today's cellular phones have been successfully miniaturized by these two developed technologies. However, the use of a semiconductor generator for microwave heating was limited because of its low microwave power output. Around 1997, Meredith noted that *although progress is being made in solid-state power supplies, they are far short of the power output requirement and are very expensive, and will not be considered further* [6]. The development of new microwave band devices (LDMOS and GaN) for the 2.45-GHz microwaves opened the possibility of using these semiconductor devices in microwave chemistry by applying an output power of several hundred watts using an amplifier.

A magnetron-based generator is generally used in commercial microwave organic synthesis apparatus, regardless of whether it operates in the single-mode or in the multimode procedure. Since high reproducibility has been achieved in recent years with single-mode apparatus, organic syntheses carried out using single-mode equipment have attracted much attention. As a result, a semiconductor generator has been used to obtain high resonance inside the single-mode applicator. However, high cost and limited output power have been disadvantages of semiconductor generators. Recently, semiconductor generators with maximum power of 500 W and with highly accurate microwave frequency generation have recently become available at a price commensurate with the costs of magnetrons. Further usages of semiconductor microwave generators should reduce the price further. The continued sales and utilization of small cellular phones depend on the further development of microwave semiconductor oscillators. This is also having an influence on microwave chemical equipment. The features of semiconductor and magnetron generators are described next.

The frequency distribution of the microwave radiation from the magnetron generator in domestic microwave cooking ovens is shown in Fig. 6.4a, which demonstrates that the frequency of the microwaves is distributed over a large frequency range: from about 2.25 to 2.60 GHz [12]. The distribution of the microwave frequency changes depending on the characteristics of the microwave generator equipment, the applicator, and the load. Note that narrower band frequency waves can be generated with high accuracy with a magnetron. By contrast, the semiconductor generator produces microwaves within a very narrow frequency range of 2.45000 ± 0.00250 GHz (Fig. 6.4b). In the case of the magnetron generator, the microwave output power distribution indicates a widely dispersed frequency distribution. Therefore, the output of the actual 2.45-GHz microwaves is smaller than the output power. On the other hand, when using a semiconductor microwave generator, microwave heating can evolve efficiently because the microwave output

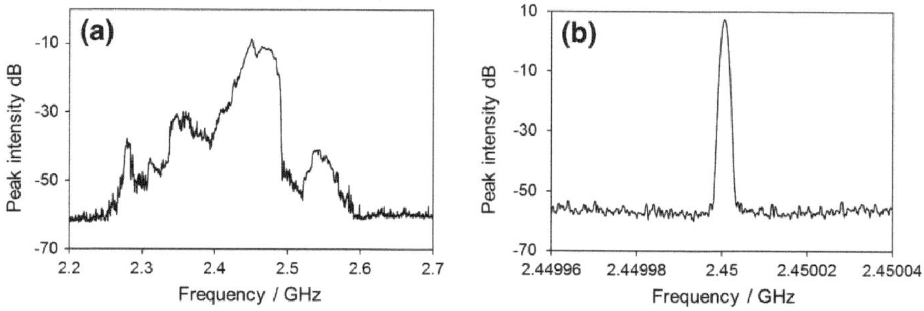

Fig. 6.4 Frequency spectral distribution of the 2.45-GHz microwave radiation emitted from (**a**) the magnetron generator and (**b**) the semiconductor generator. Reproduced from Ref. [12]. Copyright 2011 by the American Chemical Society

Table 6.3 Relative comparison of the semiconductor generator and the magnetron generator

	Magnetron generator	Semiconductor generator
Applied power	30 W to kW	~3 kW
Weight of the power supply	6 kg (for 1 kW)	0.5 kg (for 1 kW)
Applied voltage	1 kV (for 1 kW)	0.05 kV (for 1 kW)
Electric power conversion efficiency	70–80%	50–70%
Frequency control	No	Yes
Lifetime	500–1000 h	20 years
Price	Inexpensive	Expensive
Use environment	Weak in vibration	Weak for high temperature and reflected microwaves

power is concentrated at the 2.45000 GHz frequency. These microwaves resonate better in a single-mode apparatus.

A comparison of the advantages of semiconductor and magnetron microwave generation is reported in Table 6.3. The compact size of the equipment is an advantage. Photographs of a magnetron and a semiconductor generator are displayed in Fig. 6.5. Clearly, the semiconductor transmitter is smaller. It is relevant to note that microwaves generated from a semiconductor generator are transmitted by a coaxial cable, not a waveguide. A flexible coaxial cable can be used for connecting the chemical reactor to a semiconductor generator. Therefore, it is also possible to install a semiconductor generator in a small available space. Also important, the lifetime of a semiconductor generator is very long compared to a magnetron generator. An output power of hundreds of watts of microwave radiation is generally sufficient for many laboratory-scale organic syntheses.

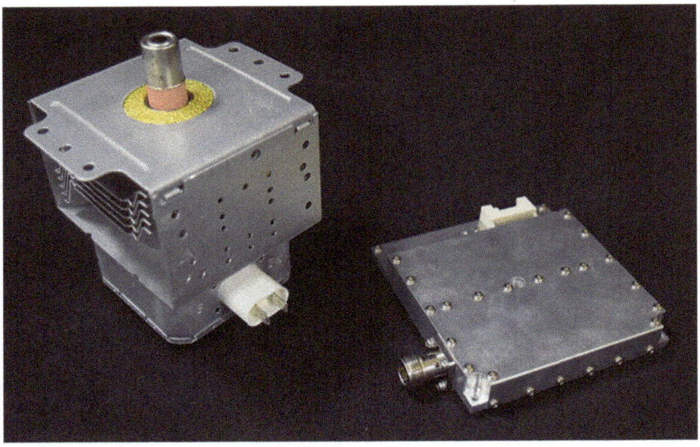

Fig. 6.5 Photograph of a magnetron generator (left) and a GaN semiconductor generator (right)

The semiconductor generator uses far less microwave output power and can produce similar temperature increases in a liquid sample when compared with a magnetron generator. Moreover, the semiconductor generator is an optimal microwave emitter in frequency-dependent reactions (processes) because the frequency can easily be adjusted and set in the semiconductor generator.

A demonstration model of a semiconductor generator using the above advantages is displayed in Fig. 6.6, which reveals the inner chamber of a commercial microwave cooking oven equipped with two GaN semiconductor generators (maximum output power, 200 W). When two waves are in phase they strengthen each other, but when they are out of phase they weaken each other. By having a food in the location of in-phase strengthened waves, heating progresses efficiently; however, if the food is at the location of out-of-phase weakened waves, heating does not progress as well. Each semiconductor generator is connected to directional antennas inside the oven (top and bottom position). The microwave radiation from the semiconductor generator consists of a wave of regular frequency, and so it is possible to synthesize the phase of the microwaves emitted from the two antennas depicted in the oven. Accordingly, selective heating of food can easily be achieved in such microwave cooking ovens.

As an example, if two different foodstuffs are placed on a dish (meat and potatoes), it is possible to selectively heat only the desired foodstuff (e.g., meat). On the other hand, selective heating can act only upon the rice in a box lunch with this microwave oven (Fig. 6.7). The temperature of the rice increases to 51.8 °C after heating for 1 min; the desired temperature of the other foods remains near room temperature. The user can now do this selective heating and ingredient temperature control using a smart phone. Moreover, even though the two foodstuffs may have different dielectric loss factors, it is possible to heat both uniformly. Such a microwave oven can also be used for chemical reactions and materials processing.

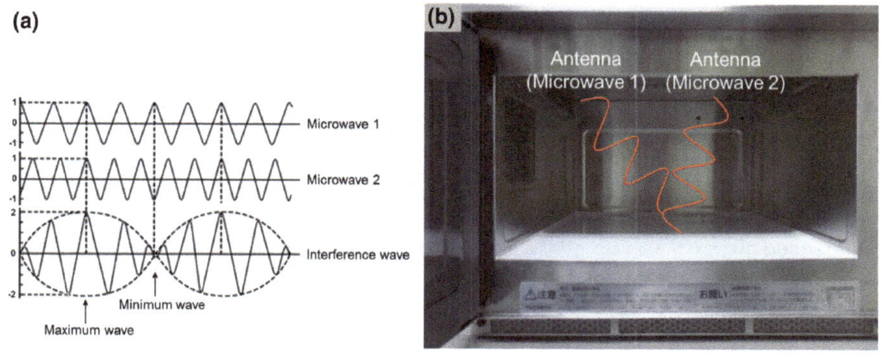

Fig. 6.6 Photograph of the inner chamber of a microwave apparatus with microwaves emitted from two GaN semiconductor generators (maximum power, 200 W). Reproduced from Ref. [13]. Copyright 2015 by Wiley-VCH Verlag GmbH

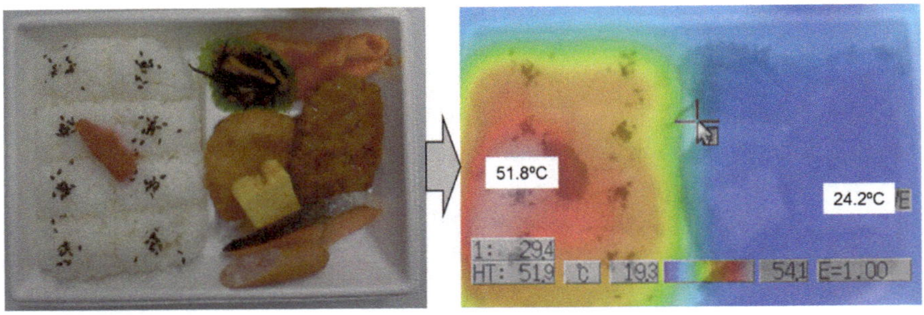

Fig. 6.7 (Left) Photograph of a lunch box in a microwave cooking oven with microwaves emitted from four GaN semiconductor generators (maximum power, 250 W); (right) thermography of the lunch box by which only rice is selectively heated

Semiconductor generators also offer other advantages for microwave chemical reactors and materials processing equipment. For example, there are many samples whose microwave absorption rate changes when the temperature of the sample solution changes. In addition, as a chemical synthesis progresses, the composition of the sample may change greatly depending on the product substances. There are also reaction systems for which the extent of microwave absorption by the samples varies largely with reaction time. It will then be necessary to tune with a tuner unit when performing such a chemical reaction. Generally, researchers will use the tuner introduced earlier (Sect. 6.3.5); however, if a semiconductor generator were used, it may change the microwave radiation according to changes of the dielectric factors of the samples without using the tuner. Because the semiconductor generator can vary the frequency, tuning can be done by changing the microwave frequency. For example, where a sensor that monitors the microwave reflected waves from a sample is linked to a device that changes the microwave frequency in a microwave continuous flow chemical synthesis system (Fig. 27a in Sect. 10.2.4), tuning can then be performed automatically by changing the microwave frequency according to changes in the dielectric factors of the samples.

A variable frequency by the semiconductor oscillator has another advantage. When a solid material containing a metal or a conductive substance is microwave-heated, a hotspot is generated at the location where electrolysis is concentrated (localized uncontrolled heating). This phenomenon also occurs in a liquid chemical reaction using a solid metal catalyst (Sect. 4.4). This problem can be resolved using a semiconductor-type oscillator. For instance, by oscillating a wideband microwave (5.850–6.650 GHz) continuously while varying the frequency every 25 μs in a multimode applicator, the standing wave instantaneously changes and uniform microwave irradiation is obtained. Figure 6.8 shows a variable frequency microwave (VFM) oven that enables high uniformity microwave heating and curing without hotspots; a rapid frequency sweeping eliminates conditions that can cause arcing to metal parts.

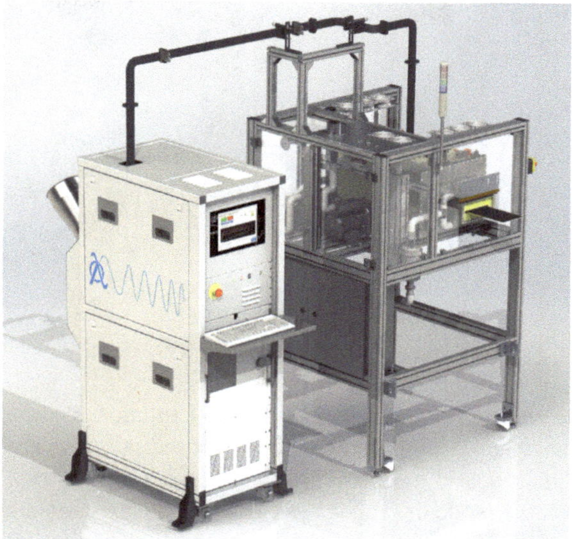

Fig. 6.8 A variable frequency microwave (VFM) oven using a semiconductor generator. Photographs provided by courtesy of Lambda Technologies Inc.

6.3 Waveguides, Isolators, Power Monitors, Tuners, Iris, and Short Plungers

6.3.1 Waveguides

In electromagnetics and communications engineering, the term waveguide may refer to any linear structure that conveys electromagnetic waves from their initiation to an endpoint. However, the original meaning is a hollow metal pipe used to carry radio waves [14]. During the 1890s, theorists performed the first analyses of electromagnetic waves in ducts [15]. Around 1893, J.J. Thomson determined the electromagnetic modes inside a cylindrical metal cavity [15], while in 1897 Lord Rayleigh provided a definitive analysis of waveguides by solving the boundary value problem of electromagnetic waves propagating through both conducting tubes and dielectric rods of arbitrary shape [15].

These transmission lines couple the energy of the microwave source to the applicator. When it is a metal or conductor pipe, the shape of the cross sections is variable. The most commonly used cross sections are rectangular because this facilitates microwave transmission most efficiently. Two modes of microwave propagation are possible in waveguides: the transverse electric mode (TE) and the transverse magnetic mode (TM). Microwaves are transmitted by transverse electromagnetic (TEM) modes in free space. The meanings of transmission modes are described in Sect. 5.10. For the TE mode, the electric field intensity in the direction of propagation is zero. For the TM mode, the magnetic field intensity in the

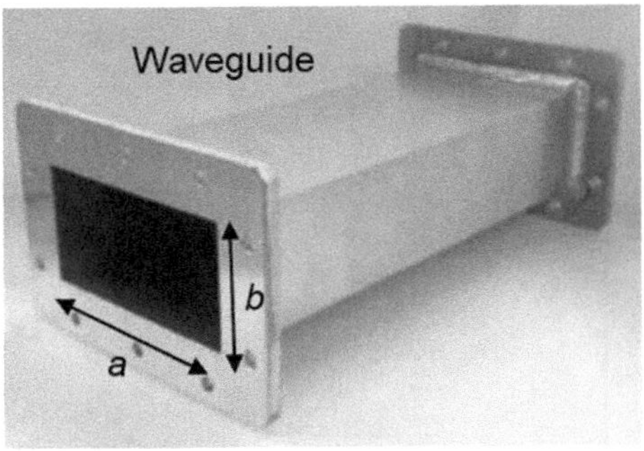

Fig. 6.9 Picture of waveguide at 2.45 GHz

propagation direction is zero. Every mathematical solution of the electromagnetic waves in a rectangular waveguide can be decomposed into a linear combination of the TE and TM modes. The most common waveguide mode is the TE_{10} mode. The subscripts specify the mode of propagation and indicate the number of maxima and minima of each field in a waveguide. In addition to waveguides, there are several other transmission line components that are used for equipment protection, sensing purposes, and coupling microwaves with the material in the applicator.

The wavelength of the 2.45-GHz microwaves is 12.24 cm; however, the wavelength of the microwaves changes to 14.78 cm inside the waveguide. This is called the *guide wavelength* (λ_g), which can be calculated using Eq. (6.1), where the term a signifies the length of the long axis of the waveguide (Fig. 6.9).

$$\lambda_g = \frac{\lambda}{\sqrt{1 + \left(\frac{\lambda}{2a}\right)^2}} \quad (6.1)$$

Not every size of a waveguide can propagate microwaves inside the waveguide. When its width is below half the wavelength, no microwaves pass. This is called the *cutoff wavelength*. For example, the long axis (a axis in Fig. 6.9) in the aspect of the waveguide cannot propagate 2.45-GHz microwaves for sizes less than 7.39 cm (=14.78/2 cm).

How are microwaves transmitted in the waveguide? In the case of free space, the microwaves' electric and magnetic fields are propagated in the same direction, but at 90° to each other (see Chap. 3, Fig. 3.5). On the other hand, microwaves from a magnetron are not propagated in a linear fashion within the waveguide. It is necessary to make the electric field move parallel to the short axis of the waveguide so that microwaves can be propagated inside the waveguide. An electric field image envisioned at some point in the waveguide is shown in Fig. 6.10a. The electric field has zero intensity along the short axis of the waveguide (i.e., at the walls), while the center of the long axis is where the electric field shows maximum strength (the sine wave distribution). The microwave's traveling wave generated from a magnetron is

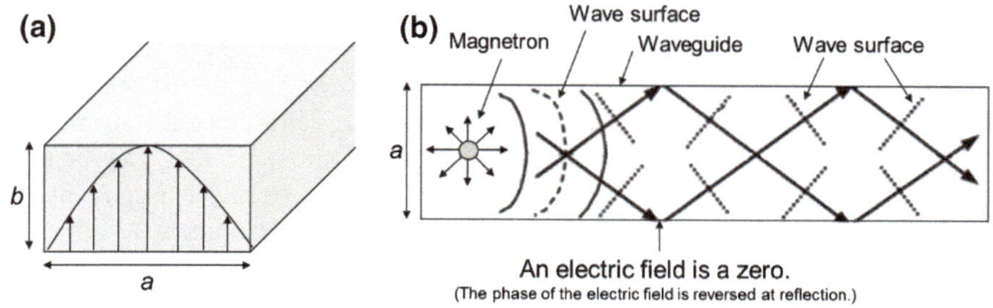

Fig. 6.10 (**a**) The distribution of the electric field inside the waveguide, and (**b**) traveling wave and wave front of the microwave seen from the upper part of a straight-line waveguide

consequently supported, while it is reflected inside the waveguide along the short axis through propagation as a plane wave (Fig. 6.10b).

A microwave in the waveguide exists as a sinusoidal synthetic wave. When two microwaves are synthesized, only that part that becomes a standing wave can propagate in the waveguide. Other microwaves in the waveguide disappear because of destructive interference of the microwaves. Therefore, a microwave in the waveguide exists only at fixed locations, and the microwave propagates while being reflected from the short axis inside a waveguide. Therefore, the phase is reversed for an electric field component parallel to the face of a conductor, and the electric field strength at the reflective location is zero. For example, when a waveguide below the cutoff wavelength is used, the reflection angle of the microwave inside the waveguide will be 90° to the waveguide wall. Consequently, microwaves just

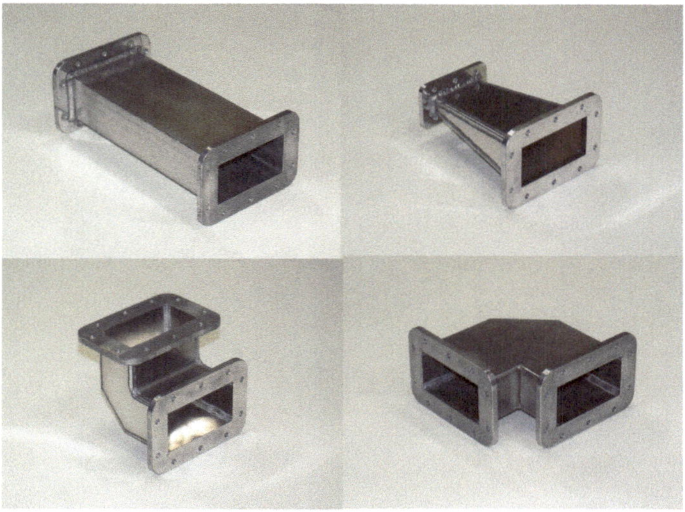

Fig. 6.11 Straight-line waveguide, tapered waveguide, electric or magnetic field corner waveguides

bounce back and forth between left and right inside the wall of the waveguide and do not advance at all.

There is a worldwide standard nomenclature for the size of the waveguide, which also comes in various shapes (Fig. 6.11). There are tapered transmission lines which link waveguides of different standard sizes, or that can bend a straight-line waveguide to match another at right angle. Hence, it is possible to combine these to meet the installation requirements of the equipment.

6.3.2 Coaxial Cables

In low-power systems, the transmission lines are often coaxial cables, which are like cables that are used in televisions. At high frequencies and output power, the losses that occur in coaxial cables are significant and damaging, making waveguides the transmission line of choice in high-power microwave heating systems. Waveguides are hollow tubes through which the electromagnetic waves propagate. Coaxial cable (concentric line) coats an insulator (polyethylene) around the center metal wire, and a metal coating structure covers the outer surfaces (Fig. 6.12). Microwaves propagate by transverse electromagnetic (TEM) modes in the space between the central metal wire (inner conductor) and the metal coating (outer conductor) separated by the insulator (polyethylene). The advantage of a coaxial cable is its flexibility. However, losses of microwaves in a coaxial cable are significantly greater than in a waveguide. When these coaxial cables are used to transmit high-power microwaves, significant heat develops that melts the cables. Therefore, coaxial cables for high-power microwave transmission tend to be very thick; under such conditions, the flexibility advantage is lost because they can no longer be bent.

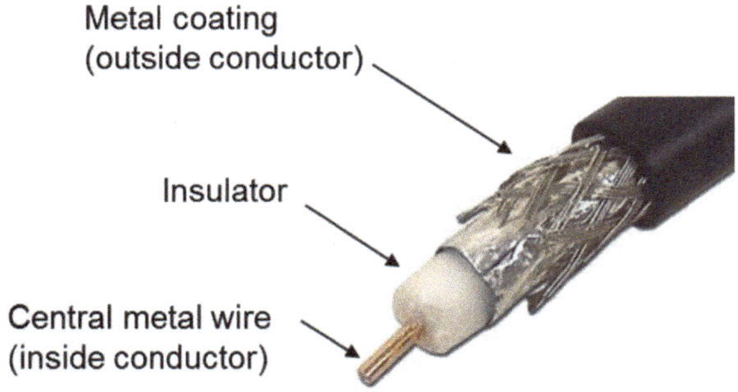

Fig. 6.12 Structure of coaxial cable

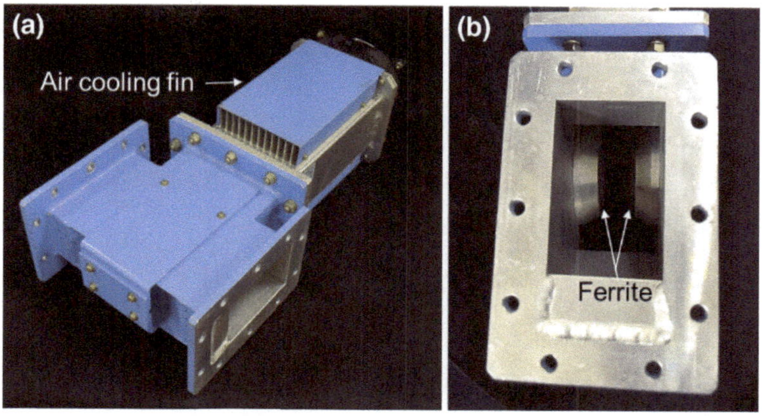

Fig. 6.13 Pictures of (**a**) isolator with an air cooling fin and (**b**) ferrite inside the waveguide

6.3.3 *Isolators*

When materials are heated that are not good absorbers of electromagnetic energy, a significant amount of power is reflected to the microwave source. Excessive reflected power can damage microwave generators (e.g., magnetrons). This reflective wave has an influence on the phase of the incident wave. The circulator or isolator protects the microwave equipment by acting as the microwave equivalent to a diode in an electrical circuit; microwaves are only allowed to pass through the circulator in one direction. In the three port circulators, one port is connected to the microwave source, another is connected to the applicator, and the third port is connected to a dummy load (Fig. 6.13). The microwave power that is reflected from the applicator is diverted into the dummy load, often water or ferrite that absorbs the reflected power so none reaches the source. The microwave absorbed by the ferrite is released into the atmosphere as heat.

6.3.4 *Power Monitors*

In microwave heating, the ability of materials to absorb electromagnetic energy depends on the dielectric properties. Thus, the magnitude of the forward and reflected power is of interest to the researcher. Power measurements are made through directional couplers, which are designed so that a small amount of forward and reflected waves are separated and measured by power meters. Power monitors, such as the one displayed in Fig. 6.14, consist of a crystal mount (diode for microwaves) for microwave detection; they digitize the incident electric and reflective electric power of the microwaves emitted by a generator. It is possible to optimize microwave absorption by a sample by using this equipment. For

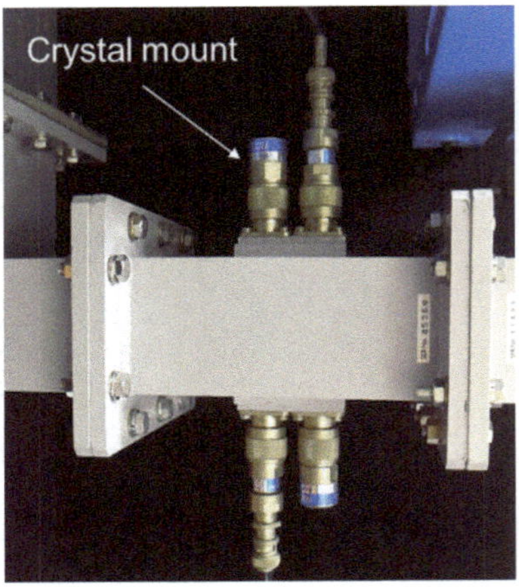

Fig. 6.14 Pictures of power monitor and power sensor

characteristic dielectric samples that change upon heating in microwave chemistry, it would be wise for the researcher to keep a close watch on these meters to make efficient use of the microwave energy to achieve best efficiency by estimating the difference between incident power and reflected power in a chemical reaction: it is best to maximize the incident and minimize the reflected powers. It is also relevant to note that the directional power indicated by a power monitor gives only an approximate value of the power consumed. It is necessary to use a power sensor (Fig. 6.14) to determine the correct microwave power consumption.

6.3.5 The Tuner

The tuner is an indispensable unit when using a standing wave single-mode applicator. Tuners are used to maximize the power absorbed by the load through impedance matching (Sect. 5.4,. i.e., microwave power transmitted is maximized if the load is *impedance-matched* to that of the generator. The microwave equipment is often required to match all manner of loads to a generator impedance of 50 Ω. When this is neglected, incident microwave power is wastefully consumed. Note that a tuner is not always needed, as when a sample of suitable size and properties absorbs the microwaves. In most cases, however, microwave heating efficiency can be improved significantly by tuning. Several tuners such as a three-stub tuner (Fig. 6.15a) and an E/H tuner, together with irises adjust differences between the impedance of the microwave source and the load. Impedance matching is adjusted by inserting each stub of a three-stub tuner (metal bar) in the waveguide (Fig. 6.15b).

The purpose of an E/H tuner is to connect the electric field side of the waveguide and the magnetic field side via a variable short through an adjustment by moving the stub in and out. This way it is possible to adjust the system to handle low

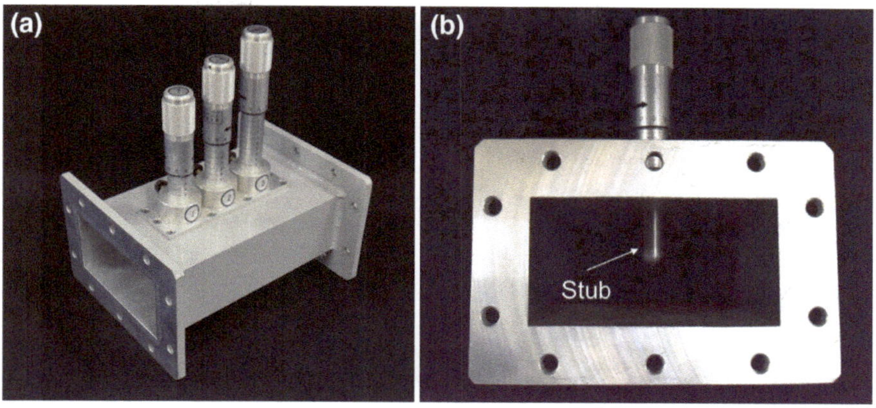

Fig. 6.15 Pictures of (a) three-stub tuner and (b) stub (metal bar) inside the waveguide

microwave absorbing samples because it is possible to change the impedance from zero to infinity.

6.3.6 Iris and Short Plungers

An iris and a short plunger are two effective units (Fig. 6.16) to control the standing wave in a single-mode applicator. The iris is installed at the entrance of the applicator by making a hole (window) in the metal plate, while the short plunger is connected at the other end of the waveguide and used to shift the location of the end plate. An illustration of the equipment necessary to focus the standing wave within a single-mode applicator is shown in Fig. 6.16. An isolator, a power monitor, and an E/H tuner are also components located between the microwave generator and the applicator.

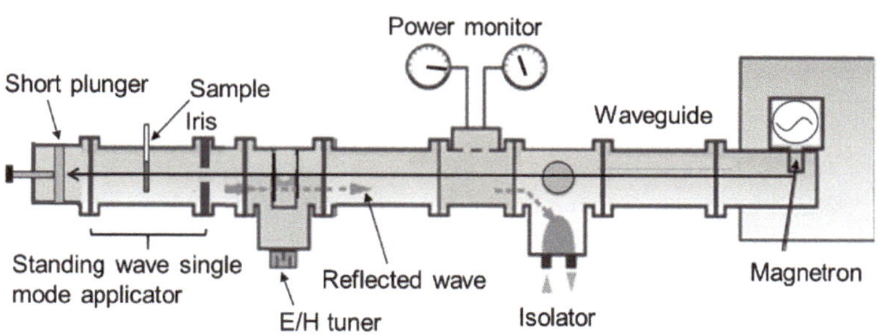

Fig. 6.16 Microwave heating system with a standing wave in the single-mode applicator

6.4 Single-Mode Versus Multimode Applicators

6.4.1 What Is an Applicator?

The design of the applicator is critical to microwave heating because the microwave energy is transmitted to materials in the applicator. The temperature dispersion within the material undergoing microwave heating is inherently linked to the distribution of the electric field within the applicator. Moreover, the role of the applicator is to concentrate uniform microwaves and to select the appropriate ratio of electric field to magnetic field to irradiate the sample. Applicators can be roughly classified as either a *traveling wave type applicator* (Fig. 6.17a) or as a *standing wave type applicator* (Fig. 6.17b). The electric field and magnetic field propagate in space at the same time and at the same position in the traveling wave type applicator. Microwaves that are not absorbed by a sample are not used so that it is difficult to effect microwave irradiation efficiently. On the other hand, the standing wave type applicator restricts the microwaves within a metallic box (i.e., the waveguide) and generates a microwave standing wave that is then applied inside the sample. The meaning of a standing wave is further explained below (Fig. 6.17b). These standing wave modes arise from the combination of reflection and interference such that the reflected waves interfere constructively with the incident waves. An important part of the condition for this constructive interference is that the waves change phase upon reflection from a fixed end. Under these conditions, the medium appears to vibrate in segments or regions and the fact that these vibrations are made up of traveling waves is not apparent—hence the term "*standing wave*" [16].

When inspecting a catalog of microwave chemical synthesis equipment, words often seen are *single-mode applicator* (or mono-mode applicator) and *multimode applicator*. Are these different from the above? Single-mode applicators have small compact cavities, in which microwave irradiation is directly focused on the reactor (Fig. 6.18a). Consequently, a high microwave field density is provided at a specific location, resulting in exceedingly fast heating rates. It is possible to control microwave output to a sample through the strength of the electric and magnetic fields. On the other hand, the size of the sample that can be put in an applicator is subject to restrictions. Multimode applicators have larger cavities, without sample

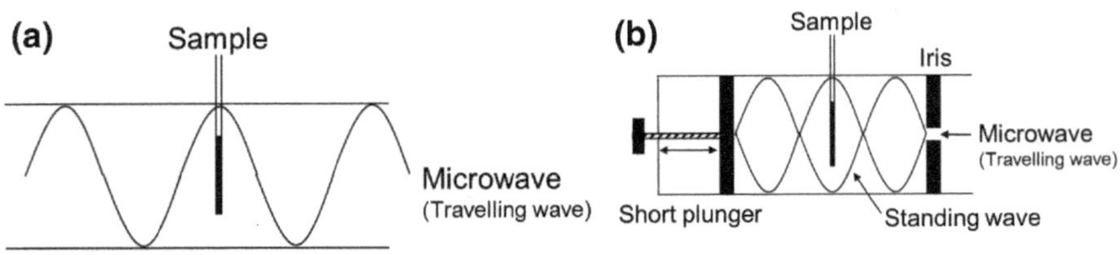

Fig. 6.17 Images of (**a**) traveling wave type single-mode applicator and (**b**) standing wave type single-mode applicator

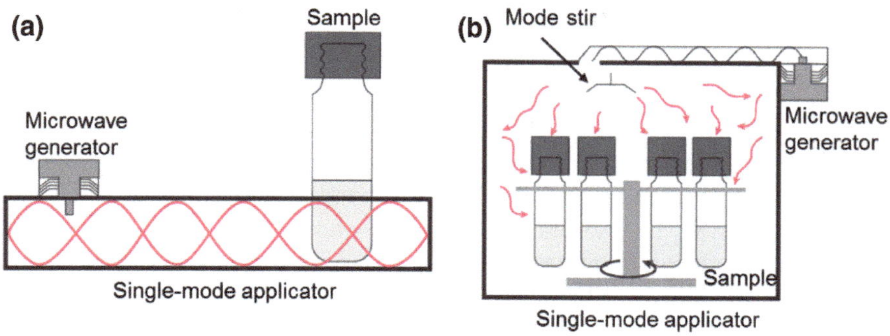

Fig. 6.18 Cartoon of microwave chemical reaction equipments with a (**a**) single-mode applicator and a (**b**) multimode applicator

size restrictions, in which the microwave field is distributed in a chaotic manner (Fig. 6.18b). In common multimode reactors, the waves are reflected from the cavity walls and multiple modes of the electromagnetic waves interact with the cavity load. Because of their size, these instruments can host many samples, rotating sample holders, and other devices. Mode stirrers and turntables are used to increase the uniformity of microwave energy dispersion within the applicator.

It is possible for a single-mode applicator to be of the traveling wave (Fig. 6.18a) type and of the standing wave type (Fig. 6.18b). A multimode form applicator can also be of the standing wave type; however, there are no traveling wave type applicators of the multimode form. What then is the difference between the single-mode applicator and multimode applicator with a standing wave? The single-mode applicator is a device in which only a single type of mode exists, whereas in a multi-mode applicator, several types of microwave modes exist. Sometimes, we see research papers that mistakenly confuse a standing wave type applicator as referring only to a single-mode applicator. The meaning of *mode* is described in Sect. 5.10.

Summarizing the above, common microwave applicators include waveguides, traveling wave applicators, single-mode cavities, and multimode cavities. For processing materials, resonant applicators such as single-mode and multimode applicators are most common because of their high field strengths. The type of applicator used in a microwave processing system often depends upon the materials to be processed. Commercially available single-mode, multimode, and variable frequency multimode processing systems are all used for microwave processing research; each of these systems has advantages and disadvantages.

6.4.2 Single-Mode Applicator

A theoretical analysis can be performed to describe the response of microwaves in a microwave applicator (or cavity). The design of single-mode applicators is based on the solution of the Maxwell equations to support one resonant mode. Consequently, the size of single-mode applicators is on the order of approximately one wavelength. To maintain the resonant mode, these cavities require a microwave source that has little variation in the frequency output, i.e., *low ripple*. Because the electromagnetic field can be determined using analytical or numerical techniques, the areas of high and low electromagnetic fields are known. Single-mode applicators have nonuniform, but predictable electromagnetic field distributions. In general, single-mode cavities have one *hot region* where the microwave field strength is high. The problem of a single-mode applicator is the limitation in sample size that can be used.

Shifting the distribution of the electric field or the magnetic field in the applicator is another different feature of the standing wave type single-mode applicator. This can be achieved when constructing the equipment with a waveguide and a standing wave single-mode applicator by also installing a short plunger after the iris and the applicator (Fig. 6.19).

The phases of the electric and magnetic fields are shifted whenever the traveling microwaves are reflected by the metal plate of the short plunger. The electric field is reflected from the center position of the short plunger, while the magnetic field is reflected from the edge position of the short plunger (Fig. 6.19). Therefore, the relation between the phase of the electric field and the magnetic field in a standing wave type single-mode applicator is shifted by 1/4 wavelength. Consequently, the magnetic field will have the smallest intensity at the location where the electric field has maximal intensity, while the electric field will have the smallest intensity at the location where the magnetic field displays maximal intensity. Images of the intensity distributions of the electric field and magnetic field inside the applicator are displayed in Fig. 6.20a, b. When considering this method, it is possible to

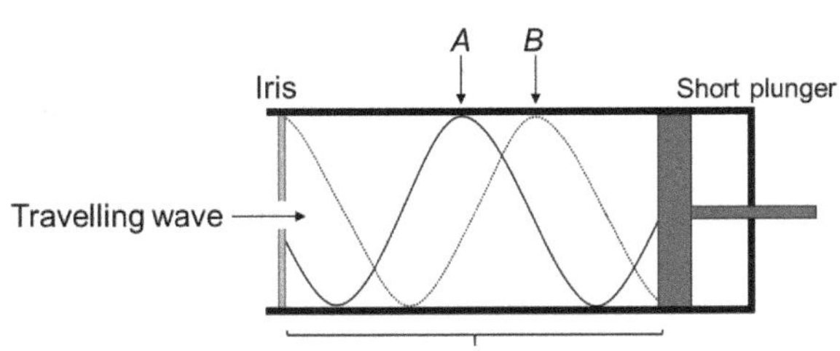

Fig. 6.19 Illustration of an electric field and magnetic field inside a standing wave type single-mode applicator. Solid line: electric field; dotted line: magnetic field; ***A***: position of maximal intensity of the electric field; ***B***: position of maximal intensity of the magnetic field

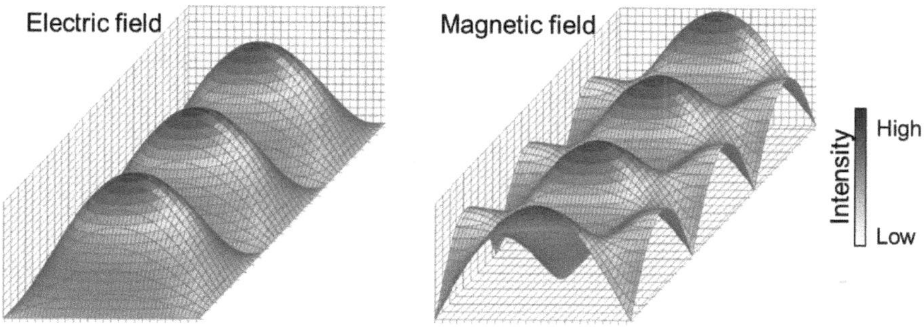

Fig. 6.20 Simulation images of the electric field distribution and the magnetic field distribution inside a standing wave type single-mode applicator (waveguide)

consider the electric and magnetic field effects upon a sample in microwave chemistry and in materials processing.

6.4.3 Multimode Applicator

This type of applicator is the type commonly used in domestic microwave ovens that have often been used both for microwave chemistry and materials processing. A multimode applicator is a metal box with a size greater than the wavelength of the microwave radiation. A multimode applicator has the advantage of being able to accommodate relatively large reactors in a variety of forms. Unlike the design of single-mode applicators, which are based on solutions of the electromagnetic field equations for a given applicator geometry, the design of multimode applicators is often based on trial and error, on experience, and often on intuition [17]. As the size of the microwave cavity increases, the number of possible resonant modes also increases. Consequently, multimode applicators are always much larger than a wavelength of the microwave radiation. For a rectangular cavity, the mode equation for the resonant frequencies is given by Eq. (6.2) [18],

$$f_{nml} = c\sqrt{\left(\frac{l}{2d}\right)^2 + \left(\frac{m}{2b}\right)^2 + \left(\frac{n}{2a}\right)^2}[\mathrm{mm}], \tag{6.2}$$

where f_{nml} is the TE_{nml} or TM_{nml} mode's (see Sect. 5.10) resonant frequency; c is the speed of light; n, m, l, denote the number of half-sinusoid variations in the standing wave pattern along the x, y, and z-axes; a, b and d are the dimensions of the cavity in the x, y, and z directions. The presence of different modes results in multiple hot regions (i.e., high $\boldsymbol{E}$ fields) within the microwave cavity. Just as in single-mode cavities, local fluctuations in the electromagnetic field results in localized overheating. Several techniques have been used to improve the field

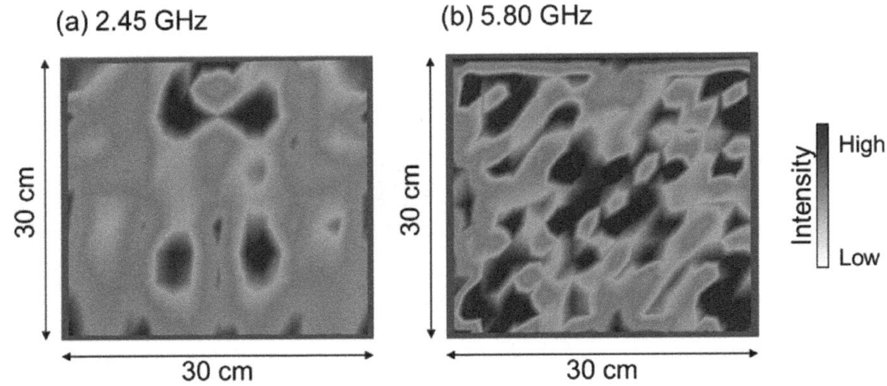

Fig. 6.21 Cross-section images for simulated intensity distribution of the electric field in a multimode applicator (30 × 30 × 30 cm) at the (**a**) 2.45-GHz and (**b**) 5.80-GHz microwave frequencies

uniformity and reduce the effect of these hot regions. The uniformity of the microwave field can be improved by increasing the size of the cavity since the number of modes within a multimode applicator increases rapidly as the dimensions of the cavity increases, and the heating patterns associated with the different resonance modes begin to overlap. The rule of thumb to achieve uniformity within an applicator is for the longest dimension to be 100 times greater than the wavelength of the operating microwave frequency [19]. Unfortunately, at the common microwave frequency of 2.45 GHz (wavelength = 12.24 cm), the largest dimension would exceed 1224 cm, which is impractical for the average laboratory. The distribution of the electric field when applying a microwave in a 30-cm^3 multimode applicator is indicated in Fig. 6.21. The intensity distribution of the electric field is relatively strong at the frequency of 2.45 GHz. A focused distribution of microwave energy, and therefore heat is generated in a sample located in the applicator. On the other hand, when the frequency is set at 5.80 GHz (wavelength = 5.17 cm), its distribution becomes more dispersed. Accordingly, it is possible to achieve greater field uniformity by operating at a higher frequency. At higher frequencies, the wavelength is shorter, and the applicator required to achieve uniformity can be reduced to a more practical size. Although higher frequency processing may seem to be the solution to create greater uniformity, the 2.45-GHz frequency can penetrate deeper into a sample to better generate volumetric heating. Consequently, many attempts have been made to achieve more uniform heating within smaller multimode ovens at 2.45 GHz [20]. To remedy this problem, a mode stirrer such as a microwave stirring device with a metallic antenna and a turntable (change of location of the sample) are typically used in multimode applicators. Moreover, the location of the sample must be at a distance of at least one wavelength from the walls of the applicator.

The size of the multimode applicator governs the sample volume. For example, when microwaves heat a 200-mL water sample, only about 60% of the microwaves are used for heating; this volume of water may not be appropriate for the size of the multimode microwave applicator. However, when a 1000-mL water sample is microwave-heated, nearly 80% of the microwaves are used for heating. Therefore, the heating of 1000 mL of water with microwave energy is more efficient than heating 200 mL of water. However, both relations are reversed from the point of view of uniformity of heating because of the penetration depth and its relationship to sample size and dimensions; uniform heating of the 200-mL water sample is far greater than that of the 1000-mL water sample. Accordingly, the choice is for the most suitable balance of *uniform heating = large size applicator* versus *high-efficient heating = small size applicator*, a decision that will depend upon the size of the multimode applicator.

6.5 Temperature Measurements

6.5.1 Thermometers

Initially, thermometers used in chemical research were alcohol-in-glass or liquid mercury thermometer types. Digital thermometers are a recent development. Significantly inaccurate temperature measurements are often the result when a liquid thermometer or a thermocouple is used in microwave heating. This is the reason why optical fiber thermometers, radiation thermometers, and/or thermography are now being used for temperature measurement. The next section examines which thermometer type is best whenever microwaves are used as a heat source in chemical reactions.

6.5.2 Temperature Measurement in a Solution Sample

An optical fiber thermometer is now used in microwave chemistry whenever the reaction is carried out in a liquid sample. For its operation, the thermometer typically employs either Raman scattering or the fluorescence phenomenon. Those that use fluorescence are generally less expensive and hence are used more frequently.

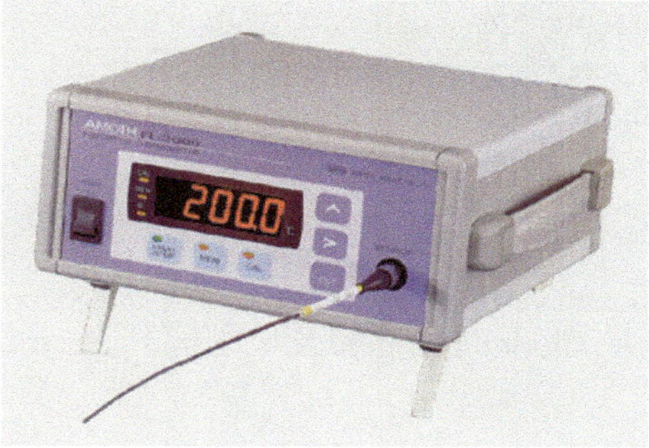

Fig. 6.22 A photograph of an optical fiber thermometer that uses a fluorescence detection system

The fiber material is usually a low microwave absorption material: for example, a silicon-based material so that the thermometer is not heated by the microwave radiation—it is *non-perturbing*. The fluorescence substance in the optic fiber may consist of a magnesium-based substrate (Fig. 6.22) whose fluorescence relaxation time changes with changes in temperature. A detector senses the pulse shape through the optical fiber and calculates the temperature. Other fluorescence substances (organic and inorganic compounds) are also used for the tip of the optical fibers, which are covered with Teflon or ceramics whenever they are immersed in strongly acidic or alkali media, or otherwise used to measure high temperatures. One disadvantage of a thermocouple over an optical fiber thermometer is that the latter can be easily bent; however, when totally covered with Teflon or a ceramic, some of the flexibility is lost.

Previously, internal heating caused by microwaves was little understood, and led to significant errors in temperature measurements. However, recent years have seen the use of equipment that monitors the temperature at the center of a solution using the optical fiber thermometer, while the temperature at the reactor's walls is monitored externally by an infrared (IR) radiation thermometer. With these two components, it is then possible to monitor whether there is a distribution of temperature within the solution.

6.5.3 Temperature Measurements in Solid Samples

The temperature range of microwave heating in a solid sample will often be in the hundreds to thousands of degrees Celsius (°C). Since the use of optical fiber thermometers is limited to approximately 250 °C, it is not possible to measure these higher temperatures with an optical fiber thermometer as is used in liquid samples. Generally, a radiation (IR) thermometer and a thermocouple are used most often. The radiation thermometer detects the radiant heat at the sample surface and converts it into temperature. By contrast, thermography expresses radiant heat at the sample surface by color changes over a wide area. However, these two types of thermometers are not without problems. A temperature gradient between the surface and the inner parts of the solid sample occurs since there is no convection possible in a solid sample. Such thermometers fail to measure the temperature in the interior of the sample. A microscopic temperature gradient often forms upon microwave heating of the solid surface (note that these thermometers cannot sense very small temperature gradients), and whenever a hotspot (Sect. 4.4) forms, the correct overall temperature cannot be measured. Although these thermometers can measure the temperature of the sample, it is necessary for the experimentalist to understand beforehand the operation of these thermometric devices and their advantages and disadvantages.

A thermocouple is usually used as a thermometer to measure the internal temperatures of a solid sample. The thermocouple is an electrical device consisting of two dissimilar conductors forming electrical junctions of differing temperatures. It

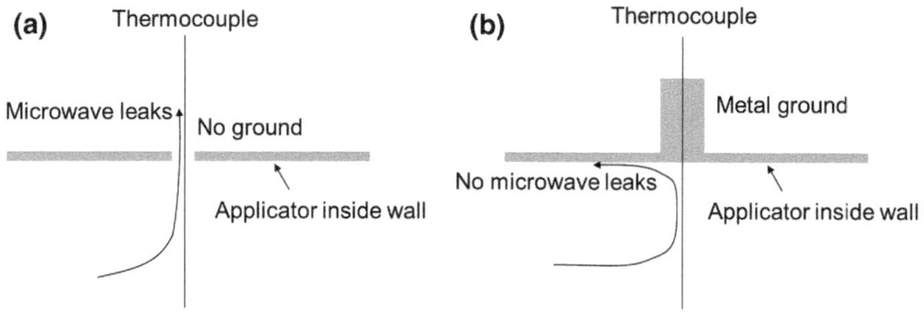

Fig. 6.23 Images of the method of grounding a thermocouple to a microwave applicator for prevention of a microwave leak with (**a**) no ground and (**b**) ground

produces a temperature-dependent voltage because of the thermoelectric effect; this voltage is then interpreted as a measure of temperature using the Seebeck effect, so-called because it converts heat directly into electricity at the junction of different types of metallic wires. This effect was named for the Baltic German physicist Thomas Johann Seebeck who, in 1821, discovered that a compass needle could be deflected by a closed loop formed by two different metals joined in two places, with a temperature difference between the joints. This was because the electron energy levels in each metal shifted differently and a voltage difference between the junctions created an electrical current and therefore a magnetic field around the wires.

Thermocouples are a widely used type of temperature sensor in industrial fields, and are also used in homes, offices, and businesses as the temperature sensors in thermostats, and as flame sensors in safety devices for gas-powered major appliances. There are several issues to consider before one uses a thermocouple to measure temperatures during microwave heating. One of these is the possibility of microwave leaks at the point of entry of the thermocouple, i.e., electric field leaking through the wall of the cavity (Fig. 6.23a). To remedy this problem, the thermocouple is electrically grounded to the microwave applicator (Fig. 6.22b).

Yet another nontrivial problem of thermocouples is that the microwave radiation may cause them to malfunction in measuring temperatures. Three kinds of thermocouples may be distinguished: (i) the non-ground junction type (the so-called sheath type), (ii) the ground junction type, and (iii) the exposed junction type

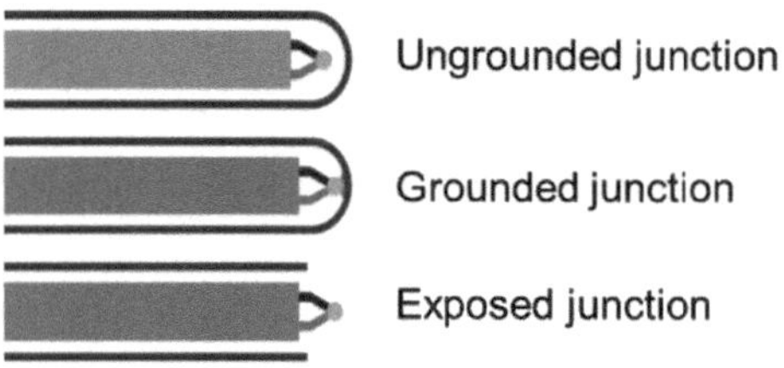

Fig. 6.24 Inner structures of the tips of the three types of thermocouples (sensor part)

(Fig. 6.24). Only the sheath type thermocouple can be used under microwave heating; malfunctions are easily observed whenever other types are used. The diameter of a thermocouple that is influenced by microwaves is empirically about 3–5 mm. A small diameter thermocouple tends to cause electric sparks, so that measures are needed to limit thermocouples to temperature measurements when heating strong microwave absorbing samples. Even if these measures are taken, they would not be enough in the case of heating nonpolar samples. It cannot be overemphasized that the use of a thermocouple within this context can have dangerous consequences; hence, it is best to consult a microwave specialist prior to beginning any microwave-assisted process.

6.6 Prevention of Microwave Leakages

6.6.1 Choke Trap

Radio waves cannot easily penetrate metals, thereby eliminating any concerns about microwaves leaking even as the microwave *door* is opened and shut. Close examination of the door of a microwave oven reveals the presence of a gap around the inner edge, whose principal function is to prevent microwave leaks (Fig. 6.25a), i.e., a choke trap that is designed into this gap (Fig. 6.25b). The length of the gap is 1/4 of the microwave wavelength; the phase of the incident microwave and that of the reflected microwave in the gap have a gap length of 1/2 the microwave wavelength. Thus, the incidence and reflective microwaves interfere destructively at the gap, thereby preventing leakage. However, it is not possible to completely prevent microwave leakage only by the choke trap when heating low-loss, poorly microwave absorbing foodstuff or other materials under high-power microwave radiation. When a WiFi device is used near a microwave domestic oven, it will not

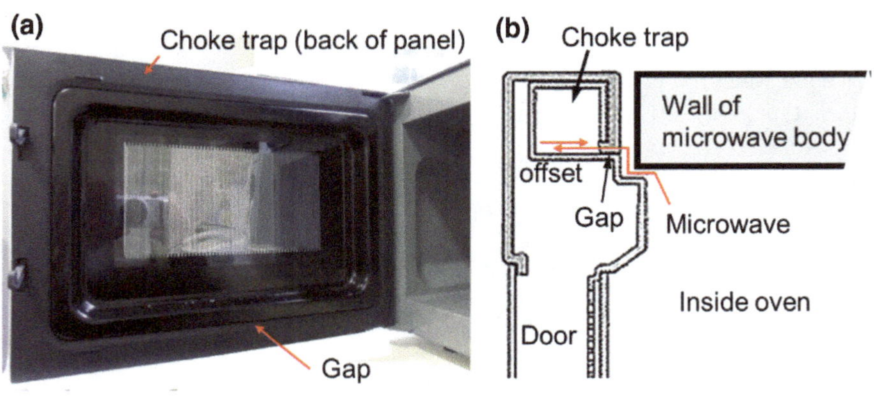

Fig. 6.25 (**a**) Photograph of a door of a microwave domestic oven, and (**b**) schematic of the choke trap on the door

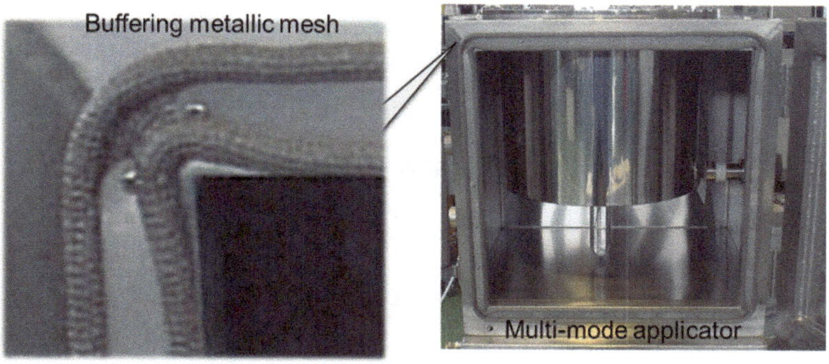

Fig. 6.26 Prevention of microwave leakage at the door using a buffering metallic mesh on the door of a multimode applicator

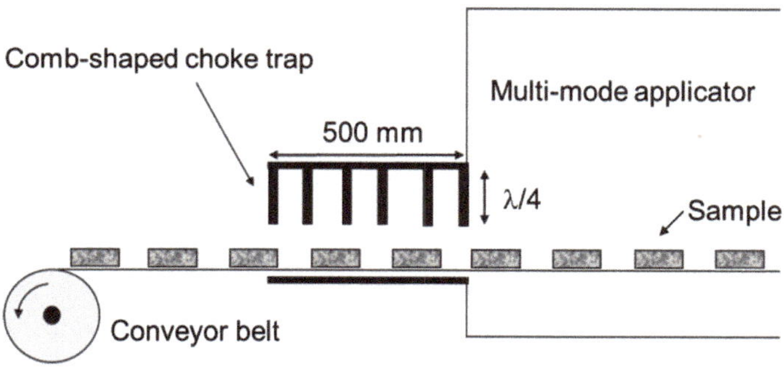

Fig. 6.27 Schematic image of a comb-shaped choke trap connected to a continuous microwave heating system with a belt conveyor

connect to the WiFi; this is because there are weak radio waves leaking from the edge of the door. Note that the amount of microwave radiation leaking from the edge at the door is very small (equal to or less than 5 mW cm^{-2}), and as such will not have any negative impact on humans.

A metallic mesh (analogous to a Faraday cage) plus a foam buffer metal mesh (Fig. 6.26) with metallic spring are additional devices used as other leak prevention methods at the door gap. These metallic materials are easily coupled to the door, inexpensive, and facilitate prevention of microwave leakages; they are often used on the access doors of industrial microwave systems. However, usage over a long time period requires periodic maintenance because of deterioration of the metal and the sponge through corrosion and occasional sparks.

A comb-shaped choke trap is used for entry and exits into the applicator of a continuous microwave heating system with the samples placed on a conveyor belt (Fig. 6.27). When using this type of choke trap, it is possible to have openings at the entry and exit ports that are 3–4 cm in height without significant leakage; when several comb-shaped choke traps are lined up contiguously, the opening size can be even larger. There exist other choke designs for similar uses.

6.6.2 Prevention of Microwave Leakages at the Sample Observation Window

When carrying out chemical reactions or, for that matter, any scientific process, a most basic experimental desire is observation of the sample during the reaction or process, especially in microwave chemistry. In the latter case, however, unlike typical chemical reactions performed at the bench, the sample cannot be manipulated or easily observed through a window in the door because of possible microwave leakage. A metallic open mesh (a Faraday cage) is used to observe heating or cooking of food through the window of a microwave domestic oven. The required size and distance of the holes in the metallic mesh can be calculated by the cutoff frequency. Figure 6.28 illustrates a schematic of the sizes of holes and distances between the holes (right-hand side of figure), together with the microwave shield ratio as a function of the distances between the holes (hole interval; left-hand side of Fig. 6.28). Clearly, the shield ratio improves by making the hole size small and separating the holes at greater distances. Also revealing is that it is possible to further suppress microwave leakages by using a thicker metallic plate.

An important facet in microwave chemistry is the need to measure temperatures continuously at the samples during a microwave-assisted process. Accordingly, it is necessary to have a hole in the applicator for making such temperature measurements of the samples, as well as for allowing a flow-type reactor to be inserted through the hole for microwave-assisted syntheses through which the reacting

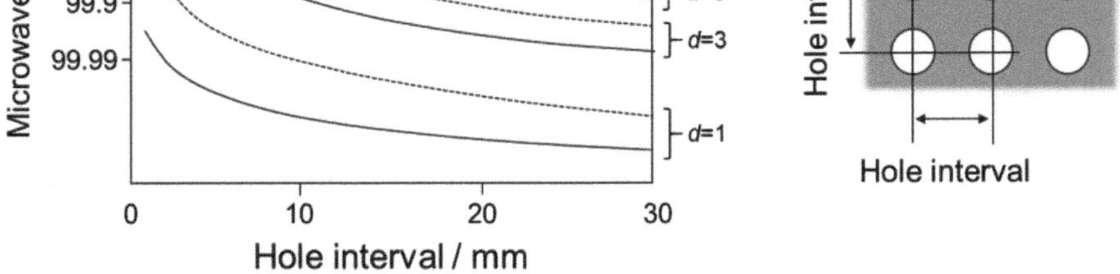

Fig. 6.28 Size and distance of the holes in a metallic Faraday cage against the microwave shield ratio for 2.45-GHz microwaves (note the log scale used for the shield ratio). Dotted lines: board thickness of 0.2 mm; solid lines: board thickness, 1.0 mm

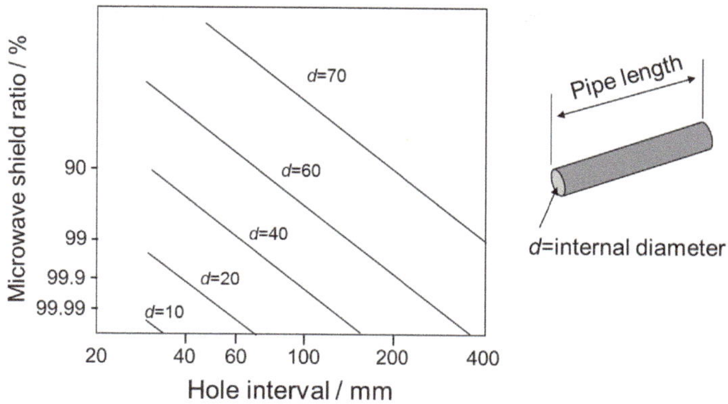

Fig. 6.29 Plot showing the microwave shield ratio against the distances between the holes (i.e., between the pipes) for various hole sizes (internal diameter of pipes). Note the log-log scale of the plots

sample is continuously circulated. Recall that, in nature, light leaks out from a hole in a straight direction. On the other hand, microwaves leak out from a hole in a manner reminiscent of a water fountain. To prevent microwave leakage in such cases, round metallic pipes are installed in a multimode applicator; the hole diameters and pipe lengths are summarized in Fig. 6.29, which displays a graphic of the microwave shield ratio as a function of the distances between the holes (hole intervals; i.e., distances between the pipes). Here again, the smaller the distances between the holes (pipe location) and the smaller the hole sizes (i.e., internal diameter of the pipes), the greater the shield ratio and the lesser the chances of any microwave leaks. For example, Fig. 6.30 shows a photograph in which a reflux

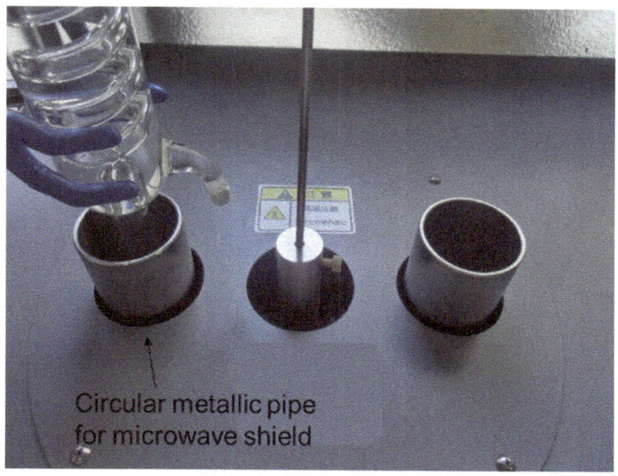

Fig. 6.30 Photograph showing circular metallic tubes used as microwave shields in a multimode applicator

condenser and a thermometer were introduced into the applicator's interior. Note that the leakage rate changes with the conductivity and the size of the pipe introduced into a hole. It cannot be overemphasized, however, that microwave leakages can lead to problems for the user of microwave radiation. Therefore, it is always necessary to consult with a microwave specialist when making changes or remodeling an applicator.

6.7 Visualization of Microwaves

Microwaves can neither be seen, heard, nor smelled. Hence, it is very difficult to ascertain their existence. Consequently, it is necessary to visualize their existence using a sensor; they can also be *visualized* through a simulation. A power monitor with a crystal mount is the component used to detect microwave power. A leak detector is used to detect possible microwave leaks (Fig. 6.31). The crystal mount and half-wave dipole are designed for low-power microwave detection and so should not be put into the microwave applicator to detect high-power microwaves.

An optical electric field sensor is the equipment with which an electric field of a microwave can be gauged correctly. The head of the sensor contains a lithium niobate ($LiNbO_3$) optical crystal substrate and a small dipole antenna. It is possible to monitor a strong electric field over a wide frequency range by the changes of the optical signal that are then propagated to the receiver (Fig. 6.32). The strength of the electric field can be calculated by the changes in the optical power. It is also possible to install more than one sensor in the applicator and thus image the electric field distribution of the microwaves.

There is also a visualization sensor with which a light-emitting diode (light-emitting diode visualization sensor), an LED and a detection diode are connected at a point of a loop antenna (Fig. 6.33). The sensors can be arranged on a printed circuit board. The sensor can then determine the image of the electric field in the applicator. However, it cannot be used under conditions of high-power microwave fields in the applicator because of heating and damage to the printed

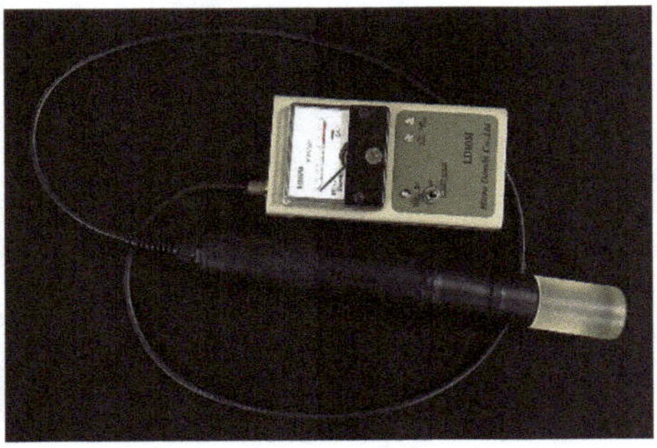

Fig. 6.31 A detector with a crystal mount to detect microwave leaks

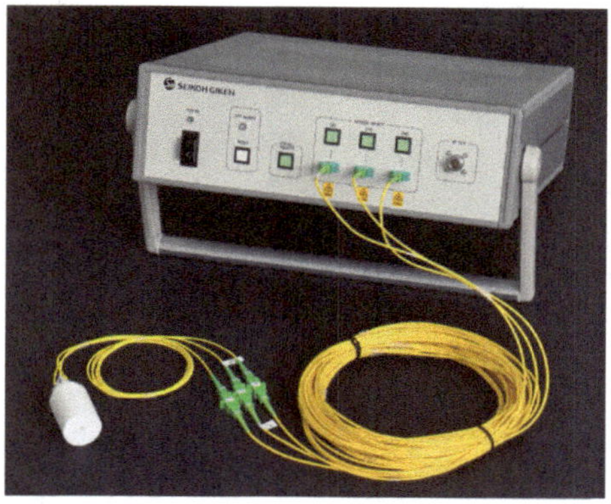

Fig. 6.32 Photograph of a photoelectric sensor. Photograph provided by courtesy of Seikoh Giken Co. Ltd.

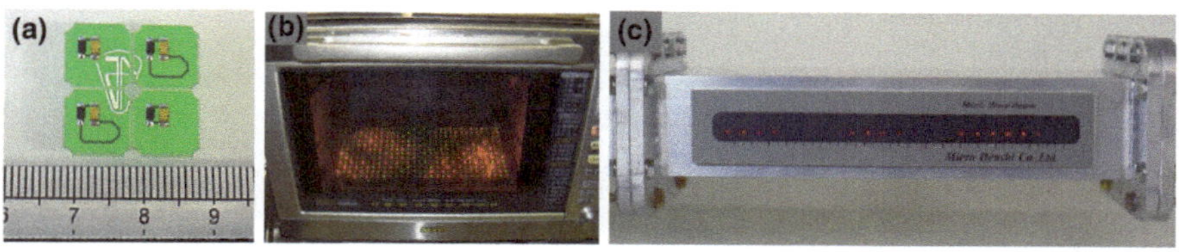

Fig. 6.33 Photographs of (**a**) a light-emitting diode visualization sensor, (**b**) a light-emitting diode visualization sensor in a microwave oven (an electric field exists in the brilliant part), and (**c**) the electric field inside the waveguide seen by a light-emitting diode visualization sensor. Photographs provided by courtesy of Micro Denshi Co. Ltd.

circuit board, but can nevertheless be used to check the electric field distribution inside the applicator in the presence of a sample under low-power microwave field conditions.

Yet another simple method to detect microwaves is the use of a dye contained in agar that responds to heat; this causes the color of the dye to change under microwave heating. Thermal paper can also be placed in the applicator to detect microwaves since it changes color under microwave heating. Note that this type of paper is often used in thermal printers and particularly in inexpensive or lightweight devices, as well as in adding machines, cash registers, and credit card terminals (Figs. 6.33 and 6.34).

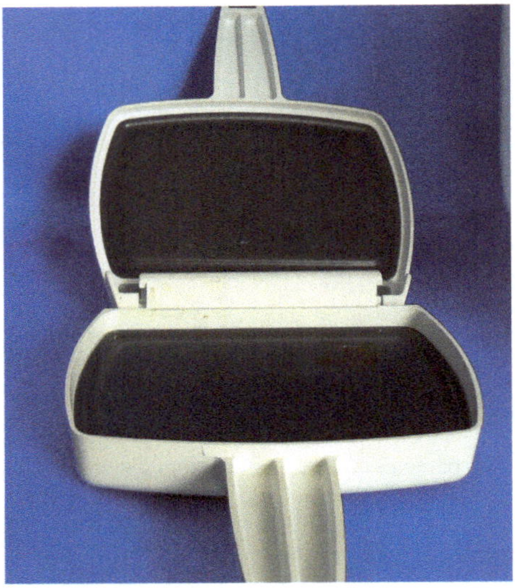

Fig. 6.34 A Raytheon "Steakmaker" has a silicon/ferrite coating, bonded to the opposite sides of the black surfaces, that heat when exposed to microwave energy for about 4–5 min, creating a very hot surface to broil a steak or hamburger when the two halves of the clamshell are closed. Copyright 2017 by R.F. Schiffmann

6.8 Coffee Break 6: Browning and Crisping in a Microwave Oven

How can you achieve browning and crisping in a microwave oven that does not have steam, hot air, or a grill element? Browning and crisping of foods are different phenomena and have different requirements in order to occur. Browning is a surface phenomenon that depends upon the Maillard Reaction, i.e., a reaction between a reducing sugar such as fructose and an amino acid such as glycine. The reaction runs best at a temperature of approximately 100 °C and in slightly alkaline conditions. Higher temperatures can also cause browning of meats: for example, by caramelization. Crisping depends upon dryness, i.e., a water activity of approximately 0.1 and temperatures on the order of 175–200 °C. The crisping problem in the microwave oven is twofold: (1) the air is cold so moisture condenses on the food surface, and (2) when microwaves heat a food, they raise the internal vapor pressure and water may be pumped to the surface where the cool air causes it to condense, raising the surface water activity to close to 1.0 (this may also dilute the reactants required for browning). There are three basic ways in which browning and crisping can be achieved; these are described in Table 6.4.

How does a microwave susceptor work? A microwave susceptor consists of a thin film of oriented polyester onto which is coated an ultrathin layer of vacuum-deposited aluminum; this is then bonded with an adhesive onto paper or paperboard (Fig. 6.36). Without this bonding, the polyester film would shrivel. The

Table 6.4 Methods to achieve browning and crisping in a microwave oven

Item	Description
Food formulation	This is possible only with browning, and uses ingredient technology, including coatings to color the surface, and blends of the amino acids and reducing sugars to stimulate the Maillard reaction
Cookware	Several cookware products available in the United States and elsewhere are useful for browning and crisping surfaces. Two major techniques are involved: (i) **Ferrites**: these are ferri and ferromagnetic materials, blends of various metals that are embedded in a silicone carrier and then coated onto an exterior metallic surface of a browning/crisping device. Ferrites can be blended to achieve and maintain a specific temperature, e.g., 200 °C, that is useful for both crisping and browning; note that the ferrite never touches the food itself, but rather heats the device, always made of metal, that becomes hot and then transfers its heat to the food (Fig. 6.34). (ii) **Tin oxide**: beginning in the 1950s, the Corning Glass Company manufactured high temperature dishes and cookware made of a glass ceramic, Pyroceram. A coating of tin oxide was embedded into the exterior bottom of some dishes to create browning dishes that the consumer would preheat to create a high interior bottom surface that could brown meats and other foods. The weakness of this design is that, in contrast with ferrites, it had no *roll off* temperature, i.e., it would simply become hotter the longer it received microwave energy
Glaze coatings	Properly formulated glazes coated onto ceramic plates or cook wares create a hot browning surface when exposed to microwave energy. Any foods placed thereon, such as meat, can be browned. But, again, the surface temperature cannot be controlled
Packaging	Packaging structures that cause browning and crisping are known as susceptors. These are thin films of aluminum vacuum-deposited onto an oriented polyester film that is then bonded onto paper or paperboard. When exposed to microwave energy, the temperature of the susceptors may reach 190–200 °C at which point the polyester degrades and the film stops heating. These susceptors are formed into trays or cartons, and usually packaged with a food such as microwavable pizza, to provide a hot surface onto which the pizza crust is placed and crisped and browned while the microwave is energized. They do not require preheating. While they work to some extent, the major problem is their low thermal mass so they are not able to deliver many calories (or joules) of heat to the surface that requires browning for crisping and may contain much water (Figs. 6.35 and 6.36)

orientation of the film is critical to controlling the heating—the film splits at approximately 190 °C so it stops heating, otherwise the paper would brown and even catch fire should the temperature reach 233 °C. The mechanism of heating of the susceptor depends upon the properties described in the equation: $\boldsymbol{P = I^2R}$, where P = Power (W), I = Current (A), and R = Resistance (Ω). If a current flows against a resistance to that flow, it will generate power, in this case: heat. For a susceptor, the aluminum is deposited in sub-microscopically tiny islands across which the current generated by the microwave energy flows and meets a resistance in the gaps between the aluminum islands. Both the size of the islands and the gaps

Fig. 6.35 Various susceptor designs used to brown and crisp foods. These are used for pizza, Hot Pockets®, microwavable popcorn pouches, and many more. Copyright by R.F. Schiffmann

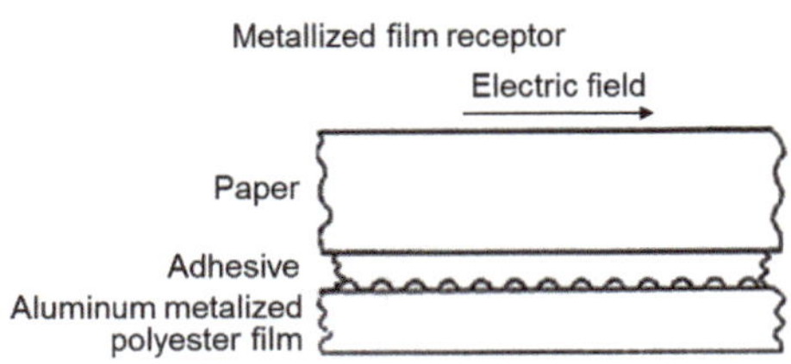

Fig. 6.36 Construction of a typical susceptor. Reproduced by permission from Ref. [21]. Copyright 1987 by the International Microwave Power Institute

are critical to achieving heating, measured by the surface resistivity of the susceptor film (Fig. 6.37); this illustrates that there will be transmission and reflection of the microwaves as well as absorption, i.e., heating. Hence, the proper surface resistivity is important to have a properly heating susceptor.

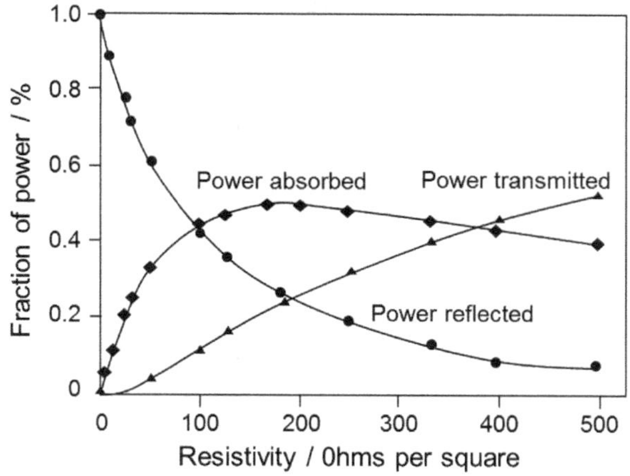

Fig. 6.37 Theoretical susceptor characteristics. Reproduced with permission from [22]. Copyright 1991 by the International Microwave Power Institute

References

1. R. Meredith, *Engineers' Handbook of Industrial Microwave Heating (Energy Engineering)* (Institution of Engineering and Technology, The Institution of Electrical Engineers, London, UK, 1998)
2. W.E. Willshaw, L. Rushforth, A.G. Stainsby, R. Latham, A.W. Balls, A.H. King, The high-power pulsed magnetron: development and design for radar applications. J. Inst. Electr. Eng. Part IIIA Radiol. **93**(5) (1946). doi:10.1049/ji-3a-1.1946.0188
3. http://www.hokuto.co.jp/hokuto/eng/products/magnetron/index.htm
4. R.J. Meredith, *Engineers' Handbook of Industrial Microwave Heating (IEE Power series)* (The Institution of Engineering and Technology, Stevenage, UK, 1998) (Chapter 9)
5. D. Varian, *The Inventor and the Pilot* (Pacific Books, Santa Barbara, CA, USA, 1983), p. 189
6. R.H. Varian, S.F. Varian, A high frequency oscillator and amplifier. J. Appl. Phys. **10**, 321 (1939)
7. A.S. Gilmour, *Klystrons, Traveling Wave Tubes, Magnetrons, Crossed-Field Amplifiers, and Gyrotrons* (Artech House, Boston, MA, USA, 2011), pp. 317–318
8. J.C. Whitaker, *The RF Transmission Systems Handbook* (CRC Press, Boca Raton, FL, USA, 2002), pp. 814–816
9. USA Patent No. US2064469 (A)
10. *Electron Discharge Device System*, USA Patent No. US2300052 (A)
11. https://en.wikipedia.org/wiki/Traveling-wave_tube
12. S. Horikoshi, A. Osawa, M. Abe, N. Serpone, On the generation of hot spots by microwave electric and magnetic fields and their impact on a microwave-assisted heterogeneous reaction in the presence of metallic Pd nanoparticles on an activated carbon support. J. Phys. Chem. C **115**, 23030–23035 (2011)
13. S. Horikoshi, N. Serpone, in *Microwaves in Catalysis: Methodology and Applications*, ed. by S. Horikoshi, N. Serpone, (Wiley-VCH Verlag GmbH, Weinheim, Germany, 2015) (Chapter 1)
14. Institute of Electrical and Electronics Engineers, *The IEEE Standard Dictionary of Electrical and Electronics Terms*, ed. by J. Radatz, 6th edn. (Institute of Electrical and Electronics Engineers, New York, NY, 1997). IEEE Std 100-1996. ISBN 1-55937-833-6 [ed. Standards Coordinating Committee 10, Terms and Definitions]
15. K.S. Packard, The origin of waveguides: a case of multiple rediscovery. IEEE Trans. Microw. Theory Tech. **32**(1984)
16. http://hyperphysics.phy-astr.gsu.edu/hbase/Waves/standw.html

17. D.E. Clark, W.R. Tinga, J.R. Laia Jr. (eds.) Microwaves: theory and application in materials processing II—ceramic transactions, in *Proceedings of Symposium 36 held during the 95th Annual Meeting of the American Ceramic Society* (Cincinnati, Westerville, OH)
18. R.E. Collin, *Foundations for Microwave Engineering* (McGraw-Hill, New York, NY, 1966)
19. W.H. Sutton, M.H. Brooks, I.J. Chabinsky (eds.) *Microwave Processing of Materials, Materials Research Society Proceedings*, vol 124 (Materials Research Society, Pittsburgh, PA, 1988), pp. 367–372
20. P.O. Risman, T. Ohlsson, B. Wass, Principles and models of power density distribution in microwave ovens. J. Microw. Power Electromagnet. Energy **22**, 193–198 (1987)
21. H.F. Huang, New product concepts in microwavable food packaging. Microw. World **8**(6) (1987)
22. C.R. Buffler, A simple approach to the calculation of microwave absorption, transmission and reflection of microwaves from susceptor film. Microw. World **12**(3) (1991)

Chapter 7
Microwave Chemistry in Liquid Media

Abstract This chapter summarizes the features and caveats in microwave-assisted chemistry. Since many samples of chemical reactions are liquids, the handling of liquid samples in microwave heating is illustrated. Also noted are problems such as accidents, temperature distribution, heat insulation, and impurities. The chapter also introduces a method of microwave chemical synthesis while cooling. Finally, it reports on existing commercial bench-scale microwave chemical synthesis equipment and scaled-up equipment. The coffee break introduces a unique hybrid cooking system.

Keywords Microwave-assisted organic synthesis (MAOS) · Heat insulators Temperature distribution · Fixed catalyst · Dewar-like double-walled insulated reactor · Tan δ · Heating rate of common organic substances · Ionic species Radical reactions · Commercial microwave chemical synthesis equipment Bench-scale equipment · Plant-scale equipment · Hybrid cooking system

7.1 Effective Microwave Heating in Chemistry

Procedures have been established and laboratory instruments have been available for some time for performing classical chemical reactions in liquid media. In many cases, appropriate information is available in several technical manuals; chemical engineering texts can also be consulted if there is a need for scale-up of some unit operations. However, when the heat source involves microwave radiation, knowledge of microwave technology is a necessity. Also relevant is an understanding of electromagnetic waves and how the resulting heat is distributed in a sample in the reaction vessel. Some syntheses that normally cannot be performed with existing protocols in classical chemical synthesis can be carried out with microwave radiation. Nonetheless, there are some issues that must be considered so as to achieve best performance when performing chemical syntheses with microwave heating; these are summarized in Fig. 7.1 [1]. Note that even if all issues were not

S. Horikoshi et al., *Microwave Chemical and Materials Processing*,
https://doi.org/10.1007/978-981-10-6466-1_7

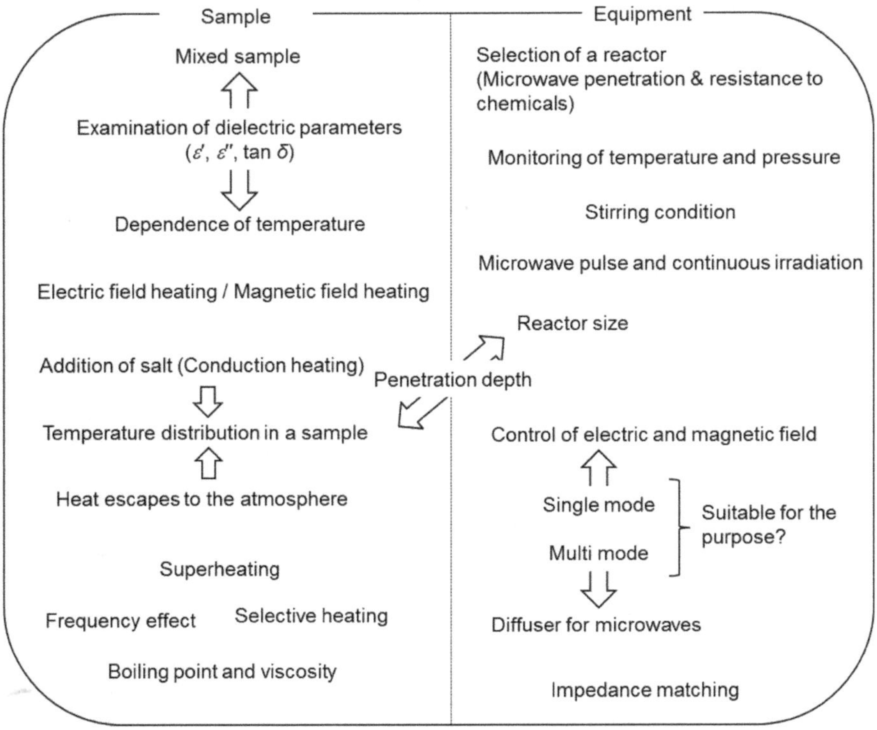

Fig. 7.1 Relevant considerations in performing microwave-assisted chemical reactions effectively. Reproduced from [1]

considered, microwave chemistry can still take place. This chapter focuses on describing reaction vessels (reactors), heat insulation, and how cooling under microwave irradiation affects chemical reactions.

7.2 Reactors

Factors that should be considered in chemical synthesis reactors are heat conduction, chemical corrosion, the melting point of the reactor material, and the potential thermal shocks that impact the reactors. In general, various materials can be used in the fabrication of chemical reactors: glass, plastics, and metals. On the other hand, an important factor in carrying out chemical syntheses with microwave heating is a determination of which material is best suited in the fabrication of reactors. This will be governed by whether the reactants will be heated directly by the microwaves or whether the microwaves will heat the reactor, which, in turn, heats the reactants. When choosing microwave heating to carry out syntheses, it is not advisable to use a metallic reactor.

If a reactor, rather than the reactants, is to be heated by microwaves, it is important that the reactor material absorb the microwave radiation to generate heat; one such material is nonpolar silicon carbide (SiC). Research reports have indicated that such reactor vessels completely absorb the microwave radiation, although the fraction of microwaves absorbed by the reactor is rather limited. This assertion can be verified by placing an electric field sensor in the reaction vessel, which will reveal the extent to which microwaves have penetrated into the vessel.

When the reactants are to be irradiated directly by microwaves, it is best to choose a reactor fabricated with low microwave absorbing materials so as to allow for the penetration of the microwave energy to heat the reactants. The specific inductive capacity (the dielectric constant, ε') and the relative physical relationship of the phase defect angle ($\tan\delta$) of the type of glass and of ceramics and plastics are shown in Fig. 7.2. Quartz glass, certain types of ceramics, Teflon (PTFE), and polyetheretherketone (PEEK) are often used as reactor materials suitable for microwave-assisted chemical reactions. Quartz glass is particularly suitable for solutions because the dissipation loss of the sample is low, *i.e.*, all the microwave energy is used for heating the reactants, not the quartz. By contrast, glass reactors are often used from the viewpoint of cost. One must consider the composition of glass reactor materials because the extent of absorption of microwaves by borosilicate glass and soda glass are 10 times and 290 times, respectively, greater than quartz glass. For example, when a nonpolar solvent is placed in an ordinary glass reactor and then microwave-heated, the glass reactor may either melt and/or break on exposure to microwaves (Fig. 7.3).

Teflon is an excellent microwave reactor material because microwave absorption is very low; even with temperature rise, the microwave absorption ratio does not change, and not least Teflon is chemically very stable. However, there may be a problem with the heat and pressure resistance of Teflon compared with quartz. Just like Teflon, PEEK is also often used as a microwave reactor material. Additionally, aluminum oxide and silicon nitride (SiN) have also proven useful reactor materials when handling special solutions and/or catalyzed syntheses because these materials do not absorb microwave radiation. One problem with these materials, however, is

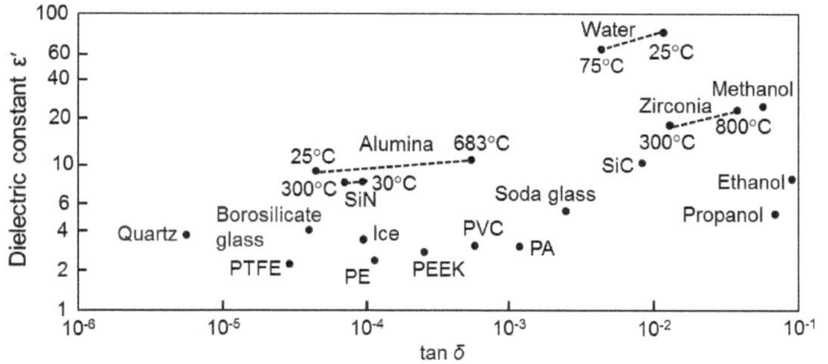

Fig. 7.2 Summary of dielectric characteristics of materials at 25 °C and at a 2.45-GHz microwave frequency. Reproduced from [2]

Fig. 7.3 Damaged glass reactors subjected to microwave irradiation (breaking and melting)

that the presence of extraneous species (*i.e.,* impurities) in the reactor can significantly enhance microwave heating, so that it is important that the reactor materials be of high purity and not the source of impurities. Hence, it is important to understand, not only the features of microwave heating and type of materials used in the fabrication of reaction vessels, but also important is the choice of the most suitable materials for the experimental objectives.

7.3 Heat Insulators

7.3.1 Differences in Heat Insulation for Classical Chemistry and Microwave Chemistry

Preventing heat loss from a solution is an important consideration in microwave chemistry. There are various ways with which heat transfer can occur: (i) by conduction, (ii) by convection, and (iii) by radiant heat transfer (Sect. 2.4), all of which are important in microwave-assisted chemical reactions. In traditional chemistry, the reactant solution is typically heated by an external heat source, for example, by placing the reactor in an oil bath. In the latter case, there is no need to heat-insulate the reactor. However, in microwave chemistry, the reactants are heated directly by the incident microwave energy, a process that brings energy directly into the reacting solution. In this case, the heat of the solution radiates outward through the reactor walls into the surrounding atmosphere within the microwave applicator. In a sense, the microwave-assisted process progresses outward from the center of the reactor to the inner reactor walls, which leads to the formation of temperature gradients and ultimately concentration gradients in the substrates (Fig. 7.4). Such thermal gradients result from the thermophoresis migration, a process known as the Ludwig–Soret effect [3]. Two problems occur

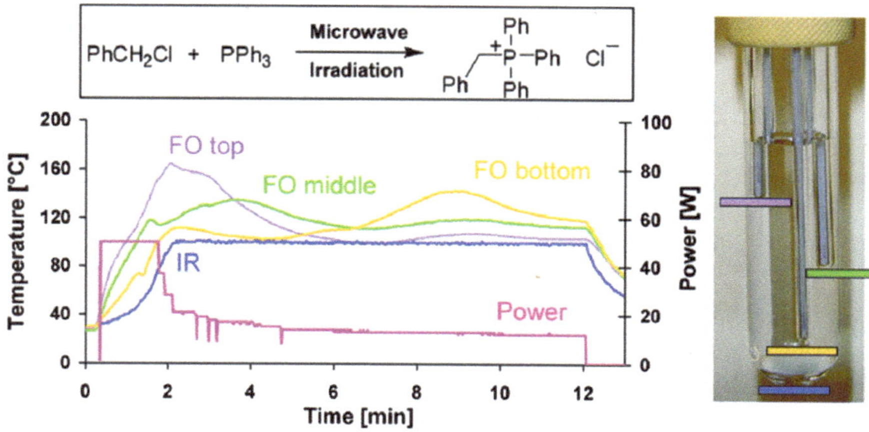

Fig. 7.4 Temperature profiles for a sample of 5 mL of *N*-methylpyrrolidone (NMP) contained in a 10-mL quartz vessel equipped with three internal fiber optic sensors. The sample was irradiated (without stirring) with constant 50-Watt magnetron output power for 40 min. Shown are the temperature profiles for the three internal fiber optic probes and a photograph depicting the external infrared sensor located at the bottom of the instrument; also reported are the experimental conditions. Reproduced from [4]. Copyright 2008 by American Chemical Society

when the heat of the solution escapes to the reactor's surrounding atmosphere: (a) microwave energy necessary to heat the solution is wasted, and (b) differences in temperature between the center of the solution and the external surroundings may be significantly large. Often the temperatures measured at different positions within the reactor will show variations as a result of thermal gradients generated within the reaction medium. In this regard, Herrero et al. [4] noted that uneven temperature distributions within a reactor result from uneven microwave heating.

To remedy the above two problems requires the reactor be insulated. This can be achieved by: (1) blowing external hot air onto the reactor surface, a method often used in industrial heating of foodstuffs—known as the microwave/hot air hybrid method; (2) maintaining the solution temperature by shielding the reactor with an insulation material; and (3) maintaining the solution temperature by performing the reaction in a vacuum-filled double-walled reactor (a Dewar-like insulated structure) [5–7].

7.3.2 *Heat Insulators*

Thermal insulation is the reduction of heat transfer between objects in thermal contact or between objects located within the range of possible radiant heat (IR) influence. Heat transfer refers to the transfer of thermal energy between objects of differing temperatures. Insulation can be achieved with special engineering methods or processes, as well as with suitable shapes of objects and suitable materials. Heat flow is an inevitable consequence of contact between objects of differing temperatures. Thermal insulation provides an insulating region in which thermal conduction is reduced or otherwise blocked, or thermal radiation is reflected from the surface of the lower temperature component rather than being absorbed by it. The insulating capability of a material is measured by its thermal conductivity (k). A low thermal conductivity is equivalent to high insulating capability (often referred to as the R-value of the insulating material). In thermal engineering, other important properties of insulating materials are product density (ρ) and specific heat capacity (c). Solid materials chosen for insulation have a low thermal conductivity k, measured in units of watts per meter per kelvin ($W\ m^{-1}\ K^{-1}$). As the thickness of insulation is increased, the thermal resistance to heat loss is also increased.

In the case of insulated reaction cylinders, a critical radius of the reaction vessel must be reached. However, before the critical radius is reached, any added insulation increases heat transfer in the reaction vessel [8]. The convective thermal resistance is inversely proportional to the surface area and therefore the radius of the cylinder, whereas the thermal resistance of a cylindrical shell (the insulation layer) depends on the ratio between the external and inner radii of the vessel wall, and not on the radius itself. An increase in the external radius of an insulated cylinder adds a fixed amount of conductive resistance equal to the expression $\ln(2)/(2\pi kL)$, where k is the thermal conductivity and L denotes the thickness of the reactor. At the same time, however, the convective resistance is reduced. This implies that adding insulation below a certain critical radius actually increases the heat transfer. For insulated cylindrical reactors, the critical radius ($r_{critical}$) is given by Eq. 7.1 [9], which shows that the critical radius depends only upon the heat transfer coefficient and the thermal conductivity of the insulation. If the radius of the insulated cylinder were smaller than the critical radius for insulation, addition of any amount of insulation would cause an increase in heat transfer.

$$r_{critical} = \frac{k}{h} \tag{7.1}$$

Insulators with low dielectric loss factors are difficult to microwave-heat from ambient conditions because of their minimal absorption of the incident microwave energy. The most suitable insulators for microwave heating are introduced and described in Sect. 6.3.3. The effect of heat insulator material has been examined by the thermal distribution in a fixed-bed flow reactor in the presence of a Pt catalyst supported on activated carbon (Pt/AC). The microwave heating efficiency of the Pt/AC catalyst is very high. However, when the Pd/AC catalyst is fixed in a flow reactor, several temperature-related problems arise. The temperature distributions of a fixed Pd/AC catalyst layer, with and without the rolled alumina fiber heat insulator, in a microwave multimode applicator are shown in Fig. 7.5. The temperatures at the upper and bottom locations of the reactor are lower when compared to temperatures measured in the middle of the reactor, particularly in a non-insulated reactor as indicated in curve (**ii**) of Fig. 7.5. By contrast, when the reactor is

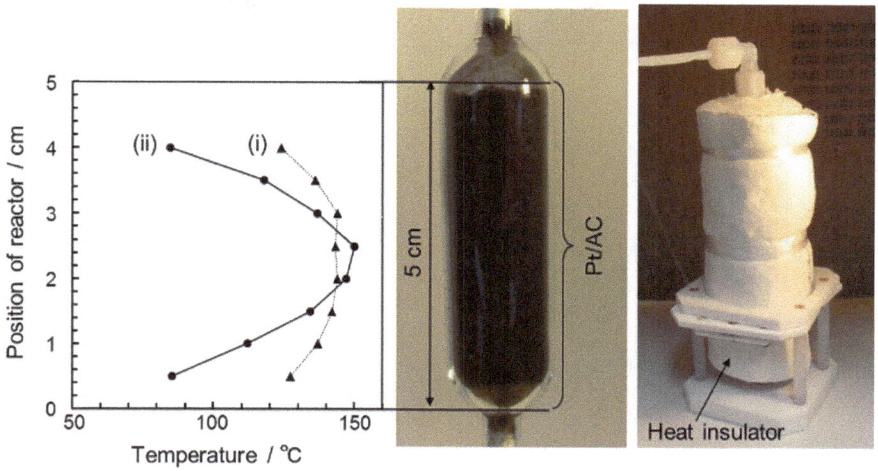

Fig. 7.5 Temperature distribution of a fixed Pd/AC catalyst layer (i) with and (ii) without a heat insulator consisting of a rolled alumina fiber material in the microwave irradiation of a fixed Pd/AC catalyst reactor located in a multimode applicator; also shown is a photograph of the rolled alumina fiber heat insulator (unpublished results)

insulated with a heat insulator (e.g., alumina fibers) surrounding the reactor, the variation in temperature between these positions is minimized (curve (**i**)).

7.3.3 Dewar-like Double-Walled Insulated Reactor

A three-neck Dewar-like vacuum-filled double-walled microwave cylindrical vessel (total capacity, 76.5 mL) is illustrated in Fig. 7.6a [7]; it consists of a Pyrex glass cylinder with a double-walled structure in which the inner pressure was 10^{-3} Torr. A photograph of the actual vessel is shown in Fig. 7.6b. The attractive features of this microwave reactor are rather evident on comparison with a conventional single-walled microwave vessel having the same volume, shape, and size.

The temperature distributions in a double-walled and a conventional single-walled vessel have been compared by recording the increase in temperature of a 20-mL water sample at various exposure times following continuous irradiation of the dielectric medium with 120-Watt microwaves in a commercial multimode microwave chemical apparatus equipped with a fiber optic thermometer. The temperature-time profiles for the water sample in these two different vessels are illustrated in Fig. 7.7. Illustrated are the temperatures of water measured at the center position B (solid curves **1, 3, 5**; inset Fig. 7.7) and at the inner wall side position A (dashed curves **2, 4, 6**; inset Fig. 7.7) of the vessels. After 1.5-min microwave irradiation, the temperatures between positions A and B in the microwave conventional vessel under non-stirring conditions varied by as much as 7 °C

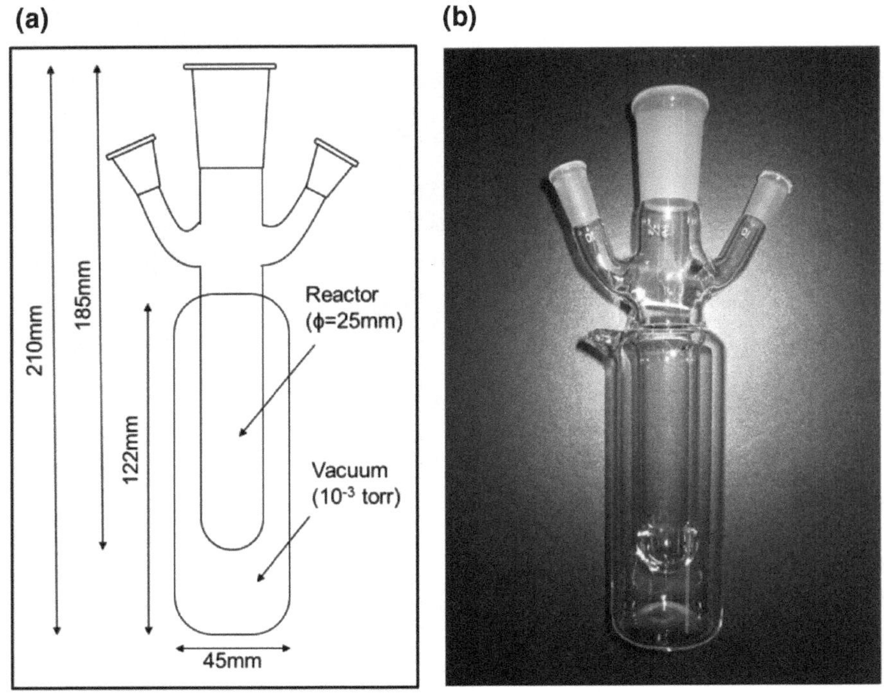

Fig. 7.6 (**a**) Illustration and dimensions of the double-walled reactor; (**b**) photograph of the actual microwave Dewar-like vacuum vessel. Reproduced from [7]. Copyright 2010 by American Chemical Society

(i.e., 49 and 56 °C, respectively; curves **4** and **3**), whereas in the Dewar-like reactor the temperatures were nearly identical (67 °C, position A; 68 °C, position B; curves **2** and **1**). The corresponding rates of increase of temperature in water for the conventional and Dewar-like reactors were, respectively, 18.7 ± 0.2 and 30.2 ± 0.3 °C min^{-1} (position A), and 25.3 ± 0.3 and 28.8 ± 0.3 °C min^{-1} (position B). The amount of heat that escapes from the reactors depends upon the temperature difference between the atmosphere surrounding the reactor vessels and the heated water. Interestingly, when the aqueous sample was stirred (magnetic stirrer), the temperatures between positions A and B in the microwave conventional vessel varied by as much as 4 °C at the end of the 1.5-min microwave irradiation period. Evidently, stirring the sample narrowed the temperature gradient between the two positions, with temperature increasing at position A and decreasing at position B. Under these conditions, the rates of temperature rise for the conventional reactor were 21.8 ± 0.2 °C min^{-1} (A; dashed curve **6**) and 24.3 ± 0.2 °C min^{-1} (B; solid curve **5**) indicating relatively little difference at the center position B, but significantly different when determined at the inner wall position A by comparison with the rates under non-stirring conditions. Temperature fluctuations in the microwave conventional vessel showed a similar tendency. By contrast, the microwave

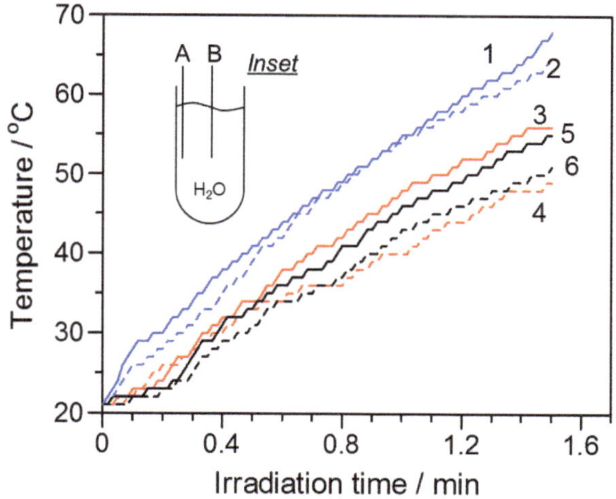

Fig. 7.7 Temperature-time profiles for a water sample in the microwave Dewar-like vacuum vessel under non-stirring conditions (lines **1, 2**) and in the conventional microwave vessel also under non-stirring conditions (lines **3, 4**) and under conditions where the aqueous medium was stirred (lines **5, 6**). Inset shows the positions where temperatures were measured in water (A: inner wall positions—dotted lines **2, 4, 6**; B: center positions—solid lines **1, 3, 5**). Reproduced from [7]. Copyright 2010 by the American Chemical Society

Dewar-like vacuum-filled double-glassed vessel achieved constant uniform heating owing to the insulation provided by the double-walled structure. Clearly, an attractive feature of the proposed microwave Dewar-like reaction vessel for a high dielectric medium, such as water, is not so much an improvement in heating efficiency, but rather the maintenance of sustained higher temperature throughout the whole medium [7].

To demonstrate the advantages of the microwave Dewar-like reactor vessel over the microwave conventional single-wall reactor, Horikoshi and coworkers [7] carried out the synthesis of biphenyl via the Suzuki–Miyaura coupling reaction. The procedure involved adding and mixing, under an argon atmosphere, the Pd/AC catalyst, phenylboronic acid, bromobenzene, and K_2CO_3 as the base in a nonpolar toluene solvent in the two types of microwave vessels; the microwave applied power was kept constant at 80 W. The initial rates of increase of temperature in the stirred dispersions were ca. 8 °C min^{-1} for the double-walled reactor and ca. 4.5 °C min^{-1} for the conventional single-walled vessel, a nearly twofold greater rate of increase of temperature with the additional heat insulation provided by the double-walled vacuum design. Continued microwave irradiation beyond 10 min for both reactor vessels led the heterogeneous system to reach a temperature of 106–107 °C for the Dewar-like vessel, but only 63–65 °C for the conventional vessel. The main absorber of microwave radiation in this heterogeneous system was the activated carbon (AC), which subsequently imparted most of its heat to the surrounding toluene solvent. For the single-walled conventional vessel, the heat was also dissipated into the surrounding atmosphere, contrary to the constant heat

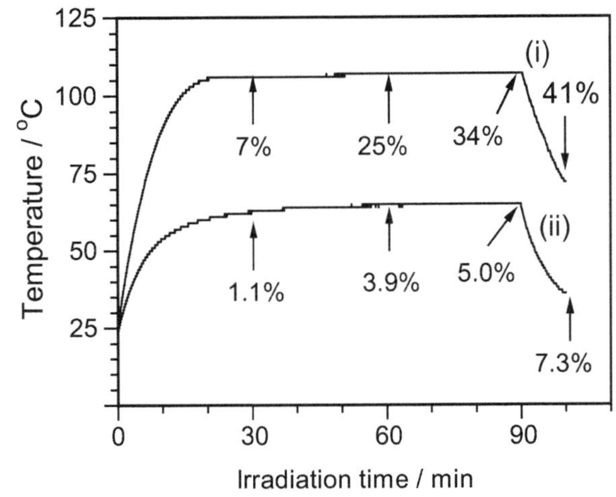

Fig. 7.8 Temperature-time profiles and percent reaction product yields of biphenyl at various times in (i) the microwave Dewar-like vacuum-filled vessel, and (ii) in the microwave conventional vessel; temperatures were measured at the center position of the reactor. Reproduced from [7]. Copyright 2010 by American Chemical Society

maintained within the double-glassed reactor. At approximately 90 min into the reaction, microwave irradiation was terminated resulting in a drop of temperature in both vessels.

The chemical yields of biphenyl from the Suzuki–Miyaura coupling reaction carried out in the microwave Dewar-like vacuum vessel were 7% after a 30-min, 25% after 60–min, and 34% after 90-min irradiation time (Fig. 7.8). The corresponding chemical yields of biphenyl in the conventional vessel were 1.1% (30 min), 3.9% (60 min), and 5.0% (90 min) under otherwise identical irradiation conditions. No doubt, the large difference in the reaction temperature between the Dewar-like vessel and the conventional vessel was the principal cause for the different yields. The chemical yields continued to increase to 41 and 7.3%, respectively, even 10 min after the termination of the microwave irradiation. Most importantly, the product yields were greater by as much as a six- to sevenfold margin with the double-walled reactor.

Another interesting application of the microwave Dewar-like vacuum-filled vessel is as a catalyst fixed-bed flow reactor displayed in Fig. 7.9. This reactor was specially designed to insulate and thereby prevent the microwave-generated heat from being transmitted to the surroundings during the Pd/AC catalyzed organic hydride process involving the dehydrogenation of methylcyclohexane (MCH) to toluene [10].

Temperatures were measured with an optical fiber thermometer at various locations from the bottom to the top and at the center of the reactor, with and without the presence of the liquid MCH organic hydride (flow rate, 0.2 mL min^{-1}). The heat distribution along the vertical locations of the Pd/AC catalyst bed in the reactor was examined under microwave heating. Because the fiber optic thermometer was limited to a maximum temperature of 220 °C; the heat distributions were determined at microwave output powers of 75 W with MCH present and 71 W without MCH. The temperature recorded under these conditions at the

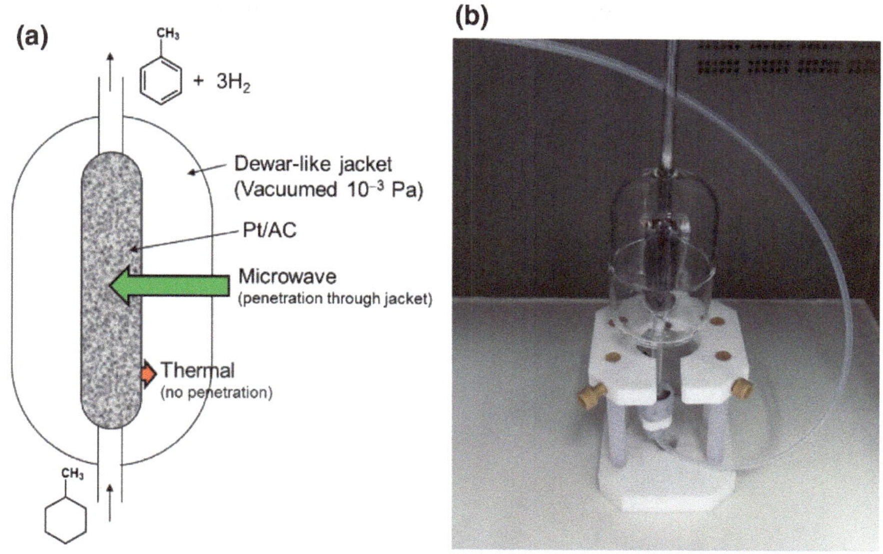

Fig. 7.9 (**a**) Schematic of the continuous flow, vacuum-filled, Dewar-like double-walled microwave reactor with packed Pd/AC catalyst powder; also shown is the addition of the methyl cyclohexane through the lower part of the reactor. (**b**) Photograph of the vacuum-filled Dewar-like double-walled microwave reactor positioned in the microwave multimode applicator. Reproduced from [10]. Copyright 2016 by Elsevier B.V

half-way position (2.5-cm from the bottom) with the classical single-walled reactor was 172 ± 2 °C (Fig. 7.10). By contrast, when the Dewar-like double-walled reactor was used, microwave output powers of only 48 W (MCH present) and 47 W (without MCH) were required to reach and maintain a temperature of 172 ± 2 °C at the 2.5-cm location. As a result, more than 33–36% of microwave energy saving was achieved using the Dewar-like reactor.

Also displayed in Fig. 7.10 are the temperature distributions within the catalyst particles under microwave heating for both types of reactor designs. Introducing the liquid MCH hydride at ambient temperature into the heated Pd/AC catalyst bed caused some cooling of the particulates at the entry point (lower part of the reactor). Using the classical reactor (left side in Fig. 7.10) led to a rather large temperature distribution within the catalyst bed: 98.9–173.5 °C (without MCH) and 86.7–189.4 °C (MCH present) indicating maximum temperature differences in the catalyst bed of about 75 °C without MCH, and 103 °C with MCH present. By comparison, using the Dewar-like reactor led to a narrower temperature distribution of 128.3–171.6 °C (gradient, ~43 °C) without MCH and 126.3–172.0 °C (gradient, ~46 °C) with MCH present. Hence, with the latter Dewar-like reactor, the heat distribution was essentially identical (right side in Fig. 7.10). In the presence of MCH, the average temperatures in the catalyst bed were 142 ± 43 °C (classical

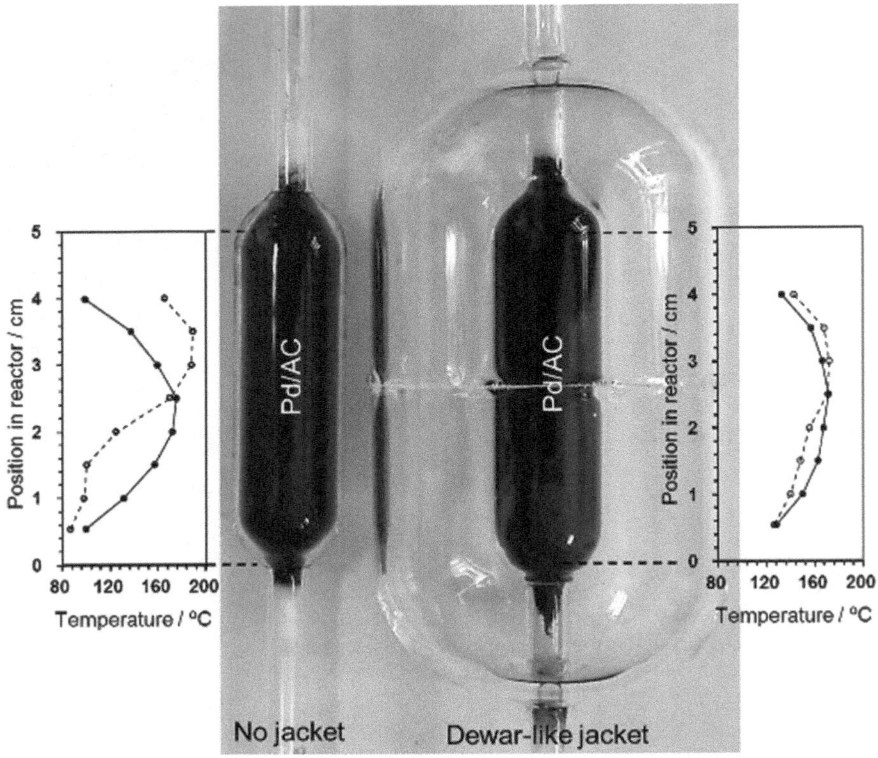

Fig. 7.10 Temperature distributions along the Pd/AC catalyst within the reactor with and without methyl cyclohexane: (left) traditional flow-through reactor; (right) Dewar-like flow-through reactor; full circles, without MCH; empty circles, with MCH present. Reproduced from [10]. Copyright 2016 by Elsevier B.V

reactor) and 153 ± 17 °C (Dewar-like reactor), a 2.6-fold lesser temperature spread in the latter reactor. The Pd/AC catalyst bed in the classical reactor and in the Dewar-like reactor reached a temperature of 340 ± 10 °C (measured with an Inconel thermometer) within 2 min under microwave heating, and with a continuous flow of MCH (0.2 mL min^{-1}); the dehydrogenation yield of MCH was 96 ± 3% after 60 min of irradiation. To reach the reaction temperature of 340 ± 10 °C from 172 ± 2 °C, the microwave power input for the classical reactor was increased from 75 to 217 W (a 65% increase), while with the Dewar-like reactor, the input microwave power needed was increased from 48 W to 141 W (a 66% increase). Accordingly, the microwave power input was 217 W for the classical reactor versus 141 W for the Dewar-like reactor, yielding an energy saving of 35% when using the latter.

7.4 Effects of Samples in a Microwave-Assisted Process

The experimentalist needs to choose appropriate samples for microwave heating. The choice of organic liquids is particularly important. Many published articles report that an indication of microwave heating efficiency depends on the dielectric loss tangent (tan δ), a parameter often measured at room temperature. The experimentalist has to understand the value of tan δ in the reaction temperature. Dielectric parameters of common organic substances typically used in industry have been determined at various temperatures in intervals of 10 °C using a conventional plate heater. The temperatures of the solutions were measured with an optical fiber thermometer. The temperature-dependent dielectric loss tan δ of these substrates at 20, 50, and 90 °C are summarized in Table 7.1. It is understood that the heating efficiency of some substances greatly changes with the increase of temperature.

Even though organic syntheses are performed using the same microwave chemical equipment, different chemical yields are likely obtained on different experimental days because of the possible influence of extraneous species (impurities) that may be present in the reaction medium. Although these species may not directly impinge on the reaction proper, they may affect the efficiency of microwave heating. For example, extraneous species such as ionic salts present in a reactor may have a significant impact on the behavior of microwaves, specifically in enhancing microwave heating efficiency [11].

The dielectric loss (ε'') in microwave heating consists of the sum of two terms (Eq. 7.2): dielectric heating (first term) and conduction heating (second term) [12]. Increasing the conductivity of the solution by addition of ions increases the heating efficiency. An important factor that also impacts the heating efficiency in microwave-assisted organic syntheses is the quantity of ionic species present in the aqueous media [13].

$$\varepsilon'' = \frac{\varepsilon_s - n^2}{1 + \omega^2\tau^2}\omega\tau + \frac{\sigma}{\omega} \tag{7.2}$$

As an illustration of the above assertions, the heating efficiency of tap water exposed to 2.45-GHz microwaves has been shown to be greater than for ultrapure water, which was attributed to the presence of various ionic impurities in tap water as evidenced by the higher electrical conductivity of 0.22 S m^{-1} for tap water *versus* 0.57×10^{-5} S m^{-1} for ultrapure water (see Table 7.2) [13]. Moreover, the heating rate of physiological saline under irradiation with the 2.45-GHz microwaves' electric field was more than twice the rate for ultrapure water. Clearly, the presence of electrolytes in microwave-assisted chemical reactions is an important factor that impinges on microwave-induced heating.

Addition of ionic species is sometimes done purposely as a method for promoting heating efficiency in microwave chemistry; however, it is important to pay attention to the quantity added. In this regard, the influence of electric field (***E***-field) and magnetic field (***H***-field) heating was examined by Horikoshi and coworkers,

Table 7.1 Temperature-dependent dielectric loss tangent (tan δ) of common organic substances at temperatures of 20, 50, and 80 °C

Organic substance	tan δ at 20 °C	Organic substance	tan δ at 50 °C	Organic substance	tan δ at 80 °C
ethyleneglycol	1.0137	1-propanol	0.8832	1-butanol	0.6943
ethanol	0.9499	2-propanol	0.8707	1-propanol	0.6237
1-propanol	0.7097	glyciline	0.8130	1-pentanol	0.5995
2-propanol	0.6660	ethanol	0.8027	α-lonone	0.5339
aminomethoxyethanol	0.6527	2-butanol	0.7089	1-hexanol	0.4998
methanol	0.5537	ethyleneglycol	0.6994	2-propanol	0.4792
β-lonone	0.5125	1-butanol	0.6835	2-butanol	0.4659
1-butanol	0.4992	aminomethoxyethanol	0.6047	ethanol	0.4450
tetrahydroxyfurfuryl alcohol	0.4608	α-lonone	0.5345	β-lonone	0.4441
glyciline	0.4442	β-lonone	0.5237	aminomethoxyethanol	0.4240
2-butanol	0.4338	1-pentanol	0.4968	1-heptanol	0.4164
ethyllactate	0.4171	methanol	0.3958	ethyleneglycol	0.3606
α-lonone	0.4071	1-hexanol	0.3773	benzylsaliciate	0.3136
o-methylphenol	0.4038	*o*-methylphenol	0.3698	benzylbenzoate	0.2902
butylbenzoate	0.3637	tetrahydroxyfurfuryl alcohol	0.3350	*o*-methylphenol	0.2692
p-methylphenol	0.3383	benzylbenzoate	0.3279	butylbenzoate	0.2204
benzylbenzoate	0.3052	1-heptanol	0.2850	2-butanone	0.1980
1-pentanol	0.3036	diethylhydroxyamine	0.2837	tetrahydroxyfurfuryl alcohol	0.1909
diethylhydroxyamine	0.3025	butylbenzoate	0.2818	p-methylphenol	0.1865
ethylbenzoate	0.2860	benzylsaliciate	0.2647	diethylhydroxyamine	0.1704
dimethylsurfoxide	0.2576	*p*-methylphenol	0.2579	ethylbenzoate	0.1549
o-dichlorobenzene	0.2451	ethyllactate	0.2479	methylbenzoate	0.1446
methylbenzoate	0.2411	acetic acid	0.2403	ethyllactate	0.1319
acetic acid	0.2409	ethylbenzoate	0.2196	*o*-dichlorobenzene	0.1289

(continued)

Table 7.1 (continued)

Organic substance	tan δ at 20 °C	Organic substance	tan δ at 50 °C	Organic substance	tan δ at 80 °C
1-hexanol	0.2325	*o*-dichlorobenzene	0.1839	dimethylsurfoxide	0.1282
1-heptanol	0.1763	dimethylsurfoxide	0.1837	bromobenzene	0.0821
benzylsaliciate	0.1666	methylbenzoate	0.1793	m-dichlorobenzene	0.0812
N,N-dimethylformamide	0.1370	bromobenzene	0.1060	octane	0.0763
m-dichlorobenzene	0.1339	m-dichlorobenzene	0.1039	propionic acid	0.0733
bromobenzene	0.1264	*N,N*-dimethylformamide	0.0971	*o*-xylene	0.0725
chlorobenzene	0.0704	nonane	0.0714	*N,N*-dimethylformamide	0.0688
2-chlorobutane	0.0654	*m*-xylene	0.0688	chlorobenzene	0.0476
butylacetate	0.0639	*p*-xylene	0.0670	butylacetate	0.0428
m-xylene	0.0498	2-chlorobutane	0.0666	propyl benzene	0.0282
p-xylene	0.0475	propionic acid	0.0650	acetonitrile	0.0128
o-xylene	0.0469	chlorobenzene	0.0628	propylacetate	0.0075
propionic acid	0.0466	*o*-xylene	0.0610	benzylamine	0.0059
1-bromopropane	0.0439	octane	0.0535	benzene	0
nonane	0.0435	butylacetate	0.0469	toluene	0
2-butanone	0.0366	1-bromopropane	0.0276	triethylamine	0
acetonitrile	0.0329	acetonitrile	0.0217	methanol	over b.p.
2-bromopropane	0.0303	2-butanone	0.0213	glyciline	over b.p.
benzylamine	0.0280	2-bromopropane	0.0189	acetic acid	over b.p.
acetone	0.0256	acetone	0.0181	2-chlorobutane	over b.p.
propylacetate	0.0256	propyl benzene	0.0181	*m*-xylene	over b.p.
octane	0.0197	benzylamine	0.0141	*p*-xylene	over b.p.
ethylacetate	0.0106	propylacetate	0.0136	1-bromopropane	over b.p.
propyl benzene	0.0095	tetrahydrofurane	0.0064	nonane	over b.p.

(continued)

Table 7.1 (continued)

Organic substance	tan δ at 20 °C	Organic substance	tan δ at 50 °C	Organic substance	tan δ at 80 °C
tetrahydrofurane	0.0057	ethylacetate	0.0045	2-bromopropane	over b.p.
benzene	0	benzene	0	acetone	over b.p.
dichloromethane	0	*n*-hexane	0	ethylacetate	over b.p.
monopropylamine	0	toluene	0	tetrahydrofurane	over b.p.
n-hexane	0	triethylamine	0	dichloromethane	over b.p.
toluene	0	dichloromethane	over b.p.	monopropylamine	over b.p.
triethylamine	0	monopropylamine	over b.p.	*n*-hexane	over b.p.

Table 7.2 Electrical conductivity (S m^{-1}) and heating rates (°C sec^{-1}) of ultrapure water, tap water, heavy water, and physiological saline solution in a quartz tube on exposure to the microwaves' electric field (***E***-field) irradiation at a microwave frequency of 2.45 GHz. Reproduced from [13]. Copyright 2012 by Wiley-VCH Verlag GmbH

Sample	Electrical conductivity/S m^{-1}	Heating rate/°C s^{-1}
Ultrapure water	0.57×10^{-5}	0.40
Tap water	0.22	0.56
Heavy water	–	0.38
Physiological saline	1.30	0.91

[14] using an ultrapure water sample whose density changed with the presence of various electrolytes (NaCl, KCl, $CaCl_2$, and $NaBF_4$). Compared to ultrapure water, the heating rate under ***E***-field irradiation with the 2.45-GHz microwaves, under otherwise identical irradiation conditions, improved significantly when NaCl was added to the water (Fig. 7.11), reaching a maximum at ca. 0.50 M in NaCl. Addition of 0.125 M NaCl enhanced the heating rate nearly 13-fold over that of pure water, whereas for the solution of 0.50 M NaCl, the heating rate enhancement was about 6-fold relative to pure water. By contrast, under ***H***-field irradiation, the heating rate showed a remarkable increase when NaCl was added to the ultrapure water sample. In general, heating of water does not occur under the microwaves' magnetic field component, yet heating aqueous electrolyte solutions is considerably enhanced. The heating rates tended to increase when increasing the quantity of NaCl. Clearly, the heating behavior of the aqueous electrolyte solutions under ***H***-field conditions is remarkably different from ***E***-field conditions with 2.45-GHz microwaves (Fig. 7.11). The relevant heating mechanisms pertaining to microwave electric and magnetic fields have been discussed by Horikoshi and coworkers [15].

With regard to the results presented in Fig. 7.11, examination of the heating behavior of the aqueous electrolyte solutions under mixed ***E***-field/***H***-field irradiation was performed by placing the samples between the positions within the waveguide for maximum ***E***-field and ***H***-field densities. The heating rates of the NaCl solutions also increased under simultaneous ***E***-field/***H***-field irradiation in a manner otherwise similar to that of ***E***-field irradiation only, albeit somewhat smaller. The heating behavior of aqueous NaCl solutions under ***E***-field/***H***-field irradiation was also examined under multimode conditions. In this case, the heating rates showed no changes regardless of whether or not the electrolyte was present. In the single-mode method, the microwaves' ***E***/***H*** fields irradiated the sample uniformly, whereas for the multimode method the ***E*** and ***H*** fields did not irradiate the sample uniformly so that any distinguishing features between ***E***-field and ***H***-field heating could not be easily discerned in the multimode method. Thus, the influence of the electrolyte's presence differed greatly under the microwave's ***E***-field/***H***-field conditions, depending on the irradiation method (single-mode versus multimode). Accordingly, Horikoshi and coworkers deduced that different heating mechanisms likely operate in the single-mode versus the multimode methods [15]. These authors also inferred that when a single-mode applicator is used in performing chemical

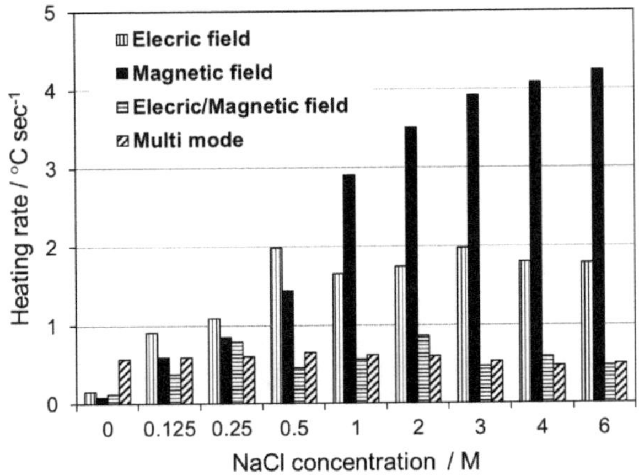

Fig. 7.11 Heating rates of an ultrapure water sample and aqueous electrolyte solutions of NaCl obtained for a 30-s irradiation period in a quartz reactor under electric field (***E***-field) and magnetic field (***H***-field) irradiation conditions using the single-mode method for 2.45-GHz microwave irradiation; the multimode irradiation method is also shown for comparison of mixed ***E***/***H***-field irradiation between the single-mode and multimode methods. Reproduced from [15]. Copyright 2012 by the International Microwave Power Institute

reactions, the heating behaviors are likely different as they depend on the electrical conductivity of the samples.

It is rather bizarre, however, that electric field heating of an electrolyte solution should be less effective than magnetic field heating. When an electrolyte is added to water, microwave heating appears to be dominated by Joule heating under ***E***-field and ***H***-field irradiation conditions. However, the results showed that ***E***-field heating is significantly different from ***H***-field heating. For example, the heating rate reached saturation at concentrations greater than 0.50 M in NaCl under ***E***-field irradiation, whereas the heating rate was significantly enhanced under ***H***-field irradiation as the concentration of the electrolytes increased (Fig. 7.11). Nonetheless, the mechanism remains Joule heating. The contrast between Joule heating by ***E***-field *versus* ***H***-field irradiation can be considered on the basis of the penetration depth (D_p; in cgs units) of the microwaves into the aqueous medium, *i.e.*, the depth at which microwaves penetrate into the sample when the power flux has fallen to 1/*e* (= 36.8%) of its original value (Sect. 5.7.4). The depth at which the 2.45-GHz microwaves penetrate into the ultrapure water sample in the reactor increased from ca. 21 to 55 mm with increases in temperature from 30 to 90 °C (Fig. 7.12). With the addition of the electrolyte NaCl at different concentrations and an increase in solution temperature, the depth of penetration of the microwaves decreased from 8.8 mm at 30 °C to 6.1 mm (0.125 M) at 90 °C (a 31% decrease); from 5.3 mm to 3.6 mm (0.50 M; a 32% decrease); from 2.7 to 1.7 mm (1.0 M; a 37% decrease); and from 1.0 to 0.8 mm (6.0 M; a 20% decrease). These penetration depths should be compared to the internal diameter of the quartz reactor: 5.0 mm. As heating of the solution progressed, the penetration depth (8.8–6.1 mm) of the microwaves into the

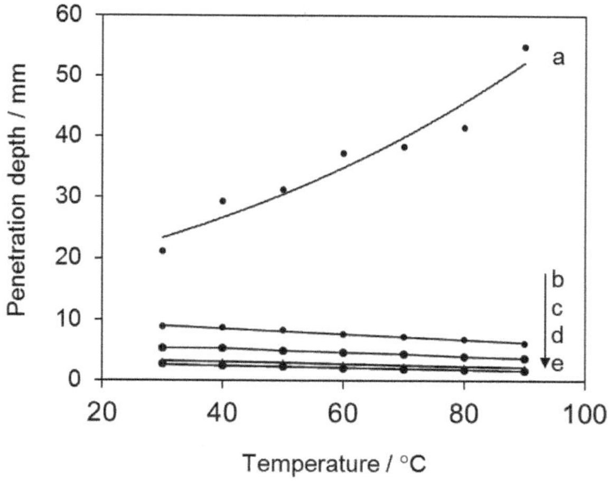

Fig. 7.12 Temperature profiles of the changes in penetration depth (mm) of the 2.45-GHz microwaves into an ultrapure water sample containing NaCl at various concentrations: **a** 0 M, **b** 0.125 M, **c** 0.25 M, **d** 0.5 M and **e** 1.0 M. Reproduced from [15]. Copyright 2012 by the International Microwave Power Institute

electrolyte solution (0.125 M) remained large compared to the quartz reactor diameter (left image of Fig. 7.13).

The optimal penetration depth for the aqueous 0.50 M NaCl solution was 5.3–3.6 mm (versus 5.0 mm of reactor size), so that the entire solution was bathed with microwaves at this concentration, which is precisely the concentration at which the maximum heating rate was observed under ***E***-field irradiation. The penetration depth of the microwaves under ***E***-field irradiation becomes shallower on increasing the electrolyte concentration; thus, heating the entire sample did not occur as demonstrated by the 6.0 M solution in Fig. 7.13. Only the aqueous solution at the interface between the quartz walls and the solution (1.0–0.8 mm) was heated at this 6.0 M concentration of electrolyte (NaCl)—see the right-hand image of Fig. 7.13. Therefore, the overall heating rate is expected to be smaller because the inner bulk solution was not heated. Such a phenomenon is reminiscent of the skin effect displayed by microwave heating of a metal.

7.5 Temperature Control by Cooling

The solution involved in a chemical reaction is heated directly under microwave irradiation. Simultaneous microwave heating and conventional cooling of a reaction can also be performed. This hybrid method presents two advantages: (a) the temperature of the solution can be adjusted by appropriate cooling under microwave heating—that is, the temperature distribution can be minimized by the hybrid conventional cooling method; and (b) only the electromagnetic wave energy is applied to the chemical reaction so that there is no need for an external thermal energy source. Such a heat/cool methodology might be particularly useful when performing organic radical and biological reactions (*e.g.*, enzyme reactions; see Sect. 9.3) under microwave irradiation.

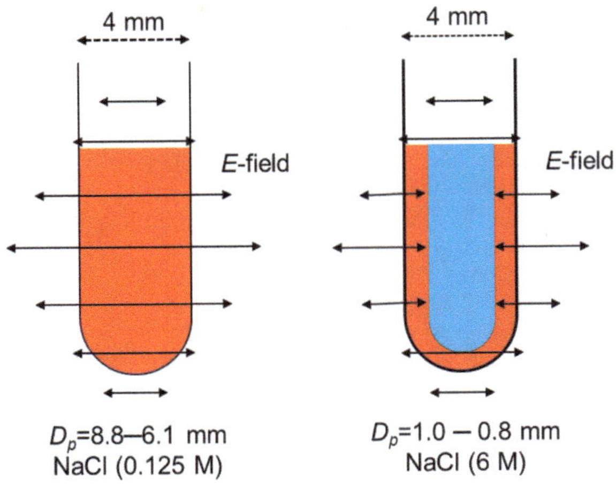

Fig. 7.13 Cartoons illustrating the heating model of aqueous electrolyte solutions (NaCl: 0.125 M and 6.0 M) under ***E***-field (electric field) irradiation; internal diameter of reactor = 4 mm. Reproduced from [15]. Copyright 2012 by International Microwave Power Institute

Several researchers have reported the simultaneous combination of cooling and microwave heating, in which care was taken to control microwave dielectric heating through a cooling system. Authors claimed that this can lead to high product yields under ambient temperature conditions (see, for example, references [16–21]. Along similar lines, the photo-assisted TiO_2 degradation of the endocrine disruptor bisphenol-A is significantly enhanced by microwave irradiation under cooling conditions and efficient magnetic stirring of the reaction mixture (temperatures measured by fiber optic probes) [22]. For the few reactions tested, combining microwave irradiation with cooling to control the reaction temperature has thus far proven advantageous, although understanding the exact effect(s) of cooling coupled to microwave heating remains somewhat elusive. This will require additional studies to establish a library of reactions that may ultimately lead to the identification of factors that lead to improved yields and shorter reaction times [20]. In this regard, the few studies that examined microwave-enhanced radical reactions ascribed improved product yields and faster kinetics solely to microwave dielectric heating effects [23–26]. To further understand microwave effects, other than dielectric heating, Horikoshi and coworkers [27] synthesized 3-cyclohexyl-1-phenyl-1-butanone (CPB) by reacting *tert*-butylhydroperoxide, triethylborane, iodocyclohexane, and crotonophenone in THF; this is a radical reaction that was enhanced by microwave dielectric heating under ambient temperature conditions. These authors proposed a simple single-modal experimental microwave reactor that involved an internal cooling system. A three-pronged microwave double cylindrical cooling reactor with a reflux condenser in the waveguide is shown in Fig. 7.14.

The single-mode reactor illustrated in Fig. 7.14 provides a reliable homogeneous electric field (wave focusing) and accurate control of temperature measured with a fiber optic probe; it makes it possible to operate under temperature profiles otherwise identical in both kinds of activation, and microwave and conventional thermal activation [28]. The coolant in the inner cold finger consisted of hexane and dry ice

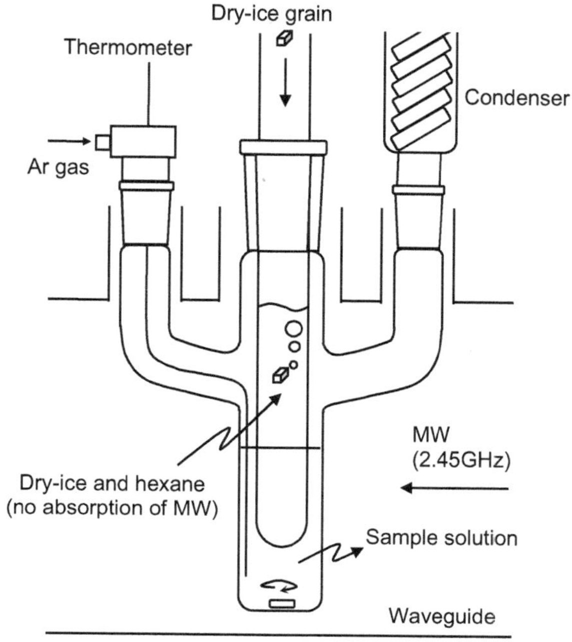

Fig. 7.14 Details of the experimental setup with a three-pronged microwave double cylindrical cooling reactor single-mode system. Reproduced from [27]. Copyright 2008 by Royal Society of Chemistry

grains with circulated cold hexane as the refrigerant. Note that hexane and dry ice grains absorb no microwave radiation; control of the addition of the dry ice achieved the desired solution temperature. Although internal heating of a reaction mixture was described by Kappe and Stadler [29] on the characterization of microwave dielectric heating, irregular temperature zones in the reacting solution due to internal heating and external cooling could not be precluded in a system such as the one displayed in Fig. 7.15a–i. However, no such irregular temperature zones were caused by an internal cooling/microwave heating system, such as the one illustrated in Fig. 7.15b–i. The advantage of internal cooling is related to preventing condensation (Fig. 7.15a–ii, b–ii) that may develop on the reactor surface by externally cooling of the sample, which can significantly attenuate the microwave radiation reaching the reaction mixture and cause a drop in reaction efficiency. For a microwave applied power of 60 W, the addition of dry ice grains easily controlled the temperature as measured by an optical fiber thermometer. Variations in temperature readings at various positions in the reactor with both probes were less than ca. ±2 °C; no significant temperature differences were noted.

The radical reaction expressed by reaction 1 was carried out by a gradual addition of a THF solution of triethylborane under an inert argon atmosphere to a vigorously magnetically stirred mixture comprised of iodocyclohexane and crotonophenone in the cylindrical reactor of Fig. 7.13, following which *tert*-butylhydroperoxide (70% aqueous solution; t-BuOOH) was added slowly, also under argon inert conditions.

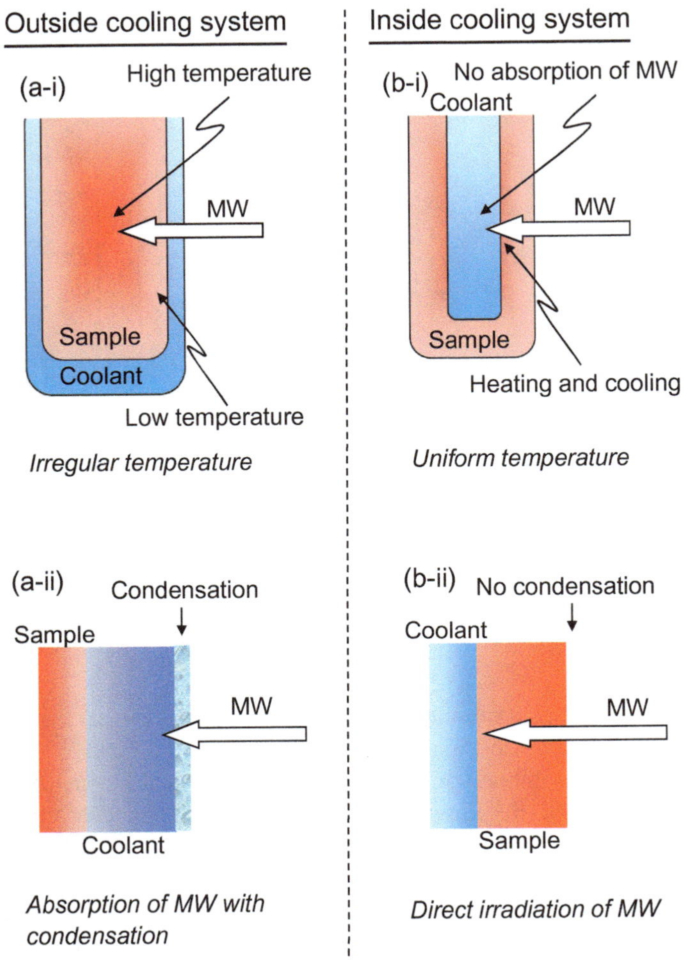

Fig. 7.15 Illustration of the possible temperature distribution and microwave irradiation conditions that might make the difference in efficiency between internal and external cooling conditions. Reproduced from [27]. Copyright 2008 by Royal Society of Chemistry

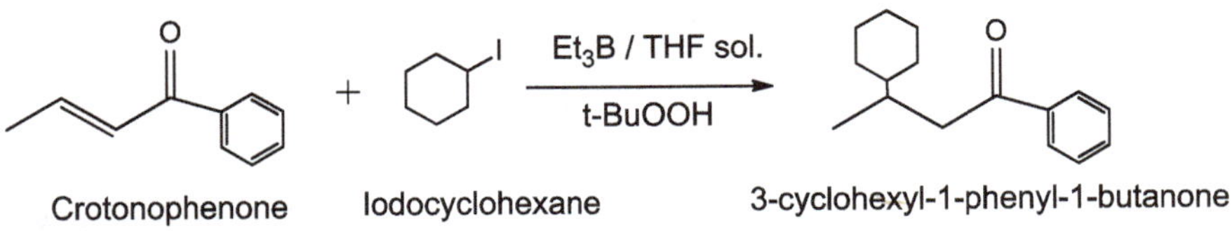

(reaction 1)

Temperature-time profiles for the sample solutions under microwave and oil bath heating, and MW/Cool protocols are reported in Fig. 7.16. An initial solution

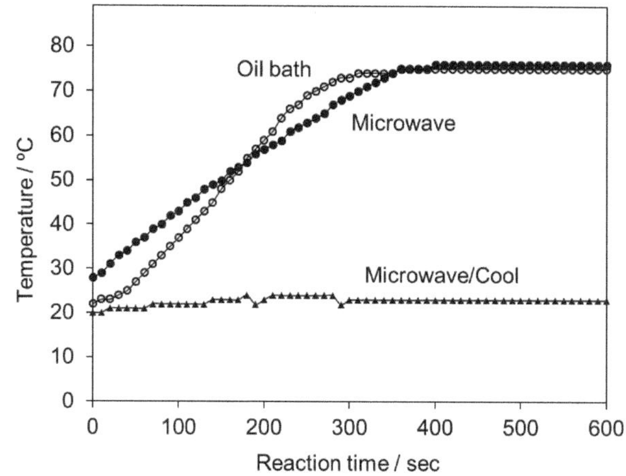

Fig. 7.16 Graph illustrating the time needed for the solution to reach maximum temperature. Reproduced from [27]. Copyright 2008 by Royal Society of Chemistry

temperature around 20–24 °C was kept relatively constant by the hexane/dry ice bath in the MW/Cool method. For the microwave protocol, the initial temperature was ca. 28 °C increasing rapidly under microwave dielectric heating to 78 °C (reflux temperature) after ca. 6 min. The corresponding profile of heating the solution by the oil bath protocol was not very different (ca. 5 min) from the microwave protocol.

The chemical yields of 3-cyclohexyl-1-phenyl-1-butanone (CPB) obtained under various experimental conditions are listed in Table 7.3. The yield of the major CPB product was nearly quantitative (93 and 95%) for the MW/Cool method, 56 and 53% for the microwave-only method, 40 and 72% for the room temperature method, and 27 and 31% for the conventional oil bath heating method after 60 and 120 min of reaction time, respectively.

Clearly, the most efficient synthesis of CPB occurred with the coupled microwave heating/internal cooling system (MW/Cool). Note the rather unusual greater yields of CPB at room temperature compared to the conventional oil bath reflux heating method. By comparison, the yield obtained under reflux by the microwave dielectric heating method (MW) was about 20–30% greater than by the conventional heating method under otherwise identical temperature conditions. In this regard, the greater yield of 53–56% from the MW method was attributed to some

Table 7.3 Chemical yields of 3-cyclohexyl-1-phenyl-1-butanone (CPB) under various experimental conditions. Reproduced from [27]. Copyright 2008 by the Royal Society of Chemistry

Method	Reaction temperature/°C	Yield/%	
		60 min	120 min
MW/Cool	20–24	93	95
MW	78 (reflux)	56	53
Room temp.	24	40	72
Oil bath	78 (reflux)	27	31

specific microwave effect, the nature of which continues to be a subject of great debate [27]. Most remarkable, however, were the greater yields of CPB by the internal cooling/MW dielectric heating of the reaction mixture (MW/Cool protocol). Moreover, cooling coupled to microwave irradiation also improved the selectivity of the radical reaction toward a single major product without losing the characteristic advantage(s) of the microwaves. The nonselectivity of the chemical reaction by either conventional or microwave dielectric heating alone could be attenuated by cooling the reaction mixture while being microwave irradiated.

7.6 Microwave Chemical Synthesis Equipment and Its Development

Since its early introduction, microwave chemical reaction equipment has caused some accidents such as the explosive destruction of a reactor. Accordingly, manufacturers of such equipment have now installed pressure and temperature sensing devices to avoid these accidents. Moreover, the malfunction of thermocouples when used to measure temperature when exposed to microwaves has now been remedied by their replacement with fiber optic thermometers. Multimode microwave chemical synthesis equipment became commercially available in the 1990s. This was followed by the availability of single-mode type microwave chemical synthesis equipment in the year 2000. A year later (2001), equipment became available combining robotic technology and automation, now widely applied in combinatorial chemistry studies. Several laboratory-scale microwave chemical synthesis equipments containing these various features are sold annually. For more than 20 years, there has been consistent equipment improvement resulting from the ideas and feedback from many researchers. For instance, special attention has been paid to improving temperature distribution in liquid samples by controlling factors such as sample stirring and the microwave mode selection in the applicator. In recent years, temperature monitoring has been further improved by direct connection to the microwave chemical synthesis equipment. For example, there is commercial equipment available that is capable of simultaneously monitoring temperature at four locations within the reaction zone (Fig. 7.17). Temperatures are measured continuously inside the solution by optical fiber thermometers, while the temperature of the reactor inner wall zone is measured by a radiant (IR) thermometer. At the same time, a pressure gauge monitors sudden pressure increases that may occur from bumping of the solution. A CCD camera may also be used to monitor the stirring or mixing conditions and the change of phase of a solution. All the data are recorded by a computer and used to confirm that a reaction is progressing properly by analyzing all these conditions.

Figure 7.18 illustrates some of the commercial laboratory-scale microwave equipment for syntheses carried out in liquid media.

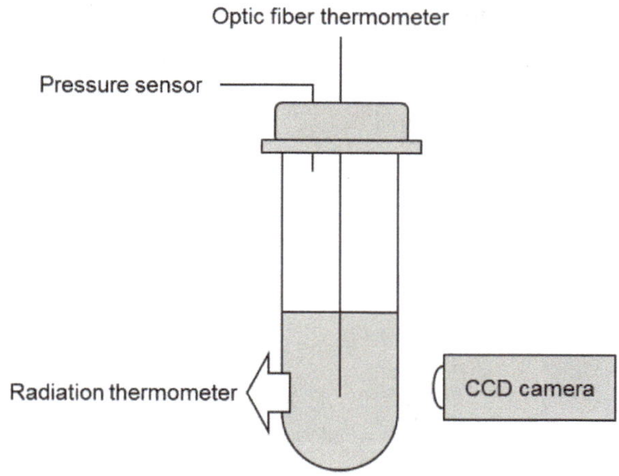

Fig. 7.17 Cartoon illustrating the means to measure pressure and temperatures, together with a solution in a microwave organic synthesis equipment. Reproduced from ref. [30]. Copyright 2013 by S. Horikoshi

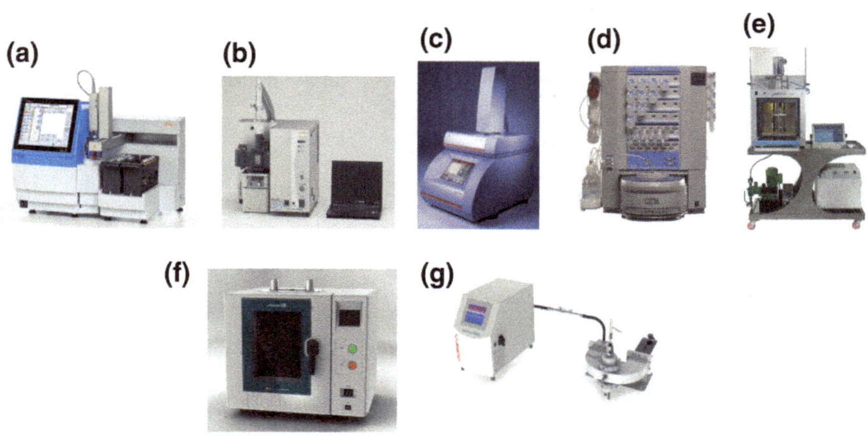

Fig. 7.18 Photographs of various commercial bench-scale microwave chemical synthesis equipment. (**a**) Initiator™ Sixty (Biotage Ltd.), (**b**) Solid phase peptide microwave system (Tokyo Rikakikai Co. Ltd.), (**c**) Monowave 300 (Anton Paar), (**d**) Discover (CEM Co.), (**e**) FlowSYNTH (Milestone General), (**f**) Shikoku Instrumentation, μ-Reactor, and (**g**) MiniFlow200SS (Sairem). All photographs are courtesy of the equipment manufacturers

Bench-scale and product-scale microwave devices have been introduced for potential industrial applications. Currently, researchers interested in pursuing microwave chemistry can choose from many equipment types to achieve the desired reaction(s). Examples of larger scale microwave chemical synthesis equipment are illustrated in Fig. 7.19.

Fig. 7.19 Photographs of various bench-scale commercial industrial microwave equipment. (**a**) Masterwave BTR with a 1 L reactor (Anton Paar), (**b**) 915 MHz chemical synthesis and extraction system (30 kW) (Sairem), (**c**) Microwave vacuum distillation drier (Shikoku Instrumentation Co.,Inc.), (**d**) Microwave polymer synthesis system (Shikoku Instrumentation Co., Inc), (**e**) High-pressure microwave chemical synthesis equipment at 6 and 2 kW power (Sairem), (**f**) Synthesis of liquid crystal material (Fuji Electronic Ind. Co.). All photographs courtesy of the equipment manufacturers

Continuous flow equipment is frequently used for industrial-scale microwave organic synthesis because of the limitations of microwave depth of penetration into many solutions. For example, the synthesis of bis(3-allyl-4-hydroxyphenyl)sulfone as a heat sensitizer was accomplished by employing industrial-scale microwave circulation flow equipment. Conventionally, it requires two steps involving classical heating for 16 h for bis(4-allyloxyphenyl)sulfone synthesis by the Claisen rearrangement reaction. Using microwave heating, the Claisen rearrangement reaction can be performed in one step in 16 min by a solvent-free technique. An image of bench-scale continuous flow synthetic equipment with the manufacturing capacities of 15 kg h^{-1} is shown in Fig. 7.20. The design of this equipment connects three reactors with tubes; the synthesis proceeds while adjusting microwave heating in each reactor. A cooling jacket surrounds each reactor; the reaction temperature is adjusted strictly by microwave heating and external cooling. Many pieces of microwave-assisted organic synthesis (MAOS) equipment are operating in industry. Many of these are custom-made and very little public information about them is available.

7.7 Coffee Break 7: Microwaves and Steam—a Unique Hybrid Cooking System

How and why is microwave energy combined with steam? Microwave ovens are essentially steamers, i.e., most foods contain water that microwave energy heats to warm the food, at the same time converting much of the water into water vapor or steam. An excellent hybrid-heating packaging system that originated in Japan combines microwave energy with the steam generated from the food. It is now also used in the United States and many other countries. This is a way of microwave cooking and reheating such foods as vegetables, pasta, rice, individually, or in combination with other items such as spices, meat, fish, chicken, etc., in a steam

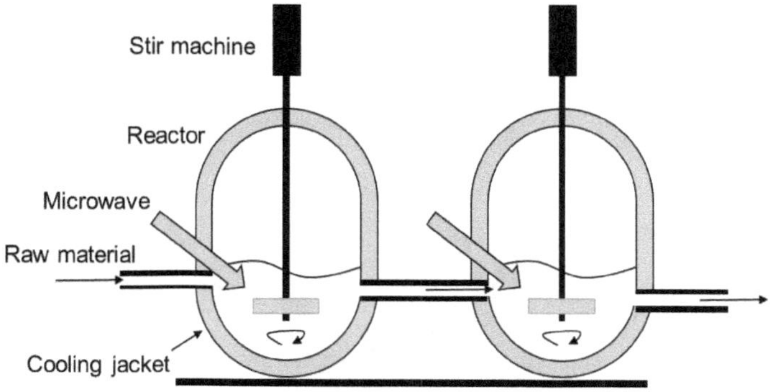

Fig. 7.20 Image of bis(3-allyl-4-hydroxyphenyl)sulfone continuous flow synthetic equipment with cooling jackets (Shikoku Instrumentation Co. Ltd.)

atmosphere, i.e., in a high moisture atmosphere; it retains the moisture in foods and produces superior eating quality at the same time. As an example, microwavable frozen pasta is very popular in Japan. The food manufacturer places the pasta and any other ingredients in a special *steamer bag* and freezes it. This flexible plastic steamer bag or pouch is usually made of some form of polyester or nylon and contains a one-way valve that opens to release excess steam when a desired pressure is reached. The combination of microwaves plus steam heats the pasta very uniformly while retaining the moisture in the pasta. The bag can be seen to inflate, but when the internal pressure reaches its maximum allowable limit, the one-way valve allows the steam to escape from the bag (Fig. 7.21). The taste and mouthfeel are usually superior to microwave-only cooking of many of these food items. Pasta, for example, is said to have a *chewier* quality and a better eating experience. In the United States, microwavable steam-packed fresh vegetables are very popular. An advantage of all these products is *portion control*, i.e., the food product can be bought in packages to feed two or three persons, even a small family. They are usually purchased and stored frozen until ready to cook. There are also some food

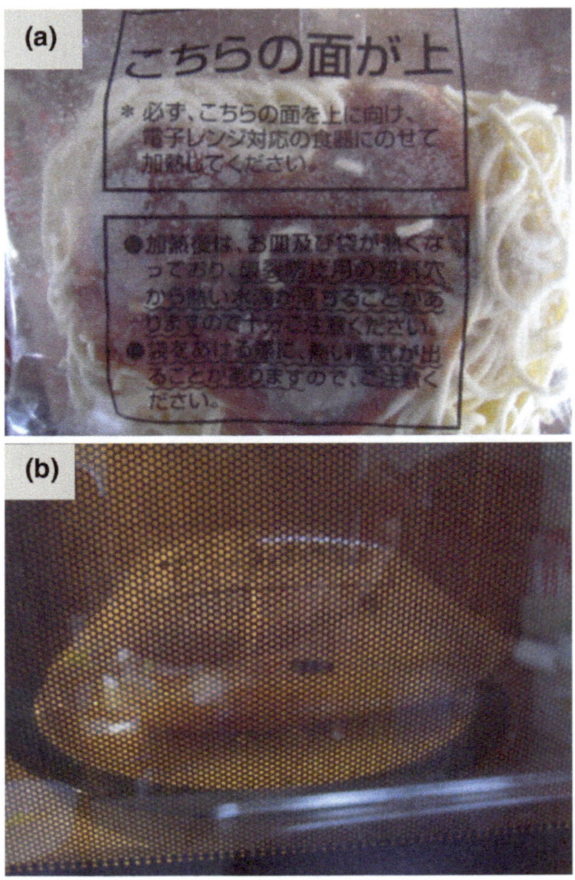

Fig. 7.21 Hybrid microwave/steam cooking using special plastic "steamer" bags. (**a**) Pasta in a steamer bag; (**b**) when microwaved the bag inflates due to steam. Reproduced by permission from S. Horikoshi. Copyright by S. Horikoshi

combinations, such as potatoes, vegetables, and meat, available in special trays overwrapped with a plastic film containing a one-way steam release valve. Despite the additional cost of steam packaging, the high quality and ease of preparation makes these products very popular.

References

1. S. Horikoshi, N. Serpone, Chapter 3, in *Microwaves in nanoparticle synthesis: fundamentals and applications*, ed. by S. Horikoshi, N. Serpone (Wiley-VCH Verlag GmbH, Weinheim, Germany, 2013)
2. http://www.jeh-center.org/mw_de.html
3. J.-S. Schanche, Microwave synthesis solutions from personal chemistry. Mol. Divers. **7**, 293–300 (2003)
4. M.A. Herrero, J.M. Kremsner, C.O. Kappe, Nonthermal microwave effects revisited: on the importance of internal temperature monitoring and agitation in microwave chemistry. J. Org. Chem. **73**, 36–47 (2008)
5. T. Sumi, R. Dillert, S. Horikoshi, A novel microwave thermodynamic model for alcohol with clustering structure in non-polar solution. J. Phys. Chem. B **119**, 14479–14485 (2015)
6. S. Horikoshi, M. Kamata, N. Serpone, Energy savings through microwave selective heating of Pd/AC catalyst particulates in a fixed-bed reactor using a vacuum-filled Dewar-like double-walled microwave continuous flow reactor. Chem. Eng. Technol. **39**, 1575–1577 (2016)
7. S. Horikoshi, A. Osawa, Y. Suttisawat, M. Abe, N. Serpone, A novel Dewar-like reactor for maintaining constant heat and enhancing product yields during microwave-assisted organic syntheses. Org. Process Res. Dev. **14**, 1453–1456 (2010)
8. http://web.mit.edu/16.unified/www/FALL/thermodynamics/notes/node123.html
9. T.L. Bergman, A.S. Lavine, F.P. Incropera, D.P. DeWitt in *Introduction to Heat Transfer*, 6th edn. (Wiley–VCH Verlag, Weinheim, USA, 2011)
10. S. Horikoshi, M. Kamata, N. Serpone, Energy savings through microwave selective heating of Pd/AC catalyst particulates in a fixed-bed reactor using a vacuum-filled Dewar-like double-walled microwave continuous flow reactor. Chem. Eng. Technol. **39**, 1575–1577 (2016)
11. H.M. Kingston, S.J. Haswell, *Microwave enhanced chemistry* (American Chemical Society, Washington DC, USA, 1997), p. 12
12. D. Stuerga, M. Delmotte, in *Microwaves in Organic Synthesis*, 2nd edn. ed. by A. Loupy (Wiley–VCH Verlag, Weinheim, Germany, 2006) Chapter 1, p. 27
13. S. Horikoshi, N. Serpone, in *Microwaves in Organic Synthesis*, 3rd edn. eds. A. de la Hoz, A. Loupy (Wiley-VCH Verlag GmbH, Weinheim, Germany, 2012)
14. S. Horikoshi, T. Sumi, N. Serpone, Unusual effect of the magnetic field component of the microwave radiation on aqueous electrolyte solutions. J. Microw. Power Electromagn. Energy **46**, 215–228 (2012)
15. S. Horikoshi, T. Sumi, N. Serpone, Unusual effect of the magnetic field component of the microwave radiation on aqueous electrolyte solutions. J. Microw. Power Electromagn. Energy **46**, 215–228 (2012)
16. K.D. Raner, C.R. Strauss, R.W. Trainor, A new microwave reactor for batch-wise organic synthesis. J. Org. Chem. **60**, 2456–2460 (1995)
17. B.K. Singh, P. Appukkuttan, S. Claerhout, V.S. Parmar, E. V. d. Eycken, Copper(II)-mediated cross-coupling of arylboronic acids and 2(1H)-pyrazinones facilitated by microwave irradiation with simultaneous cooling. Org. Lett. **8**, 1863–1866 (2006)

18. J. Kurfürstova, M. Hájek, Microwave-induced catalytic transformation of 2-tert-butylphenol at low temperatures. Res. Chem. Intermed. **30**, 673–681 (2004)
19. R.K. Arvela, N.E. Leadbeater, Suzuki coupling of aryl chlorides with phenylboronic acid in water, using microwave heating with simultaneous cooling. Org. Lett. **7**, 2101–2104 (2005)
20. S. Horikoshi, N. Ohmori, M. Kajitani, N. Serpone, Microwave-enhanced bromination of a terminal alkyne in short time at ambient temperature: Synthesis of phenylacetylene bromide. J. Photochem. Photobiol. A Chem. **189**, 374–379 (2007)
21. N.E. Leadbeater, S.J. Pillsbury, E. Shanahan, V.A. Williams, An assessment of the technique of simultaneous cooling in conjunction with microwave heating for organic synthesis. Tetrahedron **61**, 3565–3585 (2005)
22. S. Horikoshi, M. Kajitani, N. Serpone, The microwave-/photo-assisted degradation of bisphenol-A in aqueous TiO_2 dispersions revisited Re-assessment of the microwave non-thermal effect. J. Photochem. Photobiol. A Chem. **188**, 1–4 (2007)
23. M. Lamberto, D.F. Corbettb, J.D. Kilburna, Microwave assisted free radical cyclisation of alkenyl and alkynyl isocyanides with thiols. Tetrahedron Lett. **44**, 1347–1349 (2003)
24. C. Wetter, A. Studer, Microwave-assisted free radical chemistry using the persistent radical effect, Chem. Commun. 174–175 (2004)
25. C. Ericsson, L. Engman, Microwave-Assisted Group-transfer cyclization of organotellurium compounds. J. Org. Chem. **69**, 5143–5146 (2004)
26. C. Holtze, K. Tauer, Surviving radicals: promises of a microwave effect on miniemulsion polymerization for technical processes. Macromol. Rapid Commun. **28**, 428–436 (2007)
27. S. Horikoshi, J. Tsuzuki, M. Kajitani, M. Abe, N. Serpone, Microwave-enhanced radical reactions at ambient temperature. 3. Highly selective radical synthesis of 3-cyclohexyl-1-phenyl-1-butanone in a microwave double cylindrical cooled reactor. New J. Chem. **32**, 2257–2262 (2008)
28. A. Loupy, L. Perreux, M. Liagre, K. Burle, M. Moneuse, Reactivity and selectivity under microwaves in organic chemistry. Relation with medium effects and reaction mechanisms. Pure Appl. Chem. **73**, 161–166 (2001)
29. C.O. Kappe, A. Stadler, *Microwaves in Organic and Medicinal Chemistry* (Wiley-VCH, Weinhiem, 2005), pp. 9–28
30. S. Horikoshi, N. Shinohara, H. Takizawa, J. Fukushima, *Microwave Chemistry* (Sankyo Publishing Co. Ltd., 2013)

Chapter 8
Microwave Materials Processing in Solid Media

Abstract This chapter summarizes precautions and various ideas for materials processing using microwaves. With a special focus on solid samples, a method is introduced on improving microwave heating efficiency and know-how on heat dissipation to the atmosphere. Since the adiabatic method is an important factor for improving the heating efficiency and heating unevenness in the microwave heating of solids, this topic is illustrated from various angles. The latter half of the chapter introduces the microwave materials processing equipment from a laboratory scale to an industrial scale. The coffee break shows the status of the microwave food processing industry.

Keywords Microwave materials processing · Heat transfer · Conduction Convection · Radiation · Heat insulation · Isothermal adiabatic wall Microwave heating efficiency of materials · Dielectric heating · Magnetic field heating · Joule heating · Oxygen vacancies · Hybrid microwave heating Susceptor · 28 GHz frequency · Asbestos

8.1 Effective Microwave Heating in Materials Processing

Microwave heating of a solid sample is completely different from heating a liquid sample since it is not possible to equalize temperature through stirring in solids. Therefore, a large temperature gradient will be established between the solid sample and its surrounding atmosphere, so that aggressive means must be taken to prevent loss of heat to the cooler atmosphere. In addition, extra care will be required when using microwave heating to perform syntheses of solid materials and/or sintering of solids with microwaves. This chapter describes and explains a uniform heating method for solid samples from the points of view of heat transfer and heat insulation; also introduced are methods of microwave heating of metals and non-microwave heating of such materials.

Except for lasers and contrary to visible light electromagnetic waves, microwaves are coherent and polarized. Microwaves also obey the laws of optics and can

S. Horikoshi et al., *Microwave Chemical and Materials Processing*,
https://doi.org/10.1007/978-981-10-6466-1_8

be transmitted, absorbed, and reflected depending on the type of material (see Fig. 4.4 in Chap. 4). Metals are opaque to microwaves and thus are good reflectors. However, materials of interest include a broad range of ceramic dielectrics, i.e., electrically insulating materials (e.g., Al_2O_3, MgO, SiO_2, and most glasses), many of which are transparent to microwaves at ambient temperature. When heated to above a critical temperature (T_{crit}), such materials begin to absorb and couple more efficiently with microwave radiation. Other ceramic materials such as Co_2O_3, MnO_3, NiO, and CuO absorb microwaves efficiently at ambient temperatures. Absorption of microwaves by transparent ceramics is greatly enhanced by the addition of conductive or magnetic materials in the form of fibers, particles, or other additives (Fig. 4.4 in Chap. 4). In fact, conductive or magnetic additives absorb microwave energy much faster than the materials themselves and so are heated selectively and rapidly.

8.2 Useful Aspects in Carrying Out Uniform Heating

8.2.1 Heat Transfer

Generally, heating of solid samples with an electric furnace begins with a very hot external atmosphere and a very cold sample interior that is ultimately heated via heat conduction. On the other hand, the distribution of temperature is completely different in microwave heating. In liquid samples, temperature equilibration occurs by convection and by vigorously stirring the sample, whereas to the extent that there is no fluidity in solid samples, these two methods are not appropriate. Therefore, it is necessary to understand the various mechanisms of heat transfer phenomena in a solid sample undergoing microwave heating and the means with which such phenomena can be controlled. The loss of heat from microwave-heated solid samples (exothermicity) is due to three heat pathways: conduction, convection, and radiation (Fig. 8.1); their mechanisms are discussed next.

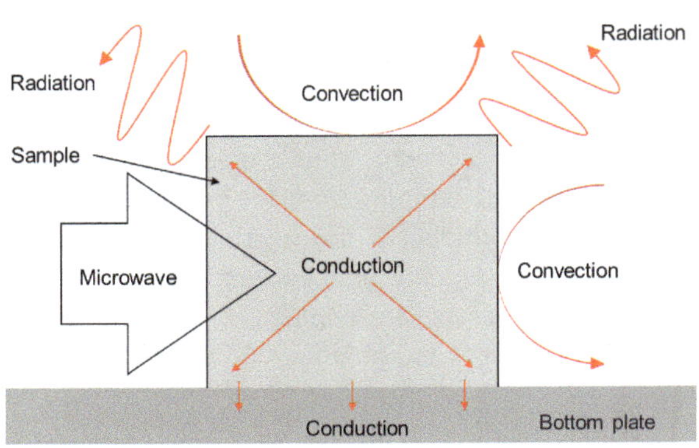

Fig. 8.1 Schematic image of heat loss (conduction, convection, and radiation) from a solid sample under microwave heating by various heat transfer mechanisms

8.2.2 *Conduction*

Even though microwaves heat unevenly **small solid samples**, uniform microwave heating can nonetheless be achieved via a combination of heat conduction and insulation by covering the sample with some insulating material. By contrast, no uniform microwave heating is possible **for large solid samples**, even if the samples were covered by an insulator since a uniform temperature distribution in the sample cannot be attained through heat conduction alone. For solid samples, it is also necessary to consider the coefficient of thermal conductivity of the samples; that is, it is necessary to establish both the heating rate of microwave heating, as well as the coefficient of thermal conductivity that describes the effectiveness that thermal conductivity will have upon the temperature distribution in the samples. Moreover, careful considerations are required whenever microwave heating is applied to a solid sample that has an unusual or complicated shape. In many of these cases, this complexity problem can be resolved by keeping the microwave heating rate very slow, which results in a slow temperature rise. This allows thermal conductivity to have a greater effect, but causes the advantage of rapid microwave heating to disappear.

8.2.3 *Convection*

Since a solid sample has no fluidity, it is not necessary to consider convection inside the sample during microwave heating. Nonetheless, heat from the sample will escape into the surrounding atmosphere by convection, thereby causing heating of the atmosphere that initially was at ambient, or at some cooler temperature—note that it is not possible to heat air directly by microwaves since air is essentially microwave-invisible. To stop the temperature of a sample to decrease by convection heat loss to the atmosphere requires that the atmosphere be heated to the same temperature as the sample. In this regard, internal heat generation by microwave heating overcomes the heat transfer limitations of conventional furnaces, and so microwave processing can be performed at much faster rates. Susceptors (materials used for their ability to absorb electromagnetic energy and convert it into heat) can accelerate microwave processing of solids by providing two-way heating with reduced heat losses from the surface of the material. Moreover, rapid initial heating with susceptors is a key factor in executing energy-efficient microwave processing of poor microwave absorbing materials. These characteristics have been applied in a variety of applications (e.g., materials processing, syntheses, and waste treatments) over the last few decades—see, for example, the review article by Bhattacharya and Basak [1].

A hybrid furnace is an industrial heating device that combines microwaves with steam and/or hot air, and can reduce the temperature gradients to a minimum by combining internal microwave heating and steam/hot air external heating. However, while this system can double the power consumption, some energy conservation is possible by employing various techniques. One way to prevent convection heat loss is by bonding the insulating material to the surface of the solid sample; that is, by maintaining good contact between the insulator and the sample's surface. This minimizes the atmosphere's effect upon the sample, and effectively stops any heat loss. Evacuating the surrounding atmosphere also reduces or prevents convection losses. It is important to note that both nitrogen and oxygen gases can exist in a plasma state upon microwave radiation in the vessel when the gas pressure reaches about 10 Pa [2]. The presence of plasma will ultimately degrade not only the sample but also the vessel.

8.2.4 Radiation

Heat escapes from the hot solid surface of a sample by radiation to a cooler solid surface (e.g., the surface of a microwave applicator and/or crucible). It is possible to calculate the radiant heat transfer by using the Stefan–Boltzmann law (Eq. 8.1). The energy of radiation I, is the product of the Stefan–Boltzmann constant (σ: 5.670367×10^{-8} W m^{-2} K^{-4}) and the fourth power of temperature. In the case of solid samples, radiant heat is the main cause of heat loss over 600 °C.

$$I = \sigma T^4 \tag{8.1}$$

8.2.5 Heat Insulation

Microwave heating of a solid sample necessitates an insulator to promote uniform heating and to conserve energy. Three properties are required for an effective insulator: (i) high melting point, (ii) low heat conduction, and (iii) low microwave absorption. For example, a magnesia carbon brick is generally used as an insulator since this magnesia carbon brick displays excellent corrosion and fire resistance. However, microwave heating of this particular insulator will occur because microwave absorption of carbon is very high. A zirconia brick is also an excellent insulator because of its high melting point and fire resistance. Unfortunately, the ion conductivity of the zirconia brick changes crystal structure under high temperature conditions, and so can develop greater microwave absorption properties. Therefore, these popular insulators cannot be used to insulate solids that are to be subjected to microwave heating. One may now correctly query what insulator materials can be used in microwave heating? A silica/alumina-based insulator has proven to be a rather suitable microwave heating material. In this case, there is almost no absorption of microwaves by the alumina fibers. Alumina fibers are white and consequently are also suitable for radiant heat insulation. When compared to other

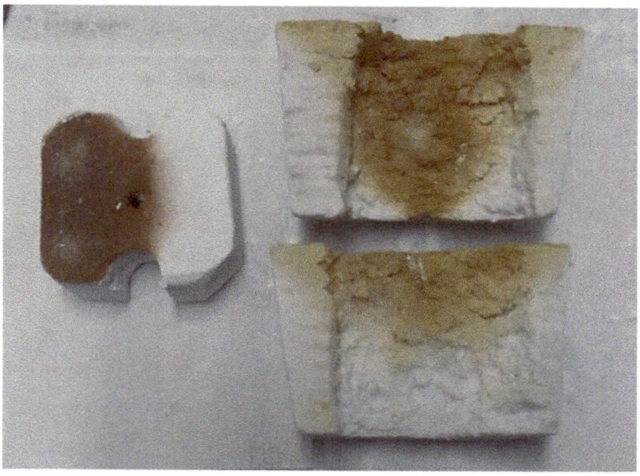

Fig. 8.2 Photograph of microwave-induced burns of impurities present in two insulator materials

insulation materials, alumina fibers are preferred because they are inexpensive and are also capable of processing. However, this material is not a perfect heat insulator. A material that at first seems to be suitable for use as a microwave insulator may, in fact, be subject to increases in microwave absorption as it becomes very hot at temperatures approaching 2000 °C. Alumina fibers fall in this category of materials since organic impurities (e.g., oil) are inherently present in alumina fibers. An insulator material that overcomes this problem is boron nitride (BN), which displays a high rate of heat conduction, but is easily oxidized under air exposure—in addition, its cost is prohibitively high. Hence, it is necessary to develop an all-round insulator for microwave heating. Figure 8.2 illustrates examples of microwave-induced burns and degradation of two insulators. Consequently, it is imperative that impurities be removed (*e.g.*, by degreasing) from a new insulator by first heating in an electric furnace.

8.2.6 Isothermal Adiabatic Wall

In the microwave sintering of ceramics, an isothermal adiabatic wall has been proposed by Sato and Takayama [3]. Such an isothermal adiabatic wall that surrounds the environment can be constructed with a substance that has the same microwave heating efficiency as the solid sample so as to prevent a temperature drop at the sample's surface via the exchange of radiant heat between the wall and the sample. Figure 8.3 shows how heat loss is prevented whenever an isothermal adiabatic wall is used in microwave heating of solids.

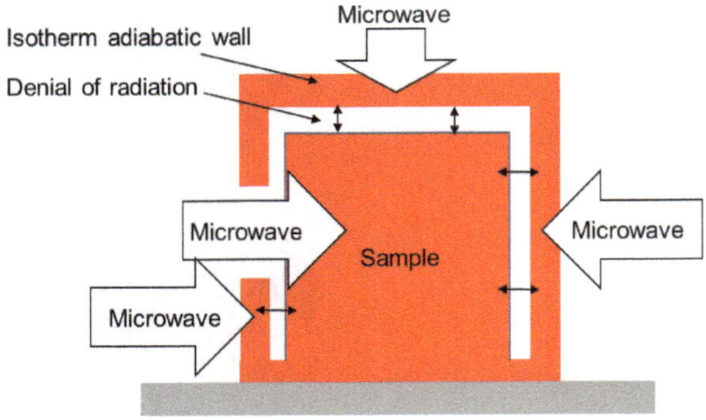

Fig. 8.3 Image of an isothermal adiabatic wall that prevents heat losses upon subjecting a solid to microwave irradiation

8.3 Purity of Samples

There are many substances that may be present as impurities in ceramics, and that affect the heating efficiency of microwaves. These impurities can lead to microwave-induced formation of hot spots that can be detrimental to solids, and so it is necessary that the solid samples be as pure as possible. Solid samples that contain high concentrations of impurities that might affect the microwave heating efficiency can still be heated, however, either by pulsed microwave irradiation or by being microwave-heated at slow heating rates.

8.4 Microwave Heating of Solid Samples

8.4.1 Heating Efficiency of Materials

Temperatures reached on microwave heating of solid materials and the heating times are summarized in Table 8.1 [4]. Solid samples that are poor microwave absorbers are SiO_2 and the alkaline earth metal oxides such as CaO; evidently, these oxides are hardly heated by microwaves. Therefore, quartz (SiO_2) is an excellent material with which to fabricate reactors and sample holders. On the other hand, microwave heating is not very efficient for such other materials as TiO_2, hematite (α-Fe_2O_3), and the metal chlorides.

The transition metal oxides NiO, CuO, and Fe_3O_4 display a very high microwave absorption ratio, and so microwave heating progresses effectively. Carbon has a high dielectric loss (ε'') in a solid sample. The relationship between dielectric loss and microwave frequency of activated carbon particles under various temperature conditions is shown in Fig. 8.4. The frequency for the largest dielectric loss is about 2 GHz, so that activated carbon is the material most suitable for 2.45-GHz microwave heating. For many solid materials, microwave heating efficiency decreases with a rise of temperature.

Table 8.1 Heating times to reach the indicated temperatures for various inorganic powders when exposed to microwave radiation

Solids with $T > 300$ °C	Temperature/°C	Heating time/min	Solids with $T < 300$ °C	Temperature/°C	Heating time/min
NiO	1305	6	$SbCl_3$	224	2
Co_2O_3	1290	3	PbO_2	182	7
C	1283	1	Pb_3O_4	122	30
CuO	701	0.5	TiO_2	122	30
V_2O_5	701	9	La_2O_3	107	30
CuCl	619	13	SnO	102	30
$ZnCl_2$	609	7	CeO_2	99	30
Al	577	6	Fe_2O_3	88	30
WO_3	532	0.5	NaCl	83	7
Fe_3O_4	510	2	CaO	83	30
$SnCl_2$	476	2	$MnCl_2$	53	2
Ni	384	1	$SnCl_4$	49	8
MnO_2	321	30	$FeCl_3$	41	4
			SiO_2	38	8

Reproduced from Ref. [4] Copyright 2013 by Horikoshi

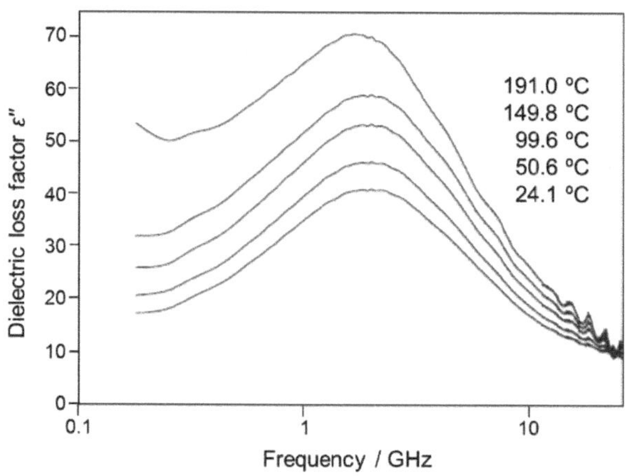

Fig. 8.4 Relationship between frequency and dielectric loss (ε'') of activated carbon under various temperature conditions. Reproduced from Ref. [5] Copyright 2004 by Springer

8.4.2 Dielectric Heating

Dielectric heating of a solid sample is different from that of a liquid sample. A dielectric solid sample typically shows some distortion of its crystal lattice that can in principle lead to microwave heating. Dielectric parameters are important factors that govern heating efficiency because molecules are highly mobile, (i.e., fluid) in liquid samples. A solid, however, is different from a liquid because atoms/

ions are held very tightly as a unit. Hence, heating of solid materials progresses only by the motion of the crystalline units, so that atomic loss and some impurities affect the microwave heating efficiency. Because the crystallinity of solid samples may be different, heating efficiencies may also be different, even though the same type of a solid substance is heated by the same microwave equipment.

The relationship between dielectric constant and dielectric loss for various inorganic substances (in powdered form) is reported in Fig. 5.4 in Chap. 5. Microwave heating efficiency is high on the upper right of the figure, which shows that carbon displays the highest microwave heating efficiency. Conversely, SiO_2 is the substance for which the heating efficiency is lowest. Note that these data are reported at room temperature and at a microwave frequency of 6.0 GHz. At present, the database of dielectric and magnetic factors has been assembled by the Japan Society of Electromagnetic Wave Energy Applications; it can be accessed via the Internet (in Japanese) [6].

8.4.3 Magnetic Field Heating

There are many solid materials with magnetic properties in which heating occurs via hysteresis loss and Eddy current loss caused by the microwaves' magnetic field. Generally, in a ferromagnetic material, magnetic domains is divided into various small zones wherein the direction of magnetization is different; these are called magnetic domains. This means that the individual magnetic moments of the atoms are aligned with one another and they point in the same direction. Hysteresis loss occurs when an alternating magnetic field is applied to a ferromagnetic material causing the direction of the magnetic field in the magnetic domains to change. Also, when microwaves are applied to a ferromagnetic material, the strength of the magnetization of a ferromagnetic material is changed. On the other hand, paramagnetic and diamagnetic materials are not subject to the microwaves' magnetic heating because there is no hysteresis loss. For example, applying a microwave magnetic field to the ferromagnetic material γ-Fe_2O_3 causes it to be heated efficiently; by contrast, α-Fe_2O_3, which is not a ferromagnetic material, but possesses the same chemical composition, is not affected by the microwaves' magnetic field, and so no heating occurs (Fig. 8.5).

Magnetic field heating is also frequency-dependent, as is dielectric heating. The frequency dependence of magnetic loss in magnetite (Fe_3O_4) particles of various particle sizes is illustrated in Fig. 8.6. Magnetic loss increases with changes in the size of the magnetite particle. For example, the biggest magnetic loss occurs at 2.59 GHz for a 38–62 μm particle, whereas at a frequency of 0.706 GHz, the biggest magnetic loss is seen for 100–180 μm particles. The behavior of magnetic field heating of powder depends on the size of the particles, which is an important factor to discuss the heating efficiency.

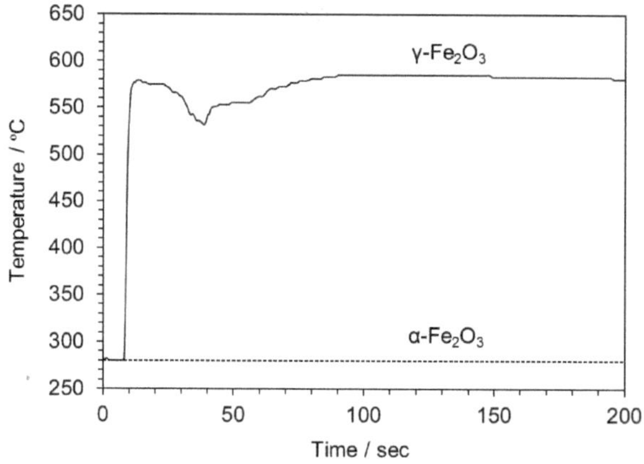

Fig. 8.5 Temperature-time profiles for α-Fe_2O_3 and γ-Fe_2O_3 powders under microwave magnetic field heating using a single-mode applicator

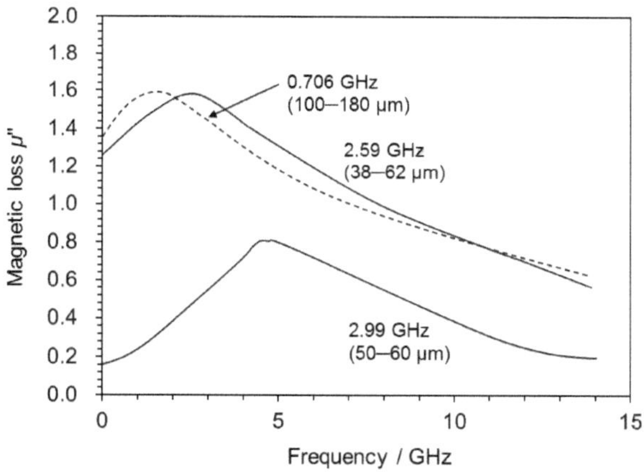

Fig. 8.6 Relationship between magnetic loss and microwave frequency for various particle sizes of a magnetite powder. Reproduced from Ref. [7] Copyright 2009 by The Iron and Steel Institute of Japan

8.4.4 Joule Heating

When a microwave magnetic field is applied to a sample, faradism (i.e., application of a faradic current of electricity) forms and heating of a sample is developed by Eddy current loss, which also occurs in metallic powders, resulting in heating. As an example, Fig. 8.7 illustrates a heating profile of a TiN powdered sample under irradiation with the microwaves' magnetic field. Evidently, TiN is a good electric conductor; however, hysteresis loss does not occur because TiN is not a magnetic material. This is evidence of magnetic field heating of TiN as a result of Joule loss heating from Eddy currents.

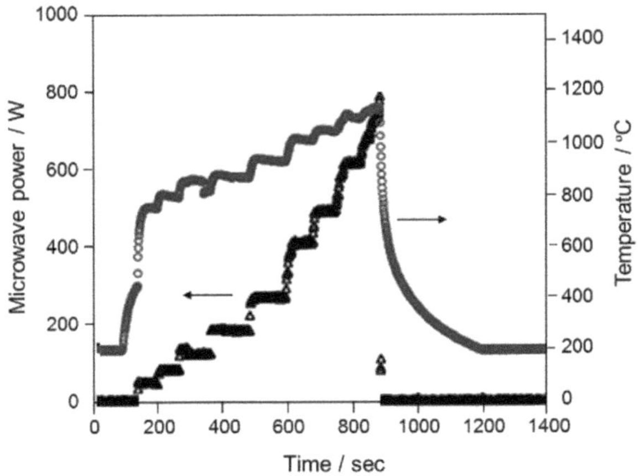

Fig. 8.7 Heating profile and microwave output power on irradiation of TiN particles with the microwaves' magnetic field. Reproduced from Ref. [8]. Copyright 2011 by Elsevier B.V

8.5 Heating of Materials Usually Unsuitable for Microwave Heating

Materials whose crystalline lattice indicates the presence of cation and/or anion vacancies are referred to as ion-defective materials. Heteroatom substitution (doping) refers to a situation wherein a regular species (atom or ion) in the lattice is substituted by a different species. Many natural crystals possess ion vacancies and/or are anion-/cation-doped, because such vacancies cause the crystals to be thermodynamically stable, more so than a perfect crystalline lattice. Even though a sample may not be suitable to be directly heated by microwaves, heating efficiency is improved by the presence of vacancy defects and dopants.

Zinc oxide (ZnO) can be heated by microwaves well especially when the sample has oxygen vacancies. A heat profile by the microwave heating (28 GHz) of ZnO is shown in Fig. 8.8 [9]. The microwave heating efficiency is very low for nearly defect-free ZnO, whereas ,for a ZnO sample which has oxygen vacancies, the heating efficiency is far greater. Such phenomenon is also observed in TiO_2 which is usually difficult to heat by microwave irradiation. THe microwave heating efficiency of $TiO_{2\text{-}x}$ particles is enhanced when oxygen vacancies are present in the TiO_2 crystal lattice [10]. Microwave heating of pure TiO_2 for a 10-min period reaches less than 200 °C, whereas $TiO_{2\text{-}x}$ can be heated over 800 °C. Clearly, oxygen vacancies in the crystal lattice of TiO2 have a great influence on the microwave heating efficiency.

The heating efficiency can also be enhanced by the addition of small amounts of some impurity substances. For example, when several mole-percent of yttrium oxide (Y_2O_3) is added to zirconium dioxide (ZrO_2), the heating efficiency by microwaves improves considerably by comparison to pure zirconia (Fig. 8.9). Note that the heating efficiency is not improved if the additive rate of the Y_2O_3 is less than 4 mol%. However, greater levels of the impurity lead to a significant enhancement of the heating efficiency, being highest for the addition of 9.4 mol%

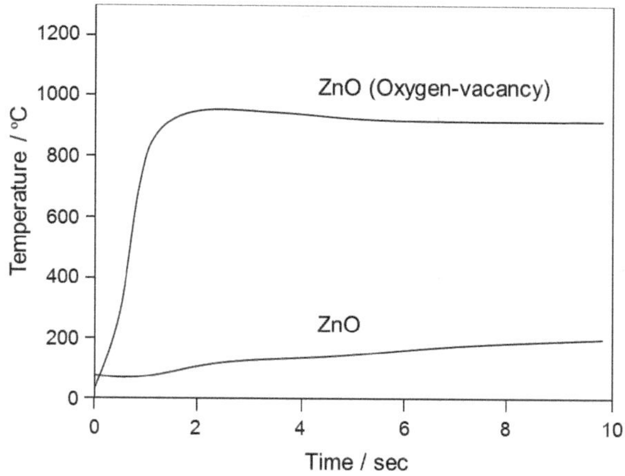

Fig. 8.8 Comparison of temperature increases of a pure ZnO powder (lower curve) and an oxygen-vacancy rich ZnO powder (upper curve) exposed to microwave heating. Reproduced from Ref. [9]. Copyright 2002 by Transactions of the Materials Research Society of Japan

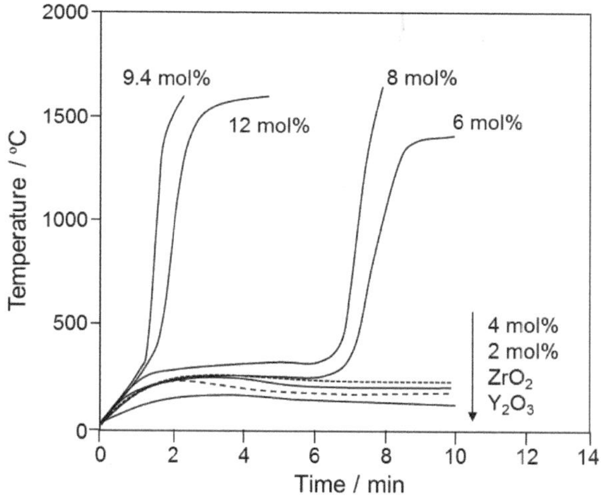

Fig. 8.9 Changes in heating behavior on addition of yttrium oxide (Y_2O_3) to a zirconium dioxide (ZrO_2) sample subjected to 28-GHz microwave irradiation. Reproduced from Ref. [11] Copyright 2001 by Iwasaki

of yttrium oxide. Interestingly, development of heating does not occur with microwave irradiation for less than 7 min following the addition of only 6–8 mol% Y_2O_3 onto the ZrO_2 sample; however, continued microwave irradiation beyond 7 min causes the temperature to rise dramatically. At even greater concentration of the yttrium oxide, the dramatic temperature rise occurs after less than 2 min of microwave irradiation.

8.6 Hybrid Microwave Heating with Susceptors

Bhattacharya and Basak summarized, in some detail, the use of susceptors in microwave processing of materials [1]. Notwithstanding a number of advantages offered by microwave heating over conventional heating, the successful application of microwave heating hinges on the careful management of some of the issues noted below. Microwave heating often leads to an uncontrolled temperature rise (or thermal runaway) arising from the continuous increase of microwave absorption as the temperature increases, via a positive feedback through dielectric loss. This uncontrolled heating not only affects product quality, but also makes the process less reproducible. Another major problem of microwave heating has to do with microwave transparent materials, which require long warm-up time to reach the effective microwave processing temperature. The long warm-up time affects the microwave processing in two ways: (a) it reduces the energy efficiency of the process and (b) introduces the additional risk of plasma formations [12], an undesirable phenomenon as it temporarily interrupts the process, in addition to inflicting abrupt changes in heating patterns. One of the routes to overcome this issue is to enhance the dielectric properties of the materials by doping the materials with secondary microwave absorbing materials or susceptors. However, this invasive technique alters the chemical composition of the original materials and, hence, restricts their application.

Alternatively, the microwave absorbing material can be incorporated as external susceptors, so that the necessary heat can be provided externally without altering the chemical composition of the material. In the review article of Bhattacharya and Basak [1], the susceptor-assisted processing referred mostly to cases of external susceptors, except for waste treatment processes. Susceptors are a class of highly lossy materials (*e.g.*, SiC and carbon, among others), which are rapidly heated by microwaves, even starting at ambient temperature. As the susceptor is heated by microwaves, the heat is transferred to the sample by conventional modes of heat transfer, which increases the temperature of the sample in a manner such that the sample becomes increasingly lossy, enabling it to interact with microwaves once the coupling temperature is reached at which point the remainder of the heating occurs via direct coupling of microwaves with the sample. The presence of a susceptor causes heating the material to occur via a combination of heating methods: (i) conventional heating during the initial phases, and (ii) microwave heating during the later stages. Hence, the susceptor-assisted microwave heating has also been referred to as ***hybrid heating*** [1]. Moreover, the material can undergo two-way heating in the presence of susceptors (Fig. 8.10).

Susceptors heat the material at the surface, while microwaves heat internally. Here, materials undergo more uniform heating compared to direct microwave heating, where the temperature at the center is higher than at the surface. In addition, in the presence of susceptors, there is reduced heat loss from the surface, which assists in maintaining temperature homogeneity during the hybrid heating process. In this regard, Lee and coworkers [13] reported a temperature difference

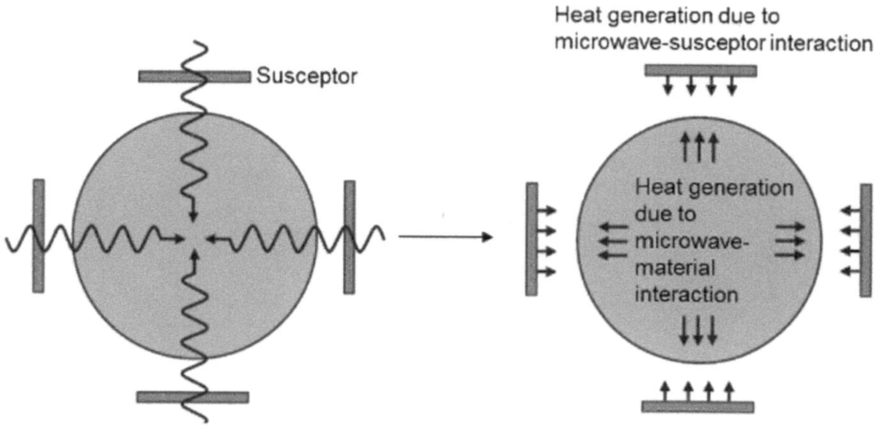

Fig. 8.10 Two-way hybrid microwave heating in the presence of susceptors. Reproduced from Ref. [1]. Copyright 2016 by Elsevier B.V

greater than 400 °C that developed between the core and the surface during the direct heating of lead zirconate titanate (PZT) ceramics with 24-GHz microwaves (Fig. 8.11a). By contrast, with the use of external susceptors (6 SiC rods), the whole sample was at the same uniform temperature (Fig. 8.11b). Hybrid microwave heating also enhances process reproducibility as heating becomes less dependent upon a material's dielectric properties.

Hybrid microwave heating has also been used successfully to perform dopant activation of arsenic-implanted Si wafers [14]. Annealing of ion-implanted wafers mounted on a SiC-coated alumina support was carried out with 2.45-GHz microwaves in an oven equipped with a 1.3 kW magnetron source (Fig. 8.12a). The microwave heating characteristics of the Si wafer, with and without the presence of a susceptor, are shown in Fig. 8.12b. However, direct microwave heating failed to activate the dopant as the wafer could not be heated beyond 100 °C. On the other

Fig. 8.11 Heating profiles of PZT samples (3 cm × 4 cm × 5 cm) during (**a**) direct (no susceptors) and (**b**) hybrid heating with 24-GHz microwaves in the presence of 6 SiC rods (the susceptors). Reproduced with permission from Ref. [13]; Copyright 2001 by Taylor & Francis

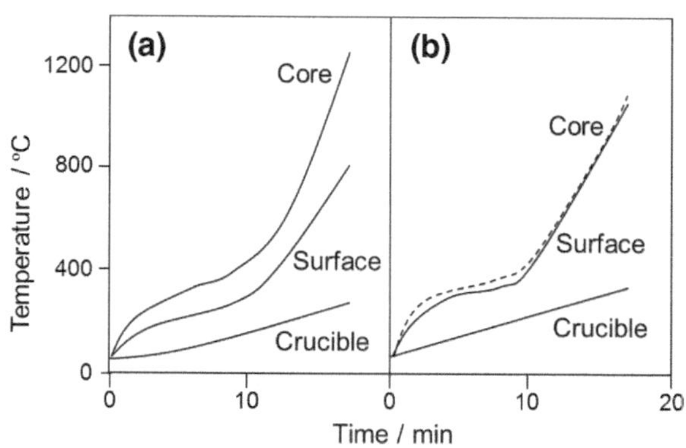

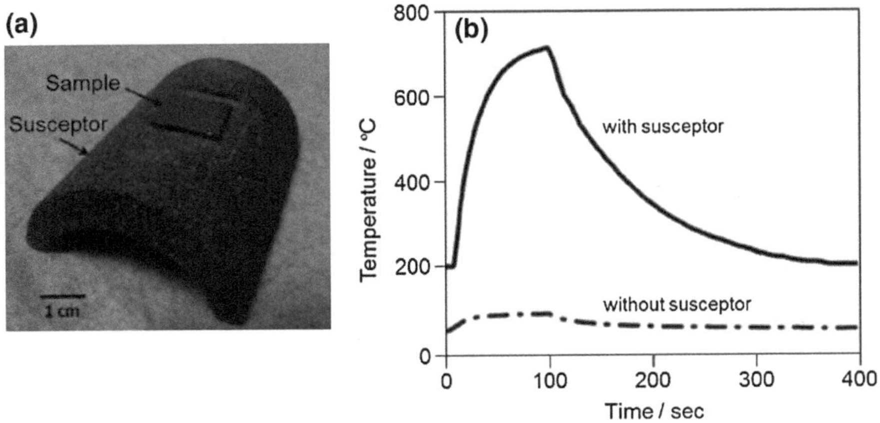

Fig. 8.12 (**a**) Sample–susceptor assembly used by Alford et al. [14] for the microwave-assisted dopant activation of arsenic-implanted silicon wafers; (**b**) surface temperature versus time profiles during direct and hybrid microwave heating of the ion-implanted wafers. Reproduced with permission from Ref. [14]; Copyright 2012 by Elsevier V. B

hand, hybrid heating with a susceptor successfully heated the wafers to temperatures greater than 600 °C; this led to near-complete dopant activation within 40 s.

Most previous studies reported a dramatic reduction of processing time (from hours or days to minutes) and improvements in energy consumption (by as much as 99%) upon the use of a susceptor for a few representative cases (Table 8.2). The processed materials were often of superior quality in terms of phase purity, uniformity of microstructure, and better mechanical/functional properties, among others. Two-way heating with the susceptors reduced the risk of crack formation, so that sintering could be performed at a much faster heating rate than direct microwave sintering. In turn, the rapid susceptor-assisted processing minimized the evaporation of volatiles and thus assisted in retaining the desired chemical composition after thermal processing. Likewise, a variety of materials involving volatiles (e.g., magnesium, lead, zinc and others) could be sintered or synthesized more easily than with other thermal processing procedures. The rapid susceptor-assisted processing also eliminated the need for a controlled environment, which reduced processing costs in many cases.

Susceptor-assisted microwave heating has resulted in various breakthrough applications in metal processing: for example, processing of bulk metals (e.g., melting, joining, or brazing) by efficient use of susceptors and masking materials that can also facilitate processing of metal powders, e.g., sintering and ignition combustion synthesis. Susceptors assist in trapping the heat within the sample during the ignition combustion synthesis of intermetallic substrates, so that a synthesis can be carried out without being terminated, even for systems with a sluggish combustion reaction. This is especially beneficial in the fabrication of functionally graded intermetallic systems containing various active and inactive layers. Moreover, the use of susceptors can extend the rapid microwave processing

Table 8.2 Representative data on the reduction of processing time and energy consumption during the hybrid microwave processing of materials

Process	Material	Microwave heating		Conventional heating	References
		Susceptor	Processing time	Processing time	
Ceramics Sintering	Al_2O_3/CeeYeZrO_2	SiC rod	2.5–3 h	>18 h	[15]
	Submicrometer alumina	SiC composite	<40 min	>11 h + cooling	[16]
	Y-TZP	2 mm thick SiC tube	1.5 h (0.4 kWh)	6 h (>4 kWh)	[17]
	Y-TZP	5 mm thick SiC tube	<1 h (heating)	>10 h (heating)	[18]
	Mullites	SiC composite	<1 h	>4.5 h + cooling	[19]
	MgO/ZrO_2-doped Mullite	SiC (box + tube)	5 h	20 h	[20]
	Porcelain	Alumina cement with up to 20 wt% ZrO_2/AlN/SiC	1 h	6 h + cooling	[21]
	Porcelain rods	SiC (casket + pieces)	20 h	120 h	[22]
	LSMO	SiC tube	<15 min	>24 h	[23]
	$La_{9.33}Si_2Ge_4O_{26}$	SiC	3.5 h	10–11 h	[24]
	Gd-CeO_2-Al_2O_3	SiC	9 h	27 h	[25]
	PZT	SiC powder	2 h (heating)	>18 h (heating)	[26]
	BZT	SiC	10 min (soaking)	24 h (soaking)	[27]
	X7R MLCC	Ferrite	2–3 h	24 h	[28]
	SBN	SiC tube	2 h	14 h	[29]
	MgCuZn/NiZn ferrites	SiC	$\leq$ 30 min (soaking)	4–6 h (soaking)	[30, 31]
	ZnO varistor	SiC	<1 h	>7 h	[32, 33]

(continued)

Table 8.2 (continued)

Process	Material	Microwave heating		Conventional heating	References
		Susceptor	Processing time	Processing time	
	ZnO bases MLV	SiC rod	<3 h	>30 h	[34]
Pyrolysis Metal Powder Sintering	Polyurethane	SiC baseplate	~2 h	>20 h	[35]
	Mg/Al	SiC powder	25 min (0.7 kWh)	170 min (17 kWh)	[36]
	Cu steel/CueSn bronze	Carbon coating	~60 min/ 2–2.5 h	≫200 min to >8 h	[37, 38]
	Tungsten alloy	SiC	25–30 min	4–6 h	[39]
	$AleAl_2O_3$ composites	SiC baseplate	1 h	10 h	[40]
Glass Sintering	SiO_2 glass	SiC grit	25 min	>10 h	[41]
	SiC whisker reinforced CaeMg-aluminosilicate glass	SiC lining	1.5 h	12 h	[42]
	Borosilicate/phosphate glass	SiC blocks	2–2.5 h	6–7 h	[43, 44]
	Metal-doped borosilicate glass	SiC baseplate	7 min	7 h	[45]
Crystallization	Silicon films	Carbon overcoat	90 s at 200 W	8–10 h	[46]
	In_2O_3	SiC baseplate	5 min at 500 °C	2 h at 500 °C	[47]
Solid-State Synthesis	$LiMn_2O_4$	CuO powder	4 × 180 s	5 days at 800 °C	[48]
	LiV_3O_8	Graphite powder	<100 min	>10 h	[49]
	$LiFePO_4/C$	Graphite powder	4–7 min	>8 h	[50]

(continued)

Table 8.2 (continued)

Process	Material	Microwave heating		Conventional heating	References
		Susceptor	Processing time	Processing time	
	$YBa_2Cu_3O_7$	SiC powder	3.5 h	1–2 days	[51]
	Eu^{2+} doped Akermanite $Ca_2MgSi_2O_7$, Ce^{3+}-substituted YAG and LuAG	Granular t powdered carbon	<1 h	>26 h	[52, 53]
	$SrAl_2O_4:Eu^{2+}/Dy^{3+}$	Activated carbon granules	30 min	5 h at 1350 °C	[54]
	$MgAl_2O_4$	SiC baseplate	60 min at 0.7 kW	>24 h	[55]
	$MgAl_2O_4$	Graphite powder	4.5 min	96 h	[56]
	$BaZr_xTi_{1-x}O_3$	SiC rod	<1.5 h	>11 h	[57]
	BMT	$MoSi_2$ rods	<100 min	>700 min	[58]

Reproduced with permission from Ref. [1]; Copyright 2016 by Elsevier V. B.

of SiO_2-based glasses and silicon-based semiconducting materials transparent to microwaves. A variety of ion-doped semiconductors can be activated simultaneously and repaired in short time through susceptor-assisted microwave heating, which facilitates the processing by suppressing the diffusion of dopants within the semiconductor. Various waste treatments have been facilitated by the use of susceptors [59]. In this regard, shortening of processing time, biofuel quality and yield patterns can be enhanced/tailored on using appropriate susceptors.

8.7 Question: Can Microwaves Heat Metals?

Generally, when an electromagnetic wave is applied to a metal substrate, that radiation is reflected. Therefore, metals cannot be heated with microwaves. However, when the metal substrate is in the form of small particulates, it can be heated with microwave radiation as demonstrated by Roy and coworkers [60]. For rubber to absorb microwave radiation, either carbon and/or metallic particles are added to rubber. Therefore, microwave heating with metallic powders was only understood empirically for a long while. It appears that microwave heating of a metallic sample proceeds via Joule loss heating. A magnetic metal is heated when subjected to the microwaves' magnetic field via formation of Eddy currents on the metal surface. When microwaves are applied to a metal, they are absorbed to a depth of several μm; this is referred to as the skin depth (Sect. 5.6.4). Hence, when metallic particle sizes are commensurate with the skin depth, it is possible to microwave-heat these particles. For example, the skin depth of copper is about 1.3 μm under 2.45-GHz microwaves, and even if only this thickness were heated by Joule heating, the whole block of copper would be heated since the heat developed in the metal skin would be distributed throughout the copper by heat conduction. The size of metal particles for the most effective heating can be calculated from an electromagnetic field simulation, which shows that the particle size could be 2.5 times the skin depth [61, 62].

When exposed to air, many metallic substrates form a thin oxide layer (ca. 100 nm) on the surface; this leads to a condition that facilitates microwave heating. An analysis of the electromagnetic field can reveal the thickness of the surface oxide on a metal substrate (e.g., iron) and the microwave electric field strength. The electric field does not penetrate into the interior of the metallic particle if it has no oxide formed on the surface. On the other hand, when the surface of the sample is covered by an oxide layer, the electric field penetrates into the interior as illustrated in Fig. 8.13. Therefore, heating of metallic particles by microwaves shows that the thickness of the surface oxidation is the important factor that influences heating efficiency.

Temperature changes occurring subsequent to microwave absorption by iron particles are reported in Fig. 8.14. When microwave heating of iron particles occurs under vacuum conditions, absorption of microwaves tends to decrease; in fact, when they are heated to 700 °C, nearly all the absorption of microwaves

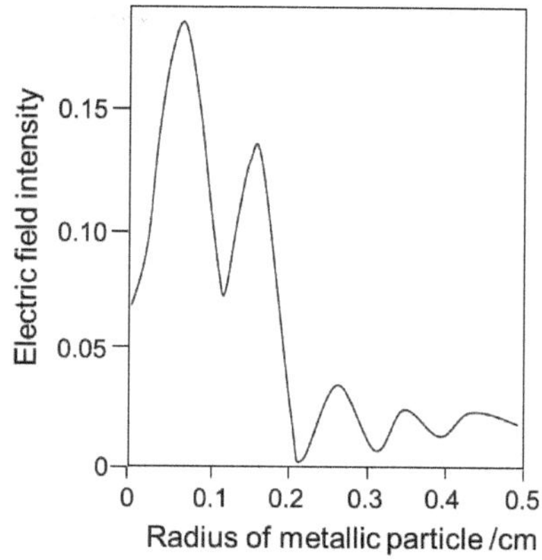

Fig. 8.13 Permeation image of an electric field on a metal substrate having a surface oxide layer. Reproduced from Ref. [63] Copyright 2006 by The Japan Society of Applied Physics

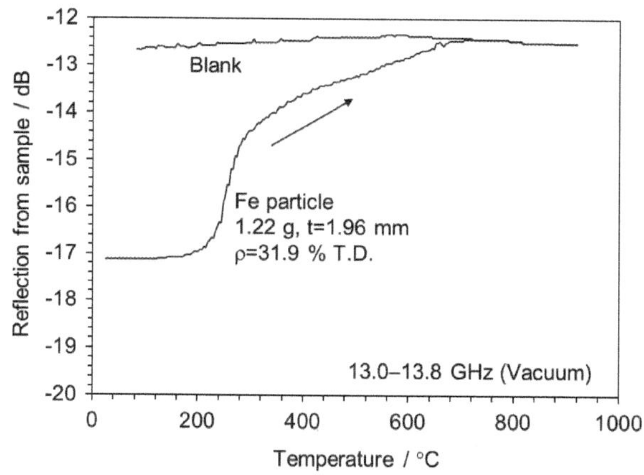

Fig. 8.14 Plots illustrating microwave absorption of iron powders as a function of temperature. Reproduced from Ref. [64]. Copyright 2015 by AGNE Gijutsu Center Inc

disappears, and even though the iron particles are cooled, absorption of microwaves does not increase. If absorption of microwaves does occur, it is because the surface of the iron particles has been oxidized.

The microwave absorption behavior of various metallic powders has been a topic of extensive and systematic research [64]. For example, when microwaves irradiated a nickel-powdered sample, 62% of the microwaves were reflected, 24% penetrated the sample, and only 14% of the microwaves were absorbed.

8.8 Microwave Sintering Equipment

Early type microwave heating devices for solid samples were used to dry ceramic powders. To reach high temperatures over 1000 °C for solid samples requires high microwave power irradiation. Sintering equipment using a klystron and a gyrotron

Fig. 8.15 Ceramic equipment of laboratory-scale using 28 GHz gyrotron generator. Photograph provided by courtesy of Professor H. Takizawa, Tohoku University, Japan

was developed for that purpose. Figure 8.15 depicts laboratory-scale ceramic sintering equipment (28 GHz; 10 kW) that uses a gyrotron generator. The temperature distribution becomes significant because the solid cannot be stirred. The shape of the applicator is also very important; the one displayed in Fig. 8.15 has a cylindrical shape. Millimeter waves uniformly irradiate the sample at the center. On the other hand, to achieve uniform microwave illumination by multi-reflection, there also exists equipment whose applicator bears an internal shape like a polygon (Fig. 8.16).

The solid samples may have various shapes, different physical/chemical properties, and no fluidity. Accordingly, the applicator was mainly developed to be used as a multimode applicator. Therefore, if a sample is to be sintered successively, the sample is passed in the applicator with a conveyor belt. On the other hand, microwave heating a solid sample or a highly viscous slurry material or a granular sample uses a multimode applicator with an extrusion screw. A multimode applicator of the revolving drum type rotary kiln has also been reported [65]. The solid sample entered the applicator on a conveyor belt, inside sample is a stir and a

Fig. 8.16 Microwave heating device for material sintering with a hexagonal shaped multimode applicator. Photograph provided by courtesy of Vötsch Industrietechnik

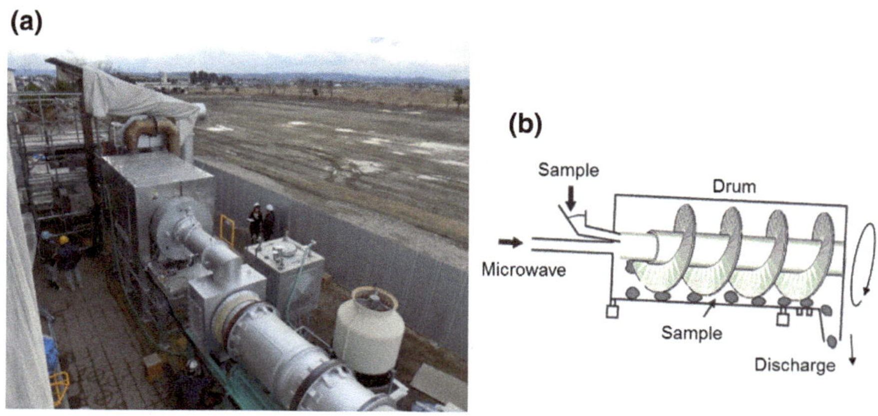

Fig. 8.17 (**a**) Picture and (**b**) image for equipment of microwave pyrolytic processing of asbestos with a revolving drum type rotary kiln multimode applicator. Reproduced from [65]. Copyright 2014 by the American Chemical Society

revolution, and microwave heating progresses. A picture of the pyrolytic processing of asbestos in the rubble is indicated on Fig. 8.17.

A component of the microwave equipment for continuous heating of films in a waveguide is sketched in Fig. 8.18; note the opened slit in the waveguide. The surface area to volume ratio for a film is significant, and so heating using a multimode applicator is inefficient. The film sample is successively passed parallel to the biggest electric field in the waveguide, and is uniformly heated by irradiation with a uniform electric field. There are many kinds of commercially available microwave heating equipment for solids; they are displayed in Fig. 8.19.

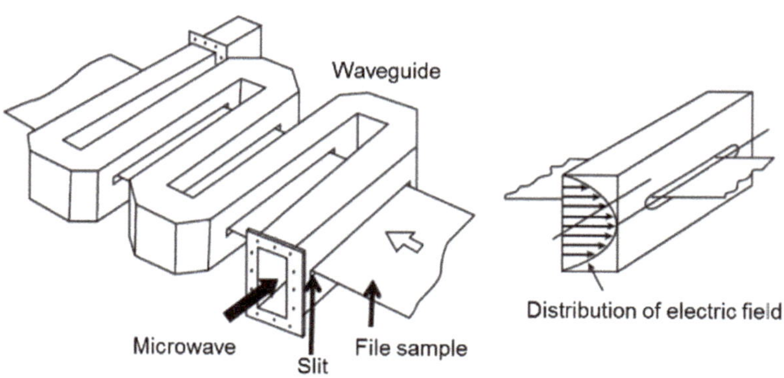

Fig. 8.18 Image of a film heating system with a microwave applicator of the waveguide type. Reproduced from Ref. [4]. Copyright 2013 by S. Horikoshi

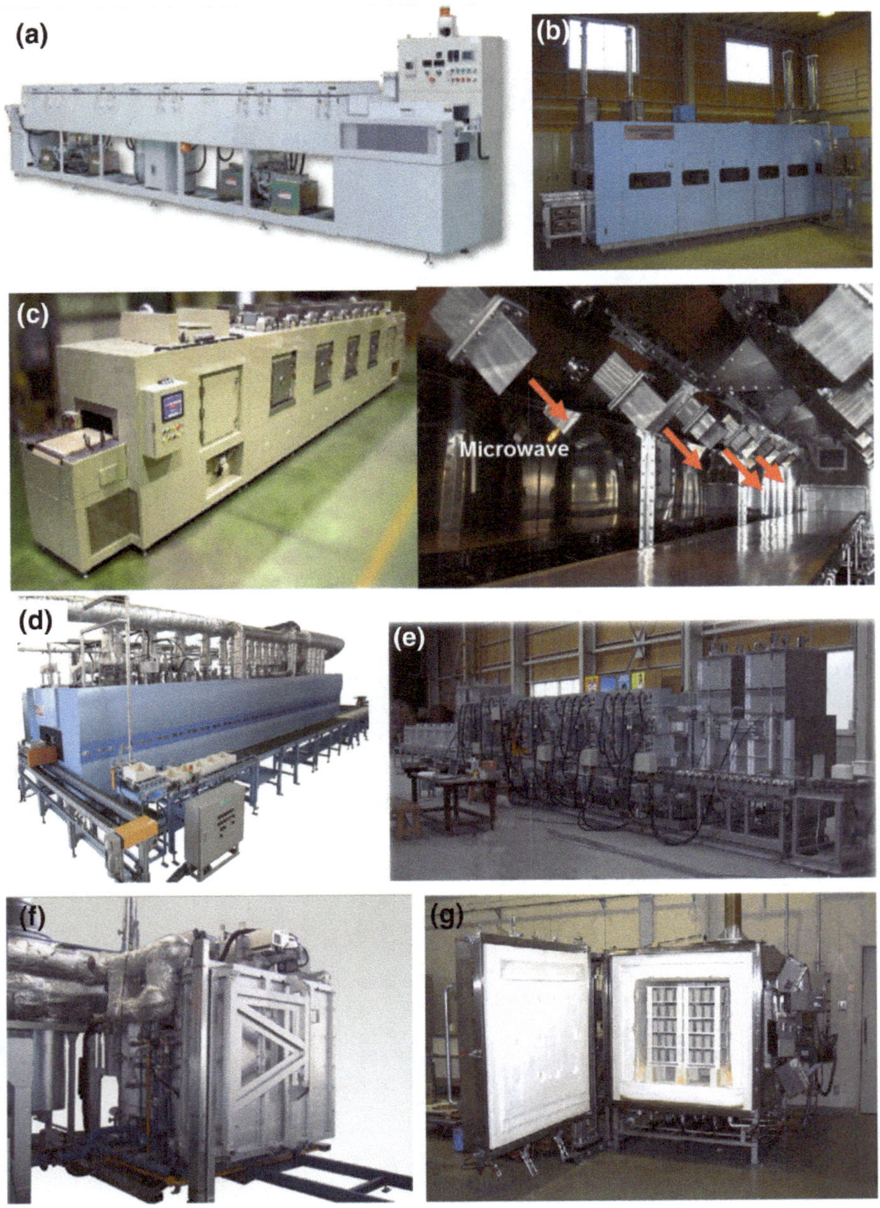

Fig. 8.19 Picture of a microwave solid sample heating equipment: Microwave solid continuous heating equipment with a conveyor belt (**a**) Micro Denshi Co., Ltd., (**b**) Mino Ceramic Co., Ltd., (**c**) Fuji Electronic Ind. Co.; Hybrid microwave/electric furnace sintering system, (**d**) Takasago Ind. Co., (**e**) Mino Ceramic Co., Ltd.; Hybrid microwave/gas sintering system, (**f**) Ceratechno Co., (**g**) Takasago Ind. Co., (**h**) solid catalyst calcination equipment at 360 kW (Fuji Electronic Ind. Co. ltd.), (**i**) large ceramics board sintering equipment (National Institute for Fusion Science, Japan), (**j**) Industrial scale device for thermal degradation of dioxins in the slag or waste (Nihon Spindle Manufacturing Co., Ltd.). All photographs were provided courtesy of the manufacturers

Fig. 8.19 (continued)

8.9 Coffee Break 8: Microwave Food Processing Industry

How is microwave energy used in the food processing industry? Within several years after the discovery of microwave heating in 1945, numerous companies investigated the potential for heating foods and other materials with microwave energy. In the beginning, most research was focused upon developing industrial microwave processes rather than microwave ovens. What intrigued these researchers was the unusual heating ability offered by microwave energy. They saw potential benefits such as those shown in the following points:

1. ***product cannot be produced by any other method***,
2. ***cost is reduced***,
3. ***quality is improved***,

4. ***yield is higher*** and
5. ***raw material can be cheaper***.

There are other criteria that influence the adoption of a new microwave processing system: e.g., reduction in floor space, reduced manpower requirements, cleaner operation and easier cleanup, and improved process efficiency and/or safety. But the major decision will be made based on return on investment (ROI)—however, can the same or better quality be achieved at less cost? What is missing from these lists is *convenience*, a major driving force behind the success of domestic microwave ovens. In 1968, a Vice President at a major manufacturer of microwave equipment (Litton Industries) predicted that domestic microwave ovens would become so popular that they would be in the homes of most consumers [66]. There was disbelief among industrialists because, apparently, *convenience* plays no significant role in industrial processing. How wrong they all were! Today, the US Dollar or Japanese Yen value of the sale of microwave ovens dwarfs that of industrial processing systems by a factor of at least 1000–1.

During the period from the 1950s through the 1970s, there were several large microwave food processing systems installed and operating, including in the USA (Table 8.3). One of these early systems is the microwave post-drying of potato chips shown in Fig. 8.20. The chips are first fried to a color slightly lighter than the desired final chip color, at which point they are too wet, so they are finish-dried with microwave energy adding the right amount of additional color while driving out all the excess water. Today, none of these systems are still operating despite all having met their goals, some producing superior products and impressive ROIs. Some of the systems shown in Table 8.3 were plagued with technical problems that were later overcome; some came too soon into the marketplace, while some were supplanted by other technologies [67].

However, several advances in microwave equipment and technology, and the common use of microwave ovens have spurred the invention and adoption of new microwave processing systems. Two food processing systems dominate in terms of numbers of units installed:

1. ***Microwave tempering of meat, fish, vegetables, and fruit:*** there are hundreds of these systems operating throughout the world. Tempering is a step in which frozen food products, such as meat, are treated to raise their temperatures to just

Table 8.3 Major commercial microwave food processing systems during the 1960s and 1970s

Process	Total number of units installed
Post-drying of potato chips	100+
Precooking of chicken parts	3
Donut proofing	24
Donut frying	6
Pasta drying	20

Reproduced with permission from Ref. [68]. Copyright 2001 by R.F. Schiffmann

Fig. 8.20 A 915-MHz microwave potato chip dryer: note 915 MHz generators and the large waveguides bringing microwave energy to the dryer-applicator. The chips are first dried to the desired color then microwave-dried with very little further color development. Reproduced with permission from Ref. [69]. Copyright 1968 by the International Microwave Power Institute

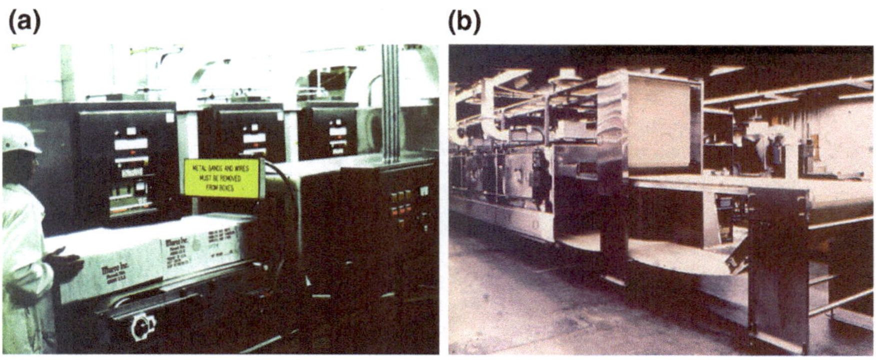

Fig. 8.21 (**a**) Microwave tempering of beef patties in overwrap carton. Note the large tunnel opening—the leakage is suppressed to safe levels. Also, note the three 915-MHz generators and waveguides. (**b**) Microwave bacon cooker looking at exit end and with the leakage suppression tunnel open in order to show two polypropylene belts: conveyor and hold-down belt to keep bacon flat; note large waveguides being fed by numerous 915-MHz generators located above the floor. Reproduced with permission; Copyright 2002 by the Raytheon Company

below the freezing temperature so they can be further processed: sliced, diced, ground, etc. The reason the temperature is maintained below the freezing temperature is to avoid thawing that would cause a dramatic increase in the dielectric properties causing the meat or other products to heat unevenly, i.e., some parts might be partially cooked while other would still be frozen. Conventional tempering can take 1–4 days and involves placing pallets of meat, or other frozen product, in a room, which may or may not be temperature-controlled to allow it to temper. Microwave energy does not thaw frozen foods evenly but it does temper extremely well and fast, in 10–20 min in many cases. It can also temper such things as hamburger patties in the shipping carton (Fig. 8.21a).

2. ***Microwave precooking of bacon*** (over 25 systems installed): bacon slices lend themselves very well to being precooked with microwave energy; this is done at 915 MHz in large tunnel-type applicators that can be 10 or more meters long and with microwave power of 500–1000 kW in a system. The grease rendered from the bacon, although not fit for human consumption, is valuable for industrial applications (Fig. 8.21b). There are or have been other highly successful microwave food processing systems such as: (i) microwave cooking of sausage patties, (ii) microwave cooking and drying of muesli, (iii) microwave drying of tea leaves, and (iv) microwave snack food drying, but these are often single or very few units.

References

1. M. Bhattacharya, T. Basak, A review on the susceptor-assisted microwave processing of materials. Energy **97**, 306–338 (2016)
2. A.D. MacDonald, *Microwave Breakdown in Gases* (Wiley, New York, 1966)
3. M. Sato, S. Takayama, Insulation Blankets of 2.45 GHz Microwave Sintering of Traditional Ceramics, *Microwaves: Theory and application in Materials Processing V (Ceramic Transactions)* , vol. 111, 2011, p. 277–285
4. S. Horikoshi, M. Shinohara, H. Takizawa, J. Fukushima, *Microwave Chemistry—Reaction, Processing and Application* (Sankyo Publishing, Japan, 2013)
5. J.E. Atwater, R.R. Wheeler Jr., Microwave permittivity and dielectric relaxation of a high surface area activated carbon. Appl. Phys. A **79**, 125–129 (2004)
6. https://permittivity.jp
7. M. Hotta, M. Hayashi, A. Nishikata, K. Nagata, Complex permittivity and permeability of SiO_2 and Fe_3O_4 powders in microwave frequency range between 0.2 and 13.5 GHz. ISIJ Intern. **49**, 1443–1448 (2009)
8. J. Fukushima, K. Kashimura, M. Sato, Chemical bond cleavage induced by electron heating - Gas emission behavior of titanium-metalloid compounds (titanium nitride and oxide) in a microwave field. Mater. Chem. Phys. **131**, 178–183 (2011)
9. H. Takizawa, M. Iwasaki, T. Kimura, A. Fujiwara, N. Haze, T. Endo, Synthesis of inorganic materials by 28 GHz microwave irradiation. Trans. Mater. Res. Soc. Jpn **27**, 51–54 (2002)
10. B. Vaidhyanathan, A.P. Singh, D.K. Agrawal, T.R. Shrout, R. Roy, Microwave effects in lead zirconium titanate synthesis: enhanced kinetics and changed mechanisms. J. Am. Ceram. Soc. **84**, 1197–1202 (2001)
11. M. Iwasaki, Microwave processing of conductive oxide ceramics, Ph. D. thesis, Tohoku University, 2001
12. C. Zhao, J. Vleugels, C. Groffils, P.J. Luypaert, O. Van der Biest, Hybrid sintering with a tubular susceptor in a cylindrical single mode microwave furnace. Acta Mater. **48**, 3795–3801 (2000)
13. W.C. Lee, K.S. Liu, I.N. Lin, Microwave sintering $Pb(Zr_{0.52}Ti_{0.48})_{03}$ piezoelectric ceramics. Ferroelectrics **262**, 293–298 (2001)
14. T.L. Alford, M.J. Gadre, R.N.P. Vemuri, N.D. Theodore, Susceptor-assisted microwave annealing for activation of arsenic dopants in silicon. Thin Solid Films **520**, 4314–4320 (2012)
15. Z. Xie, J. Yang, X. Huang, Y. Huang, Microwave processing and properties of ceramics with different dielectric loss. J. Eur. Ceram. Soc. **19**, 381–387 (1999)

16. R.R. Menezes, P.M. Souto, R.H.G.A. Kiminami, Microwave fast sintering of submicrometer alumina. Mater. Res.-Ibero.-Am. J. Mater. **13**, 345–350 (2010)
17. C. Zhao, J. Vleugels, C. Groffils, P.J. Luypaert, O. Van der Biest, Hybrid sintering with a tubular susceptor in a cylindrical single mode microwave furnace. Acta Mater. **48**, 3795–3801 (2000)
18. S. Charmond, C.P. Carry, D. Bouvard, Densification and microstructure evolution of Y-tetragonal zirconia polycrystal powder during direct and hybrid microwave sintering in a single mode cavity. J. Eur. Ceram. Soc. **30**, 1211–1221 (2010)
19. B.G. Ravi, V. Praveen, M.P. Selvam, K.J. Rao, Microwave-assisted preparation and sintering of mullite and mullite-zirconia composites from metal organics. Mater. Res. Bull. **33**, 1527–1536 (1998)
20. S. Bodhak, S. Bose, A. Bandyopadhyay, densification study and mechanical properties of microwave-sintered mullite and mullite-zirconia composites. J. Am. Ceram. Soc. **94**, 32–41 (2011)
21. R.R. Menezes, P.M. Souto, R.H.G.A. Kiminami, Microwave hybrid fast sintering of porcelain bodies. J. Mater. Process. Technol. **190**, 223–229 (2007)
22. L.N. Satapathy, Microwave assisted sintering of high voltage porcelain material and its characterization. J. Ceram. Process. Res. **10**, 637–642 (2009)
23. D. Grossin, S. Marinel, J.G. Noudem, Materials processed by indirect microwave heating in a single mode cavity. Ceram. Int. **32**, 911–915 (2006)
24. C. Alves, T. Marcelo, F.A.C. Oliveira, L.C. Alves, J. Mascarenhas, B. Trindade, On the influence of silica type on the structural integrity of dense $La_{9.33}Si_2Ge_4O_{26}$ electrolytes for SOFCs. J. Eur. Ceram. Soc. **33**, 2251–2258 (2013)
25. R. Chockalingam, S. Chockalingam, V.R.W. Amarakoon, The electrical properties of microwave sintered gadolinia doped ceria-alumina nano-composite electrolyte. J. Power Sources. **196**, 1808–1817 (2011)
26. M.V. Ramana, S.R. Kiran, N.R. Reddy, K.V.S. Kumar, V.R.K. Murthy, B.S. Murty, Investigation and characterization of $Pb(Zr_{0.52}Ti_{0.48}])O_3$ nanocrystalline ferroelectric ceramics, by conventional and microwave sintering methods. Mater. Chem. Phys. **126**, 295–300 (2011)
27. C.Y. Fang, C.A. Randal, M.T. Lanagan, D.K. Agrawal, Microwave processing of electroceramic materials and devices. J. Electroceram. **22**, 125–130 (2009)
28. Y. Fang, M.T. Lanagan, D.K. Agrawal, G.Y. Yang, C.A. Randall, T.R. Shrout, A. Henderson, M. Randall, A. Tajuddin, An investigation demonstrating the feasibility of microwave sintering of basemetal-electrode multilayer capacitors. J. Electroceram. **15**, 13–19 (2005)
29. J. Li, Y. Pu, Z. Wang, J. Dai, A comparative study of $Sr_{0.7}Ba_{0.3}Nb_2O_6$ relaxor ferroelectric ceramics prepared by conventional and microwave sintering techniques. Ceram. Int. **39**, 5069–5075 (2013)
30. C.Y. Tsay, K.S. Liu, T.F. Lin, I.N. Lin, Microwave sintering of NiCuZn ferrites and multilayer chip inductors. J. Magn. Magn. Mater. **209**, 189–192 (2000)
31. M. Yan, J. Hu, Microwave sintering of high-permeability $(Ni_{0.20}Zn_{0.60}Cu_{0.20})\ Fe_{1.98}O_4$ ferrite at low sintering temperatures. J. Magn. Magn. Mater. **305**, 171–176 (2006)
32. I.N. Lin, W.C. Lee, K.S. Liu, H.F. Cheng, M.W. Wu, On the microwave sintering technology for improving the properties of semiconducting electronic ceramics. J. Eur. Ceram. Soc. **21**, 2085–2088 (2001)
33. R.F.K. Gunnewiek, R.H.G.A. Kiminami, Effect of heating rate on microwave sintering of nanocrystalline zinc oxide. Ceram. Int. **40**, 10667–10675 (2014)
34. B. Vaidhyanathan, K. Annapoorani, J. Binner, R. Raghavendra, Microwave sintering of multilayer integrated passive devices. J. Am. Ceram. Soc. **93**, 2274–2280 (2010)
35. K. Wu, H.S. Park, M. Willert-Porada, Pyrolysis of polyurethane by microwave hybrid heating for the processing of NiCr foams. J. Mater. Process. Tech. **212**, 1481–1487 (2012)
36. W.L.E. Wong, M. Gupta, Improving overall mechanical performance of magnesium using nano-alumina reinforcement and energy efficient microwave assisted processing route. Adv. Eng. Mater. **9**, 902–909 (2007)

37. R.M. Anklekar, D.K. Agrawal, R. Roy, Microwave sintering and mechanical properties of PM copper steel. Powder Metall. **44**, 355–362 (2001)
38. G. Sethi, A. Upadhyaya, D. Agrawal, Microwave and conventional sintering of premixed and prealloyed Cu-12Sn bronze. Sci. Sinter. **35**, 49–65 (2003)
39. C.S. Zhou, J.H. Yi, S.D. Luo, Y.D. Peng, L.Y. Li, G. Chen, Effect of heating rate on the microwave sintered W-Ni-Fe heavy alloys. J. Alloy. Compd. **482**, L6–8 (2009)
40. J.H. Park, Z.S. Ahn, The production of low shrinkage porous alumina by using microwave (hybrid) heating. J. Ceram. Soc. Jpn. **103**, 211–216 (1995)
41. A. Goldstein, R. Ruginets, Y. Geffen, Microwave sintering of amorphous silica powders. J. Mater. Sci. Lett. **16**, 310–312 (1997)
42. L. Chen, C. Leonelli, T. Manfredini, C. Siligardi, Processing of a silicon-carbidewhisker-reinforced glass-ceramic composite by microwave heating. J. Am. Ceram. Soc. **80**, 3245–329 (1997)
43. A.K. Mandal, D. Agrawal, R. Sen, Preparation of homogeneous barium borosilicate glass using microwave energy. J. Non-Cryst. Solids **371–372**, 41–46 (2013)
44. A.K. Mandal, K. Biswas, K. Annapurna, C. Guha, R. Sen, Preparation of alumino-phosphate glass by microwave radiation. J. Mater. Res. **28**, 1955–1961 (2013)
45. A.R. Boccaccini, P. Veronesi, C. Leonelli, Microwave processing of glass matrix composites containing controlled isolated porosity. J. Eur. Ceram. Soc. **21**, 1073–1080 (2001)
46. S.C. Fong, H.W. Chao, T.H. Chang, H.J. Leu, I.S. Tsai, S.Y. Cheng, C.Y. Wang, T.S. Chin, Microwavecrystallization of amorphous silicon film using carbon-overcoat as susceptor. Thin Solid Films **519**, 4196–4200 (2011)
47. F.V. Motta, R.C. Lima, A.P.A. Marques, E.R. Leite, J.A. Varela, E. Longo, In_2O_3 microcrystals obtained from rapid calcination in domestic microwave oven. Mater. Res. Bull. **45**, 1703–1706 (2010)
48. M. Nakayama, K. Watanabe, H. Ikuta, Y. Uchimoto, M. Wakihara, Grain size control of $LiMn_2O_4$ cathode material using microwave synthesis method. Solid State Ion **164**, 35–42 (2003)
49. G. Yang, Y. Kong, W. Hou, Q. Yan, Heating behavior and crystal growth mechanism in microwave field. J. Phys. Chem., B **109**, 1371–1379 (2005)
50. X.F. Guo, H. Zhan, Y.H. Zhou, Rapid synthesis of $LiFePO_4$/C composite by microwave method. Solid State Ion **180**, 386–391 (2009)
51. A. Agostino, P. Benzi, M. Castiglioni, N. Rizzi, P. Volpe, $YBa_2Cu_3O_7$ synthesis using microwave heating. Supercond Sci. Tech. **17**, 685–688 (2004)
52. A. Birkel, L.E. Darago, A. Morrison, L. Lory, N.C. George, A.A. Mikhailovsky, S.B. Christina, R. Seshadria, Microwave assisted preparation of Eu^{2+}-doped Akermanite $Ca_2MgSi_2O_7$. Solid State Sci. **14**, 739–745 (2012)
53. A. Birkel, K.A. Denault, N.C. George, C.E. Doll, B. Hery, A.A. Mikhailovsky, C.S. Birkel, B.-C. Hong, R. Seshadri, Rapid microwave preparation of highly efficient Ce^{3+}-substituted garnet phosphors for solid state white lighting. Chem. Mater. **24**, 1198–1204 (2012)
54. T. Ishigaki, H. Mizushina, K. Uematsu, N. Matsushita, M. Yoshimura, K. Toda, M. Sato, Microwave synthesis technique for long phosphorescence phosphor $SrAl_2O_4$, Eu^{2+}, Dy^{3+} using carbon reduction. Mater. Sci. Eng., B **173**, 109–112 (2010)
55. I. Ganesh, B. Srinivas, R. Johnson, B.P. Saha, Y.R. Mahajan, Microwave assisted solid state reaction synthesis of $MgAl_2O_4$ spinel powders. J. Eur. Ceram. Soc. **24**, 201–207 (2004)
56. I. Gomez, M. Hernandez, J. Aguilar, M. Hinojosa, Comparative study of microwave and conventional processing of $MgAl_2O_4$-based materials. Ceram. Int. **30**, 893–900 (2004)
57. V. Vinothini, B. Vaidhyanathan, J. Binner, Microwave assisted synthesis of barium zirconium titanate nanopowders. J. Mater. Sci. **46**, 2155–2161 (2011)
58. B. Vaidhyanathan, D.K. Agrawal, T.R. Shrout, Y. Fang, Microwave synthesis and sintering of $Ba(Mg_{1/3}Ta_{2/3})O_3$. Mater. Lett. **42**, 207–211 (2000)
59. T.J. Appleton, R.I. Colder, S.W. Kingman, I.S. Lowndes, A.G. Read, Microwave technology for energy-efficient processing of waste. Appl. Energy **81**, 85–113 (2005)

60. R. Roy, D. Agrawal, J. Cheng, S. Gedevanishvlli, Full sintering of powdered-metal bodies in a microwave field. Nature **399**, 668–670 (1999)
61. M. Suzuki, M. Ignatenko, M. Yamashiro, M. Tanak, M. Sato, Numerical study of microwave heating of micrometer size metal particles. ISIJ Intern. **48**, 681–684 (2008)
62. M. Ignatenko, M. Tanaka, M. Sato, Absorption of microwave energy by spherical nonmagnetic metal particle. Jpn. J. Appl. Phys. **48**, 067001 (2009)
63. K.I. Rybakov, V.E. Semenov, S.V. Egorov, A.G. Eremeev, I.V. Plotnikov, Y.V. Bykov, Microwave heating of conductive powder materials. J. Appl. Phys. **99**, 023506 (2006)
64. S. Sano, Heating behavior of metallic powder. Kinzoku Mater. Sci. Technol. **85**, 4–9 (2015)
65. S. Horikoshi, T. Sumi, S. Ito, R. Dillert, K. Kashimura, N. Yoshikawa, M. Sato, N. Shinohara, Microwave-driven asbestos treatment and its scale-up for use after natural disasters. Environ. Sci. Technol. **48**, 6882–6890 (2014)
66. J. Ellman, Commercial microwave ovens: the market today and tomorrow. J. Microwave Power Electromagnetic Energy **3**, 81–84 (1968)
67. J. O'Meara, Why did they fail? J. Microwave Power Electromagnetic Energy **8**, 167–172 (1973)
68. R.F. Schiffmann, State of the art of microwave applications in the food industry in the USA, in *Advances in Microwave and Radiofrequency Processing: Proceedings of the 8th AMPERE Conference* (Springer, Berlin, 2001)
69. M.R. Jeppson, The evolution of industrial microwave processing in the United States. J. Microwave Power Electromagnetic Energy **3**, 29–38 (1968)

Chapter 9
Microwave-Assisted Chemistry

Abstract This chapter introduces the research situation of microwave-assisted organic synthesis (MAOS) and some characteristic reaction cases. It also explains why MAOS is a reaction suitable for Green Chemistry (no solvent or water solvent condition). The effectiveness of removal of dissolved oxygen is discussed. In addition, the advantages and features of the microwave method in polymer synthesis, enzymatic reactions, heterogeneous catalytic reactions, and photocatalytic reactions are introduced. The coffee break talks about the future of microwave processing of foods.

Keywords Microwave-assisted organic synthesis (MAOS) · Green Chemistry
Solvent-free · Water solvent · Labeling with isotope element · Dissolved oxygen
Scaling up · Polymer syntheses · Enzymatic reactions · Catalytic reactions
Hydrogen storage · Air pollution · Thermodynamics simulation
Photocatalytic reactions · Wastewater treatment · Catalyzed synthesis

9.1 Microwave-Assisted Organic Synthesis

9.1.1 Heat Sources in Organic Synthesis

Ancient *chemists* were drawn to a flame. Whether transforming dull rocks into gleaming metals, converting sand into glass, or separating mixtures into intoxicating vapors, a hot flame always played a starring role in the laboratory [1]. Well into the nineteenth century, chemists had a problem regarding heating. Furnaces and charcoal burners were fine for calcinations, but delicate experiments such as distillation required something craftier. Oil-based heaters were finicky: the sootiness of the flame rapidly blackened flasks and reduced the amount of heat the lamp could deliver. A better option was to use a spirit lamp—a squat glass bottle filled with alcohol tapered to a narrow neck that held a cotton wick. As pure ethanol burns with a luminous sooty flame, the percentage of alcohol in the spirit was crucial; adding water reduced not only the soot but also the combustion

S. Horikoshi et al., *Microwave Chemical and Materials Processing*,
https://doi.org/10.1007/978-981-10-6466-1_9

temperature; the flame could only be varied by changing the thickness of the wick. British chemists faced a further difficulty, as J.J. Griffin bemoaned in the 1830s in his **Chemical Recreations**: *In France and Germany spirit of wine is very cheap, but in England it is very dear, in (as a) consequence of the excise duty* [2]. After that, as coal gas became available and began to be used for illumination lamps, [3], many chemists used it as the heat source for their reactions. Named after its inventor, the Bunsen burner was invented in 1855 (Robert Bunsen, a German chemist who taught and performed research at the University of Heidelberg). A new laboratory at the university was designed with coal-gas lines running to it. To make use of these gas lines, Bunsen worked with the university's mechanic, Peter Desaga, to create a new and improved burner design. By the time the laboratory opened, the first 50 Bunsen burners were ready for use [4].

Later, electrical heating from the use of impedance enabled chemists to adjust the heating temperature when performing chemical reactions. It was then possible to uniformly heat a sample over a wide temperature range through a heat carrier because the very high caloric content could be maintained constant for some time. However, the use of an insulator was required for the reactants to be heated efficiently by a smaller calorie input. As a result, today's laboratories use such heat sources as oil, water, or sand baths, aluminum block heaters, or mantle heaters for performing chemical reactions. A gas burner is the only presently used heat source for the processing of glass. Electric resistance heaters can provide uniform heating to a sample over a wide range of temperatures. The use of a flame heater may result in spot (local) heating, because the distribution of temperature is large and nonuniform. Heating by electrical resistance often turns liquid and solid matter into heat carriers. However, when gas-flame heating is used as the gas heat carrier, the sample can be shaken or stirred. A far better means to provide thermal energy to reactions or processes is microwave radiation; Table 9.1 lists some of the principal features of microwave heating in organic synthesis.

9.1.2 Overview of Microwave-Assisted Organic Syntheses

Microwave radiation is a nonconventional energy source. Its popularity and its utility in synthetic organic chemistry have increased and developed considerably

Table 9.1 Features of microwave heating in organic synthesis

Evaluation	Features of microwave heating in organic synthesis
Excellent	Microwave heating has the features of both electric resistance and flame heating
Inferior	It is possible to do local heating, but control is difficult
Inferior	It is possible to do uniform heating, but control is difficult. However, it is possible with the use of a susceptor
Inferior	An insulator can be used, but it is limited

during the last two decades. Since the first published reports on the use of microwave radiation to accelerate organic chemical transformations, more than 3500 articles have been published in this field, which today is generally referred to as microwave-assisted organic synthesis (MAOS). Results of searches of publications, books, and journals by SciFinder (American Chemical Society) with the keywords *microwave* and *organic synthesis* are reported in Fig. 9.1, which also includes the synthesis of polymers. Overall sales of microwave chemical synthesis equipment have also been added to Fig. 9.1. Early research in MAOS, which had a significant impact on researchers, was reported in a USA patent by Vanderhoff [5]. The effect of dielectric heating at the liquid–liquid interface by microwaves and radio waves was discovered by researchers at the Dow Chemical Company. This phenomenon was applied to the emulsion polymerization of a vinyl monomer [5]. Emulsions of styrene, methacrylic acid, and acryl ester (among others) in the presence of emulsifying agents were polymerized under pulsed microwave and radio wave irradiation (reaction 1). Latex could be produced in high yields in short time (several minutes) compared with the more traditional heating methods. The advantage of pulse irradiation was to contain simple thermal polymerization. In addition, the product *N*-{4-[[(aminoiminomethyl)amino]sulphonyl]phenyl}acetamide (ASG) was synthesized by Halmos at the American Cyanamid Company with the use of microwaves (reaction 2) [6]. ASG is a chemical intermediate in the synthesis of sulphaguanidine useful in the treatment of gut infections; high yields were also obtained using microwave irradiation.

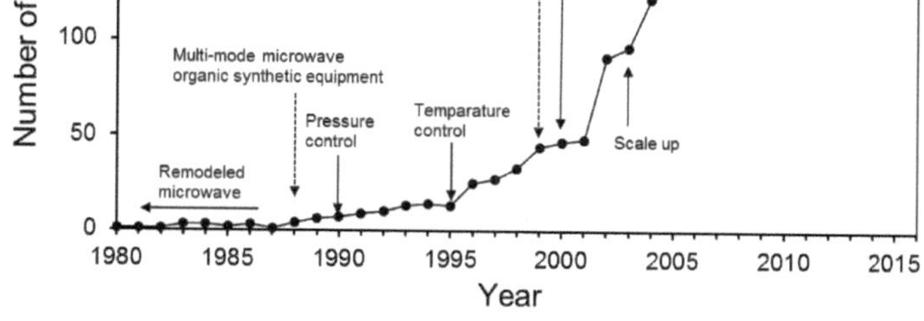

Fig. 9.1 Development of microwave organic synthesis equipment and increased comparison of the number of articles (journals and books) on keywords *microwave* and *organic synthesis* from 1980 to 2016 in SciFinder

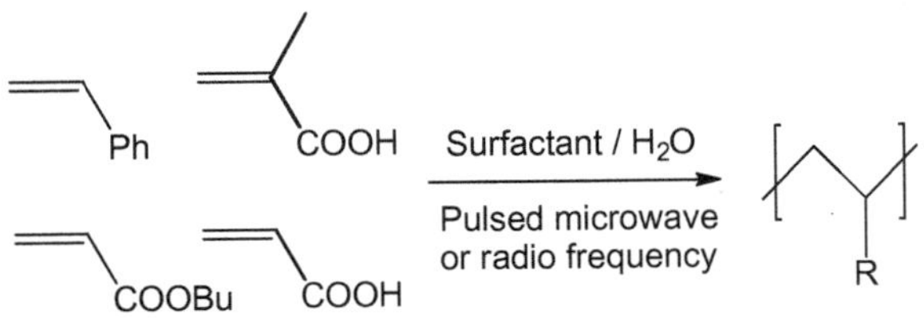

(reaction 1)

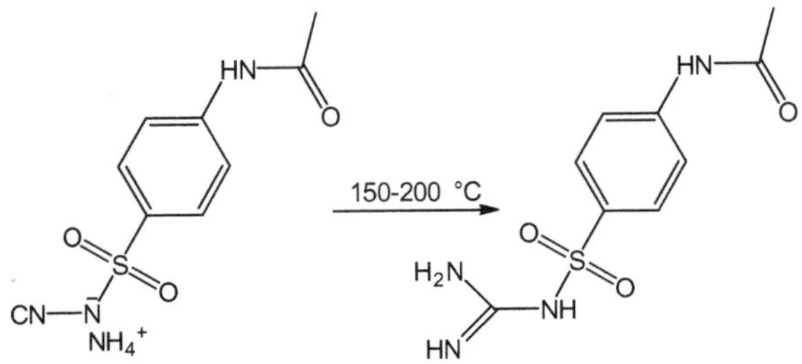

(reaction 2)

Three decades ago (1986), Gedye and Giguere independently reported two pioneering studies carried out in academic environments that made use of microwaves in organic synthesis. For instance, Gedye et al. [7] performed organic syntheses using a Teflon reactor and a commercial/domestic microwave oven; this resulted in a remarkable decrease in reaction time compared to conventional heating methods. In the other study, Giguere et al. [8] pointed out that reactor, solvent, and temperature management were important factors in microwave-assisted organic synthesis. Nonetheless, prior to these two studies, organic syntheses had been carried out that made use of microwave heating; however, problems with the required microwave equipment failed to attract researchers to use microwaves to carry out organic reactions, and so the use of microwave radiation as a potential heat source failed to proliferate. Curiously, microwave technology had been used in inorganic chemistry since the late 1970s; yet, it was only implemented in organic chemistry in the mid-1980s.

The development of microwave technology for organic chemistry has been rather slow, compared to combinatorial chemistry and computational chemistry. This slow acceptance of the technology has been principally attributed to its lack of controllability and reproducibility, safety concerns, and a generally low degree of understanding of the essentials of microwave heating of dielectric materials. The number of published research papers began to increase beginning at the end of the 1980s. Since then, multimode applicators and microwave organic synthesis equipment have become commercially available. The equipment evolved further as control of microwave power by pressure and/or temperature was added. The number of publications has increased significantly since the year 2000 as attested in the data reported in Fig. 9.1, the reason being that single-mode applicator microwave organic chemical synthesis equipment also became available, while the technology for multimode applicators increased substantially. Safety, reproducibility, simple microwave power control, sensitive temperature, and pressure controls all improved remarkably during this period. Consequently, many researchers with no knowledge of microwaves could still use this equipment as part of their tools. Currently, robotic microwave organic synthesis equipment is available and has become an active player in the ever-increasing number of available types of chemical synthesis production equipment. Moreover, microwave organic synthesis equipment with large and bench-scale reactors is being sold.

The current popularity of the microwave technique is characterized by the spectacular increase in reaction speed achieved in many reactions. Further, it has been shown that by reducing unwanted side reactions, milder reaction conditions, shorter reaction times, higher yields, and enhanced product purity can all be achieved. The combination of solvent-free conditions and microwave irradiation [9] has several advantages in the eco-friendly approach in *Green Chemistry*. The selectivity (chemo-, regio-, and stereo-selectivity) of a reaction could be modified in relation to that obtained with conventional heating. These results showed that microwave irradiation can be used as an alternative to conventional heating, as a means to obtain results different from those obtained by classical methods. The use of microwave radiation in chemistry has increased in the scientific community, and now many chemists use microwave energy as a heat source for chemical reactions on a laboratory scale [10]. Despite the relatively large body of published work in this area, however, the exact reasons why microwave radiation enhances chemical processes remain somewhat elusive.

When performing MOAS, an issue in which a researcher is mostly interested is the specific effects (not purely thermal; i.e., *nonthermal effects*) that microwave radiation might impart on reactions/processes. Results obtained by microwave exposure arise from material/wave interactions, with the claim that these interactions lead to both ***thermal*** effects and specific ***nonthermal*** effects. A combination of these two contributions is thought to be responsible for the effects observed. The main debate is the question of what actually alters the outcome of the synthesis: is it merely an effect of the thermal heat generated by the microwaves, or is it a nonthermal effect caused by the microwave radiation? Thermal effects can easily be ascertained from temperature measurements. These effects are a consequence of a different microwave heating profile, which arises from the overheating of polar solvents that may lead to the presence of ***hotspots*** particularly in heterogeneous systems, together with the selective heating of solvents, catalysts and/or reagents. In most examples, the claimed specific nonthermal microwave effects can be attributed exclusively to thermal effects, as reactions are reproducible under conventional heating when the temperature is accurately measured and controlled. Some researchers have reinvestigated previous reports of nonthermal effects employing more accurate measurement systems, and all seem to agree that specific microwave nonthermal effects were not present.

Germane to the above issue of thermal versus nonthermal effects, Schmink and Leadbeater have shown by in situ Raman spectroscopy that while microwave energy may interact with polar molecules to a greater extent than with nonpolar molecules, the conversion of electromagnetic energy into kinetic energy is slower than conversion of kinetic energy into thermal energy [11]. They concluded that localized superheating did not exist. Kappe and coworkers further described a new technique that used reactors made of silicon carbide (SiC) to separately evaluate thermal and nonthermal effects [12]. SiC absorbs microwave irradiation very efficiently and can heat the sample while blocking the electromagnetic field within the reaction mixture. In all the reactions studied, only bulk temperature effects were shown to be responsible for the observed rate enhancements, and the electromagnetic field had no direct influence on the reaction pathway. They also reported that using commercial microwave organic synthesis equipment causes temperature gradients to form in the reactor under microwave heating. This is commonsense knowledge from the viewpoint of a microwave engineer. However, the number of engineers and researchers who took notice of this point was small because microwave chemical equipment after the year 2000 had become a common tool. Many chemists were using inexpensive modified domestic microwave cooking ovens, and not the more expensive and more desirable microwave organic synthesis equipment. As a result, the measurement of temperature was often inaccurate. Hence, many research results were doubtful and researchers, rightfully so, began to question the microwave effects. As a result, today's articles in which domestic microwave ovens were used to carry out reactions are no longer accepted by several journals. For instance, guidelines from the Journal of Organic Chemistry [13] directed to the attention of submitting authors stated that *...reports of syntheses conducted in microwave reactors must indicate whether sealed or open reaction vessels were used, how the reaction temperature was monitored (external surface sensor or internal probe type), and the temperature reached or maintained in each experiment. The Journal does not publish reports of studies conducted with domestic (kitchen) microwave ovens in which yields or selectivities observed using microwave irradiation are compared with results obtained using conventional heating*. Today's chemists have come to use suitable microwave organic synthesis equipment for their studies, and microwave heating is now firmly established in organic synthesis as a result of the many repeated investigations that have used the latest microwave equipment. When considering modern technology, this is a necessary process in order for the technology to mature.

9.1.3 Microwave-Assisted Organic Synthesis (MAOS) in Green Chemistry

Considering the 12 principles of ***Green Chemistry*** reported by Anastas and Warner [14], the use of microwaves may be applicable to the principle of increased energy efficiency [15]. These 12 principles reiterated by the United States Environmental Protection Agency (EPA) [16] and reported below demonstrate the breadth of the concept of Green Chemistry.

1. **Prevent waste**: Design chemical syntheses to prevent waste. Leave no waste to treat or clean up.
2. **Maximize atom economy**: Design syntheses so that the final product contains the maximum proportion of the starting materials. Waste few or no atoms.
3. **Design less hazardous chemical syntheses**: Design syntheses to use and generate substances with little or no toxicity to either humans or the environment.
4. **Design safer chemicals and products**: Design chemical products that are fully effective yet have little or no toxicity.
5. **Use safer solvents and reaction conditions**: Avoid using solvents, separation agents, or other auxiliary chemicals. If you must use these chemicals, use safer ones.
6. **Increase energy efficiency**: Run chemical reactions at room temperature and pressure whenever possible.
7. **Use renewable feedstocks**: Use starting materials (also known as feedstocks) that are renewable rather than depletable. The source of renewable feedstocks is often agricultural products or the wastes of other processes; the source of depletable feedstocks is often fossil fuels (petroleum, natural gas, or coal) or mining operations.
8. **Avoid chemical derivatives**: Avoid using blocking or protecting groups or any temporary modifications if possible. Derivatives use additional reagents and generate waste.
9. **Use catalysts, not stoichiometric reagents**: Minimize waste by using catalytic reactions. Catalysts are effective in small amounts and can carry out a single reaction many times. They are preferable to stoichiometric reagents, which are used in excess and carry out a reaction only once.
10. **Design chemicals and products that degrade after use**: Design chemical products to break down to innocuous substances after use so that they do not accumulate in the environment.
11. **Analyze in real time to prevent pollution**: Include in-process, real-time monitoring and control during syntheses to minimize or eliminate the formation of byproducts.
12. **Minimize the potential for accidents**: Design chemicals and their physical forms (solid, liquid, or gas) to minimize the potential for chemical accidents including explosions, fires, and releases to the environment.

In this regard then, microwave-assisted organic synthesis is one of the most powerful and sustainable tools in synthetic chemistry. It has attracted a great deal of attention due to its specific features that include efficient atom utilization, improved temperature regulation, reaction homogeneity, and possible modifications of activation parameters. Varma and coworkers have noted that the use of water as a solvent in MAOS has evolved into a rapid and greener alternative method when compared to conventional heating because of the slow rate of conventional methods of organic synthesis to meet the demand for expeditious generation of compounds [9, 17].

As an alternative solvent, water is attractive from both an economic and environmental point of view. Organic synthesis is conventionally performed by heating conductively with an external heat source (e.g., a classical oil bath, which is a comparatively slow and inefficient method for transferring energy) dictated by the thermal conductivity of the various materials involved, which may result in the temperature of the reaction vessel being higher than that of the reaction mixture [18]. By comparison, microwave radiation produces efficient in-core volumetric heating by direct coupling of the microwave energy with the molecules (solvents, reagents, and catalysts) present in the reaction mixture [19]. The basic underlying principle behind the heating in a microwave oven resides in the interaction of the polar entities of the reaction materials with the electromagnetic waves. The phenomena of generating heat by electromagnetic radiation are either by collision or conduction, and sometimes both [20].

The efficiency of microwave flash heating has resulted in dramatic reductions in reaction times from days and hours to minutes, and is potentially important for the expedient production of fine chemicals [21]. The short reaction times and expanded breadth of reactions offered by MAOS are well suited to the increased demand of faster synthesis in the industry [see, e.g., 22]. Since the first report in 1986 [7, 8], microwave-assisted chemistry has fulfilled the promise of a fast synthesis technique, and various microwave equipment have become indispensable tools in modern organic synthesis (Fig. 9.2) [19].

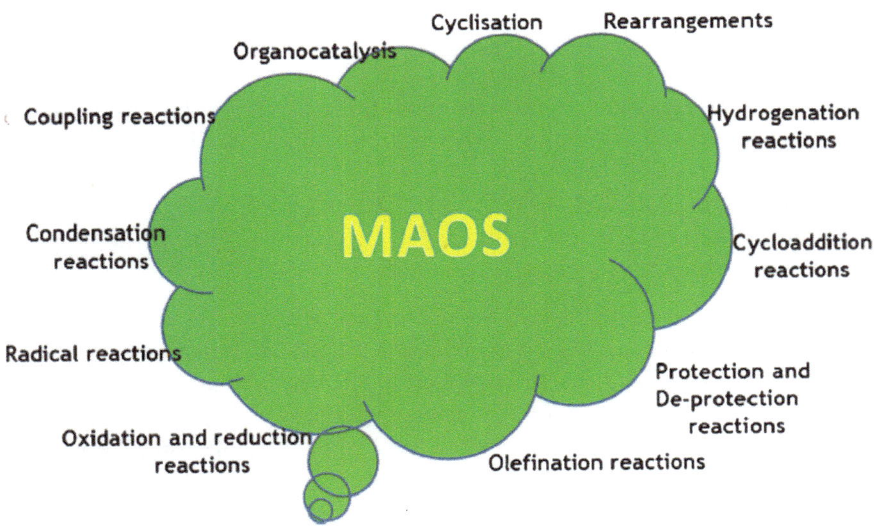

Fig. 9.2 Application of MAOS in Green Chemistry. Reprinted with permission from Ref. [9]. Copyright 2015 by Elsevier V.B.

9.1.4 Solvent-Free Microwave-Assisted Organic Syntheses

Chemical synthesis in the absence of organic solvents is a hopeful and attractive objective in Green Chemistry. The economy of atoms, [23] takes place and the occurrence of volatile organic compounds (VOC) is reduced, if not totally eliminated. Although microwave radiation is a safe source of heating, uncontrolled reaction conditions may occur that may involve volatile reactants and/or solvents at high pressure and may result in undesirable effects. This problem has been addressed and the resulting organic syntheses carried out using open-vessel solvent-free microwave conditions have become more sustainable processes [24]. The absence of organic solvents in reactions leads to a clean, efficient and economical technology; safety is increased significantly, preparation is simplified, costs are reduced, larger amounts of reactants can be employed, the reactivity is enhanced, and in some cases, the selectivity is modified without dilution. In summary, the absence of solvents, in conjunction with the high yields and short reaction times, which are characteristic of microwave-assisted processes, makes these procedures very attractive for sustainable syntheses.

Over two decades ago (1993), Loupy et al. [25] reported that potassium acetate could be alkylated in the absence of a solvent in a domestic microwave oven using equivalent amounts of salt and an alkylating agent in the presence of Aliquat 336 (10% mol) (reaction 3). Yields were practically quantitative within 1–2 min of microwave heating, regardless of the chain length, the nature of the halide leaving group, and the scale of the operation.

$$CH_3COO^- K^+ + R\text{-}X \xrightarrow[\text{Microwave}]{\text{Aliquat 336}} CH_3COOR + K^+ X^- \qquad \text{(reaction 3)}$$

Quinolines are known not only for their important biological activities but also for the formation of conjugated molecules and polymers, combining enhanced electronic or nonlinear optical properties with good mechanical properties. Song et al. [26] described the preparation of a mini-library of 12 quinoline derivatives by the Friedlander coupling condensation reaction between an acetophenone and a 2-aminoacetophenone in the presence of diphenylphosphate (0.1–0.5 equiv.), a reaction that occurred within 4 min under microwave irradiation in the absence of solvent (reaction 4). This procedure resulted in product yields of up to 85%, whereas the yield obtained with conventional heating under similar conditions did not exceed 24%.

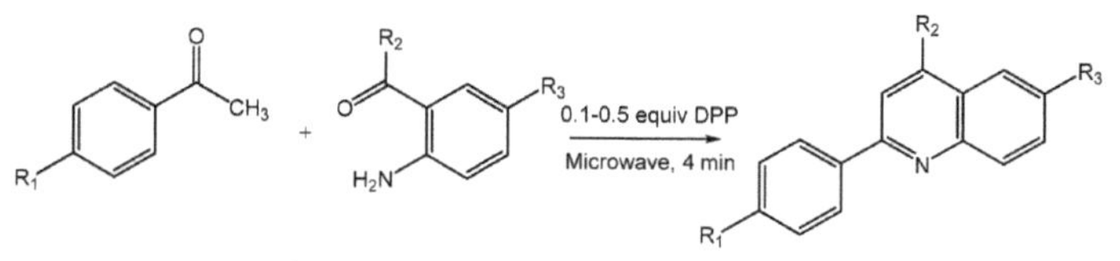

(reaction 4)

The microwave-promoted solvent-free oxidation of α-methylene ketones to 1,2-diketones involves the reaction of α-(4-nitrophenylsulfonyloxy) ketone intermediates with pyridine-1-oxide. α-(4-Nitrophenylsulfonyloxy) ketones are produced by reaction of ketones with [hydroxy(4-nitrophenylsulfonyloxy)iodo] benzene (HNIB) under pulsed microwave irradiation (3 times for 30 s ON, and 10 s OFF), that then undergo substitution reactions with pyridine-1-oxide to give the 1,2-diketones (at 30 s microwave irradiation) in fair to good yields (reaction 5) [27]. Microwaves are not applied continuously in this aforementioned experiment, but rather are pulsed. The reason is that the synthesis problem that exists under solvent-free conditions is the highly viscous raw materials used. If a solvent were used, the viscosity could easily be lowered; however, stirring of the sample is hindered under solvent-free conditions. Consequently, under microwave heating, a nonuniform temperature distribution forms in the sample, so microwave irradiation is stopped, the sample is mixed or stirred, and then microwave irradiation is resumed.

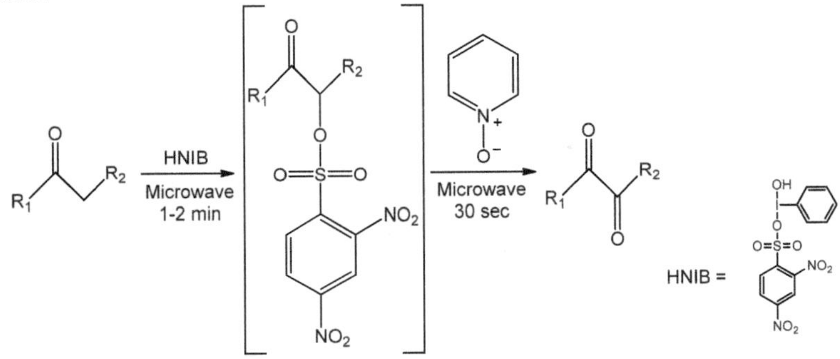

(reaction 5)

9.1.5 Water Solvent System

Organic syntheses carried out in aqueous media are as important as when performed under solvent-free conditions. Water is considered as a paradigm green solvent. It is readily available and is nontoxic and nonflammable. Unfortunately, the use of water in organic synthesis has been hampered by the low solubility of many organic compounds. Further, there are other problems regarding the use of water as a solvent in microwave organic synthesis: water has a high tanδ at room temperature and so is heated rapidly under microwave irradiation, and as the water becomes hotter, the rate of temperature rise becomes slower. This is because the tanδ of water decreases with temperature increase. To correct this problem, salts are added to the sample; for example, the synthesis of indole derivatives with only water as the solvent is performed via the intramolecular cyclization of 2-alkynyl aniline (reaction 6) [28]. With water as the solvent, the sample would not reach the required high temperatures under microwave heating. Consequently, the addition of either KCl or $NaHCO_3$ to the sample in the aqueous medium results in the sample temperature to easily reach 200 °C for the reaction to occur.

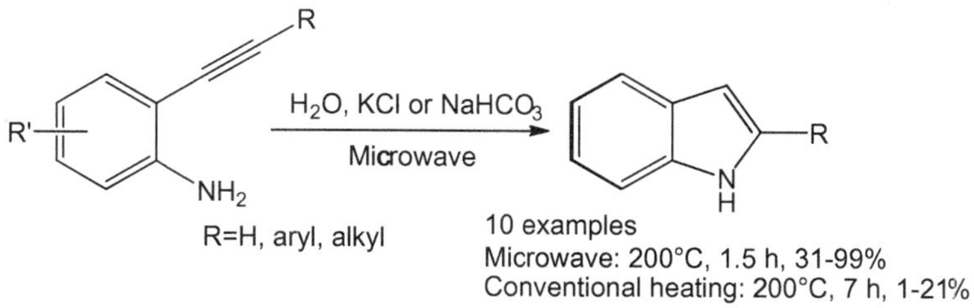

(reaction 6)

The Suzuki coupling reactions in water occur under microwave irradiation (reaction 7). The addition of tetrabutylammonium bromide (TBAB) as a phase transfer agent facilitates the reaction because it enhances the solubility of the organic substrate in water, and increases the rate of the coupling reaction through the formation of a complex with the boronate system; the reaction could be performed without the use of a palladium catalyst [29]. In a second paper, they [30] reported that ultralow levels of palladium (50 ppb), found in the sodium carbonate base, were responsible for the reaction. In all cases, on using microwave irradiation, a clear reduction in the reaction time was observed (from 120 to 5 min). In situ Raman detection showed that the reaction had reached completion after 135 s [31].

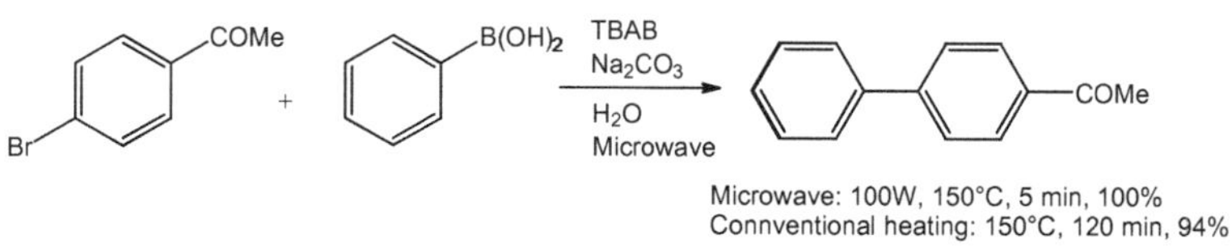

(reaction 7)

The Tour reaction of single-walled carbon nanotubes (SWCNTs) in water performed under microwave irradiation was described for the functionality of CNTs [32]. Pristine SWCNTs were dispersed in water with aniline derivatives in a microwave glass vessel. After application of sonic energy for a few minutes, isoamyl nitrite was added and a condenser was attached. The mixture was irradiated for 90 min at 80 °C in a focused microwave reactor (reaction 8). The presence of one functional group for each 68 carbon atoms (approximately) was analyzed with a thermogravimetric analyzer (TGA).

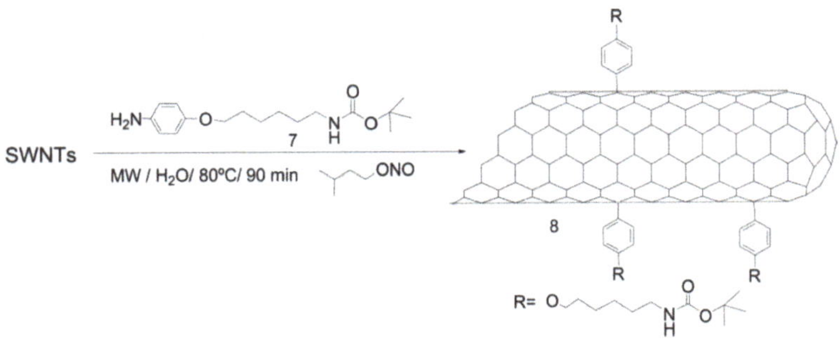

(reaction 8)

In aqueous solid-phase peptide synthesis, researchers have focused on developing novel hydrophilic N-protecting groups for amino acids in order to improve their water solubility, as the conventional protecting groups deployed (e.g., Boc and Fmoc groups) are hydrophobic [33]. In this regard, Hojo et al. [34] reported on the water-based microwave-assisted solid-phase synthesis using Boc-amino acid nanoparticles (Fig. 9.3) and used that for the syntheses, in high yield and purity, of Leu-enkephalin, Tyr-Gly-Gly-Phe-Leu-OH, and the difficult sequence model peptide, Val-Ala-Val-Ala-Gly-OH. The *tert*-butyloxycarbonyl protection group was identified [35] to be the most suitable for industrial and sustainable synthesis

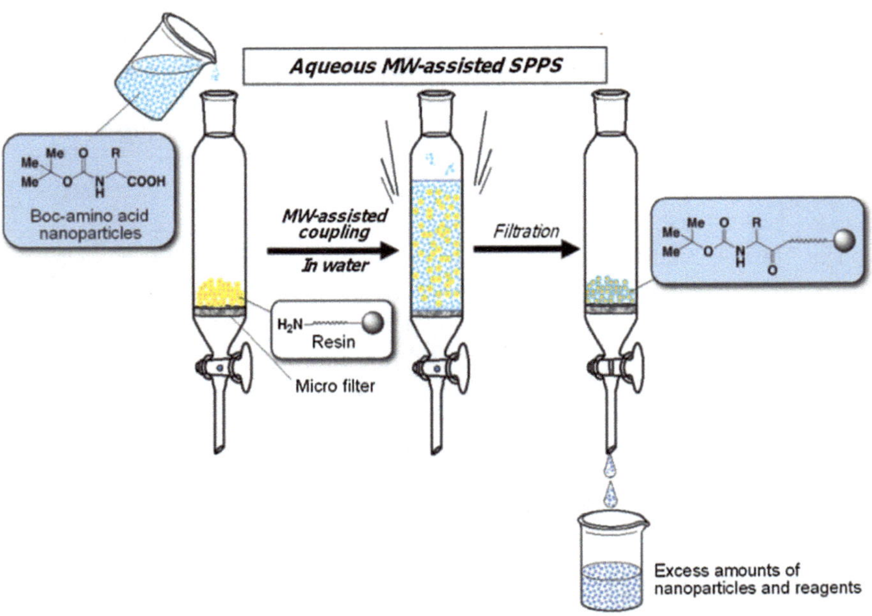

Fig. 9.3 Aqueous microwave-assisted solid-phase synthesis using Boc-amino acid nanoparticles. Reprinted with permission from Ref. [34]. Copyright 2013 by MDPI

because only gases are generated, without any other byproducts produced during de-protection.

9.1.6 Labeling by an Isotope Element

Unique properties such as heat, electrical conductivity, and durability are known and actively studied by deuterium labeling for discovering and developing functional materials. Organic electroluminescence materials are an example of improving luminescence efficiency and durability by deuterium labeling. Deuterated compounds are used as internal standards in quantification of trace compounds, such as agricultural chemical residues in foods, and samples in pharmacokinetic studies by mass spectrometry. In general, however, the use of internal standards labeled with isotopes is limited because of their price and availability. Thus far, there are but a few reports on deuteration of organic compounds. One of the conventional methods of deuteration is the metal-catalyzed exchange of hydrogen isotopes in a compressed D_2 gaseous atmosphere [36]. The processes generally require long reaction times and complicated handling of expensive compressed gases. A novel and facile H-D exchange method has been determined for several organic compounds, especially heterocyclic aromatic compounds, under microwave irradiation in deuterium oxide, as the deuterium source, and in the presence of aluminum powder as the metal catalyst [37]. To prepare phenazine-d_8, the following procedure was used: phenazine (10 mmol), deuterium oxide (30 ml), 5% Pt/C (10%w/w), and aluminum powder (10%w/w) were stirred in the reaction vessel and heated at 200 °C in an autoclave apparatus or microwave reactor. In the conventional autoclave heating method, the deuterium exchange reaction of phenazine required 24 h to achieve 64 atom % D enrichment at best; upon microwave irradiation, the reaction required only 3 h to achieve enrichment as high as 99 atom % D (reaction 9).

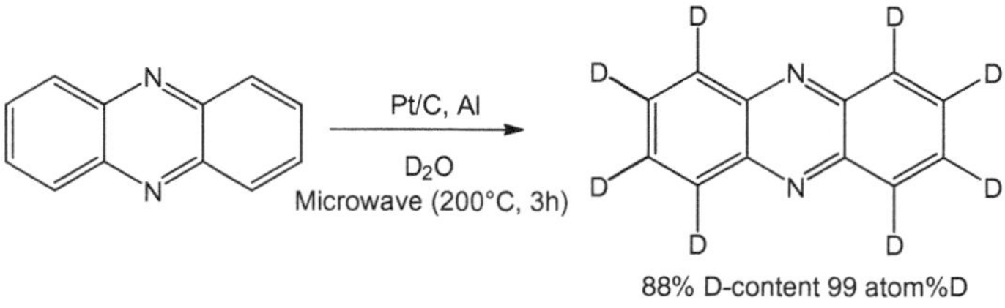

(reaction 9)

9.1.7 Removal of Dissolved Oxygen

The next example does not directly address the microwave influence on organic synthesis; rather, the example relates to the removal of dissolved oxygen from a solution. The relationship between oxygen and microwaves is important in solution chemistry. To the extent that some reactions are oxygen-sensitive, degassing of dissolved oxygen in water and in 2-propanol solvent has been examined under microwave heating by Horikoshi et al. [38]. Degassing dissolved oxygen from a water sample was more efficient when the solution was exposed to microwave irradiation (microwave heating) than when heated by conventional means (oil bath) under otherwise identical temperature conditions. Degassing was carried out by exposing the water sample to 2.45-GHz and 915-MHz microwaves with the sample located within a waveguide, at positions rich in either the electric field or magnetic field densities (T = 80 °C). The efficacy of the microwaves was greater at the lower microwave frequency of 915 MHz than with the more commonly used frequency of 2.45 GHz because of the greater penetration depth into the reactor by the 915-MHz microwaves. Microwaves were again more effective than conventional heating (38% vs. 30%) in degassing dissolved oxygen in 2-propanol solvent. Figure 9.4 is a photograph of the experimental solutions degassed with argon purging the solutions using a Pasteur pipette after microwave heating for a 10-min period in the synthesis of cyclohexanol from cyclohexanone in the presence of 2-propanol and Wilkinson's catalyst [38].

The initial light yellow solution (Fig. 9.4a) changed to a brownish color after this 10-min time period, with the color intensity differing according to the heating method used. The intensity of the brownish color increased in the order: 915 MHz (Fig. 9.4b) < 2.45 GHz (Fig. 9.4c) << oil bath (Fig. 9.4d). The solution heated with the oil bath displayed the most intense color change (Fig. 9.4d), suggesting some oxidative decomposition of Wilkinson's catalyst present in the synthesis of cyclohexanol, from cyclohexanone in 2-propanol solvent. Similar observations of color change were made when the initial solution, which had been purged with oxygen gas, was heated. Evidently, degassing the oxygen-rich 2-propanol solution was slow, compared to the inactivation of Wilkinson's catalyst. With regard to the effects on product yields of microwave irradiation versus conventional heating in the conversion of cyclohexanone to cyclohexanol in 2-propanol media and in the

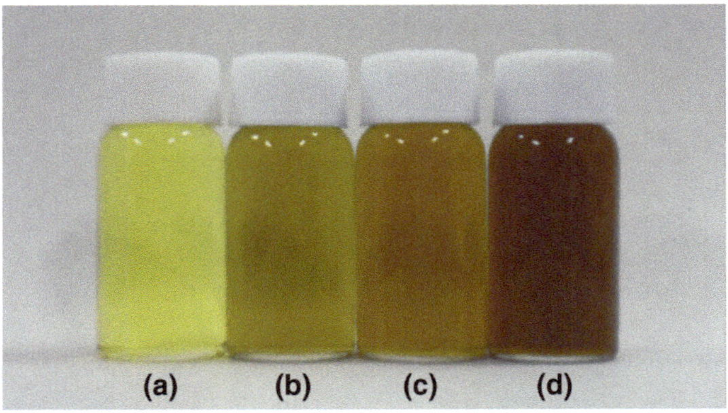

Fig. 9.4 Photograph of the solutions before (**a**) and after (**b**)–(**d**) the synthesis of cyclohexanol from cyclohexanone in the presence of 2-propanol and Wilkinson's catalyst on removing dissolved oxygen by the pipette degassing method following a 10-min heating period: (**b**) 915 MHz, (**c**) 2.45 GHz, and (**d**) oil bath. Reproduced from Ref. [38]. Copyright 2014 by Elsevier B.V.

presence of Wilkinson's catalyst $ClRh(PH_3)_3$, little difference was found irrespective of whether the reaction was activated by microwave irradiation or by conventional heating from an oil bath (81–86% after 10 min). However, argon purging the solution with a Pasteur pipette did reveal some changes in product yields: ca. 13% by conventional heating, 37% by 2.45-GHz microwaves, and 41% with 915-MHz microwaves, also after 10 min. Wilkinson's catalyst appeared more stable under microwave irradiation than under conventional heating.

9.1.8 Scaling Up of Microwave-Assisted Organic Syntheses (MAOS)

Let us now consider the experiment of a scaled-up MAOS reported by Gronnow et al. [39]: the experiment involved installing a wattmeter in a desktop microwave organic synthetic equipment (CEM Discover; 2.45 GHz; 300 W). For the Suzuki coupling reaction (reaction 10) performed using a Pd catalyst the calculated energy saving per product mol was 99% with microwave heating against oil bath heating.

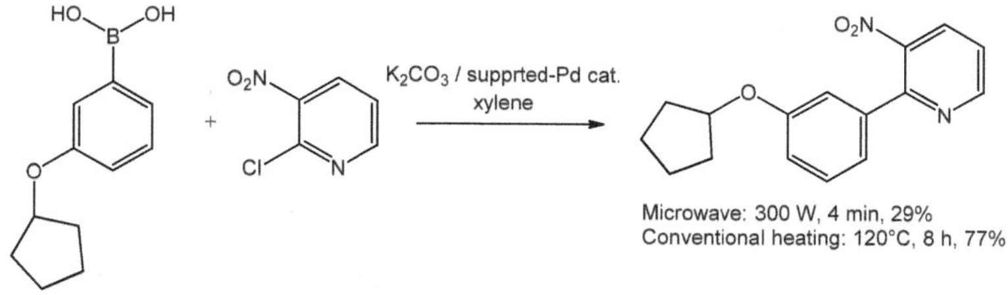

(reaction 10)

The important technology for scaling up microwave-assisted reactions is using continuous flow conditions in open- or closed-loop modes [40]. The idea was initially developed by Strauss and Trainor [41] at the end of the 1980s, which led to commercialized industrial applications among which is the microwave-assisted synthesis of Laurydone®, a moisturizer found in lipsticks and other make-up products (Fig. 9.5) [42]. The procedure involved a 1-m^3 tank connected to a circulation system; the circulated solution was microwave-heated. It was inefficient to heat the solution in the tank with the microwaves because of issues with the microwave penetration depth (Sect. 5.6); however, if the circulating solution was heated by this microwave system, it was possible to heat it uniformly. The reaction time could be reduced by 1/5 and energy consumption to 1/2.6 when compared to a conventional method. A catalyst (*p*-toluene sulfonic acid) and a solvent (toluene) became unnecessary for the microwave-assisted process.

An ester exchange reaction of discarded cooking oil has been performed using large-scale microwave equipment; this process resulted in the synthesis of biodiesel from the discarded cooking oil. A manufacturing daily throughput of 2000 L was

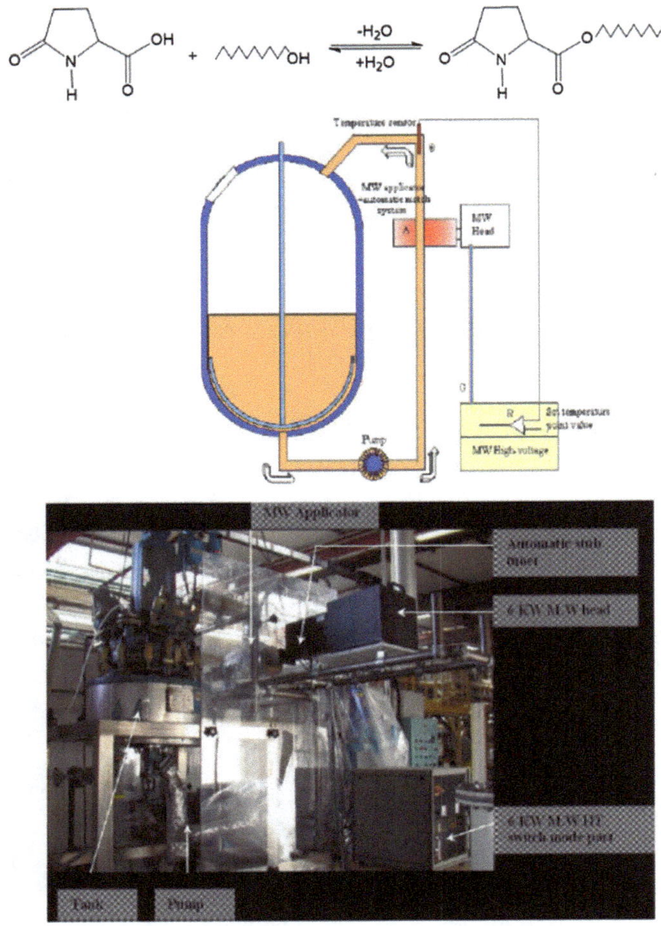

Fig. 9.5 Sairem microwave-assisted reactor for Laurydone® production (courtesy of Sairem SAS). Reproduced from Ref. [42]. Copyright 2014 by MDPI

possible; this equipment has the advantages of a speeded-up process, plus significant energy conservation (Fig. 9.6) [42]. Additionally, while esterification of the isolated fatty acid was not possible with conventional heating, it was possible with microwave heating.

It is not possible, in a single page, to list all the cases of microwave-assisted organic syntheses that have been studied, and what can be listed is limited to a single page in this book. The many reference books related to MAOS are summarized in Table 9.2. Various examples of chemical syntheses may be found by the interested reader.

Fig. 9.6 Microwave esterification reactor production equipment. Photograph provided by courtesy of Microwave Chemistry Co. Ltd

Table 9.2 List of books devoted to microwave-assisted organic synthesis (MAOS)

Book title	Editor or Author	Ref.
Microwave-enhanced chemistry: Fundamentals, sample preparation, and applications	H.M. Kingston, S.J. Haswell	[43]
Microwaves in organic synthesis	A. de la Hoz, A. Loupy	[44]
Practical microwave synthesis for organic chemists: Strategies, instruments, and protocols	C.O. Kappe, D. Dallinger, S. S. Murphree	[45]
Microwaves in organic and medicinal chemistry, (Methods and principles in medicinal chemistry)	C.O. Kappe, A. Stadler, D. Dallinger	[46]
Microwave-assisted organic synthesis: One hundred reaction procedures	D. Bogdal	[47]
Microwave-assisted organic synthesis: A green chemical approach	S.C. Ameta, P.B. Punjabi, R. Ameta, C. Ameta	[48]
Microwave assisted organic synthesis	J. Tierney, P. Lidström	[49]
Microwave methods in organic synthesis (Topics in current chemistry)	M. Larhed, K. Olofsson	[50]
Solid-phase organic syntheses, Volume 3: Microwave-assisted solid-phase organic synthesis	P. H. Scott	[51]
Microwaves in catalysis: Methodology and applications	S. Horikoshi, N. Serpone	[52]
Microwave-assisted synthesis of heterocycles	E. van der Eycken, C.O. Kappe	[53]
Aqueous microwave assisted chemistry: Synthesis and catalysis	V. Polshettiwar, R. S. Varma	[54]
Laboratory experiments using microwave heating	N.E. Leadbeater, C.B. McGowan	[55]
Microwave chemistry	G. Cravotto, D. Carnaroglio	[56]
Microwave heating as a tool for sustainable chemistry (Sustainability: contributions through science and technology)	N.E. Leadbeater	[57]

9.2 Microwave-Assisted Polymerization

9.2.1 Overview of Microwave-Assisted Polymer Syntheses

Application of microwaves in process development has a long tradition, since they often provide workable solutions to technical problems. An early example (late 1960s) is the development of a resin polymerized with microwaves for the manufacture of dental plates [58]; the method has become increasingly popular as an alternative to conventional water bath processing [59]. In addition, microwaves have been used successfully in the vulcanization of rubber used for tires and other automobile parts, as well as insulating weather stripping, door seals, and more. Another example is the microwave-assisted radical cross-linking of an unsaturated polyester with styrene, as reported by Gourdenne and coworkers in the late 1970s, [60], which was followed, soon thereafter by the bulk polymerization of 2-hydroxyethyl methacrylate under microwave and conventional heating conditions [61]. All these studies were performed earlier than the research on the use of microwaves in organic syntheses. In current polymer science, the use of microwave-assisted polymer synthesis has been investigated largely by several research groups; reviews summarizing studies carried out prior to the mid-2000s have been published [see, e.g., 62, 63]. General remarks on the use of microwaves in polymer chemistry appear in a review article dealing with biopolymers [64]. Also reported are polymerization experiments that involve microwaves and an amino acid in regard to a model of the effect of the primitive atmosphere to the origin of life [65]. The polymerization of aqueous glycine, alanine and aspartic acid solution after several minutes of heating under microwave irradiation produced a polyamino acid with a molecular weight of at most 4000. In ancient geological times, it is believed that the protein that gave rise to the origin of life was synthesized by the microwave background radiation present in the universe; this represents the first example in which microwaves were applied in a condensation polymerization reaction.

Research studies of polymerization processes involving radical species have been abundant since the early 1980s. In fact, free radical polymerization has become the most important industrial polymerization technique. Polymers such as polyethylene, polystyrene, polyvinyl chloride, polyvinyl acetate, polyacrylonitrile, polymethyl methacrylate (PMMA), and polytetrafluoroethylene are produced by radical polymerization. In some early examples, the free radical polymerizations, employing microwaves, of styrene, vinyl acetate, methyl methacrylate (MMA), and acrylonitrile, were reported by Porto et al. [66]. Also investigated have been copolymerizations, emulsion polymerizations, preparation of composite materials, and controlled radical polymerizations. Some of these studies are summarized in Table 9.3 [67–116].

Table 9.3 Examples of homopolymerization, copolymerization, graft copolymers, ring-opening, and the use of composites as initiators performed by the microwave heating method

	Raw material/initiator	Ref.
Homopolymerization	Methyl methacrylate (MMA)/Benzoyl peroxide	[67]
	(Meth)acrylic acid/Aliphatic and aromatic amines	[68]
	(Meth)acrylic acid/(*R*)-1-phenylethylamine	[69]
	Methacrylic acid/3-(dimethylamino)-1-propylamine	[70]
	N-alkylacrylamides/Azobisisobutyronitrile (AIBN)	[71]
	Diisopropyl fumarate	[72]
	Dialkyl fumarates/Benzoyl peroxide	[73]
	Acrylamide/2-ethylhexyl acrylate	[74]
	N-substituted maleimides	[75–77]
Copolymerization	Methyl methacrylate (MMA)/Styrene	[78, 79]
	Methyl methacrylate (MMA)/Styrene/Azobisisobutyronitrile (AIBN)	[80]
	2,3,4,5,6-pentafluorostyrene/*N*-phenylmaleimide/Azobisisobutyronitrile (AIBN)	[81]
	2-(dimethylamino)ethyl methacrylate/1-allylthiourea	[82]
Graft copolymers	Methyl methacrylate (MMA)/Chitosan	[83]
	Acrylamide/Chitosan	[84]
	Acrylonitrile/Chitosan	[85]
	Acrylamide/Guar gum	[86]
	Acrylonitrile/Guar gum	[87]
	Acrylonitrile/cassia siamea seed gum	[88]
	Acrylamide/Potato starch	[89]
	Methyl methacrylate (MMA)/κ-carrageenan	[90]
	Acrylic acid/2-(acrylamido)-2-methylpropane-1-sulfonic acid/*N*,*N*-methylenebisacrylamide	[91]
Ring-opening	ε-caprolactone/Metal catalyst	[92–100]
	ε-caprolactone/(Meth)acrylic acid/$Sn(Oct)_2$	[101]
	ε-caprolactone/Carboxylic acid catalyst	[102–105]
	ε-caprolactone/Poly(ethylene glycol)	[106]
	ε-caprolactone/Poly(vinyl alcohol)	[107]
	2-ethyl-2-oxazoline/Methyl tosylate	[108]
	2-phenyl-2-oxazoline	[109]
	2-nonyl-2-oxazoline	[110]
	2-alkyl-2-oxazolines	[111]

(continued)

Table 9.3 (continued)

	Raw material/initiator	Ref.
Composites	Polymethyl methacrylate (PMMA) coated metal fiber	[112]
	Poly(acrylamide)–metal composite	[113]
	Poly(acrylamide)–metal sulfide composite	[114]
	Poly(ε-caprolactone)/Clay nanocomposite	[115]

9.2.2 The Advantage of Microwaves in Macromolecular Syntheses

The microwave synthesis of *N*-isopropylacrylamide as a heat-sensitive polymer gel was reported by Shi and Liu [116]. Polymerization performed by conventional means required heating for 24 h, resulting in a product yield of about 76%. By comparison, the microwave heating method generated a product yield of 99% in about 20 min; moreover, the surface area of as-synthesized *N*-isopropylacrylamide gel, produced with the microwave polymerization method, was about tenfold greater compared to that from the conventional heating method.

Micro- or nanosize polymer particles are very attractive materials in the medical, analytical, and materials science (among others) as they have enhanced aqueous solubility, stability, and uniform size. The polymerization of reactive vinyl derivatives (e.g., styrene and methacrylic acid) results in the generation of the well-known latex particles. The pioneering work of Zhang et al. [117], carried out in the late 1990s, dealt with the production of nearly monodispersed polystyrene (PS) nanospheres, formed under mild microwave irradiation (80 W) for over 1 h, by polymerizing a surfactant-free styrene nanoemulsion. Conversion of the monomer was complete within 40 min, while conversion under conventional heating required no less than 10 h. Polystyrene nanoparticles were produced with sizes between 60 and 120 nm, the size being a linear function of the initial monomer concentration; a dramatic decrease in size polydispersity was also observed (see Fig. 9.7).

9.2.3 Scale-Up of Polymer Syntheses

Microwave irradiation hastens the emulsion polymerization more than 70-times when compared to a similar reaction under conventional heating. In addition, the molecular weight of the product by microwave heating increased nearly 1.2-fold [118]. The main feature of microwaves that leads to a more uniform size distribution probably stems from the fast and homogeneous heating of the droplet, in both the bulk of the emulsion and at the interface; homogeneity is one of the

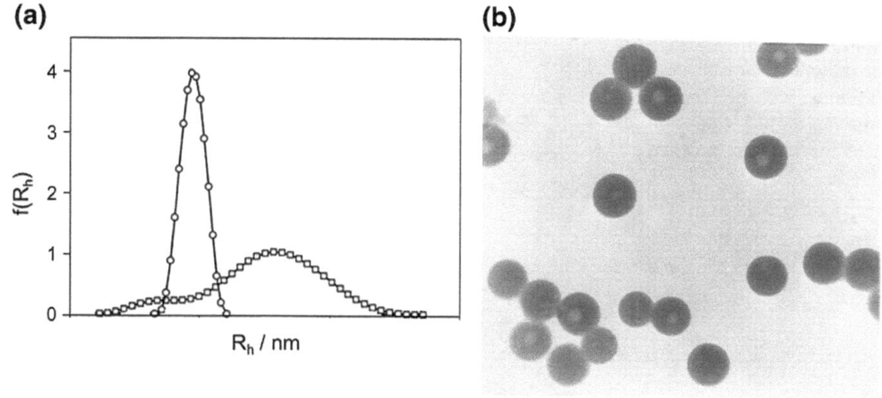

Fig. 9.7 (**a**) Comparison of typical hydrodynamic radius distributions (*f*(*R*h)) of the nanospheres prepared by microwave heating and by the conventional heating methods. (**b**) Electron microscopy micrograph of surfactant-free polystyrene nanoparticles prepared by microwave irradiation. Magnification = 50,000 times. Reprinted with permission from Ref. [117]. Copyright 1997 by the American Chemical Society

features of microwave-assisted synthesis of polymers, which leads to narrower molecular weight distributions. According to the work of Correa et al. [118], the mechanism depends upon the concentration and nature of the monomer. The microwave technique can also be applied to emulsion polymerization of other monomers such as methyl methacrylate (MMA) [119], *n*-butyl methacrylate (BMA) [120], and *N*-isopropylacrylamide (NIPAM) [121]

Pawluczyk et al. [122] attempted the scale-up of the microwave-assisted living free radical polymerization of styrene from Rasta resin beads. They reported that this heterogeneous reaction was not directly scalable from reaction conditions determined on a lower scale. However, after optimization, 40–100 g quantities of Rasta resins with loading levels greater than 3.8 mmol g^{-1} could be achieved with benchtop microwave equipment. Regardless of the remarkable advantages shown by microwave chemistry, the use of this technology in Polymer Science and Biomaterials Science has been relatively scarce [123]. A growing interest in the basic aspects of polymeric biomaterials [124] has been apparent only during the last decade. In this regard, a landmark is the construction of the first industrial plant for the microwave-assisted production of lactic acid oligomers at a commercial level by the National Institute of Advanced Industrial Science and Technology of Japan (AIST), (Fig. 9.8) [125]. Poly(lactic acid) (PLA) constitutes one the most popular and broadly used biodegradable polyesters in drug delivery. In addition to mass production of high-quality polymers in shorter manufacturing times (as claimed in the website), a reduction of 70% of CO_2 emissions was envisioned [126].

Principal plastic raw materials consist of thermoplastic and thermosetting resins, with the former being used most. Thermoplastics are ideal injection molding resins as they are easily moldable when softened by heat and harden to their original strength during the cooling process. It is possible to mass-produce them in short

Fig. 9.8 The 30-kg scale multimode microwave reactor for polymerization reactions. Reprinted with permission from Ref. [126]. Copyright 2011 by Bentham Science

Fig. 9.9 Picture of resin lens production equipment using high-frequency heating with a thermosetting resin. Photograph provided by courtesy of S. Horikoshi

time. On the other hand, thermosetting resins require the addition of a curing agent to a liquid sample and slow heating in a mold. It is also possible to make high-quality plastics with them, though the productivity is poor because polymerizing times are long. For example, thermosetting resins are used as raw materials in the fabrication of high-quality glass lenses and optical plastic lenses because of the excellent control of the optical constant, the heat resistance, chemical resistance, and frame insertion. The polymerization of thermosetting resins for high-quality glass lenses and optical plastic lenses occurs both by microwave heating or high-frequency heating (Fig. 9.9). The heating time is more than 20 h for the

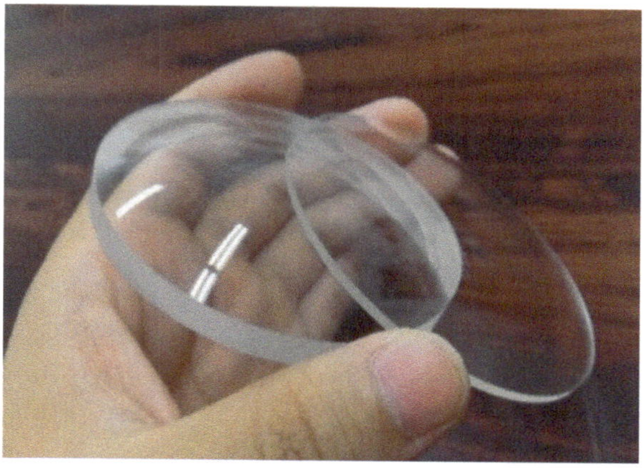

Fig. 9.10 Picture of different thickness glass lenses made from thermosetting resins using high-frequency heating in one batch. Photograph provided by courtesy of S. Horikoshi

classic heating process, which can be reduced to less than 3 h by the microwave heating process. Lenses of higher quality can be formed by the microwave process than is possible for the classical heating process. Moreover, lenses of different thickness and/or size can be polymerized at the same time in one batch (Fig. 9.10); larger size lenses can also be produced that cannot be made by the classical method.

During the course of a polymerization process, the viscosity of the solution increases and an air bubble may form in the solution. This air bubble is not a problem under conventional heating; however, under microwave heating, a microwave plasma will likely form inside the air bubble causing some damage to the sample, to the reactor, and to the equipment. Therefore, the interior of the reactor is first evacuated and any air bubbles are removed; nonetheless, the microwave plasma can also easily form under these lower pressure conditions. To avoid such a problem, microwave equipment design must be carefully scaled-up. Importantly, measurement of the dielectric properties of raw materials and of the mixed sample, calculation of the penetration depth, and monitoring of temperature during the synthesis are necessary.

9.3 Enzymatic Reactions

9.3.1 Situation of Microwave Heating in Biomaterials

Many researchers have shown interest in the microwave effect on enzymatic reactions, so investigations into the mechanism of the microwave effect are highly relevant. Accordingly, it is important to consider some of the features of microwave-assisted enzymatic reactions and on equipment problems; examples of enzyme reactions are also introduced in this section (see, e.g., Ref. [127] for more details).

Enzymatic reactions with microwaves have been researched extensively since the end of the twentieth century. Three situations are worth mentioning in regard to the use of microwave radiation for enzymatic reactions: (i) the influence of microwaves on food enzymes in microwave cooking [see, e.g., 128], (ii) the impact of microwaves on a human body [see, e.g., 129], and (iii) microwave-assisted reactions in organic chemistry. Of particular relevance are the extensive studies carried out on the impact that microwave radiation has on the human body. The first studies on the microwave irradiation of enzymes were published at the end of the 1970s. The technique of microwave irradiation for the inactivation of brain enzymes of animals has been used widely since its introduction by Stavinoha et al. [130]. Inactivation of an enzymatic reaction in a living body by microwave heating was reported nearly three decades ago by Ikarashi et al. [129] who noted that microwave heating could inactivate the enzyme in the brain of the rat (see block diagram in Fig. 9.11 for the model NJE 2603 microwave instrument used). Because it is a sacrificial method, the microwave procedure has gained widespread acceptance. Inactivating the enzymes makes possible the sampling and measurement of

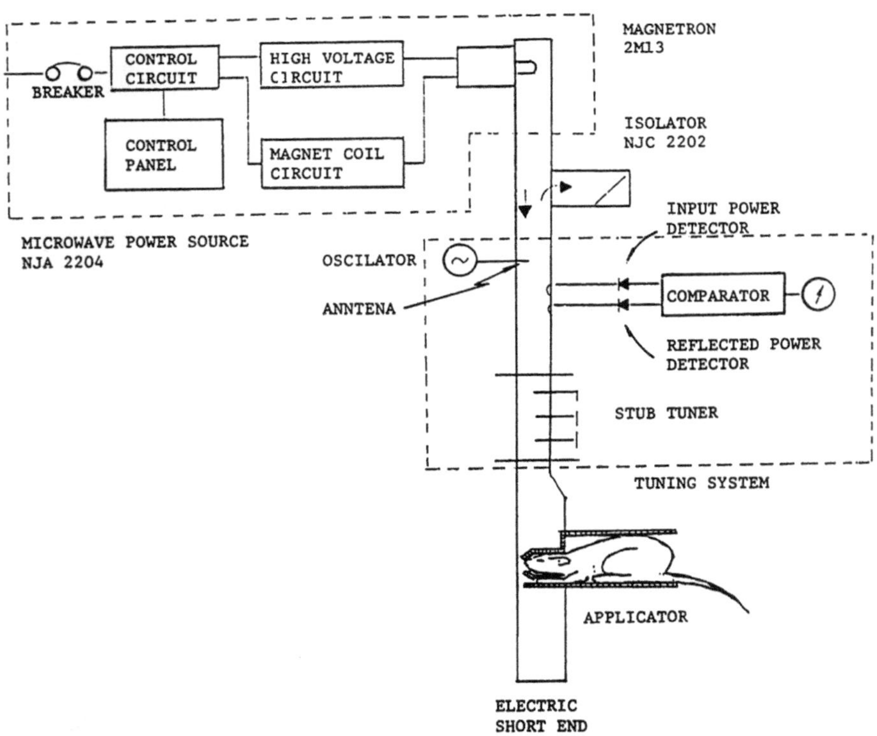

Fig. 9.11 Block diagram of the model NJE 2603 microwave instrument for inactivation of brain enzymes in a rat. Reprinted with permission from Ref. [129]. Copyright 1984 by the Japanese Pharmacological Society

many enzymatically destroyed brain neurochemicals, and reduces the influence of postmortem changes [131].

The heating effect of microwaves was well known, but doubts remained as to the existence of a nonthermal effect. Isolated enzymes in aqueous media were then used as models to determine if such an effect of the radiation on biological processes existed. The first studies were performed with various microwave systems designed and constructed in the test laboratories. Consequently, from one study to another, microwave frequencies and irradiation powers employed were different (Table 9.4). Initially, researchers were interested in the influence of microwave radiation on the specific enzymatic activity. Experiments consisted in heating two independent volumes of an enzyme solution at the same rate and to the same temperature; one volume was heated by microwave energy, the other by conventional resistive heating. The two volumes were maintained at the same temperature for the same length of time, and then assayed to compare the results. In most cases, similar enzymatic activity was found [132, 133]. In only one study, was a significant inactivation of the enzyme, horseradish peroxidase, observed when applying microwave irradiation at room temperature [134]. However, considering the high irradiation power used, local heat enzyme denaturation could not be ruled out, (Table 9.2). Thereafter, the enzyme activity was measured simultaneously with microwave exposure in order to observe possible reversible effects [see, e.g., 135]. Again, no influence of the irradiation was observed.

9.3.2 Microwave Effect in Enzymatic Reactions

Proteins and peptides have significant dipole moments, and so may be especially susceptible to microwave radiation [144]. Microwave promotion of enzymatic reactions has been considered; however, it is not clear whether specific microwave effects are important in biocatalysis [145]. Such effects are difficult to measure because rapid heating of aqueous solutions under high-power microwave irradiation can result in protein denaturation and inactivation [141]. Moreover, the intrinsic catalytic ability of enzymes at a relatively low temperature can obscure the potential rate enhancement resulting from microwave irradiation. There have been attempts to conduct enzymatic reactions under microwave irradiation in nonaqueous solvents; however, this has yielded inconclusive results because of the greatly reduced enzymatic activity [146].

Microwave effects on enzymatic digestion could, in principle, be due to increased dipole moments of the α-helices of proteins. Collins and Leadbeater [145] have questioned why microwave heating has such a profound effect on the desensitization and coupling reactions in the synthesis of peptides, even at relatively low microwave powers and bulk temperatures. Optimal coupling conditions require a fully solvated peptide–polymer matrix, which allows for efficient reagent penetration. During the difficult synthesis of peptides, the reaction matrix becomes partially inaccessible typically involving 6–12 residues into a chain assembly [147].

Table 9.4 Studies of microwave/enzyme interactions in aqueous media

Enzyme	Frequency/GHz	Power input/mW g^{-1}	Temperature/°C	Subsequent/simultaneous radiation	Effect	Ref.
G6P dehydrogenase	2.8	160–300	37–50	Subsequent	No effect	[132]
Trypsin	2.45	100–600	30–95	Subsequent	No effect	[133]
Horseradish peroxidase	2.45	62,500–375,000	25	Subsequent	Inactivation	[134]
Adenylate kinase	2.45	42	25–60	Simultaneous	No effect	[135]
Lactate dehydrogenase	3	30–1800	25–60	Simultaneous	No effect	[137]
Acetylcholinesterase	2.45	1–100	37	Simultaneous	No effect	[138]
Acetylcholinesterase	2.45	2.46–4.29	25	Simultaneous	No effect	[139]
Acid Phosphatase	9.375	160–480	60	Subsequent	Stabilization	[140]
Thermophilic enzymes	10.4	1500–3100	70–90	Subsequent	Inactivation	[141]
Thermophilic enzymes	10.4	800–1700	30–70	Subsequent	Inactivation	[142, 143]

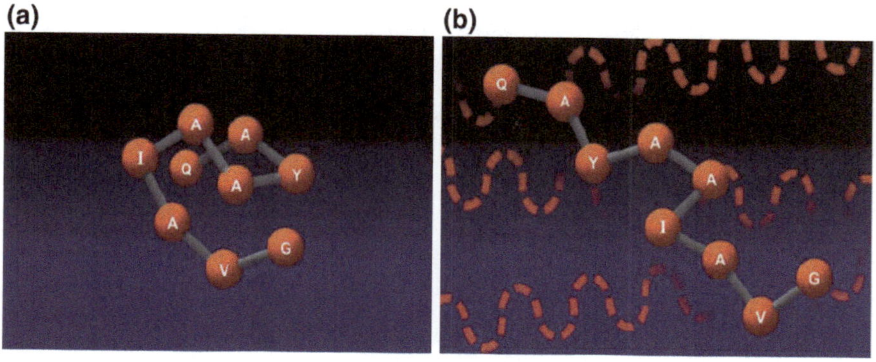

Fig. 9.12 A peptide backbone (scaffold) (**a**) before and (**b**) during microwave irradiation. Reprinted with permission from Ref. [145]. Copyright 2007 by The Royal Society of Chemistry

This could be attributed to the formation of secondary structures, resulting in poor solvation of the peptide–polymer matrix. As a peptide is built stepwise on a resin bead, it can form aggregates with itself or with neighboring chains because of hydrogen bonding between the peptide molecular scaffolds (backbones). One hypothesis is that irradiation with microwave energy leads to de-aggregation of the peptide scaffolds, thus allowing for reagents to more easily reach the reaction sites at the ends of the growing chains (Fig. 9.12).

The amide group of amino acids in a peptide has two resonance forms, which lead to several important properties, one of which is an unusually high dipole moment of roughly 3.5 Debye. When molecules possessing such a high dipole moment are exposed to microwave irradiation, the dipoles attempt to align with the applied electric field. Since the electric field is oscillating, the dipoles constantly attempt to realign and follow the oscillating polarity. At the microwave frequency of 2.45 GHz, molecules attempt to align with the oscillating electric field but cannot follow it exactly. This continual reorientation of the molecules is one of the accepted mechanisms by which microwaves produce localized heating of reaction mixtures [148]. In a polypeptide, this movement could feasibly lead to de-aggregation of the peptide scaffold and to localized superheating, which can have an accelerating effect on the reaction rate for desensitization and coupling steps.

It is relevant to query why microwave heating has such a profound effect on the rate of enzymatic digestion. Together, α-helixes and β-conformations compose a major portion of the secondary structure of large proteins. In the α-helix, the polypeptide scaffold is tightly wound around the molecular axis with each amino acid side chain pointing outwards and downwards from this backbone [145]. The α-helix reveals the highly optimized use of internal hydrogen bonding and creates a stacking of peptide bond dipoles that are added across the hydrogen bonds in the helix. As such, it leads to a large net dipole from one end of the helix to the other. The presence of a large net dipole moment across α-helices may be, at least in part,

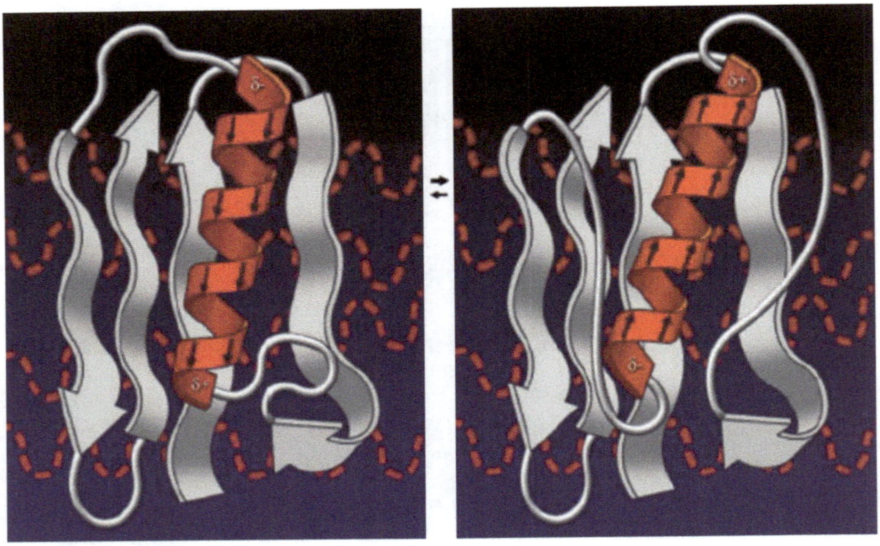

Fig. 9.13 Dipole movement across α-helices and interaction with microwave radiation. Reprinted with permission from Ref. [145]. Copyright 2007 by the Royal Society of Chemistry

responsible for the increased digestion rates of certain proteins upon microwave irradiation. If the microwave energy interacts with the dipole of the α-helix, this may result in perturbation of the three-dimensional structure of the protein as illustrated in Fig. 9.13 [145], thereby facilitating digestion of previously enclosed areas of the protein.

To understand enzyme stabilization and activity in ionic liquids, Zhao and coworkers [149] examined the effect of ionic liquid properties on lipase stabilization under microwave irradiation. The transesterification of ethyl butyrate and 1-butanol was performed by use of Novozyme435, a thermal lipase B obtained from *C. Antarctica* (CaLB) immobilized on acrylic resin in ionic liquids under microwave heating (Fig. 9.14). A Discover Lab-Mate apparatus equipped with both an infrared temperature probe and a pressure monitoring sensor was used as the microwave reactor. Thus, no superheating of the enzyme's surface and no enhanced rate were observed under microwave irradiation. The authors suggested a model of water-induced superheating on the free or immobilized enzyme in dry media (Fig. 9.14a) or a small amount of dispersed water (Fig. 9.14b); they concluded that the effect of microwave irradiation for an enzyme could be explained as originating from superheating of the water layer near the enzyme's surface.

In a 2012 study on proteomics, the microwave technology was used for enzymatic protein digestions and also examined nonthermal microwave effects in proteomics [150]. These authors reevaluated the tryptic (trypsin) digestion of three proteins (bovine serum albumin, cytochrome-c, and β-casein) at 37 and 50 °C under both microwave and conventional heating. They further simulated the tertiary structure of trypsin and bovine serum albumin (BSA) in the electromagnetic field.

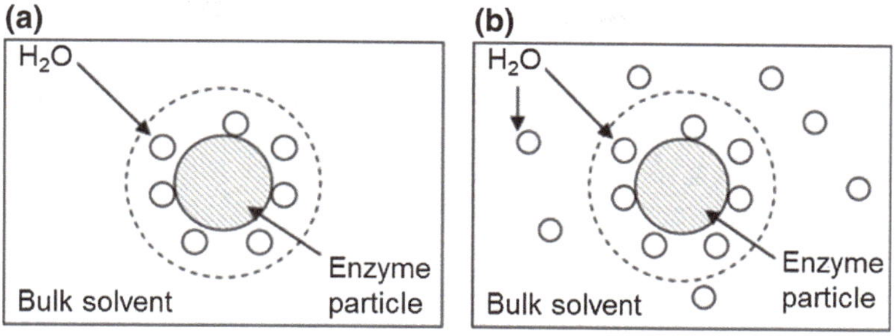

Fig. 9.14 Illustrations of water-induced superheating: (**a**) the free or immobilized enzyme particle is surrounded by a layer of water molecules, while the bulk hydrophobic solvent is dry; (**b**) the enzyme particle is surrounded by a layer of water molecules, while the bulk solvent contains a small amount of dispersed water. Reprinted with permission from Ref. [149]. Copyright 2009 by Elsevier V.B.

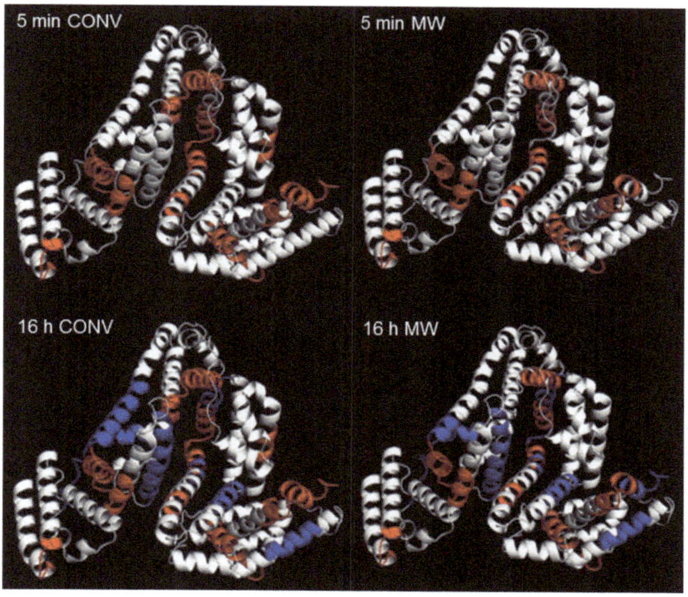

Fig. 9.15 Comparison of the time-dependent BSA digestion using the PyMOL software. Tryptic peptides obtained after a 5 min heating period at 37 °C are shown in red while additional fragment generated after 16 h at 37 °C are highlighted in blue (*CONV* Conventional heating; *MW* Microwave heating). Reprinted with permission from Ref. [150]. Copyright 2012 by Elsevier V.B.

A comparison of microwave irradiation with conventional heating for the digestion of BSA was analyzed using the PyMOL software (see Fig. 9.15). Results of their combined experimental and computational studies revealed that neither the enzyme stability itself nor the trypsin digestion was directly affected by the electromagnetic field. They concluded that incorrect temperature measurements led to erroneous conclusions concerning the nonthermal effects in microwave-assisted proteomics experiments.

A method using microwave dielectric measurements was reported by Suzuki et al. [151] for evaluating protein hydration in aqueous media using a precision microwave network analyzer and a thermostatic glass cell at 20.0 ± 0.01 °C with an open-end flat-surface coaxial probe. Measurements were done on such proteins as cytochrome-*c*, myoglobin, ovalbumin, bovine serum albumin, and hemoglobin. The microwave frequency dependence on the relative complex permittivity of bovine serum albumin (BSA) was measured in the range from 0.2 to 20 GHz. The plotted curves of the relative complex permittivity, changed with changes in frequency.

Control of temperature in microwave-assisted enzymatic reactions is very important, since inactivation of an enzyme can occur with a very small difference in temperature, in most cases within 1 °C. The temperature effect resulting from microwave heating is very severe and is different for organic syntheses, more so than for inorganic material processes. As an example of such an occurrence, Horikoshi et al. [152] examined the decomposition of hydrogen peroxide ($2H_2O_2 \rightarrow 2H_2O + O_2$) using the common metalloenzyme catalase; details of the reaction protocol follow. The enzymatic reaction was performed separately using the microwaves' magnetic field (***H***-field) and electric field (***E***-field) in a single-mode apparatus. Commercial multimode microwave chemical equipment was also used in their study. The temperature profile of microwave heating was nearly identical with conventional heating. A microwave input power of 1–3 W was used to achieve 25 °C, and 37 °C reaction conditions in the single-mode system. Ambient temperature was maintained at 24 °C for the reaction carried out at 37 °C, and 15 °C for the 25 °C reaction. The microwave input power was similar for both ***E***-field and ***H***-field irradiation. The same reactor (with magnetic stirring) was used for the reactions performed in either the single-mode or multimode microwave heating systems, and for the conventional heating system. All experiments were repeated at least six times for microwave and water bath heating. The decrease in the concentration of hydrogen peroxide (H_2O_2) at 37 °C originating from the catalase-assisted reaction subjected to various conditions is displayed in Fig. 9.16a (each experiment was repeated 8 times). Compared to conventional heating, the degradation of H_2O_2 at 37 °C was enhanced in the first 3 min: 1.2-times and 1.3-times under the microwaves' ***E***-field heating and ***H***-field heating, respectively. Under microwave heating (***E***-field and ***H***-field), the degradation of H_2O_2 stopped after the 3-min period; however, under conventional heating, the degradation of H_2O_2 progressed continuously after that period, This suggested that the activity of catalase was considerably affected by the microwave irradiation at times longer than 3 min. The degradation of H_2O_2, with multimode microwave heating, tended to be less efficient than with conventional heating throughout the 12-min heating period, indicating a decreased activity of the catalase with the microwave heating method, while catalase seemed less affected by conventional heating.

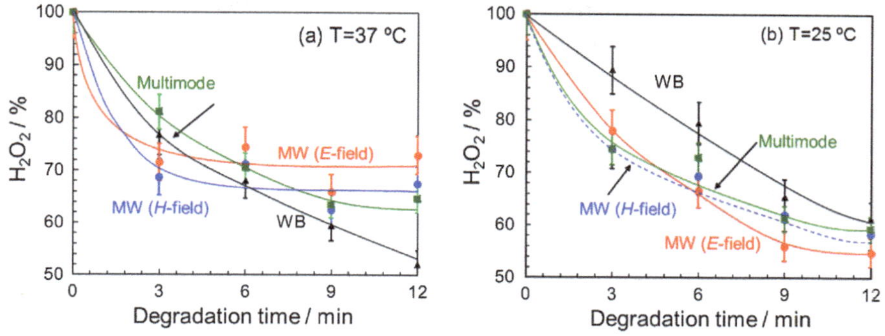

Fig. 9.16 Residual quantity of H_2O_2 in the catalase-assisted degradation reaction under microwave heating (electric field: *E*-field and magnetic field: *H*-field) and conventional heating (water bath: WB), at temperatures of (**a**) 37 °C and (**b**) 25 °C at which the enzyme is most active. Reprinted with permission from Ref. [152]. Copyright 2016 by the Royal Society of Chemistry

9.3.3 Some Relevant Issues Regarding the Equipment Used in Enzymatic Reactions

Regardless of the above inferences (Sect. 9.3.2), it was important to further investigate the degradation of H_2O_2 in the presence of catalase so as to compare microwave heating with conventional heating; the obvious question was why the H_2O_2 reaction was enhanced by catalase in the first 3 min, but considerably reduced at times longer than 3 min. In this regard, enzyme inactivation can be achieved by physicochemical and such environmental factors as pH, temperature, pressure, concentration of the enzyme, and concentration of the substrate; the temperature factor plays an important role that may be changed by either the microwave and/or conventional (water bath) heating. It is important to note that the activity of an enzyme tends to decrease when the enzyme is subjected to heating beyond the optimal temperature at which the enzyme displays its greatest activity. Accordingly, Horikoshi et al. [153] examined the catalase-assisted degradation of H_2O_2 at both 37 °C (optimal temperature) and at a slightly higher temperature (39 °C) under conventional heating. Results reported in Fig. 9.16b showed that at the latter temperature the degradation efficiency was significantly lower after a heating period of 12 min (0.58 times that at 37 °C). However, since there was no temperature variation between microwave heating with a single-mode system and conventional heating, this called into question the concept of selective heating of catalase in the aqueous sample by the microwave heating method. Results at the lower temperature of 25 °C showed that after 3 min into the catalase-assisted reaction the degradation of hydrogen peroxide was 2.1-times and 2.4-times more efficient, respectively, under the microwaves' ***E***-field and ***H***-field heating, in contrast to conventional heating (water bath). Indeed, the microwaves' enhancing effect at 25 °C was greater than at 37 °C. Parallel observations were made on comparing the degradation of H_2O_2 using the microwave multimode system to those from single-mode

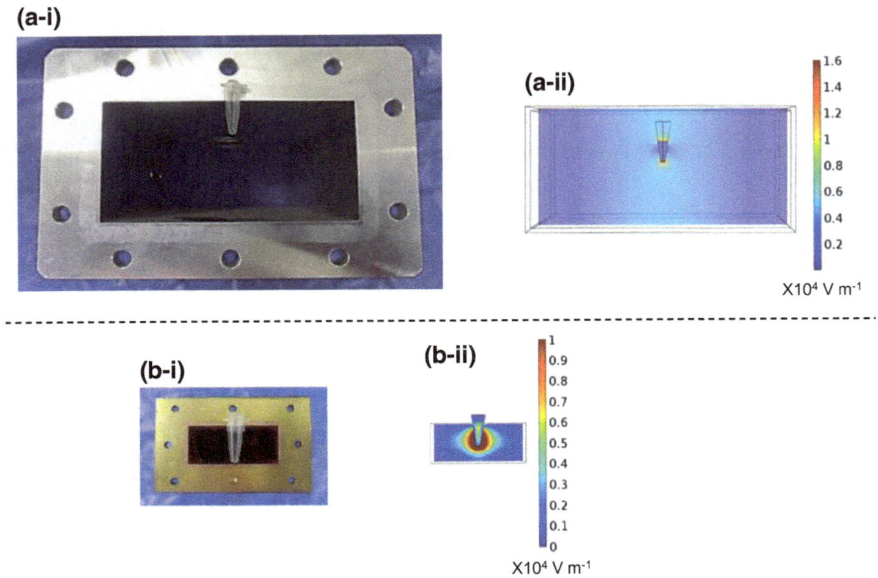

Fig. 9.17 Illustrations of 2.45 and 5.8 GHz waveguides with PCR tube (**a-i**) and (**b-i**), and the distribution (**a-ii**) and (**b-ii**) of the microwaves' electric fields (2.45 and 5.8 GHz) calculated with COMSOL Multiphysics software. Reprinted with permission from Ref. [153]. Copyright 2015 by Elsevier V.B.

microwave irradiation. At the lower temperature (25 °C), the enhancing effect of microwave heating continued and catalase activity was little affected, at least up to 12 min into the reaction.

The choice of microwave equipment to carry out enzymatic reactions is a very important factor to consider, since the equipment design may often be different from equipment used in organic syntheses. Also, it is important to note that the sample used in a biomaterial-related reaction is of a remarkably small volume capacity, compared with sample volumes typically used in chemical reactions (e.g., organic syntheses). Moreover, there are significantly large differences between the common microwave wavelength (12.24 cm for 2.45 GHz microwaves) and the volume of a sample at the microscale (i.e., microliters contained in very small reactors). Figure 9.17 shows a comparison of the cross sections of the waveguides used for 2.45 and 5.8-GHz microwaves and the polymerase chain reaction (PCR) propylene tube reactor generally used in carrying out such reactions [153]. Differences in the sizes of the waveguides employed are responsible for the variations in the input power density of the microwave radiation. The cross-sectional area of the waveguide commonly used for the 2.45-GHz frequency is 59.65 cm^2 (10.92 × 5.46 cm), while it is 8.0 cm^2 for the 5.8-GHz frequency (4.0 × 2.0 cm) so that the ratio of the power density (W cm^{-1}) at the latter frequency is expected to be 7.5-times greater relative to that at the 2.45-GHz frequency; the corresponding wavelengths at these frequencies are 12.24 cm (2.45 GHz) and 5.17 cm (5.8 GHz), respectively.

A photograph comparing the sizes of the 2.45-GHz microwave waveguide and the PCR tube is displayed in Fig. 9.17a-i. The PCR tube (height: 20.0 mm) is very small relative to the sectional area of the 2.45-GHz waveguide. Accordingly, since only a small part of the inner 2.45-GHz microwave waveguide is used for heating, a microliter (μL) solution sample would be difficult to heat since it is difficult to adjust the microwave phase control with the tuner and the short plunger. Hence, an exploratory experiment was performed using a microliter solution sample in the PCR tube inserted in the 2.45-GHz single-mode apparatus equipped with a semi-conductor microwave generator. Horikoshi et al. [153] confirmed that, in this situation, it was difficult to obtain reproducible temperature measurements; a 2.45-GHz semiconductor generator and a single-mode microwave applicator were also used. The electric field (***E***-field) distribution about the PCR tube containing the sample solution in the 2.45-GHz waveguide is illustrated in Fig. 9.17a-ii, which shows that the distribution of the ***E***-field is concentrated in a narrow area near the PCR tube. Consequently, the ***E***-field distribution (microwave phase state) is sensitive to the sample position in the waveguide. The photograph that compares the size of the 5.8-GHz microwave waveguide and the PCR tube is shown in Fig. 9.17b-i, which reveals that the sectional size of the 5.8-GHz microwave waveguide (2 cm × 4 cm) is more appropriate to the size of the PCR tube. Thus, the microwaves in the 5.8-GHz waveguide efficiently irradiated the sample in the PCR tube. The computer-calculated ***E***-field distribution displayed in Fig. 9.17b-ii reveals a large distribution around the PCR tube and confirms the suitability of 5.8-GHz microwaves to effectively heat microliter-size samples.

Temperature control in a small sample can be rather difficult. The highly reproducible temperature of the sample solution in the PCR tube was monitored using a fiber optic thermometer with the aid of a Teflon guide (see Fig. 9.18) [153]. The fiber optic thermometer was inserted through a hole at the center of the Teflon guide and was fixed in position with a silicone O-ring (note that microwaves are not absorbed by Teflon or by silicone). The silicone guide was connected to the upper part of the Teflon guide, while the reactor was installed at the appropriate positions within the waveguide as required by the experimental conditions. Precise temperature measurement is possible when using such a device.

Another study designed to examine bio-related reactions was reported by Shaw et al. [154] who used a custom-built re-entrant microwave cavity operating at 8 GHz to directly heat the glass microfluidic device illustrated in Fig. 9.19. The copper cavity was connected via a coaxial coupling loop to the microwave power source delivering power levels up to 10 W. The mean power consumption of the cavity during a typical heating cycle was under 500 mW. The microwave heating system response time was orders of magnitude faster than typical commercially available systems. The fast-thermal transition capability enabled 28 cycles to be performed in ca. 42 min, a considerable time saving over previously reported microwave PCR systems where 33 cycles took 127 min. Further work was aimed at reducing the hold times at each temperature to achieve DNA amplification that

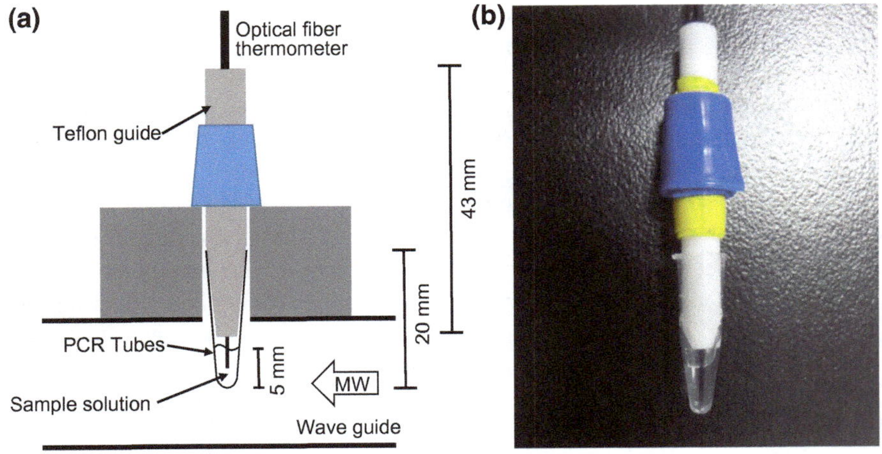

Fig. 9.18 Experimental setup of the PCR tube reactor and optical fiber thermometer with the Teflon guide: (**a**) schematic and (**b**) photograph of the actual PCR reactor/thermometer. Reprinted with permission from Ref. [153]. Copyright 2015 by Elsevier V.B.

would no longer be thermally limited, but restricted by the process efficiency of the DNA polymerase where an average DNA polymerase enzyme adds 60–100 nucleotides per second.

9.3.4 Summary of Data in Microwave-Assisted Enzymatic Reactions

Table 9.5 lists various enzymatic reactions that have been performed in aqueous media using microwaves. Studies concerning the influence of microwave irradiation on enzymatic properties in nonaqueous media are summarized in Table 9.6.

9.4 Catalysts

9.4.1 History of Microwave-Assisted Reactions

The development of chemistry in the twentieth century was essentially based on the use of substrates that acted as catalysts. For example, at the beginning of the last century, whatever soil nitrogen (mostly nitrates) was available proved insufficient for intensive production of crops. Accordingly, several commercial processes were developed to produce nitrogen-based fertilizers with the Haber–Bosch process being the best known (discovered by Fritz Haber in 1906 and produced industrially

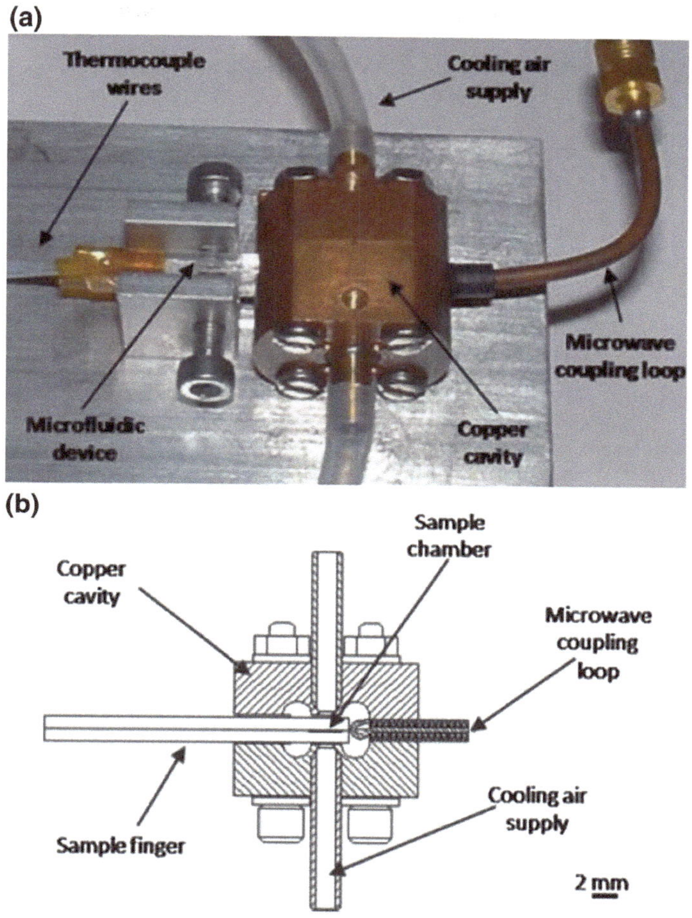

Fig. 9.19 (**a**) Photograph and (**b**) schematic cross section of the re-entrant cylindrical microwave cavity. Reprinted with permission from Ref. [154]. Copyright 2010 by the Royal Society of Chemistry

by Carl Bosch in 1913); this process converts nitrogen in the atmosphere into ammonia in the presence of iron-based catalysts [194]. The challenge of the twenty-first century will be the development of processes that release lower greenhouse gas (e.g., CO_2) emissions and the discovery or development of novel energy-saving catalysts and catalyzed reactions. A natural choice to achieve such goals might involve both catalysis and microwave radiation.

An early study on the calcination and sintering of ceramics by microwaves was reported in the 1970s; fundamental and application studies followed around 1980. The time required and the microwave conditions for heating solids were investigated, resulting in applying the information to microwave heating in the presence of solid catalysts. Several interesting review articles have appeared in the last few

Table 9.5 Studies of the microwave/enzyme interaction in an aqueous medium

Kinds of enzyme	Title	Ref.
Lipase	Microwave radiation can increase the rate of enzyme-catalyzed reactions in organic media	[155]
	Lipase-catalyzed esterification of some α-D-glucopyranosides in dry media under focused microwave irradiation	[156]
	Microwave-promoted lipase-catalyzed reactions. Resolution of (±)-1-phenylethanol	[157]
	Selective enzymic esterification of free fatty acids with n-butanol under microwave irradiation and under classical heating	[158]
	Microwave-induced enzyme-catalyzed chemoselective reduction of organic azides	[159]
	Stability improvement of immobilized Candida antarctica lipase B in an organic medium under microwave radiation	[160]
	Influence of microwave radiation on free Candida antarctica lipase B activity and stability	[161]
	Influence of microwave irradiation on enzymatic properties: applications in enzyme chemistry	[136]
	Nonthermal effects of microwaves on proteins: thermophilic enzymes as model system	[141]
	Different effects of microwave energy and conventional heat on the activity of a thermophilic ß-galactosidase from *Bacillus acidocaldarius*	[143]
Glucosidase	Microwave-assisted synthesis of galacto-oligosaccharides from lactose with immobilized β-galactosidase from Kluyveromyces lactis	[162]
	Synergism between microwave and enzyme catalysis in intensification of reactions and selectivities: transesterification of methyl acetoacetate with alcohols	[163]
	Influence of microwave irradiation on the lipase-catalyzed ring-opening polymerization of ε-caprolactone	[164]
	Intensification of enzymatic synthesis of propylene glycol monolaurate from 1,2-propanediol and lauric acid under microwave irradiation: Kinetics of forward and reverse reactions	[165]
	Kinetic modeling of microwave-assisted chemoenzymatic epoxidation of styrene	[166]
	Microwave-assisted enzyme catalysis for synthesis of n-butyl dipheyl methyl mercapto acetate in nonaqueous media	[167]
	Synergism of microwave irradiation and enzyme catalysis in synthesis of isoniazid	[168]
	Probing the effects of microwave irradiation on enzyme-catalyzed organic transformations: the case of lipase-catalyzed transesterification reactions	[145]
	Kinetic and mechanistic investigation of microwave-assisted lipase-catalyzed synthesis of citronellyl acetate	[169]

(continued)

Table 9.5 (continued)

Kinds of enzyme	Title	Ref.
	Microwave-assisted enzymatic esterification of lactic acid and ethanol in phosphonium type ionic liquids as cosolvents	[170]
	Thermal effect on the microwave-assisted biodiesel synthesis catalyzed by lipases	[171]
	Microwave-assisted lipase-catalyzed synthesis of isoamyl myristate in solvent-free system	[172]
	An efficient microwave-assisted enzyme-catalyzed regioselective synthesis of long chain acylated derivatives of flavonoid glycosides	[173]
	Microwave irradiated immobilized lipase-catalyzed synthesis of alkyl benzoate esters by transesterification: mechanism and kinetic modeling	[174]
	Microwave-assisted synthesis of ethyl 2-(4-aminophenyl) acetate using novozyme 435	[175]
Enzymatic resolution	Microwave-assisted resolution of (R,S)-2-octanol by enzymatic transesterification	[176]
	A chemo-enzymatic process for sequential kinetic resolution of (R,S)-2-octanol under microwave irradiation	[177]
	The influence of microwave irradiation on lipase-catalyzed kinetic resolution of racemic secondary alcohols	[178]
	Microwave-promoted racemization and dynamic kinetic resolution of chiral amines over Pd on alkaline earth supports and lipases	[179]
	Kinetic resolution of rac-1-phenylethanol with immobilized lipases: a critical comparison of microwave and conventional heating protocols	[180]
	Lipase-catalyzed kinetic resolution of (±)-1-(1-naphthyl) ethanol under microwave irradiation	[181]
	Insight into microwave irradiation and enzyme catalysis in enantioselective resolution of dl-(±)-3-phenyllactic acid	[182]
	Microwave-assisted enzymatic kinetic resolution of (±)-1-phenyl-2-propyn-1-ol in nonaqueous media	[183]
β-glucosidase/ β-galactosidase	Microwave activation of enzymatic catalysis	[184]
	Synthesis of UDP-activated oligosaccharides with commercial β-galactosidase from Bacillus circulans under microwave irradiation	[185]
Aldolase	Stable and efficient immobilization technique of aldolase under consecutive microwave irradiation at low temperature	[186]

Reprinted with permission from Ref. [136]. Copyright 2007 by Taylor & Francis

Table 9.6 Studies of the microwave/enzyme interactions in nonaqueous media

Enzyme	Reaction medium	Enzymatic reaction	Temperature/°C	Tested enzymatic parameter	Microwave effect	Ref.
Supported lipases	Dry	(trans) Esterification	70–100	Activity	Higher	[157]
Supported lipases	Dry	(trans) Esterification	70–100	Selectivity	Higher	[157]
Supported lipases	Dry	(trans) Esterification	70–100	Stability	No effect	[157]
Supported lipase	Dry	Esterifications	95–110	Activity	Higher	[156]
Supported glucosidase	Dry	Glucosidation	80	Activity	Smaller	[187]
Supported glucosidase	Aqueous	Hydrolysis	60	Activity	Higher	[187]
Supported glucosidase	Dry	(trans) Glucosidations	80–110	Activity	Higher	[188]
				Selectivity	No effect	[188]
Cutinase	Liquid substrates	Transesterification	50–70	Activity	Higher	[155]
Supported lipases	Organic solvent	Esterification	40–70	Activity	Higher	[155]
	Liquid substrates	Esterification	40–70	Activity	Higher	[155]
Chymotrypsin, Subtilisin	Organic solvent	(trans) Esterification	25–60	Activity	Higher	[189]
Supported lipase	Organic solvents	Transesterification	50	Activity	Higher	[163]
Supported lipase	Organic solvents	Transesterification	50	Selectivity	No effect	[163]
Lipases	Microemulsion	Hydrolysis	80	Activity	Higher	[190]
Lipases	Microemulsion	Hydrolysis	80	Selectivity	No effect	[190]
Lipases	Organic solvent	Esterification	80	Activity	Higher	[190]
Lipases	Organic solvent	Esterification	80	Selectivity	No effect	[190]
Porcine pancreas lipase	Organic solvent	Transesterification	35, reflux	Activity	Higher	[191]
Porcine pancreas lipase	Organic solvent	Transesterification	35, reflux	Selectivity	Higher	[191]
Supported lipase	Liquid substrate	Transesterification	100	Activity	No effect	[160]
Supported lipase	Liquid substrate	Transesterification	100	Stability	Higher	[160]

(continued)

Table 9.6 (continued)

Enzyme	Reaction medium	Enzymatic reaction	Temperature/°C	Tested enzymatic parameter	Microwave effect	Ref.
Lipase	Liquid substrate	Transesterification	40–110	Activity	No effect	[192]
Lipase	Liquid substrate	Transesterification	40–110	Stability	Modified	[192]
Supported galactosidase	Solvent/water (70/30, v/v)	Transglycosylation	40	Activity	No effect	[162]
Supported galactosidase	Solvent/water (70/30, v/v)	Transglycosylation	40	Second reaction	Smaller	[162]
Porcine pancreas lipase	Organic solvent	Azide reduction	50–60	Activity	Higher	[159]
Porcine pancreas lipase	Organic solvent	Azide reduction	50–60	Second reaction	Smaller	[159]
Glycosidases	Solvent/water (9/1 and 4/1, v/v)	(trans)Glycosylations	50–65	Activity	Higher	[193]
Supported lipases	Organic solvent	Esterification	30–70	Activity	Higher	[165]

Reprinted with permission from Ref. [136]. Copyright 2007 by Taylor & Francis years that describe the use of microwave radiation in several processes [195–200]. The combination of heterogeneous catalysis and microwaves was exploited in microwave-assisted organic syntheses, which provided conversion of natural products to valuable chemicals and to biomass-derived compounds; also presented was the utilization of microwaves in a series of heterogeneous acid-catalyzed and base-catalyzed reactions to redox and coupling chemistries.

The process intensification of the production of fatty acid methyl esters from waste cooking oil (WCO) using microwave reactors was reviewed by Mazubert et al. [196] who addressed the transformation of WCO feedstock by transesterification, esterification and hydrolysis reactions together with choice of catalyst and product separation. Specifically, their article examined the role of important process parameters (e.g., the presence and type of solvent, the dielectric properties of the mixture and the individual phases, the use of heterogeneous catalysts, pressure, stirring, reflux conditions, temperature measurement methods, and microwave absorbing fillers) to determine the occurrence and magnitude of the benefits enabled by microwaves during polycondensation reactions. The use of the microwave technology (including multi-feedstock) reported in a 2015 article by Soltani et al. [199] described microwave-assisted heterogeneously catalyzed processes for the production of biodiesel through the transesterification of plant oils and animal fats; it also discussed reaction variables that impact the transesterification process (e.g., heating system, microwave power, type and amount of heterogeneous catalyst, oil/methanol molar ratio, reaction time, temperature, and mixing intensity).

Therefore, the following subsections will provide a non-exhaustive introduction into the study of microwave-assisted syntheses of catalysts/carbonaceous materials, as exemplified by the synthesis of TiO_2-based catalysts using selective microwave heating and activated carbon as the support and, where available, their activity (see, e.g., Ref. [127] for more details).

9.4.2 Advantages of the Microwave-Assisted Heterogeneous Catalytic Method

Despite the large application of microwave-assisted sintering protocols [201], the use of solid catalysts subjected to microwave radiation has been limited for several years because of fears of the possible arcing phenomena that would create hazardous conditions in the presence of flammable solvents and/or gases [202]. The electric discharge phenomenon occurring in metal–solvent mixtures (arcing) was described in Sect. 4.5. Attractive specific-reaction fields are provided by the heterogeneous catalyzed reactions under microwave irradiation. When microwave selective heating is desired, it is possible to carry out organic syntheses using some of the features of microwave radiation. For example, when nonabsorbing microwave substrates such as a nonpolar solvent, or a nonpolar sample, or gas sample are involved in a catalyzed chemical process, only the solid catalyst may be heated

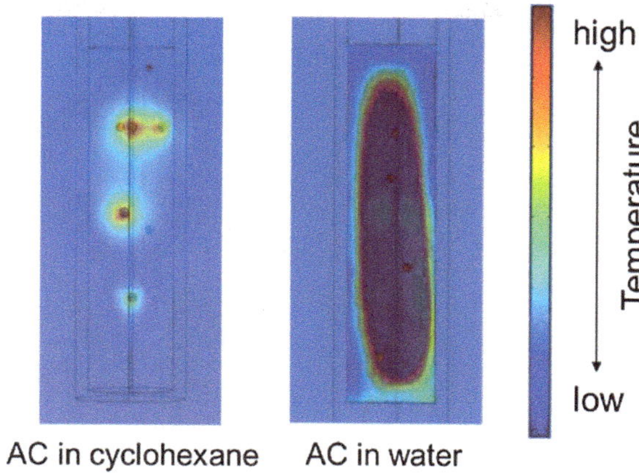

Fig. 9.20 Simulated temperature distribution in water and cyclohexane solvents in the presence of Pt/activated carbon (AC) catalyst particles subjected to microwave radiation. The simulation was carried out using an RF module and the COMSOL Multiphysics software Version 4.3a. Reproduced with permission from [203]. Copyright 2014 by the Royal Society of Chemistry

selectively. Figure 9.20 illustrates the results of a simulation experiment to determine the temperature distributions in water and in cyclohexane (a poor microwave absorber) in the presence of an activated carbon (AC) catalyst support (a strong microwave absorber) when exposed to microwave radiation [203]. The activated carbon was selectively heated in cyclohexane by Joule loss (resistive) heating; however, since water is a strong microwave absorber, selective heating of activated carbon did not occur in aqueous media. Chemical reactions typically occur at the catalyst particle surface, so that whenever microwaves selectively heat only dispersed catalyst particles, not only does the reaction take place but significant energy conservation is also achieved. A researcher who would like to use this feature, however, will need to control the formation of hotspots.

A study by Mastuzawa et al. [204] is germane to the present discussion since it examined the synthesis of 4-phenyltoluene by the Suzuki–Miyaura coupling reaction using both heterogeneous and homogeneous Pd-based catalysts. Microwaves enhanced the heterogeneous catalyzed process, while the same reaction in homogeneous media was only slightly enhanced under similar conditions. Moreover, when carried out in a heterogeneous solvent-free system, the reaction was only slightly enhanced by the microwaves.

In examining the microwave effect, another example of a catalyzed reaction involved the microwave enhancement of the oxidation of water to generate dioxygen; the α-Fe_2O_3 electrode was activated by pulsed microwave radiation at a constant applied potential [205]. Rectangular waves of current density were observed under pulsed microwave irradiation; in other words, upon introduction of microwaves, the oxidation current of water was increased instantaneously (Fig. 9.21) and remained stable at maximum current density under continuous

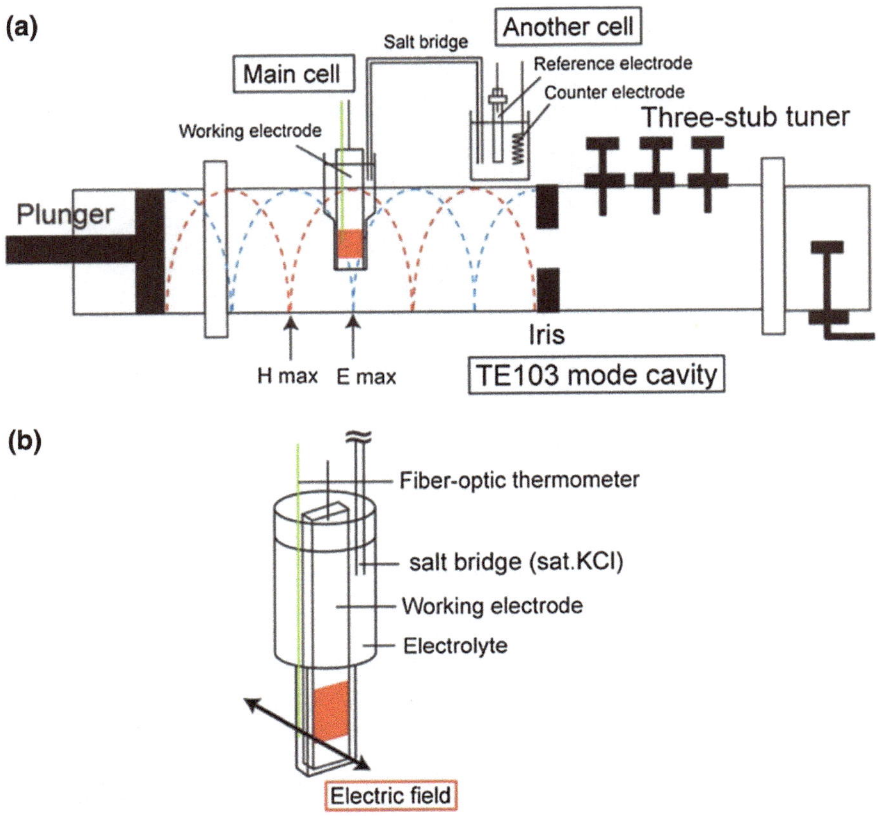

Fig. 9.21 Experimental setup of the electrochemical measurements: (**a**) microwave irradiation cavity and positions of the electrochemical cells; the main cell was introduced into the microwave cavity at the maximal ***E*** (electric) or ***H*** (magnetic) fields. (**b**) Detailed illustration of the main cell; a fiber optic thermometer was introduced close to the electrode. Reproduced with permission from [205]. Copyright 2016 by Nature

microwave irradiation. The microwave enhancement was observed only for the α-Fe_2O_3 electrode with a specific surface electronic structure that was evaluated by electrochemical impedance spectroscopy. This discovery provided firm evidence of the ***nonthermal*** microwave (special) effect on electron transfer reactions caused by the interaction of the oscillating microwaves and the irradiated samples. In summary, Kishimoto et al. [205] discovered that the anodic current of α-Fe_2O_3, electrochemically deposited on F-doped tin oxide (FTO) substrate, was steadily enhanced by 2.45-GHz microwave irradiation. The enhancement was strongly related to the interaction between the surface electronic structures of α-Fe_2O_3 and the microwave oscillating electric field. The enhancement in the anodic current on α-Fe_2O_3 was attributed to the enhanced oxidation of water into dioxygen under microwave irradiation. This was clear experimental evidence of a microwave special effect upon electron transfer reactions. Electron transfer reactions are typical chemical reactions often involving a metal catalyst; evidently, microwaves can accelerate such electron transfer reactions.

Carbonaceous materials have been widely investigated as functional materials for various applications that include battery electrodes and fuel cells, adsorbents for water purification, and as catalyst supports [206–210]. With the continued development of activated carbon technologies, new applications of microwave heating used in preparing carbon supports have witnessed successful and continued interest (see e.g., ref. [211]).

Microwave-induced chemical reactions have been investigated for use in many applications, such as the pyrolysis of highly volatile bituminous coal, and coal gasification in water vapor. A solid material (char) that remains after light gases (e.g., coal gas) and tar have been released from a carbonaceous material during the initial stage of combustion (i.e., carbonization, charring, or pyrolysis) turns out to be a good absorber of microwave energy [212]. Char has led to applying a certain amount of microwave energy for promoting char-containing syntheses performed with minimal reaction conditions. Metal catalysts on carbon supports are mostly used in microwave-assisted catalyzed reactions. Microwave heating proceeds easily with carbon materials; the ratio of microwave absorption is increased as microwave heating progresses (Sect. 8.4.1). The difference in the heating mechanism of a solution and a carbon-based solid is shown in Fig. 9.22 [213]. Microwave heating of a solution is mainly related to the dipole rotation of polar solvent molecules. However, in electron-rich solids, such as carbon-based solids which have no freely rotatable dipoles, it is the motion of electrons that generates heat through Joule (resistance) heating within the grain or causes arcing at phase boundaries. Microwave heating of solid-state carbon-based materials has distinct merits, even though they have scarcely been reported in microwave-assisted organic syntheses.

A heterogeneous catalyst system may appear to be ideal for performing a catalyzed reaction by the microwave-assisted method, yet the nature of the solvent medium remains an important factor to contend with. For instance, if a polar solvent were used in carrying out the Suzuki–Miyaura cross-coupling reaction in the presence of the palladium/activated carbon (Pd/AC) catalyst in heterogeneous media, the advantages of microwaves are diminished as the microwaves will non-selectively heat the whole of the bulk solution. Despite this, however, Arvela

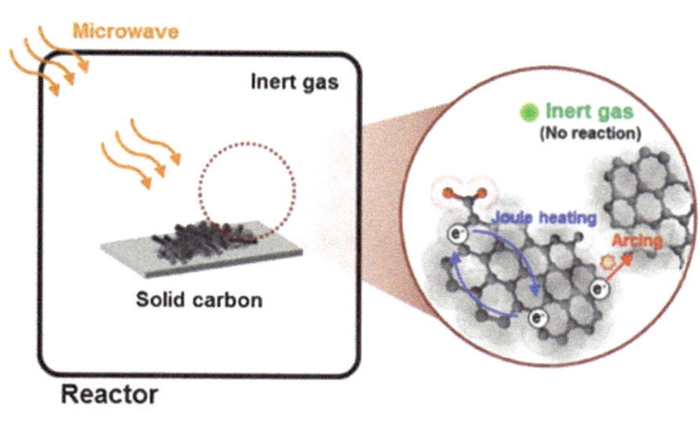

Fig. 9.22 Mechanism of microwave heating of solid carbon. Reproduced with permission from [213]. Copyright 2014 by Korean carbon society

and Leadbeater [214] witnessed significant enhancement effects by microwaves in the synthesis of biphenyl compounds from the Suzuki–Miyaura cross-coupling reaction; they used a novel method of microwave heating while simultaneously cooling the reactor externally. This led to significant product yields under such experimental conditions.

9.4.3 Gaseous Reactions with Solid Catalysts

Microwave-assisted catalysis has some benefits in systems designed for environmental detoxification; for example, elimination of VOCs, NO_x, and SO_x occurs on the solid surface of metal-oxide catalysts through oxidative and reductive reactions [215]. Using appropriate devices, heating can be concentrated on the catalyst mass resulting in reducing the size of the catalytic reactors and shortening the heating time with very little energy absorption by those reactants adsorbed on supports [216]. Application of the microwave heating method to catalytic processes for environmental purification is a desirable effort. The microwave dielectric heating effect uses the ability of some absorbers or catalysts to transform electromagnetic energy into heat via dielectric loss and thus able to drive catalytic reactions [217]. Table 9.5 reports typical temperature data measured for several pure metal-oxide catalysts subjected to microwave heating. Such metal-oxide catalysts as CuO, Co_2O_3, NiO, V_2O_5, and MnO_2 are compatible with incident microwave radiation and are rapidly heated to about 200 °C or above after 8 cycles of 5-s irradiation and 10-s rest periods (i.e., no irradiation). Mignos and Baghurst [218] have reported similar results on NiO, Co_2O_3, V_2O_5, and CuO. However, the reason why these oxides are easily heated under microwave irradiation has not been determined. In this regard, Tamura [219] has speculated that the increase in conduction with increasing temperature of these oxides may be associated with the thermal activation of electrons that migrate from the oxygen $2p$ valence band to the $3s3p$ conduction band. In addition, electrical conduction may be enhanced by the presence of defects in the oxides. Defects such as oxygen vacancies may form at higher temperatures and lead to the emergence of an *extrinsic band gap* of the metal oxide (i.e., an effective energy gap between the vacancy levels and the conduction band) in contrast to the *intrinsic band gap*.

Comparing the heating properties determined for several metal oxides reveals that most transition metal oxides absorb microwaves and are heated. By contrast, rare earth metal oxides, alkaline earth group oxides, SiO_2, and γ-Al_2O_3 are weak absorbers of microwave radiation and thus exhibit a little rise in temperature when irradiated. From a practical viewpoint, such as in the oxidation of VOCs or NO_x, the most active catalysts originated from platinum group metals deposited on suitable supports; an example is γ-Al_2O_3 that is normally used as a catalyst support and is found in automobiles for emission control. Supported platinum group metals (Pt, Rh, and Pd) are the most efficient and active catalysts. However, considering the effectiveness of microwave radiation, catalysts composed of transition metal oxides

Table 9.7 Heating profiles of pure oxides (catalysts or catalyst supports) by pulsed microwave irradiation at 2.45 GHz; 2-s irradiation and then 10-s rest periods, (8 cycles); starting temperature, 25 °C

Oxide	Temperature/°C	Oxide	Temperature/°C	Oxide	Temperature/°C
CaO	58	Ta_2O_5	55	NiO	260
BaO	57	Cr_2O_3	54	CuO	341
CeO_2	46	MoO_3	41	ZnO	131
TiO_2	40	WO_3	188	γ-Al_2O_3	85
ZrO_2	47	MnO_2	194	SiO_2	106
V_2O_5	221	Fe_2O_3	165	SnO_2	143
Nb_2O	51	Co_2O_3	272	Bi_2O_3	59

Data reproduced with permission from Ref. [215]; Copyright 2016 by the Wiley-VCH Verlag GmbH

are more suitable for use in reactors irradiated by microwaves than are metal catalysts on supports because materials normally used for supports (e.g., γ-Al_2O_3, SiO_2, and CeO_2) do not possess a high susceptibility to the incident microwave radiation (Table 9.7).

Several commercial catalysts consisting of mixed oxides, including transition metal elements, used frequently for improving or protecting the environment as a result of their high oxidation and reduction properties, have been investigated to assess their heating properties under microwave irradiation [220]. The catalysts examined were **G-3A** consisting of Fe_2O_3 (90%) and Cr_2O_3 (10%), **G-132** consisting of CuO (40%) and ZnO (60%), **G-53D** consisting of NiO (70%) and SiO_2 (30%), **N-140** consisting of CuO (40%) and MnO_2 (60%), **N-150** consisting of Fe_2O_3 (40%) and MnO_2 (60%), and **NF-9** consisting of pure NiO. Since every catalyst is typically tablet-shaped, small amounts of graphite have been added at concentrations between 0.5 and 1% to stabilize the tablets. The quantities of graphite are too small to influence microwave heating. These mixed metal-oxide catalysts used in environmental purification were repeatedly pulse-heated by microwaves at 2-s irradiation and 10-s rest periods; the resulting temperatures reached after several pulsed cycles are listed in Table 9.8—also shown are the measured dielectric constants of the catalysts at the microwave frequency of 2.45 GHz and at 25 °C.

Among the catalysts tested, **NF-9, N-140,** and **N-150** showed a rapid temperature rise, while **G-132** and **G-53D** showed a slow temperature rise. Comparison of the rates of temperature rise for the mixed oxides with those of pure oxides revealed little correlation; that is, the commercial catalysts **N-140** and **NF-9** containing CuO and NiO, respectively, showed relatively higher temperature rise, whereas even though **G-132** and **G-53D** contained CuO and NiO, their temperature increases under microwave irradiation were smaller. The **N-150** system that contained both Fe_2O_3 and MnO_2 displayed a relatively higher temperature rise, while the two separate metal oxides showed smaller increases. These results indicate that the

Table 9.8 Surface area, tanδ, and temperature reached in commercial catalysts subjected to pulsed microwave irradiation at 2.45 GHz; 2-s irradiation, 10-s rest; starting temperature, 25 °C

Catalyst	Surface area/$m^2\ g^{-1}$	tanδ	Temperature/°C
G-3A (Fe_2O_3–Cr_2O_3)	78	0.13	61
G-132 (CuO–ZnO)	35	0.12	49
G-53D (NiO–SiO_2)	193	0.15	48
N-140 (CuO–MnO_2)	133	0.19	96
N-150 (Fe_2O_3–MnO_2)	175	0.13	79
NF-9 (NiO)	152	0.18	101

Data reproduced with permission from Ref. [215]; Copyright 2016 by the Wiley-VCH Verlag GmbH

coexistence of different metal oxides in a catalyst system has different microwave susceptibility from that of separate catalyst oxides; that is, the dielectric loss of mixed oxides is significantly altered. The existence of two types of sites in perovskite-type oxides led to different dipole strengths and resulted in the formation of hotspots during microwave dielectric heating [221].

VOCs that also include odorants are major contributors to atmospheric pollution; their removal is best achieved at the source of contamination. Catalytically aided oxidation of such pollutants is preferred over thermal combustion because of the lower processing temperatures required and the higher selectivity to CO_2 evolution [222]. Lowering the temperature, at which these processes occur with great efficiency, can significantly reduce energy costs. Hence, it is important to utilize a catalyst in these microwave-assisted processes as they offer certain advantages such as volumetric and high heating rates, indirect contact, simple control of heating, and minimization of equipment size.

Microwave-heated catalytic processes for the purpose of oxidative removal of environmental pollutants have been investigated extensively. For instance, aliphatic hydrocarbons, with propane as a model substrate, are more difficult to oxidize than aromatic hydrocarbons (e.g., benzene, toluene, and xylene) and CO over such a doped perovskite catalyst as $La_{0.5}Co_{0.2}MnO_3$ [221]. There is no indication that microwave and conventional heating led to different reaction pathways because only small differences in selectivity were observed. However, the mechanisms of propane oxidation and of CO oxidation are different; that is, CO is oxidized by reaction with adsorbed oxygen at low temperatures, while propane is oxidized by reaction with lattice oxygen at high temperatures as per the Mars/van Krevelen mechanism. Microwave energy enables fast and efficient heating of the perovskite-type catalysts of complex oxides where only the catalyst is selectively heated. Direct heating with microwave irradiation is rapid, effective, and much more suitable for periodic operation than conventional heating.

Microwave-assisted catalysis has been applied to areas of environmental concern, including degradation of NO_x and SO_x [223]. NO_x and SO_x are major air pollutants emitted from fossil fuel combustion from both mobile and stationary sources. NO_x contributes significantly to photochemical smog, acid rain, ozone depletion and greenhouse effects, while SO_x is also an important precursor to acid rain [224]. Among the many technologies used to reduce the emission of NO_x and SO_x, the catalytic technology of direct and reductive decomposition of these oxides is more attractive because of the compact apparatus used and for its high efficiency; moreover, there is no need for a reducing agent. Reduction of NO_x can reach as much as 98% conversion when microwave energy is applied continuously [225]. On the other hand, where a reducing agent is used, the use of both catalyst and ammonium bicarbonate, combined with microwave energy, can significantly increase the catalytic reductions of both NO_x and SO_x. The catalytic reduction of sulfur dioxide with methane to form carbon dioxide and sulfur takes place over MoS_2/Al_2O_3 catalysts. A higher conversion of SO_2 and CH_4 was claimed when microwave heating was applied [226]. The conversion in an electromagnetic field was as high as the one obtained with conventional heating requiring temperatures around 200 K higher (Fig. 9.23). A morphological study of the catalyst showed that localized temperatures under microwave heating had to be significantly higher than the measured average temperature of the catalytic bed (1000 K), since the observed phase transition of c-alumina to a-alumina occurs only at temperatures above 1273 K. The observed phase transition was claimed to be a clear evidence of hotspot formation inside the catalytic bed. The decomposition of H_2S performed by conventional heating showed that the results obtained were in very good agreement with equilibrium data [227]. However, when microwave heating was applied, the conversion of H_2S was twice as much as that obtained under conventional heating at the same temperature. Observed effects on gas–solid catalytic reactions performed under microwave radiation are summarized in Table 9.9 [200].

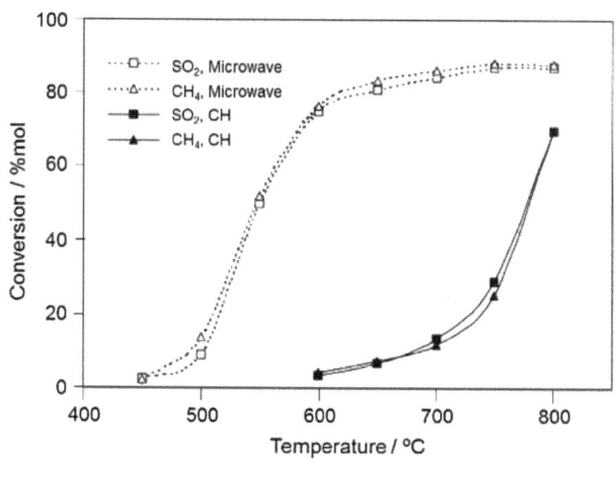

Fig. 9.23 SO_2 and CH_4 conversion as a function of temperature over catalyst MC-2. Reproduced with permission from [226]. Copyright 2001 by Elsevier V.B.

Table 9.9 Observed effects of gas–solid catalytic reactions performed under microwave radiation, and their possible explanations

Application	Observed effect	Explanation of observed effects	Ref.
Methane decomposition	Higher selectivity	Nonuniform temperature conditions	[228]
	No comparison with conventional process	n/a	[223, 229, 230]
	Higher conversion	No explanation	[231]
Methane oxidation	Higher conversion, higher selectivity	Plasma formation, arcing	[232]
	No comparison with conventional process	n/a	[233, 234]
	Higher conversion, higher selectivity, reduction of reaction temperature	Hotspot formation	[235]
	No effects	n/a	[232]
	Reduction of reaction temperature	Hotspot formation	[236–238]
Methane oligomerization	No comparison with conventional process	n/a	[239, 240]
Methane reforming	Higher conversion, higher selectivity	Hotspot formation	[241]
Methane synthesis	No comparison with conventional process	n/a	[242, 243]
HCN synthesis	Higher selectivity	Selective catalyst heating	[244]
H_2S decomposition	Higher conversion	Hotspot formation	[227]
NO_x, SO_2 reduction	Higher conversion, enhanced reaction rate	No explanation	[229, 245–249]
SO_2 reduction	Higher conversion, phase transition of catalyst	Hotspot formation	[226]
	Higher rate	Imprecise temperature measurement	[250]
CO oxidation	No effects	n/a	[251, 252]
	Negligible effect	n/a	[253, 254]
Propane oxidation	Higher conversion reaction temperature	Hotspot formation	[255]
Ethane/propane/*n*-butane dehydrogenation	No comparison with conventional process	n/a	[256]
Ethane dehydrogenation	Higher conversion	Phase transition of catalyst	[257]
o-Xylene/toluene oxidation	Higher conversion, higher selectivity	Hotspot formation, more homogeneous dispersion of metallic particle	[258]

(continued)

Table 9.9 (continued)

Application	Observed effect	Explanation of observed effects	Ref.
2-Methylpentane isomerization	Higher conversion, higher selectivity	Changed dispersion of metallic particles	[167, 259–261]
Methanol steam reforming	Higher production rate, higher selectivity	No explanation	[262]
	Reduction of the reaction temperature	Homogeneous energy dissipation	[263]
2-Propanol dehydration	Reduction of the reaction temperature	No explanation	[264]

Reproduced with permission from [200]. Copyright 2009 by Wiley-VCH

9.4.4 Applications Directed Toward Hydrogen Storage

As a clean energy carrier, hydrogen is expected to make a significant contribution to a sustainable energy society [265] as it is a regenerative and an environmental friendly energy source, which reacts cleanly with oxygen in a highly exothermic reaction to yield pure water as the only exhaust product, similar to that from a hydrogen fuel cell. However, this is not without its problems; the drawbacks of hydrogen as an energy source are its storage, its transport, and energy costs because the energy density of hydrogen on a per volume basis is relatively low compared to other commonly used fuels.

Organic hydrides are efficient, high-density materials used for hydrogen storage, transportation, and supply to fuel cell systems in a Hydrogen Society. The organic hydride storage method (no need for low temperatures and high pressures) can make use of the infrastructure currently used for transport (Lorries and ships) and to store (reservoir tanks) gasoline. After storing the hydrogen gas, the aromatic hydrocarbon is collected by existing infrastructure equipment; otherwise, the infrastructure equipment used to store the hydrogen energy can share the equipment used to store gasoline. Aromatic hydrocarbons can also be used as fuels in times of natural disasters. The cyclohexane ↔ benzene cycle is a system capable of storing and releasing hydrogen. However, benzene is a major carcinogenic material; as such, the benzene present in gasoline is regulated to less than 1 vol.%. Consequently, the use of the cyclohexane ↔ benzene cycle is not desirable as an organic hydride carrier. By comparison, the decalin ↔ naphthalene cycle is highly efficient for hydrogen evolution, making it is possible to extract most of the hydrogen from the decalin molecule. Unfortunately, the melting point of naphthalene is 80 °C, so that when it is cooled in the reactor or in the flow pipe after the reaction it tends to solidify, which can clog either the reactor and/or the flow pipes. Unfortunately, the large quantities of these materials required for industrial application are not available. Moreover, a higher temperature reaction field is needed because the boiling point of ring compounds tends to be rather high. The presence of a methyl group in naphthalene lowers the melting point when compared with pure naphthalene. Accordingly, the methylcyclohexane (MCH) ↔ toluene cycle may well be the most suitable carrier of hydrogen for the organic hydride method in the industry.

Hydrogen production from MCH as the source of hydrogen was investigated by Horikoshi and coworkers, by subjecting MCH to microwave heating [266]. Pt supported on activated carbon particulates (Pt/AC) is the most common catalyst used in dehydrogenation reactions. Current research focuses on enhancing the efficiency of hydrogen evolution using conditions that require low applied energy for the thermal activation of the catalysts. Energy conservation for hydrogen evolution can be achieved by microwave energy selectively heating the solid catalysts (e.g., Pt/AC). The features pertaining to hydrogen evolution from MCH with the microwave method are worth noting; they are described next.

The reaction temperature for the dehydrogenation of MCH begins at ca. 200 °C [267], following which the yield of hydrogen is saturated at ca. 330 °C. Platinum (5 wt%) deposited as particulates onto activated carbon as the support (Pt/AC) was used as the catalyst in the dehydrogenation reactions of MCH performed under atmospheric pressure in a quartz fixed-bed tubular reactor loaded with 2.5 g of Pt/AC catalyst. Results from the dehydrogenation of MCH under microwave heating and conventional heating with a ceramics heater and insulator are displayed in Fig. 9.24 [268]. The packed Pt/AC catalyst particulates, microwave-heated to 330 °C within 1 min, resulted in ca. 94% dehydrogenation of MCH that occurred in less than 90 s. On the other hand, when using a ceramics heater, dehydrogenation needed more than 30 min to reach 330 °C, even though an insulator was used. Microwave heating led to hydrogen evolution in much shorter time. On interruption of microwave irradiation, the catalyst particulates promptly cooled and hydrogen evolution stopped, whereas interruption of conventional heating led to continuous hydrogen gas evolution for more than 30 min. Total power consumption of the microwave system was 36 W (microwave applied power, 17 W), while the ceramics heater consumed 115 W. Hence, microwave heating is to be favored in terms of power consumption, and therefore is an attractive profitable energy-saving

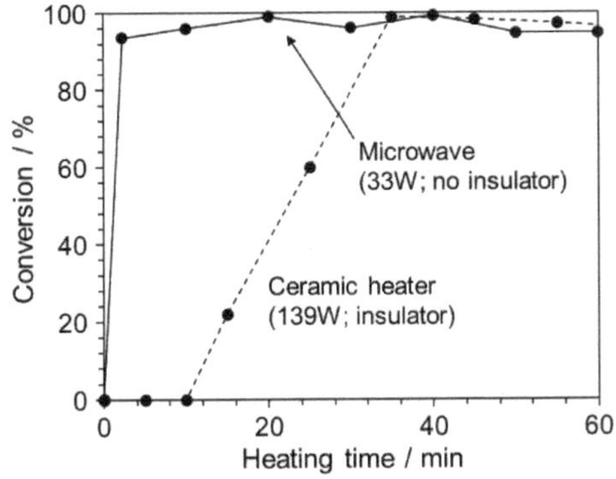

Fig. 9.24 Dependence of the conversion of methylcyclohexane (MCH) to hydrogen gas and toluene over the Pt/AC catalyst, under microwave heating and conventional heating with a ceramics heater (temperature, 330 °C). Reproduced with permission from Ref. [268]. Copyright 2016 by Elsevier B.V.

system; in addition, there are benefits of shorter times for hydrogen evolution. It is worth recalling that electrical energy is being used to produce hydrogen energy.

Heat convection from Pt/AC in methylcyclohexane solution was also investigated by simulation using the COMSOL Multiphysics software package (Version 4.3a) [268]. For the simulation, selective heating of 180 particles of Pt/AC dispersed in methylcyclohexane was considered; the model geometry is displayed in Fig. 9.25a. The correlation between bulk temperature and the temperature of the Pt/AC particles is illustrated in Fig. 9.25b. Results of the simulation demonstrated that to bring MCH from room temperature (25 °C) to 330 °C required the temperature

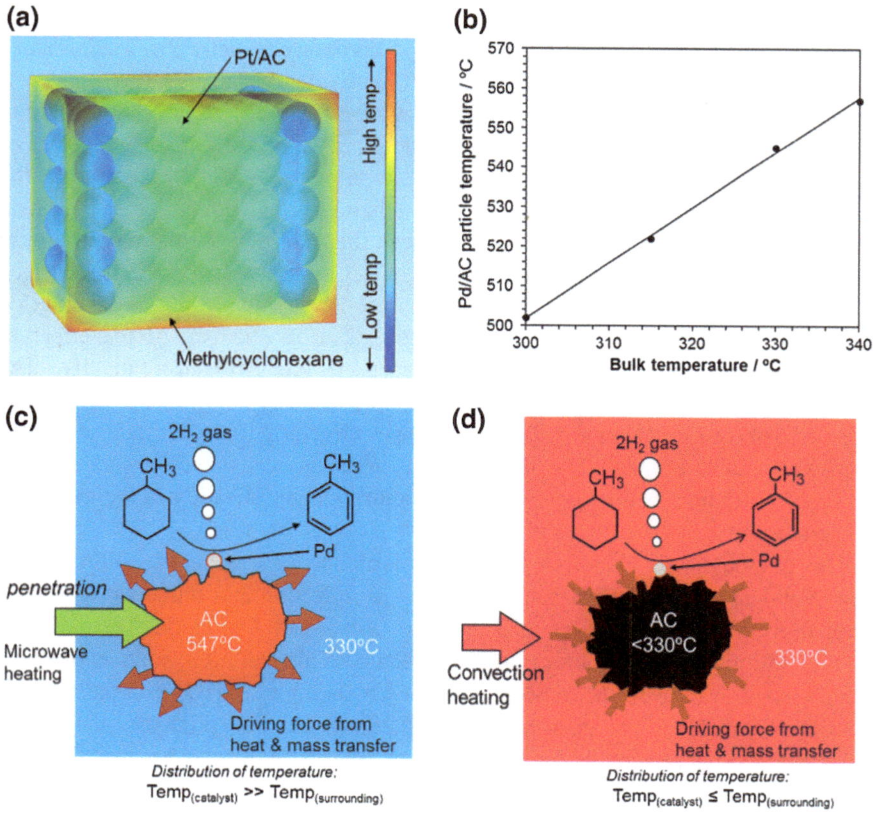

Fig. 9.25 Relationship between the Pt/AC particle temperatures and bulk temperatures. (**a**) Simulation image; (**b**) relationship between the Pt/AC particle temperatures and bulk temperatures. Model of heat transfer direction: (**c**) under microwave heating (MWH): $T_{catalyst} > T_{surrounding}$, heat transfers from the catalyst to the surroundings; mass also transfers in the same direction because of the coupling vector, and (**d**) under classical convection heating (CH): $T_{catalyst} \leq T_{surrounding}$, heat transfers from the surroundings to the catalyst. Reproduced with permission from Ref. [268]. Copyright 2016 by Elsevier B.V.

of the Pt/AC particles to be 547 °C. Under microwave irradiation, the microwave radiation field continuously heated the Pt/AC particles, with much of the resulting thermal energy spent to heat MCH.

9.4.5 Microwave-/Photo-Driven Catalytic Treatment of Wastewaters

Several years have passed since the discovery that led to the improvement in photocatalyst activity on exposure to microwave radiation. A photocatalyst is a material that absorbs the energy of electromagnetic waves, such as ultraviolet (UV light), and changes it into chemical energy. Microwave radiation also consists of electromagnetic waves and so can interact with photocatalysts. The notion of irradiating TiO_2 with microwaves may appear strange at first since the photon energy (1×10^{-5} eV) of the microwaves of frequency 2.45 GHz is several orders of magnitude lower than the band-gap energy required (3.0–3.2 eV) to activate the TiO_2 semiconductor. However, microwave radiation brings effects other than heat to bear on a photocatalyst. It has been deduced that microwave nonthermal effects contribute significantly to the enhancement of TiO_2-photoassisted reactions, since they may sensitize both the surface and the crystalline structure of the metal oxide to reactions taking place at the surface [269]. The mechanism of the effect of microwave radiation on photocatalyzed reactions has evolved gradually. The enhanced treatment of wastewaters through improvements in photocatalyst activity when exposed to microwave radiation is now a clear possibility. In fact, coupling microwave radiation with UV light in TiO_2-photoassisted processes as a novel advanced oxidation technology (AOT) can contribute significantly to the treatment of wastewaters.

The photoassisted oxidative (and reductive) decomposition of pollutants by means of TiO_2 photocatalyst nanoparticles is an effective and attractive oxidation (reduction) method in the general area of Advanced Oxidation Technologies (AOTs). Applications of photoassisted treatments to air pollution, such as in filters found in air conditioners, have been developed by fixation of TiO_2 nanoparticulates on several suitable substrate supports [270]. However, this photoassisted degradation method is not suitable for large-scale wastewater treatment because the rates of degradation of organic compounds dissolved in wastewaters tend to be rather slow. Relatively little has been done in this area in the last few years, as large-scale treatment of organic pollutants in aquatic environments, have been problematical, including low photodegradation efficiency. Accordingly, Horikoshi and coworkers [271] have proposed to couple microwave radiation with UV radiation to enhance the activity of photocatalysts.

Some dyes are poorly photodecomposed, and so are not useful in assessing the photoactivity of a metal-oxide photocatalyst. For example, the rate of photodegradation of the cationic rhodamine-B (RhB) dye is slow in acidic aqueous

Fig. 9.26 Visual comparison of color fading in the degradation of RhB solutions (0.05 mM) subsequent to being subjected to various degradation methods for 150 min. From left to right: initial RhB solution; RhB subjected to photoassisted degradation (TiO_2/UV); RhB subjected to integrated microwave/photoassisted degradation (TiO_2/UV/MW); and RhB subjected to thermal- and photoassisted degradation (TiO_2/UV/CH). Reproduced with permission from Ref. [272]. Copyright 2009 by Elsevier B.V.

media because the surface of the TiO_2 particles is positively charged (Ti–OH_2^+; pI (isoelectric point) = 6.3). Nonetheless, RhB has proven an interesting model compound to examine the microwave effect. In earlier studies, the major focus was on the degradation of organic pollutants as exemplified by the degradation of this dye catalyzed by TiO_2 semiconductor particles under both UV and microwave irradiation [271]. Changes in color intensity of the RhB dye solutions occurring under various conditions are illustrated in Fig. 9.26 [272]. The photodegradation of RhB is unambiguous in the use of the TiO_2-assisted UV/MW method. These observations demonstrated that a method is conceivable that can treat large quantities of pollutants in wastewaters by a hybrid combination of microwaves and TiO_2 photoassisted technologies. The photodegradation of RhB by this metal oxide is unaffected by conventional heating (CH)—compare, for example, the results from the UV and the UV/CH methods in the presence of TiO_2 (Fig. 9.26).

The possible enhancement of the photoactivity of TiO_2 photocatalysts subsequent to being exposed to microwave radiation has also been investigated from the viewpoint of the amount of $^{\bullet}OH$ radicals generated [273]. Formation of $^{\bullet}OH$ radicals during TiO_2-assisted photooxidations driven simultaneously by UV light and microwave radiation has been investigated by electron spin resonance spectroscopy; a novel setup was employed in which the ESR sample (containing the DMPO spin-trap agent and various TiO_2 particles in aqueous media) was irradiated by both UV light and microwave radiation [273]. In this case, microwave radiation was produced using a magnetron-excited microwave generator (frequency, 2.45 GHz), a three-stub tuner, a power monitor, and an isolator (Fig. 9.27). The UV irradiation source was an Ushio 250-W mercury lamp; the emitted UV light irradiated the sample at an angle to the horizontal plane using a fiber optic light guide.

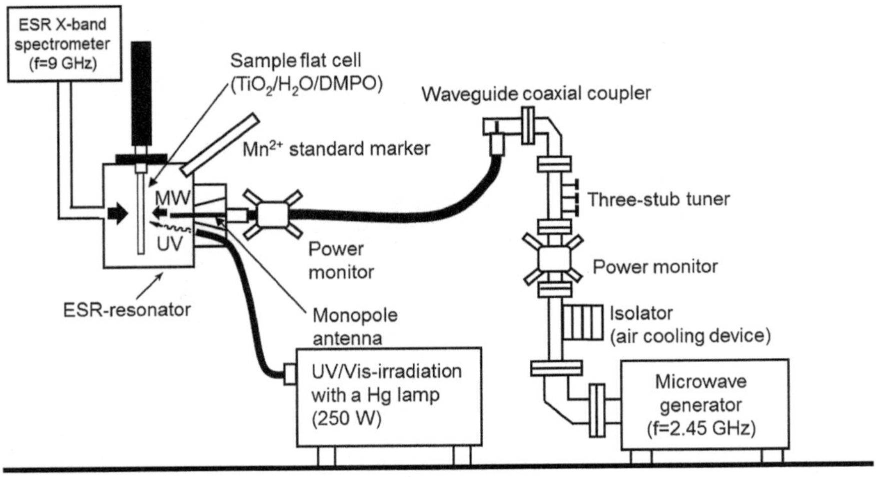

Fig. 9.27 System used to generate ${}^{\bullet}OH$ radicals in aqueous TiO_2 dispersions by MW irradiation alone, and by the UV and UV/MW methods, together with assessing the quantity of radicals formed. Reproduced with permission from Ref. [273]. Copyright 2003 by Elsevier B.V.

Table 9.10 Number of DMPO-${}^{\bullet}OH$ spin adducts produced in the various heterogeneous systems under microwave irradiation, UV irradiation, and MW/UV irradiation relative to those formed in the rutile TiO_2 specimen for the TiO_2/H_2O/MW heterogeneous system; data reproduced with permission from Ref. [273]

Methodology	P25 TiO_2	UV100 TiO_2	Anatase	Rutile
UV	182	45	110	110
UV/MW (3 W)	259	51	92	76
UV/MW (16 W)	369	–	–	–

Copyright 2003 by Elsevier B.V.

The number of ${}^{\bullet}OH$ radicals generated under various experimental conditions is given in Table 9.10. For P25 titanium dioxide, the number of ${}^{\bullet}OH$ radicals produced by the UV/MW method was nearly 30% greater than the quantity generated by the UV method alone [273]. A fivefold increase in incident microwave power from 3 to 16 W led to a significant increase (ca. 40%) in the number of ${}^{\bullet}OH$ radicals. Such an increase was sufficient to raise the efficiency of the photooxidation of organic pollutants in water. However, for the UV100 TiO_2 sample, the increase in the number of ${}^{\bullet}OH$ radicals produced by increasing the MW power five times was only 10% greater. On the other hand, the number of ${}^{\bullet}OH$ radicals generated for the pure anatase and rutile TiO_2 samples decreased under microwave irradiation. The P25 specimen was clearly affected by the microwaves, generating a significant number of ${}^{\bullet}OH$ radicals. Therefore, under such conditions, the rate of decomposition of pollutants can be enhanced by the combined UV/MW method when P25 is used to detoxify wastewaters. On the other hand, to the extent that the quantity of

$^{\bullet}$OH radicals produced by the other three TiO_2 systems does not increase, even when irradiated with microwaves, the rate of decomposition is not expected to be greatly affected by the microwaves.

The photoactivity of the TiO_2 photocatalyst appears hardly affected when conventional heating is substituted for the microwave radiation, Using various approaches, the nonthermal effects of microwaves have been examined since they have been suggested to be factors that might affect organic/inorganic reactions: for instance, (i) the influence of nonthermal microwave effects upon improving the affinity of the TiO_2 surface toward organic pollutants [274], (ii) the increase in the number of $^{\bullet}$OH radicals generated by the photooxidation of water (Table 9.7) [273], (iii) the influence of the microwaves' magnetic and electric fields on photocatalyst activity [275], (iv) the effect of frequency [276], and (v) the enhanced photoactivity of TiO_2 by microwave irradiation under cooling to ambient temperature with the heating attributed not only to a microwave thermal effect but also to a significant nonthermal effect that involved hotspots generated on the TiO_2 particle surface [277]. Additionally, microwaves likely influence the electrovalence inside the light-activated catalyst; that is, electron recombination in the photocatalyst TiO_2 particle with an oxygen vacancy is obstructed by microwave irradiation. To assess this possibility, transient decay kinetics at 550 nm were determined in situ for all three samples using a 150 ps Nd-YAG pulsed laser system (10 Hz) and a Xe flash lamp (2-μs pulses) probe while samples were being microwave irradiated (2.45 GHz) (Fig. 9.28). The transient(s) absorbed at the probe wavelength displayed double exponential decay kinetics: (i) a fast decay that occurred within ca. 5–12 ns ascribed to recombination of photogenerated shallow-trapped (or free conduction band) electrons with valence band holes, and (ii) a slower decay that occurred from hundreds of nanoseconds to several microseconds, attributable to recombination of electrons caught in deep traps (e.g., either as Ti^{3+} or as F color centers) with free holes.

9.4.6 Synthesis of Metal Catalysts on Carbonaceous Material Supports

Rapid and even heating applications of microwaves open new opportunities in chemistry, materials science, and in other areas. Microwave heating is now developing into an effective method for the preparation of catalysts [279] and can provide (i) rapid drying, (ii) moisture leveling leading to a uniform distribution of metal in the support, (iii) physically strong pellets compared to conventionally heated catalysts, and (iv) a change in the activity of the catalyst compared to its conventionally prepared analogs [280]. Thus, microwave heating offers the possibility of developing catalysts with uniform pellet strength, enhanced reaction rates, and selectivity.

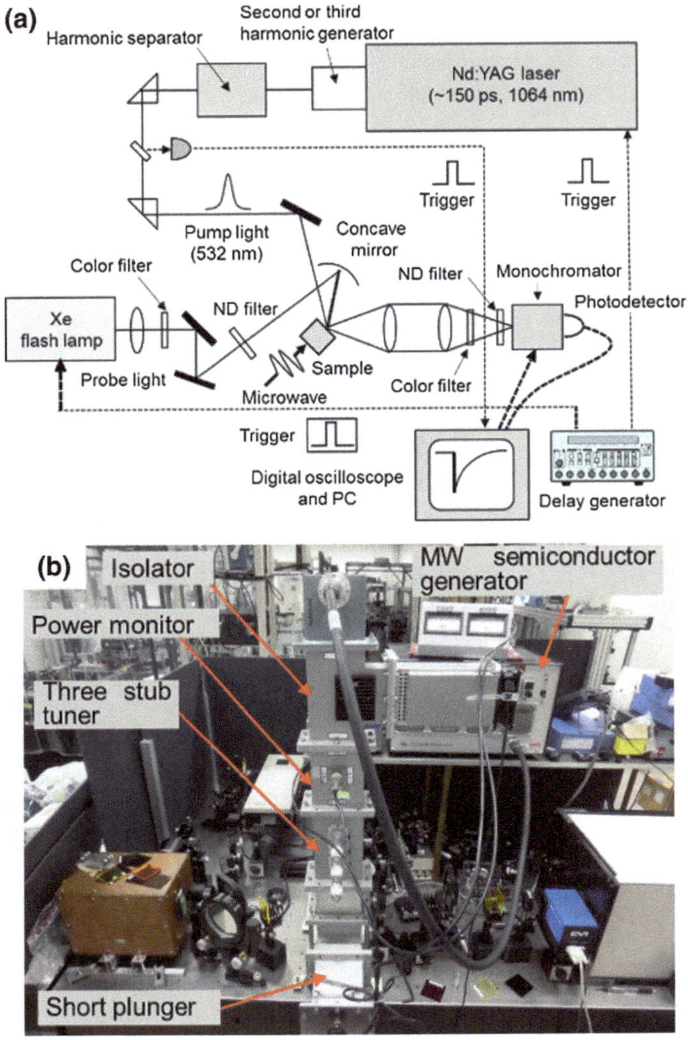

Fig. 9.28 (**a**) Schematic illustrating the overall diffuse reflectance setup; (**b**) photograph of the microwave irradiation setup. Reproduced with permission from Ref. [278]. Copyright 2015 by Royal Society of Chemistry

According to van Wyk et al. [281], microwave enhancement of activated carbon particles produces a product that performs better than standard activated carbon As such, microwaves are now being used, where heating dielectric materials is required, in various technological and scientific areas. Palladium nanoparticles deposited on carbon nanotubes (CNTs) find applications that range from hydrogen storage to catalysis [282–284]. The one-step linker-free synthesis of hybrid materials consisting of palladium nanoparticles and multiwall carbon nanotubes (Pd-NP/

MWCNTs) has been described, [285]. The method uses microwave radiation for the effective decomposition of the tris(dibenzylideneacetone)dipalladium(0), $Pd_2(dba)_3$, complex in the presence of MWCNTs. High loadings of Pd nanoparticles (up to 40 wt%), having sizes between 3 and 5 nm, could be deposited on the surface of MWCNTs within a very short time (2 min). Moreover, the microwave-assisted procedure to synthesize graphene-supported metallic catalysts has become a widely documented method [286, 287]. Although graphene-supported metal catalysts have high surface areas and stabilities, only a few studies have examined their application for organic reactions; greater attention has been focused on applications in fuel cells, hydrogen storage, sensors and solar cells [288, 289].

To prevent a solution of graphene from being agglomerated through van der Waals attractive forces and restacked to form graphite owing to the absence of O groups, some efforts have been expended on synthesizing graphene-supported Pt nanostructures by the microwave heating method [290, 291]. Aggregation of graphene sheets reduces the high surface area of graphene and so limits the effective dispersion of Pt nanoparticles being deposited onto graphene sheets [292]. Microwave-assisted syntheses have also been reported on the preparation of graphene-supported Pt/Pd nanostructures, in which graphene oxide (or graphitic oxide, GO), and Pt precursors (K_2PtCl_4) or Pd precursors ($PdCl_2$) were reduced through microwave irradiation using ascorbic acid in the presence of hexadecyltrimethylammonium bromide [293]. Poly (diallydimethyl ammonium chloride) is effective in modifying Pd/Pt nanoparticles through the functionalization of graphene oxide in ethylene glycol [294]. Other reports have described the deposition of Pt and Pd/Pt nanocrystals on the few exfoliated layers of graphene by solvothermal reactions [295]. Also, the preparation and the deposition of Pd/Pt nanoparticles in the cubic form on graphene have been performed in *N,N*-dimethylformamide solvent [296].

Microwaves can be used to synthesize graphitic carbon particles at temperatures lower than a conventional method, for which the most common synthesis route is chemical vapor deposition at temperatures around 800 °C in the presence of a transition metal catalyst [297]. Microwave irradiation of polyethylene glycol at temperatures between 160 and 220 °C for 40 min in the absence of a catalyst yields chains of graphitic carbon particles. Ruthenium nanocatalysts have been deposited on a γ-Al_2O_3 catalyst support using a microwave-assisted solvothermal method [298]. The resulting material can hydrogenate phenol to cyclohexanone. A feature of this method is the rapid synthesis of Ru nanoparticles with a small diameter and a narrow size distribution. Similarly, Glaspell et al. [299] reported a simple method for the microwave-assisted synthesis and characterization of Au and Pd nanoparticle catalysts supported on CeO_2 nanoparticles for CO oxidation; catalysts so produced, displayed high activity and stability. The method offers extremely short reaction times, and efficiently produces high yields of very pure nanoparticle catalysts.

In their 2003 article, Shi et al. [300] reported some microwave-assisted wet chemical syntheses of various catalytic materials, in which they described the advantages, the significance, and steps needed toward possible industrialization. The steps identified and discussed to bring the microwave-assisted syntheses of catalysts to industrial manufacturing were: (i) identify and evaluate the microwave nonthermal effect, (ii) examine the microwave effects at different synthesis stages, and (iii) examine the temperature influence on the microwave effects. In this context, the influence of microwave radiation on the properties and structure of the superacid Fe_2O_3/SO_4^{2-} solid catalyst was examined by Ning et al. [301] who noted that, compared to the traditional method of preparing such superacids, the noncrystalline Fe_2O_3/SO_4^{2-} superacid, prepared in a microwave field, displayed the highest magnetic susceptibility and catalytic activity when applied to esterification reactions.

The characterization and catalytic properties of nickel aluminate nanoparticles prepared by both a conventional combustion method (CCM) and a microwave combustion method (MCM) using nickel nitrate and aluminum nitrate as precursors in aqueous media were reported by Ragupathi et al. [302]. The preparation involved the *Opuntia dillenii haw* plant extract, which simplified the process, provided an alternative procedure for a simple and economical synthesis that required no surfactant, and led to a rapid method for the synthesis of $NiAl_2O_4$ nanoparticles. The MCM method produced $NiAl_2O_4$ with uniform size and well-defined crystals; optical property was determined by diffuse reflectance spectroscopy. The $NiAl_2O_4$ prepared by the microwave combustion method possessed a higher surface area and a lower crystallite size than the $NiAl_2O_4$ nanoparticles prepared by the conventional combustion method; as a result, this led to improved performance for the selective oxidation of benzyl alcohol to benzaldehyde.

Microwave hydrothermal processing has been used [303] to prepare two bismuth oxides: ABi_2O_6 (A = Mg, Zn) with the unusual oxidation state of Bi^{5+} that could previously only be prepared under limited hydrothermal conditions. The authors discovered that microwave hydrothermal processing for soft-chemical and dissolution-precipitation reactions was far more advantageous to the kinetics of the process than conventional hydrothermal processing. The microwave method has also been used [304] to synthesize solids with the saponite structure, with Mg^{2+} or Ni^{2+} as octahedral cations and Fe^{3+} in the tetrahedral sheet; the formulae of the resulting solids were calculated from elementary chemical analysis data, resulting in SNiFe: $\{[Si_{7.136}Fe_{0.864}][Ni_{5.813}][Na_{0.810}]O_{20}(OH)_4{\cdot}13H_2O\}$ and SMgFe $\{[Si_{7.200}Fe_{0.800}][Mg_{5.820}][Na_{0.7020}]O_{20}(OH)_4{\cdot}9H_2O\}$. A special type of silico-alumino-phosphate with an opal structure (Opal-SAPO) has been prepared under microwave irradiation conditions using tetraethylorthosilicate, triethanolamine, aluminum isopropoxide and phosphoric acid (85%) [305]. The catalytic behavior of the Opal-SAPO on the regeneration of methylethylketone, benzaldehyde, and cyclohexanone from the corresponding semicarbazone derivatives was examined by Derikvand and Farzaneh [305]. Results indicated that the OH groups on the Opal-SAPO surface provided the Bronsted acidity to the substrates. The first step in the regeneration of aldehydes or ketones from semicarbazone derivatives involved the chemisorption of semicarbazones on the Bronsted acid sites. Cleavage of the C–N bond was catalyzed by H^+ ions, followed by hydrolysis by small amounts of water; no reaction occurred without the presence of water. The reactivity of Opal-SAPO for the regeneration of carbonyls was methylethylketone > benzaldehyde > cyclohexanone.

9.4.7 Catalyst Synthesis Using Features of Microwave Heating

In the field of heterogeneous photocatalysis, recent years have seen the immobilization of titanium dioxide particles on support materials such as glass, silica beads, polymers, and zeolites [306] as the reuse of dispersed TiO_2 nanoparticles can cause undue problems of filtration. Accordingly, it is important to fix TiO_2 particles on supports for possible recycling after the photocatalytic events. Note that free nanoparticles may have important consequences on human health and on ecological systems [307]. In this regard, the U.S. Food and Drug Administration (FDA) permits up to 1% TiO_2 as an inactive ingredient in food products. While there are no known health effects associated with the use of TiO_2, a recent study found that 3–6-year-old children are the most affected group consuming TiO_2 particles from food products. A regulatory framework for the use of TiO_2 in food products has yet to be firmly established in many countries, especially in developing nations. Many new properties of TiO_2 have been explored during the past few years [308]. On the other hand, the degree of affinity of TiO_2 and carbon materials (e.g., activated carbon and graphite, among others) is very low so that it is difficult to immobilize TiO_2 particles on the surface of carbon-based materials.

Accordingly, it is instructive to examine the preparation of TiO_2 particles supported on activated carbon (AC) particulates using the microwave (MW) heating method, subsequently compared to the more traditional hydrothermal (oil bath) heating method. Scanning Electron Microscopic (SEM) photographs revealed an average particle size of TiO_2 on the AC surface of around 415 nm (diameter) under microwave heating (Fig. 9.29a); when the same heating rate and reaction temperature were used by the oil bath heating procedure, the TiO_2 formed on the AC surface was different from that of the microwave method in that with oil bath heating only a thin TiO_2 film, not nanoparticles, formed on the AC support (Fig. 9.29b). No growth of TiO_2 particles on activated carbon by the microwave

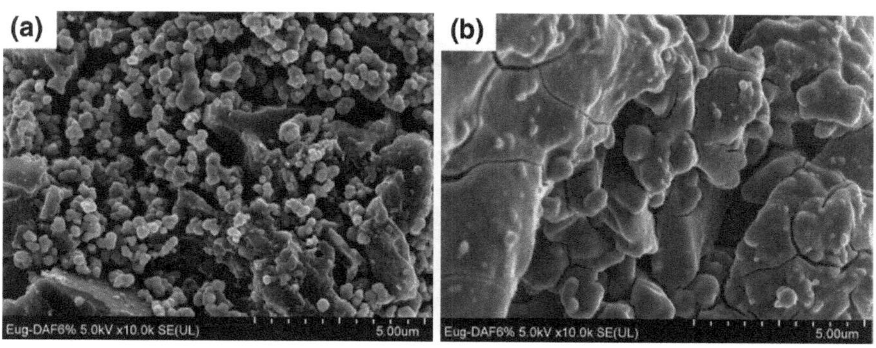

Fig. 9.29 Scanning electron microscopic images of TiO_2/activated carbon (AC) particulates under various synthesis conditions: reaction temperatures 90 °C under (**a**) microwave heating and (**b**) oil bath heating. Reproduced from Ref. [309]. Copyright 2013 by Elsevier B.V.

method had been hitherto been reported. Obviously, the formation mechanisms of TiO_2 on the AC surface differ between the microwave and the oil bath heating methods as evident from the results observed by the SEM technique.

The proposed growth mechanism of TiO_2 under microwave heating and oil bath heating is seen in the illustrations of Fig. 9.30 [309]. At temperatures greater than 80 °C, the microwave heating efficiency of the water decreased, whereas microwave direct heating of the AC particles increased (Fig. 9.30a). Evidently, direct heating of the AC particles produced TiO_2 particles on the AC surface, together with smaller TiO_2 particles that were formed at the higher reaction temperatures. While the dielectric loss (ε'') of water decreases at higher temperatures, the AC particles are heated selectively. In the case of the oil bath heating method, formation of TiO_2 on the AC surface occurred only through heat conduction from the oil heating the reactor externally. Therefore, under high-temperature conditions, the temperature of the AC particle was lower than that of the solution and caused a thin TiO_2 coating to form efficiently on the AC surface by oil bath heating (Fig. 9.30b). Clearly, the growth of the TiO_2 thin film took place through heat conduction from the solution.

A novel microwave-assisted synthesis of Ni/C nanocomposites was reported by Gunawan et al. [310]; the synthesis required only a few minutes starting from nickel salts and a renewable high-content carbon source (tannin). The carbon precursor was *Quebracho tannin*, a renewable resource material obtained from the hot-water extraction from *Schinopsis lorentzii* and *Schinopsis balansae,* two products indigenous to Argentina and Paraguay. The process involved a simultaneous carbonization of the carbon precursor, as well as the reduction of nickel ions to elemental nickel nanoparticles in an ambient atmosphere. The technique provided a fast, easy, and economical way to produce nickel/carbon nanocomposites, requiring neither hydrogen nor an inert gas during the transformation. The resulting nanocomposites possessed a high surface area, thus suitable as high-efficiency

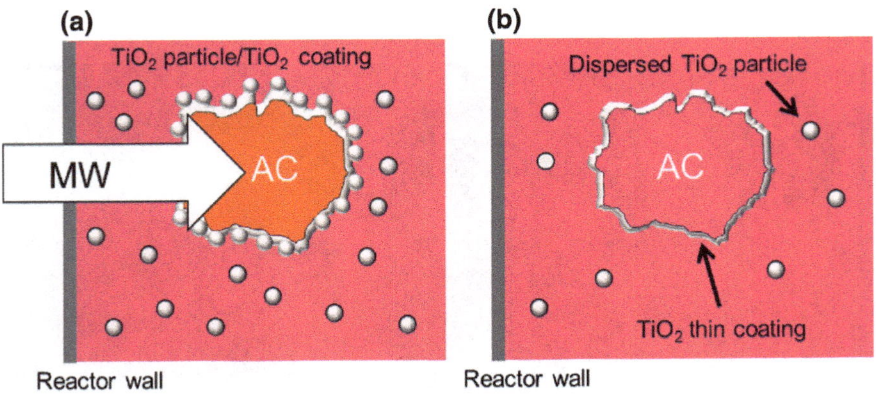

Fig. 9.30 Illustration of the growth mechanism of TiO_2/AC particles produced by the microwave heating method at (**a**) a temperature of 70 °C and (**b**) at temperatures greater than 80 °C. Reproduced from Ref. [309]. Copyright 2013 by Elsevier B.V.

Fig. 9.31 Scanning Electron Microscopic images of the Ni/C nanocomposites prepared by solution dispersion at: (**a**) 5000× magnification, (**b**) 10,000× magnification, (**c**) 25,000× magnification, and (**d**) 75,000× magnification. Reproduced from Ref. [310]. Copyright 2011 by the Taylor & Francis Publisher, London, UK

catalysts. SEM images (Fig. 9.31) showed that the solution dispersion method gave a better dispersion of nickel nanoparticles than the solid-state dispersion method. The surface area of the nanocomposites was high, indicating a highly porous nanocomposite material.

9.5 Coffee Break 9: Future of Microwave Processing of Foods

What are the major barriers to adopting new microwave food processes? There are several but the major problems are:

1. Microwave processing systems are expensive: this is electrically generated energy that must compete in cost with natural gas and steam. Also, the capital cost can be high. For example, to dry 1 kg of water in an hour requires 1 kW of microwave energy that may cost from $5000 to $10,000 per installed kilowatt; this includes the generators, the applicators, the conveyor, controls, etc. While removing many kilograms of water is probably not commercially feasible, removing the last few percent of water that takes a long time conventionally can be done very rapidly with microwaves, which saves not only time but also floor space, extensive equipment, and more (Fig. 9.32) [311].

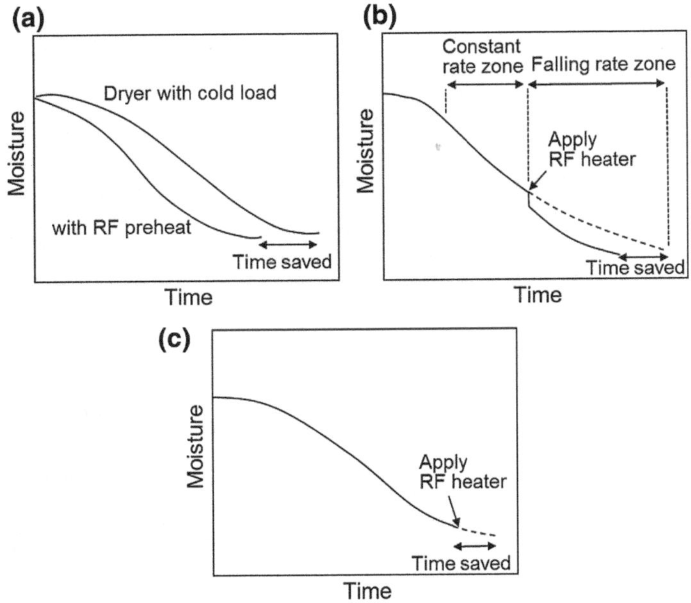

Fig. 9.32 Drying with microwaves: three methods all combining microwave energy with conventional hot-air drying: (**a**) preheating—microwaves applied throughout drying process; (**b**) booster drying—microwaves applied in the falling drying rate zone; and (**c**) finish drying—microwaves applied as hot-air drier efficiency becomes low: it may shorten drying time by 2/3. Reproduced with permission from Ref. [311]. Copyright 2014 by the CRC Press

2. Very few companies want to be first! The industry is often reluctant to change from its tried-and-true old processes to something so different and perhaps be the first to change.
3. The major microwave equipment manufacturers are small companies unable to do much process research on the users' products, making it necessary for the user to do the research for which it may be ill-equipped.

What are some of the major opportunities for processing of foods with microwave energy? There are a number of new developing systems, as well as some opportunities waiting to be tried and developed:

1. **Microwave pasteurization and sterilization of packaged consumer products**: Two major processes are under investigation, one of which, Micvac®, is marketing ambient packaged meals with extended shelf life. The other is a system developed by Washington State University [312]. Both have the advantage of lower temperature and shorter processing times, providing better flavor, color, and texture of their food products.
2. **Baking**: In the mid-1970s, Schiffmann [312] developed a highly successful microwave bread baking system that produced excellent quality bread in 1/3 the

time and reduced the energy required for baking by more than 50%. All these systems are ideal for processing with microwave energy.

3. **Microwave frying**: Combining microwave energy with frying can dramatically increase the production throughput, i.e., the number of pounds or kilograms produced in a given time. This can be applied to chicken, fish, doughnuts, pastries, and more. It was technically and economically successful in the 1970s and should be resurrected [312].

References

1. S. Andrea, Classic kit: bunsen burner. *Chem. World* **4**(10) (2007) (Royal Society of Chemistry)
2. J.J. Griffith, *Chemical Reactions—A Compendium of Experimental Chemistry*, 8th edn. (R Griffin and Co., Glasgow, 1838)
3. https://en.wikipedia.org/wiki/Gas_lighting#cite_note-18
4. M. Kohn, Remarks on the history of laboratory burners. J. Chem. Educ. **27**, 514–516 (1950)
5. J.W. Vanderhoff, Method for carrying out chemical reactions using microwave energy. Patent US 3,432,413 (1969)
6. I.A. Halmos, Process for preparing acetylsulfaguanidine. Patent US 4,210,593 (1980)
7. R. Gedye, F. Smith, K. Westaway, H. Ali, L. Baldisera, L. Laberge, J.R. Rousell, The use of microwave ovens for rapid organic synthesis. Tetrahedron Lett. **27**, 279–282 (1986)
8. R.J. Giguere, T.L. Bray, S.M. Duncan, G. Majetich, Application of commercial microwave ovens to organic synthesis. Tetrahedron Lett. **27**, 4945–4948 (1986)
9. A.K. Rathi, M.B. Gawande, R. Zboril, R.S. Varma, Microwave-assisted synthesis—catalytic applications in aqueous media. Coord. Chem. Rev. **291**, 68–94 (2015)
10. J.M. Kremsner, A. Stadler, C.O. Kappe, The scale-up of microwave-assisted organic synthesis. Top. Curr. Chem. **266**, 233–278 (2006)
11. J.R. Schmink, N.E. Leadbeater, Probing microwave effects using Raman spectroscopy. Org. Biomol. Chem. **7**, 3842–3846 (2009)
12. D. Obermayer, B. Gutmann, O. Kappe, Microwave chemistry in silicon carbide reaction vials: separating thermal from nonthermal effects. Angew. Chem. Int. Ed. **48**, 8321–8324 (2009)
13. http://pubs.acs.org/paragonplus/submission/joceah/joceah_authguide.pdf
14. P.T. Anastas, J.C. Warner, in *Green Chemistry: Theory and Practice* (Oxford University Press, New York, 1988)
15. T. Razzaq, C.O. Kappe, On the energy efficiency of microwave-assisted organic reactions. Chemsuschem **1**, 123–132 (2008)
16. https://www.epa.gov/greenchemistry/basics-green-chemistry#twelve
17. R.S. Varma, Solvent-free organic syntheses using supported reagents and microwave irradiation. Green Chem. **1**, 43–55 (1999)
18. B.A. Roberts, C.R. Strauss, Toward rapid, "Green", Predictable microwave-assisted synthesis. Acc. Chem. Res. **38**, 653–661 (2005)
19. C.O. Kappe, Controlled microwave heating in modern organic synthesis. Angew. Chem. Int. Ed. **43**, 6250–6284 (2004)
20. D. Adam, Microwave chemistry: out of the kitchen. Nature **421**, 571–572 (2003)
21. V. Polshettiwar, R.S. Varma, Greener and expeditious synthesis of bioactive heterocycles using microwave irradiation. Pure Appl. Chem. **80**, 777–790 (2008)
22. F. Mavandadi, A. Pilotti, The impact of microwave-assisted organic synthesis in drug discovery. Drug Discov. Today **11**, 165–174 (2006)

23. https://www.acs.org/content/acs/en/greenchemistry/what-is-green-chemistry/principles/gc-principle-of-the-month-2.html
24. M.B. Gawande, V.D.B. Bonifácio, R. Luque, P.S. Branco, R.S. Varma, Solvent-free and catalysts-free chemistry: a benign pathway to sustainability. Chemsuschem **7**, 24–44 (2014)
25. A. Loupy, A. Petit, M. Ramdani, C. Yvanaef, M. Majdoub, B. Labiad, D. Villemin, The synthesis of esters under microwave irradiation using dry-media conditions. Can. J. Chem. **71**, 90–95 (1993)
26. S.J. Song, S.J. Cho, D.K. Park, T.W. Kwon, S.A. Jenekhe, Microwave enhanced solvent-free synthesis of a library of quinoline derivatives. Tetrahedron Lett. **44**, 255–257 (2003)
27. J.C. Lee, H.J. Park, J.Y. Park, Rapid microwave-promoted solvent-free oxidation of α-methylene ketones to α-diketones. Tetrahedron Lett. **43**, 5661–5663 (2002)
28. A. Carpita, A. Ribecai, P. Stabile, Microwave-assisted synthesis of indole- and azaindole-derivatives in water via cycloisomerization of 2-alkynylanilines and alkynylpyridinamines promoted by amines or catalytic amounts of neutral or basic salts. Tetrahedron **66**, 7169–7178 (2010)
29. N.E. Leadbeater, M. Marco, Transition-metal-free Suzuki-type coupling reactions. Angew. Chem. Int. Ed. **42**, 1407–1409 (2003)
30. R.K. Arvela, N.E. Leadbeater, M.S. Sangi, V.A. Williams, P. Granados, R.D. Singer, A reassessment of the transition-metal free Suzuki-type coupling methodology. J. Org. Chem. **70**, 161–168 (2005)
31. N.E. Leadbeater, R.J. Smith, Real-time monitoring of microwave-promoted Suzuki coupling reactions using *in situ* Raman spectroscopy. Org. Lett. **8**, 4589–4591 (2006)
32. N. Rubio, M.A. Herrero, A. de la Hoz, M. Meneghetti, M. Prato, E. Vázquez, Versatile microwave-induced reactions for the multiple functionalization of carbon nanotubes. Org. Biomol. Chem. **8**, 1936–1942 (2010)
33. A.S. Galanis, F. Albericio, M. Grøtli, Solid-phase peptide synthesis in water using microwave-assisted heating. Org. Lett. **11**, 4488–4491 (2009)
34. K. Hojo, N. Shinozaki, Y. Nozawa, Y. Fukumori, H. Ichikawa, Aqueous microwave-assisted solid-phase synthesis using boc-amino acid nanoparticles. Appl. Sci. **3**, 614–623 (2013)
35. M. Raju, S. Maeorg, O. Tsubrik, U. Maeorg, Efficient solventless technique for Boc-protection of hydrazines and amines. Arkivoc **6**, 291–297 (2009)
36. H. Sajiki, F. Aoki, H. Esaki, T. Maegawa, K. Hirota, Efficient C–H/C–D exchange reaction on the alkyl side chain of aromatic compounds using heterogeneous Pd/C in D_2O. Org. Lett. **6**, 1485–1487 (2004)
37. H. Shimodaira, T. Abe, T. Abe, Y. Kawanishi, A. Miyazawa, in *Proceedings of Global Congress on Microwave Energy Applications*, 2008, pp. 627–628
38. S. Horikoshi, S. Matsuzaki, S. Sakamoto, N. Serpone, Efficient degassing of dissolved oxygen in aqueous media by microwave irradiation and the effect of microwaves on a reaction catalyzed by Wilkinson's catalyst. Rad. Phys. Chem. **97**, 48–55 (2014)
39. M.J. Gronnow, R.J. White, J.H. Clark, D.J. Macquarrie, Energy efficiency in chemical reactions: a comparative study of different reaction techniques. Org. Proc. Res. Dev. **9**, 516–518 (2005)
40. I.R. Baxendale, J.J. Hayward, S.V. Ley, Microwave reactions under continuous flow conditions. Comb. Chem. High T. Scr. **10**, 802–836 (2007)
41. C.R. Strauss, R.W. Trainor, Developments in microwave-assisted organic chemistry. Aust. J. Chem. **48**, 1665–1692 (1995)
42. S. Christiaens, X. Vantyghem, M. Radoiu, J.J.V. Eynde, Batch and continuous flow preparation of hantzsch 1,4-dihydropyridines under microwave heating and simultaneous real-time monitoring by Raman spectroscopy. An exploratory study. Molecules **19**, 9986–9998 (2014)
43. H.M. Kingston, S.J. Haswell (eds.), *Microwave-Enhanced Chemistry: Fundamentals, Sample Preparation, And Applications* (American Chemical Society, USA, 1997)

44. A. de la Hoz, A. Loupy (eds.), *Microwaves in Organic Synthesis* (Wiley-VCH, Germany, 2012)
45. C.O. Kappe, D. Dallinger, S.S. Murphree, *Practical Microwave Synthesis for Organic Chemists: Strategies, Instruments, and Protocols* (Wiley-VCH, Germany, 2009)
46. C.O. Kappe, A. Stadler, D. Dallinger, *Microwaves in Organic and Medicinal Chemistry* (Methods and principles in medicinal chemistry) (Wiley-VCH, Germany, 2012)
47. D. Bogdal, *Microwave-Assisted Organic Synthesis: One Hundred Reaction Procedures* (Elsevier, UK, 2005)
48. S.C. Ameta, P.B. Punjabi, R. Ameta, C. Ameta (eds.), *Microwave-Assisted Organic Synthesis: A Green Chemical Approach* (CRC Press, NW, USA, 2015)
49. J. Tierney, P. Lidström (eds.), *Microwave Assisted Organic Synthesis* (Wiley-VCH, Germany, 2009)
50. M. Larhed, K. Olofsson (eds.), *Microwave Methods in Organic Synthesis* (Topics in current chemistry) (Springer, Germany, 2010)
51. P.J.H. Scott, *Solid-Phase Organic Syntheses, Volume 3: Microwave-Assisted Solid-Phase Organic Synthesis* (Wiley-VCH, Germany, will be published 2018)
52. S. Horikoshi, N. Serpone (eds.), *Microwaves in Catalysis: Methodology and Applications* (Wiley-VCH, Germany, 2015)
53. E. van der Eycken, C.O. Kappe (eds.), *Microwave-Assisted Synthesis of Heterocycles* (Springer, Germany, 2006)
54. V. Polshettiwar, R.S. Varma (eds.), *Aqueous Microwave Assisted Chemistry: Synthesis and Catalysis* (Royal Society of Chemistry, UK, 2010)
55. N.E. Leadbeater, C.B. McGowan (eds.), *Laboratory Experiments Using Microwave Heating* (CRC Press, NW, USA, 2013)
56. G. Cravotto, D. Carnaroglio (eds.), *Microwave Chemistry* (de Gruyter, Germany, 2017)
57. N.E. Leadbeater, *Microwave Heating as a Tool for Sustainable Chemistry* (Sustainability: contributions through science and technology) (CRC Press, NW, USA, 2010)
58. M. Nishii, Studies on the curing of denture base resins with microwave irradiation. J. Osaka Dent. Univ. **2**, 23–40 (1968)
59. B. Levin, J.L. Sanders, P.V. Reitz, The use of microwave energy for processing acrylic resins. J. Prosthet. Dent. **61**, 381–384 (1989)
60. A. Gourdenne, A. Maassarani, P. Monchaux, S. Aussudre, L. Thourel, *Polym. Prepr. (Am. Chem. Soc., Div. Polym. Chem.)* **20**, 471 (1979)
61. M. Teffal, A. Gourdenne, Activation of radical polymerization by microwaves—I. Polymerization of 2-hydroxyethyl methacrylate. Eur. Polym. J. **19**, 543–549 (1983)
62. C. Zhang, L. Liao, S. Gong, Recent developments in microwave-assisted polymerization with a focus on ring-opening polymerization. Green Chem. **9**, 303–314 (2007)
63. S. Sinnwell, H. RitterA, Recent advances in microwave-assisted polymer synthesis. Aust. J. Chem. **60**, 729–743 (2007)
64. A. Sosnika, G. Gotelli, G.A. Abraham, Microwave-assisted polymer synthesis (MAPS) as a tool in biomaterials science: how new and how powerful. Progress Poly. Sci. **36**, 1050–1078 (2011)
65. H. Yanagawa, K. Kojima, M. Ito, N. Handa, Synthesis of polypeptides by microwave-heatin. 1. Formation of polypeptides during repeated hydration- dehydration cycles and their characterization. J. Mol. Evol. **31**, 180–186 (1990)
66. A.F. Porto, B.L. Sadicoff, M.C.V. Amorim, M.C.S. de Mattos, Microwave- assisted free radical bulk-polyaddition reactions in a domestic microwave oven. Polym. Test. **21**, 145–148 (2002)
67. J. Jovanovic, B. Adnadjevic, Comparison of the kinetics of conventional and microwave methyl methacrylate polymerization. J. Appl. Polym. Sci. **104**, 1775–1782 (2007)
68. C. Goretzki, A. Krlej, C. Steffens, H. Ritter, Green polymer chemistry: Microwave-assisted single-step synthesis of various (meth)acrylamides and poly(meth)acrylamides directly from (meth)acrylic acid and amines. Macromol. Rapid Commun. **25**, 513–516 (2004)

69. M. Iannelli, H. Ritter, Microwave-assisted direct synthesis and polymerization of chiral acrylamide. Macromol. Chem. Phys. **206**, 349–353 (2005)
70. S. Schmitz, H. Ritter, Access to poly{N-[3-(dimethylamino)propyl](meth)acrylamide} via microwave-assisted synthesis and control of LCST-behavior in water. *Macromol.* Rapid Commun. **28**, 2080–2083 (2007)
71. F. Fischer, R. Tabib, R. Freitag, Chain transfer polymerisation of poly-N-alkylacrylamides in superheated methanol and by microwave induction. Eur. Polym. J. **41**, 403–408 (2005)
72. M.S. Cortizo, Polymerization of diisopropyl fumarate under microwave irradiation. J. Appl. Polym. Sci. **103**, 3785–3791 (2007)
73. M.S. Cortizo, S. Laurella, J.L. Alessandrini, Radiolysis study of actinide complexing agent by irradiation with helium ion beam. Radiat. Phys. Chem. **76**, 1140–1144 (2007)
74. J. Lu, S.-J. Ji, J. Wu, F. Guo, H. Zhou, X. Zu, Study of microwave irradiation polymerization mechanism in the presence of carriers.J. Appl. Polym. Sci. **91**, 1519–1524 (2004)
75. E. Bezdushna, H. Ritter, Microwave accelerated synthesis of *N*-phenylmaleimide in a single step and polymerization in bulk. Macromol. Rapid Commun. **26**, 1087–1092 (2005)
76. P. Eckstein, H. Ritter, Microwave-assisted synthesis, transesterification and polymerization of N-(2-acetoxy-ethyl-)maleimide. Des. Monom. Polym. **8**, 601–607 (2005)
77. M. Iannelli, E. Bezdushna, H. Ritter, Microwave-assisted bulk synthesis and polymerization of *N*-benzenesulfonamide maleimide. J. Macromol. Sci. Part A Pure Appl. Chem. **44**, 7–10 (2007)
78. H. Stange, M. Ishaque, N. Niessner, M. Pepers, A. Greiner, Microwave-assisted free radical polymerizations and copolymerizations of styrene and methyl ethacrylate. Macromol. Rapid Commun. **27**, 156–161 (2006)
79. H. Stange, A. Greiner, Microwave-assisted free radical copolymerizations of styrene and methyl methacrylate. Macromol. Rapid Commun. **28**, 504–508 (2007)
80. C.M. Fellows, Preliminary observations on the copolymerisation of acceptor monomer: donor monomer systems under microwave irradiation. Cent. Eur. J. Chem. **3**, 40–52 (2005)
81. S. Agarwal, M. Becker, F. Tewes, Synthesis, characterization and properties evaluation of copolymers of 2,3,4,5,6-pentafluorostyrene and *N*-phenylmaleimide. Polym. Int. **54**, 1620–1625 (2005)
82. J.M. Lu, J.F. Wu, L.H. Wang, S.H. Ya, The synthesis of copolymer–Cu coordinates and its catalysis to methyl methacrylate. J. Appl. Polym. Sci. **97**, 2072–2075 (2005)
83. V. Singh, D.N. Tripathi, A. Tiwari, R. Sanghi, Carbohydr. Microwave synthesized chitosan-*graft*-poly(methylmethacrylate): An efficient Zn^{2+} ion binder. Polym. **65**, 35–41 (2006)
84. V. Singh, A. Tiwari, D.N. Tripathi, R. Sanghi, Microwave enhanced synthesis of chitosan-*graft*-polyacrylamide. Polymer **47**, 254–260 (2006)
85. V. Singh, D.N. Tripathi, A. Tiwari, R. Sanghi, Microwave promoted synthesis of chitosan-*graft*-poly(acrylonitrile). J. Appl. Polym. Sci. **95**, 820–825 (2005)
86. V. Singh, A. Tiwari, D.N. Tripathi, R. Sanghi, Carbohydr. Synthesis of amino acid based covalently cross-linked polymeric gels using tetrakis(hydroxymethyl) phosphonium chloride as a cross-linker. Polym. **58**, 1–8 (2004)
87. V. Singh, A. Tiwari, D.N. Tripathi, R. Sanghi, Grafting of polyacrylonitrile onto guar gum under microwave irradiation. J. Appl. Polym. Sci. **92**, 1569–1575 (2004)
88. V. Singh, D.N. Tripathi, Microwave promoted grafting of acrylonitrile onto Cassia siamea seed gum. J. Appl. Polym. Sci. **101**, 2384–2390 (2006)
89. V. Singh, A. Tiwari, S. Pandey, S.K. Singh, Microwave-accelerated Synthesis and Characterization of Potato Starch-*g*-poly(acrylamide). Starch **58**, 536–543 (2006)
90. K. Prasad, R. Meena, A.K. Siddhanta, Microwave-induced rapid one-pot synthesis of κ-carrageenan-g-PMMA copolymer by potassium persulphate initiating system. J. Appl. Polym. Sci. **101**, 161–166 (2006)
91. K. Xu, W.D. Zhang, Y.M. Yue, P.X. Wang, Swelling behaviors of a three-component copolymer (starch graft sodium acrylate and 2-acrylamido-2-methyl-propanosulfonic acid) synthesized by microwave polymerization. J. Appl. Polym. Sci. **98**, 1050–1054 (2005)

92. P. Albert, H. Warth, R. Mulhaupt, R. Janda, Comparison of thermal and microwave-activated polymerization of ε-caprolactone with titanium tetrabutylate as catalyst. Macromol. Chem. Phys. **197**, 1633–1641 (1996)
93. X.M. Fang, R. Hutcheon, D.A. Scola, Microwave syntheses of poly(ε-caprolactam-co-ε-caprolactone). J. Polym. Sci. Part A Polym. Chem. **38**, 1379–1390 (2000)
94. D. Barbier-Baudry, C.H. Brachais, A. Cretu, A. Loupy, D. Stuerga, An Easy Way Toward ε-Caprolactone Macromonomers by Microwave Irradiation Using Early Lanthanide Halides as Catalysts. Macromol. Rapid Commun. **23**, 200–204 (2002)
95. X.M. Fang, C.D. Simone, E. Vaccaro, S.J. Huang, D.A. Scola, Ring-opening polymerization of ∈-caprolactam and ∈-caprolactone via microwave irradiation. J. Polym. Sci. Part A Polym. Chem. **40**, 2264–2275 (2002)
96. B. Koroskenyi, S.P. McCarthy, Microwave-Assisted Solvent-Free or Aqueous-Based Synthesis of Biodegradable Polymers. J. Polym. Environ. **10**, 93–104 (2002)
97. L.Q. Liao, L.J. Liu, C. Zhang, F. He, R.X. Zhuo, K. Wan, Microwave-assisted ring-opening polymerization of ∈-caprolactone. J. Polym. Sci. Part A Polym. Chem. **40**, 1749–1755 (2002)
98. L.Q. Liao, L.J. Liu, C. Zhang, F. He, R.X. Zhuo, Physical and mechanical behavior of single-walled carbon nanotube/polypropylene/ethylene–propylene–diene rubber nanocomposites. J. Appl. Polym. Sci. **90**, 2657–2663 (2003)
99. D. Barbier-Baudry, A. Brachais, A. Cretu, R. Gattin, A. Loupy, D. Stuerga, Synthesis of polycaprolactone by microwave irradiation—an interesting route to synthesize this polymer via green chemistry. Environ. Chem. Lett. **1**, 19–23 (2003)
100. G. Sivalingam, N. Agarwal, G. Madras, Kinetics of microwave-assisted polymerization of ∈-caprolactone. J. Appl. Polym. Sci. **91**, 1450–1456 (2004)
101. S. Sinnwell, A.M. Schmidt, H. Ritter, Direct Synthesis of (meth-)acrylate poly (∈-caprolactone) macromonomers. J. Macromol. Sci. Part A Pure Appl. Chem. **43**, 469–479 (2006)
102. Y. Song, L.J. Liu, X.C. Weng, R.X. Zhuo, Acid-initiated polymerization of ε-caprolactone under microwave irradiation and its application in the preparation of drug controlled release system. J. Biomater. Sci. Polym. Ed. **14**, 241–253 (2003)
103. Y. Song, L.J. Liu, R.X. Zhuo, Microwave-assisted polymerization of ε-caprolactone with maleic acid as initiator and drug release behavior of ibuprofen-poly(ε-caprolactone)system. Chin. Chem. Lett. **14**, 32–34 (2003)
104. Z.J. Yu, L.J. Liu, R.X. Zhuo, Microwave-improved polymerization of ∈-caprolactone initiated by carboxylic acids. J. Polym. Sci. Part A Polym. Chem. **41**, 13–21 (2003)
105. Z.J. Yu, L.J. Liu, Effect of microwave energy on chain propagation of poly(ε-caprolactone) in benzoic acid-initiated ring opening polymerization of ε-caprolactone. Eur. Polym. J. **40**, 2213–2220 (2004)
106. Z.J. Yu, L.J. Liu, ArticleMicrowave-assisted synthesis of poly(ε-caprolactone)-poly (ethylene glycol)-poly-(ε-caprolactone) tri-block co-polymers and use as matrices for sustained delivery of ibuprofen taken as model drug. J. Biomater. Sci. Polym. Ed. **16**, 957–971 (2005)
107. Z.J. Yu, L.J. Liu, Biodegradable poly(vinyl alcohol)-*graft*-poly(ε-caprolactone) comb-like polyester: Microwave synthesis and its characterization. J. Appl. Polym. Sci. **104**, 3973–3979 (2007)
108. F. Wiesbrock, R. Hoogenboom, C.H. Abeln, U.S. Schubert, Single-mode microwave ovens as new reaction devices: Accelerating the living polymerization of 2-ethyl-2-oxazoline. Macromol. Rapid Commun. **25**, 1895–1899 (2004)
109. S. Sinnwell, H. Ritter, Microwave Accelerated Polymerization of 2-Phenyl-2-oxazoline. Macromol. Rapid Commun. **26**, 160–163 (2005)
110. R. Hoogenboom, F. Wiesbrock, M.A.M. Leenen, M.A.R. Meier, U.S. Schubert, Accelerating the living polymerization of 2-nonyl-2-oxazoline by implementing a microwave synthesizer into a high-throughput experimentation workflow. J. Comb. Chem. **7**, 10 (2005)

111. R. Hoogenboom, M.W.M. Fijten, H.M.L. Thijs, B.M. Van Lankvelt, U.S. Schubert, Microwave-assisted synthesis and properties of a series of poly(2-alkyl-2-oxazoline)s. Des. Monomers Polym. **8**, 659–671 (2005)
112. M. Klink, U. Kolb, H. Ritter, Microwave assisted synthesis of channel-containing polymeric materials. E-Polym. Art. **069**, 1–6 (2005)
113. J.-F. Zhu, Y.-J. Zhu, Microwave-assisted one-step synthesis of polyacrylamide−metal (M=Ag, Pt, Cu) nanocomposites in ethylene glycol. J. Phys. Chem. B **110**, 8593–8597 (2006)
114. J.F. Zhu, Y.J. Zhu, M.G. Ma, L.X. Yang, L. Gao, Simultaneous and rapid microwave synthesis of polyacrylamide−metal sulfide (Ag_2S, Cu_2S, HgS) nanocomposites. J. Phys. Chem. C **111**, 3920–3926 (2007)
115. L. Liao, C. Zhang, S. Gong, Preparation of poly(ε-caprolactone)/clay nanocomposites by microwave-assisted *in situ* ring-opening polymerization. Macromol. Rapid Commun. **28**, 1148–1154 (2007)
116. S. Shi, L. Liu, Microwave-assisted preparation of temperature sensitive poly(*N*-isopropylacryl-amide) hydrogels in poly(ethylene oxide)-600. J. Appl. Poly. Sci. **102**, 4177–4184 (2006)
117. W. Zhang, J. Gao, C. Wu, Microwave preparation of narrowly distributed surfactant-free stable polystyrene nanospheres. Macromolecules **30**, 6388–6390 (1997)
118. R. Correa, G. Gonzalez, V. Dougar, Emulsion polymerization in a microwave reactor. Polymer **39**, 1471–1474 (1998)
119. Z. Cheng, X. Zhu, M. Chen, J. Chen, L. Zhang, Atom transfer radical polymerization of methyl methacrylate with low concentration of initiating system under microwave irradiation. Polymer **44**, 2243–2247 (2003)
120. W.D. He, C.Y. Pan, T. Lu, Soapless emulsion polymerization of butyl methacrylate through microwave heating. J. Appl. Polym. Sci. **80**, 2455–2459 (2001)
121. Z. Deng, X. Hu, L. Li, C. Yi, Z. Xu, Thermosensitive P(St-co-NIPAM) particles prepared by emulsifier-free emulsion polymerization with microwave irradiation. Acta. Polym. Sin. **2**, 293–296 (2005)
122. J.M. Pawluczyk, R.T. McClain, C. Denicola, J.J. Mulhearn, D.J. Rudda, C.W. Lindsley, Microwave-initiated living free radical polymerization: optimization of the preparative scale synthesis of Rasta resins. Tetrahedron Lett. **48**, 1497–1501 (2007)
123. R. Hoogenboom, U.S. Schubert, Microwave-assisted polymers synthesis: recent developments in a rapidly expanding field of research. Macromol. Rapid Commun. **28**, 368–386 (2007)
124. F. Wiesbrock, R. Hoogenboom, U.S. Schubert, Microwave-assisted polymer synthesis: state-of-the-art and future perspectivas. Macromol. Rapid Commun. **25**, 1739–1764 (2004)
125. T. Nakamura, R. Nagahata, K. Kunii, H. Soga, S. Sugimoto, K. Takeuchi, Large-scale polycondensation of lactic acid using microwave batch reactors. Org. Proc. Res. Develop. **14**, 781–786 (2010)
126. T. Nakamura, R. Nagahata, K. Takeuchi, Microwave-assisted polyester and polyamide synthesis. Mini-Rev. Org. Chem. **8**, 306–314 (2011)
127. S. Horikoshi, N. Serpone (eds.), *Microwaves in Nanoparticle Synthesis: Fundamentals and Applications* (Wiley-VCH Verlag GmbH & Co., Germany, 2013)
128. M. Regier, K. Knoerzer (eds.), *The Microwave Processing of Foods* (Woodhead Publishing, England, 2005)
129. Y. Ikarashi, Y. Maruyama, W.B. Stavinoha, Study of the use of the microwave magnetic field for the rapid inactivation of brain enzymes. Jpn. J. Pharmacol. **35**, 371–387 (1984)
130. W.B. Stavinoha, B. Pepelko, P.W. Smith, Microwave radiation to inactivate cholinesterase in the rat brain prior to analysis for acetylcholine. *Pharmacologist* **12**, 257 (1970)
131. W.B. Stavinoha, S.T. Weintraub, A.T. Modak, The use of microwave heating to inactivate cholinesterase in the rat brain prior to analysis for acetylcholine. J. Neurochem. **20**, 361–371 (1973)

132. M.L. Belkhode, D.L. Johnson, A.M. Muc, Thermal and athermal effects of microwave radiation on the activity of glucose-6-phosphate dehydrogenase in human blood. Health Phys. **26**, 45–51 (1974)
133. E.K. Yeagers, A.P. Langley, A.P. Sheppard, G.K. Huddleston, Effects of microwave radiation on enzymes. Ann. N Y Acad. Sci. **28**, 301–304 (1975)
134. H.M. Henderson, K. Hergenroeder, S.S. Stuchly, Effect of 2450 MHz microwave radiation on horseradish peroxidase. J. Microwave Power **10**, 27–35 (1975)
135. T.R. Ward, J.W. Allis, J.A. Elder, Measure of enzymatic activity coincident with 2450 MHz microwave exposure. J. Microwave Power **10**, 315–320 (1975)
136. B. Réjasse, S. Lamare, M.D. Legoy, T.J. Besson, Influence of microwave irradiation on enzymatic properties: applications in enzyme chemistry. Enzym. Inhib. **22**, 519–527 (2007)
137. M. Bini, A. Checcucci, A. Ignesti, L. Millanta, N. Rubino, G. Camici, G. Manao, G. Ramponi, Analysis of the effects of microwave energy on enzymatic activity of lactate dehydrogenase. J. Microwave Power **13**, 95–99 (1978)
138. M.J. Galvin, D.L. Parks, D.I. McRee, Influence of 2.45 GHz microwave radiation on enzyme activity. Radiat. Environ. Biophys. **19**, 149–156 (1981)
139. D.B. Millar, J.P. Christopher, J. Hunter, S.S. Yeandle, The effect of exposure of acetylcholinesterase to 2.45-MHz microwave radiation. Bioelectromagnetics **5**, 165–172 (1984)
140. G. d'Ambrioso, R. Massa, L. Gianfreda, G. Greco, A. Scaglione, Influence of microwave radiation on acid phosphatase deactivation process. Alta Frequenza **58**, 361–364 (1989)
141. M. Porcelli, G. Cacciapuoti, S. Fusco, R. Massa, G. d'Ambrosio, C. Bertoldo, M. de Rosa, V. Zappia, Non-thermal effects of microwaves on proteins: thermophilic enzymes as model system. FEBS Lett. **402**, 102–106 (1997)
142. F. La Cara, S. d'Auria, M.R. Scarfi, O. Zeni, R. Massa, G. d'Ambrioso, G. Franceschetti, M. De Rosa, M. Rossi, Microwave exposure effect on a thermophilic alcohol deshydrogenase. Protein Pept. Lett. **6**, 155–162 (1999)
143. F. La Cara, M.R. Scarfi, S. d'Auria, R. Massa, G. d'Ambrioso, G. Franceschetti, M. Rossi, M. De Rosa, Different effects of microwave energy and conventional heat on the activity of a thermophilic b-galactosidase from Bacillus acidocaldarius. Bioelectromagnetics **20**, 172–176 (1999)
144. J.M. Collins, N.E. Leadbeater, Microwave energy: a versatile tool for the biosciences. Enzym. Inhib. **5**, 1141–1150 (2007)
145. J.M. Collins, N.E. Leadbeater, L.M. Stencel, E.C. Wood, Probing the effects of microwave irradiation on enzyme-catalysed organic transformations: the case of lipase-catalysed transesterification reactions. Org. Biomol. Chem. **5**, 1052–1055 (2007)
146. A.M. Klibanov, Improving enzymes by using them in organic solvents. Nature **409**, 241–246 (2001)
147. J.P. Tam, Y.A. Lu, Coupling difficulty associated with interchain clustering and phase transition in solid phase peptide synthesis. J. Am. Chem. Soc. **117**, 12058–12063 (1995)
148. C. Gabriel, S. Gabriel, E.H. Grant, B.S. Halstead, D.M.P. Mingos, Dielectric parameters relevant to microwave dielectric heating. Chem. Soc. Rev. **27**, 213–224 (1998)
149. H. Zhao, G.A. Baker, Z. Song, O. Olubajo, L. Zanders, S.M. Campbell, Effect of ionic liquid properties on lipase stabilization under microwave irradiation. J. Mol. Catal. B Enzym. **57**, 149–157 (2009)
150. M. Damm, C. Nusshold, D. Cantillo, G.N. Rechberger, K. Gruber, W. Sattler, C.O.J. Kappe, Can electromagnetic fields influence the structure and enzymatic digest of proteins? A critical evaluation of microwave-assisted proteomics protocols. Proteomics **75**, 5533–5543 (2012)
151. M. Suzuki, J. Shigematsu, T. Kodama, Hydration study of proteins in solution by microwave dielectric analysis. J. Phys. Chem. **100**, 7279–7282 (1996)
152. S. Horikoshi, K. Nakamura, M. Kawaguchi, J. Kondo, N. Serpone, Effect of microwave radiation on the activity of catalase. Decomposition of hydrogen peroxide under microwave and conventional heating. RSC Adv. **6**, 48237–48244 (2016)

153. S. Horikoshi, T. Nakamura, M. Kawaguchi, N. Serpone, Enzymatic proteolysis of peptide bonds by a metallo-endoproteinase under precise temperature control with 5.8-GHz microwave radiation. *J. Mol. Cat. B: Enzym.* **116**, 52–59 (2015)
154. K.J. Shaw, P.T. Docker, J.V. Yelland, C.E. Dyer, J. Greenman, G.M. Greenway, S. J. Haswell, Rapid PCR amplification using a microfluidic device with integrated microwave heating and air impingement cooling. Lab Chip **10**, 1725–1728 (2010)
155. M.-C. Parker, T. Besson, S. Lamare, M.-D. Legoy, Microwave radiation can increase the rate of enzyme-catalysed reactions in organic media. Tetrahedron Lett. **37**, 8383–8386 (1996)
156. M. Gelo-Pujic, E. Guibé-Jampel, A. Loupy, S.A. Galema, D. Mathé, Lipase-catalysed esterification of some α-D-glucopyranosides in dry media under focused microwave irradiation. J. Chem. Soc. Perkin Trans. **1**, 2777–2780 (1996)
157. J.-R. Carrillo-Munoz, D. Bouvet, E. Guibé-Jampel, A. Loupy, A. Petit, Microwave-promoted lipase-catalyzed reactions. Resolution of (±)-1-phenylethanol. J. Org. Chem. **61**, 7746–7749 (1996)
158. M. Vacek, M. Zarevúcka, Z. Wimmer, K. Stránský, K. Demnerová, M.-D. Legoy, Selective enzymic esterification of free fatty acids with n-butanol under microwave irradiation and under classical heating. Biotechnol. Lett. **22**, 1565–1570 (2000)
159. S. Mazumder, D.D. Laskar, D. Prajapati, M.K. Roy, Microwave-induced enzyme-catalyzed chemoselective reduction of organic azides. Chem. Biodivers. **1**, 925–930 (2004)
160. B. Réjasse, S. Lamare, M.-D. Legoy, T. Besson, Stability improvement of immobilized Candida antarctica lipase B in an organic medium under microwave radiation. Org. Biomol. Chem. **2**, 1086–1089 (2004)
161. B. Réjasse, T. Besson, M.-D. Legoy, S. Lamare, Influence of microwave radiation on free Candida antarctica lipase B activity and stability. Org. Biomol. Chem. **4**, 3703 (2006)
162. T. Maugard, D. Gaunt, M.D. Legoy, T. Besson, Microwave-assisted synthesis of galacto-oligosaccharides from lactose with immobilized β-galactosidase from Kluyveromyces lactis. Biotechnol. Lett. **25**, 623–629 (2003)
163. G.D. Yadav, P.S. Lathi, Synergism between microwave and enzyme catalysis in intensification of reactions and selectivities: transesterification of methyl acetoacetate with alcohols. J. Mol. Catal. A Chem. **223**, 51–56 (2004)
164. P. Kerep, H. Ritter, Influence of microwave irradiation on the lipase-catalyzed ring-opening polymerization of ε-caprolactone. Macromol. Rapid Commun. **27**, 707–717 (2006)
165. G.D. Yadav, P.S. Lathi, Intensification of enzymatic synthesis of propylene glycol monolaurate from 1,2-propanediol and lauric acid under microwave irradiation: kinetics of forward and reverse reactions. Enzyme Microb. Technol. **38**, 814–820 (2006)
166. G.D. Yadav, I.V. Borkar, Kinetic modeling of microwave-assisted chemoenzymatic epoxidation of styrene. AIChE J. **52**, 1235–1247 (2006)
167. G.D. Yadav, P.S. Lathi, Microwave assisted enzyme catalysis for synthesis of n-butyl diphenyl methyl mercapto acetate in non-aqueous media. Clean Technol. Environ. Policy **9**, 281–287 (2007)
168. G.D. Yadav, A.D. Sajgure, Synergism of microwave irradiation and enzyme catalysis in synthesis of isoniazid. J. Chem. Technol. Biotechnol. **82**, 964–970 (2007)
169. G.D. Yadav, I.V. Borkar, Kinetic and mechanistic investigation of microwave-assisted lipase catalyzed synthesis of citronellyl acetate. Ind. Eng. Chem. Res. **48**, 7915–7922 (2008)
170. B. Major, I. Kelemen-Horváth, Z. Csanádi, K. Bélafi-Bakó, L. Gubicza, Microwave assisted enzymatic esterification of lactic acid and ethanol in phosphonium type ionic liquids as co-solvents. Green Chem. **11**, 614–616 (2009)
171. I.C.R. Costa, S.G.F. Leite, C.R. Leal, L.S.M. Miranda, R.O.M.A. de Souza, Thermal effect on the microwave assisted biodiesel synthesis catalyzed by lipases. J. Braz. Chem. Soc. **22**, 1993–1998 (2011)
172. G.D. Yadav, P.A. Thorat, Microwave assisted lipase catalyzed synthesis of isoamyl myristate in solvent-free system. J. Mol. Catal. B Enzym. **83**, 16–22 (2012)

173. Z.H.P.V. Rupasinghe, An efficient microwave-assisted enzyme-catalyzed regioselective synthesis of long chain acylated derivatives of flavonoid glycosides. Tetrahedron Lett. **54**, 1933–1937 (2013)
174. S.D. Shinde, G.D. Yadav, Microwave irradiated immobilized lipase catalyzed synthesis of alkyl benzoate esters by transesterification: mechanism and kinetic modeling. Ind. Eng. Chem. Res. **53**, 8706–8713 (2014)
175. B. Sontakke, G.D. Yadav, Microwave assisted synthesis of ethyl 2-(4-aminophenyl) acetate using novozyme 435. Current Catal. **3**, 27–34 (2014)
176. D. Yu, Z. Wang, P. Chen, L. Jin, Y. Cheng, J. Zhou, S. Cao, Microwave-assisted resolution of (R, S)-2-octanol by enzymatic transesterification. J. Mol. Catal. B Enzym. **48**, 51–57 (2007)
177. D. Yu, P. Chen, L. Wang, Q. Gu, Y. Li, Z. Wang, S. Cao, A chemo-enzymatic process for sequential kinetic resolution of (R, S)-2-octanol under microwave irradiation. Process Biochem. **42**, 1312–1318 (2007)
178. P. Bachu, J.S. Gibson, J. Sperry, M.A. Brimble, The influence of microwave irradiation on lipase-catalyzed kinetic resolution of racemic secondary alcohols. Tetrahedron Asymmetry **18**, 1618–1624 (2007)
179. A. Parvulescu, E. Vandereycken, P. Jacobs, D. Devos, Microwave-promoted racemization and dynamic kinetic resolution of chiral amines over Pd on alkaline earth supports and lipases. J. Catal. **255**, 206–212 (2008)
180. R.O. de Souza, O.A. Antunes, W. Kroutil, C.O. Kappe, Kinetic resolution of rac-1-phenylethanol with immobilized lipases: a critical comparison of microwave and conventional heating protocols. J. Org. Chem. **74**, 6157–6162 (2009)
181. G.D. Yadav, S. Devendran, Lipase catalyzed kinetic resolution of (±)-1-(1-naphthyl)ethanol under microwave irradiation. J. Mol. Catal. B Enzym. **81**, 58–65 (2012)
182. G.D. Yadav, S.V. Pawar, Insight into microwave irradiation and enzyme catalysis in enantioselective resolution of dl-(±)-3-phenyllactic acid. Appl. Microbiol. Biotechnol. **96**, 69–79 (2012)
183. S. Devendran, G.D. Yadav, Microwave assisted enzymatic kinetic resolution of (±)-1-phenyl-2- propyn-1-ol in nonaqueous media. Biomed. Res. Int. **2014**, 482678–482687 (2014)
184. D.D. Young, J. Nichols, R.M. Kelly, A. Deiters, Microwave activation of enzymatic catalysis. J. Am. Chem. Soc. **130**, 10048–10049 (2008)
185. C. Kamerke, M. Pattky, C. Huhn, L. Elling, Synthesis of UDP-activated oligosaccharides with commercial β-galactosidase from Bacillus circulans under microwave irradiation. J. Mol. Catal. B Enzym. **79**, 27–34 (2012)
186. A. Wang, M. Wang, Q. Wang, F. Chen, F. Zhang, H. Li, Z. Zeng, T. Xie, Stable and efficient immobilization technique of aldolase under consecutive microwave irradiation at low temperature. Bioresource Technol. **102**, 469–474 (2011)
187. M. Gelo-Pujic, E. Guibé-Jampel, A. Loupy, Enzymatic glycosidations in dry media on mineral supports. *Tetrahedron* **53**, 17247–17252 (1997)
188. M. Gelo-Pujic, E. Guibé-Jampel, A. Loupy, A. Trincone, Enzymatic glycosidation in dry media under microwave irradiation. J. Chem. Soc. Perkin Trans. **1**, 1001–1002 (1997)
189. W. Huang, Y.M. Xia, H. Gao, Y.J. Fang, Y. Wang, Y. Fang, Enzymatic esterification between n-alcohol homologs and n-caprylic acid in non-aqueous medium under microwave irradiation. J. Mol. Catal. B Enzym. **35**, 113–116 (2005)
190. S. Bradoo, P. Rathi, R.K. Saxena, R. Gupta, Microwave-assisted rapid characterization of lipase selectivities. J. Biochem. Biophys. Meth. **51**, 115–120 (2002)
191. G. Lin, W.Y. Lin, Microwave-promoted lipase-catalyzed reactions. Tetrahedron Lett. **39**, 4333–4336 (1998)
192. B. Réjasse, T. Besson, M.D. Legoy, S. Lamare, Influence of microwave radiation on free candida antarctica lipase B activity and stability. Org. Biomol. Chem. **4**, 3704–3707 (2006)

193. M. Zarevucka, M. Vacek, Z. Wimmer, C. Brunet, M.D. Legoy, Models for glycosidic juvenogens: Enzymic formation of selected alkyl-b-D-galactopyranosides under microwave irradiation. Biotechnol. Lett. **21**, 785–790 (1999)
194. F. Haber, *Thermodynamik technischer Gasreaktionen*, 1st edn. (Salzwasser Verlag, Paderborn, R. Oldenburg, Germany, 1905) (in German)
195. R. Luque, A.M. Balu, D.J. MacQuarrie, Microwave-assisted heterogeneously catalyzed processes, in *Microwaves in Organic Synthesis*, 3rd edn. vol. 2, Chapter 18, ed. by A. De la Hoz, A. Loupy (Wiley-VCH Verlag GmbH & Co. KGaA, Germany, 2012), pp. 811–842
196. A. Mazubert, M. Poux, J. Aubin, Intensified processes for FAME production from waste cooking oil: a technological review. Chem. Engin. J. **233**, 201–223 (2013)
197. L.N. Protasova, M. Bulut, D. Ormerod, A. Buekenhoudt, J. Berton, C.V. Stevens, Latest highlights in liquid-phase reactions for organic synthesis in microreactors. Org. Proc. Res. Develop. **17**, 760–791 (2013)
198. M. Komorowska-Durka, G. Dimitrakis, D. Bogdal, A.I. Stankiewicz, C.D. Stefanidis, A concise review on microwave-assisted polycondensation reactions and curing of polycondensation polymers with focus on the effect of process conditions. Chem. Engin. J. **264**, 633–644 (2015)
199. S. Soltani, U. Rashid, R. Yunus, Y.H. Taufiq-Yap, Synthesis of biodiesel through catalytic transesterification of various feedstocks using fast solvothermal technology: a critical review. *Catal. Rev. Sci. Engin.* **57**, 407–435 (2015)
200. T. Durka, T. Van Gerven, A. Stankiewicz, Microwaves in heterogeneous gas-phase catalysis: experimental and numerical approaches. Chem. Eng. Technol. **32**, 1301–1312 (2009)
201. J. Zheng, H. Liu, B. Wu, Y. Guo, T. Wu, G. Yu, Y. Liu, D. Zhu, Production of high-quality carbon nanoscrolls with microwave spark assistance in liquid nitrogen. Adv. Mater. **23**, 2460–2463 (2011)
202. A.G. Whittaker, D.M.P. Mingos, Microwave-assisted solid-state reactions involving metal powders and gases. *J. Chem. Soc. Dalton Trans.* 2541–2543 (1993)
203. S. Horikoshi, N. Serpone, Role of microwaves in heterogeneous catalytic systems. Catal. Sci. Technol. **4**, 1197–1210 (2014)
204. M. Matsuzawa, S. Togashi, S. Hasebe, Investigation of microwave effects on transition metal catalyzed reaction using an isothermal reaction. J. Chem. Eng. Jpn. **45**, 429–435 (2012)
205. F. Kishimoto, M. Matsuhisa, S. Kawamura, S. Fujii, S. Tsubaki, M.M. Maitani, E. Suzuki, Y. Wada, Enhancement of anodic current attributed to oxygen evolution on α-Fe_2O_3 electrode by microwave oscillating electric field. Sci. Rep. **6**, 35554–35562 (2016)
206. C.-T. Hsieh, W.-Y. Chen, I.L. Chen, A.K. Roy, Deposition and activity stability of Pt–Co catalysts on carbon nanotube-based electrodes prepared by microwave-assisted synthesis. J. Power Sources **199**, 94–102 (2012)
207. D. Kalpana, S.H. Cho, S.B. Lee, Y.S. Lee, R. Misra, Recycled waste paper—a new source of raw material for electric double-layer capacitors. J. Power Sources **190**, 587–591 (2009)
208. J. Wang, M. Chen, C. Wang, J. Wang, J. Zheng, Preparation of mesoporous carbons from amphiphilic carbonaceous material for high-performance electric double-layer capacitors. J. Power Sources **196**, 550–558 (2011)
209. C.-M. Yang, Y.-J. Kim, M. Endo, H. Kanoh, M. Yudasaka, S. Iijima, K. Kaneko, Nanowindow- regulated specific capacitance of supercapacitor electrodes of single-wall carbon nanohorns. J. Am. Chem. Soc. **129**, 20–21 (2007)
210. Y. Yuan, J.A. Smith, G. Goenaga, D.-J. Liu, Z. Luo, J. Liu, Platinum decorated aligned carbon nanotubes: electrocatalyst for improved performance of proton exchange membrane fuel cells. J. Power Sources **196**, 6160–6167 (2011)
211. F.K. Yuen, B.H. Hameed, Recent developments in the preparation and regeneration of activated carbons by microwaves. Adv. Coll. Interf. Sci. **149**, 19–27 (2009)
212. Y. Kong, C.Y. Cha, Reduction of NOX adsorbed on char with microwave energy. *Carbon* **34**, 1035–1040 (1996); D. Djebabra, O. Dessaux, P. Goudmand, Coal gasification by microwave plasma in water vapor. *Fuel* **70**, 1473–1475 (1991)

213. T. Kim, J. Lee, K.-H. Lee, Microwave heating of carbon-based solid materials. Carbon Lett. **15**, 15–24 (2014)
214. R.K. Arvela, N.E. Leadbeater, Suzuki coupling of aryl chlorides with phenylboronic acid in water using microwave heating with simultaneous cooling. Org. Lett. **7**, 2101–2104 (2005)
215. T. Hirano, in *Microwaves in Catalysis*, Chapter 18, ed. by S. Horikoshi, N. Serpone (Wiley-VCH Verlag GmbH, Weinheim, Germany, 2016)
216. D.A. Jones, T.P. Leyveld, S.D. Mavrofidis, S.W. Kingman, N.J. Miles, Microwave heating applications in environmental engineering—review. Resourc. Conservat. Recycl. **34**, 75–90 (2002)
217. C.C. Liu, B.K. Na, A.B. Walters, M.A. Vannie, Microwave absorption measurements of the electrical conductivity of small particles. Catal. Lett. **26**, 9–24 (1994)
218. D.M.P. Mignos, D.R. Baghurst, Applications of microwave dielectric heating effects to synthetic problems in chemistry. Chem. Soc. Rev. **20**, 1–47 (1991)
219. H. Tamura, Amer. Ceramics Soc. Bull. **73**, 92–95 (1994)
220. T. Hirano, Y. Matsuda, in *Proceedings of the 2nd Kyushu-Taipei International Congress on Chemical Engineering*, 1997, pp. 335–338
221. Y.Z. Steenwinkel, H.L. Castricum, J. Beckers, E. Eiser, A. Blieke, Dielectric heating effects on the activity and SO_2 resistance of $La_{0.8}Ce_{0.2}MnO_3$ perovskite for methane oxidation. J. Catal. **221**, 523–531 (2004)
222. J.J. Spivery, Complete catalytic oxidation of volatile organics. Ind. Eng. Chem. Res. **26**, 2165–2180 (1987)
223. J.K.S. Wan, M.Y. Tse, H. Husby, M.C. Depew, High-power pulsed microwave catalytic processes: decomposition of methane. J. Microwave Power Electromag. Rad. **25**, 32–38 (1990)
224. J.K.S. Wan, M.S. Ioffe, Surface heating and energy transfer in pulsed microwave catalytic systems: a microwave-induced acoustic study. Res. Chem. Intermed. **20**, 115–132 (1994)
225. D.X. Zang, A.M. Yu, Q.H. Jin, Studies on microwave-carbon reduction method for the treatment of nitric oxide. Chem. J. Chinese Univ. **18**, 1271–1280 (1997)
226. X. Zhang, D.O. Hayward, C. Lee, D.M.P. Mingos, Microwave assisted catalytic reduction of sulfur dioxide with methane over MoS2 catalysts. Appl. Catal. B: Environ. **33**, 137–148 (2001)
227. X. Zhang, D.O. Hayward, D.M.P. Mingos, Apparent equilibrium shifts and hot-spot formation for catalytic reactions induced by microwave dielectric heating. Chem. Commun. **11**, 975–976 (1999)
228. M.S. Ioffe, S.D. Pollington, J.K.S. Wan, High-power pulsed radio-frequency and microwave catalytic processes: selective production of acetylene from the reaction of methane over carbon. J. Catal. **151**, 349–355 (1995)
229. S.D. Pollington, M.S. Ioffe, M. Westergaard, J.K.S. Wan, A high-power X-band pulsed microwave induced catalytic decomposition of methane with integral acoustic measurements. Res. Chem. Intermed. **21**, 59–68 (1995)
230. M.Y. Tse, M.C. Depew, J.K.S. Wan, Applications of high power microwave catalysis in chemistry. Res. Chem. Intermed. **13**, 221–236 (1990)
231. D.O. Cooney, Z. Xi, Production of hydrogen from methane and methane/team in a microwave irradiated char-loaded reactor. Fuel Sci. Technol. Inter. **14**, 1111–1141 (1996)
232. X. Zhang, C.S.M. Lee, D.M.P. Mingos, D.O. Hayward, Oxidative coupling of methane using microwave dielectric heating. Appl. Catal. A Gen. **249**, 151–164 (2003)
233. G. Roussy, J.M. Thiebaut, M. Souiri, E. Marchal, A. Kiennemann, G. Maire, Controlled oxidation of methane doped catalysts irradiated by microwaves. Catal. Today **21**, 349–355 (1994)
234. G. Roussy, E. Marchal, J.M. Thiebaut, A. Kiennemann, G. Maire, C2+ selectivity enhancement in oxidative coupling of methane over microwave-irradiated catalysts. Fuel Process. Technol. **50**, 261–274 (1997)
235. X.-J. Bi, P.-J. Hong, X.-G. Xie, S.-S. Dai, Microwave effect on partial oxidation of methane to syngas. React. Kinet. Catal. Lett. **66**, 381–386 (1999)

236. G. Bond, R.B. Moyes, D.A. Whan, Recent applications of microwave heating in catalysis. Catal. Today **17**, 427–437 (1993)
237. C. Chen, P. Hong, S. Dai, J. Kan, Microwave effects on the oxidative coupling of methane over proton conductive catalysts. J. Chem. Soc. Faraday Trans. **91**, 1179–1180 (1995)
238. C.-L. Chen, P.-J. Hong, S.-S. Dai, C.-C. Zhang, X.-Y. Yang, Microwave effects on the oxidative coupling of methane over Bi_2O_3-WO_3 oxygen ion conductive oxides. React. Kinet. Catal. Lett. **61**, 175–180 (1997)
239. C. Marun, L.D. Conde, S.L. Suib, Catalytic oligomerization of methane via microwave heating. J. Phys. Chem. A **103**, 4332–4340 (1999)
240. L.D. Conde, C. Marún, S.L. Suib, Z. Fathi, Frequency effects in the catalytic oligomerization of methane via microwave heating. J. Catal. **204**, 324–332 (2001)
241. X. Zhang, C.S.M. Lee, D.M.P. Mingos, D.O. Hayward, Carbon dioxide reforming of methane with Pt catalysts using microwave dielectric heating. Catal. Lett. **88**, 129–139 (2003)
242. J.K.S. Wan, Microwaves and chemistry: the catalysis of an exciting marriage. Res. Chem. Intermed. **19**, 147–158 (1993)
243. J.K.S. Wan, G. Bamwenda, M.C. Depew, Microwave induced catalytic reactions of carbon dioxide and water: mimicry of photosynthesis. Res. Chem. Intermed. **16**, 241–255 (1991)
244. T.A. Koch, K.R. Krause, M. Mehdizadeh, Improved safety through distributed manufacturing of hazardous chemicals. Process Saf. Prog. **16**, 23–24 (1997)
245. C.Y. Cha, Microwave induced reactions of SO_2 and NO_x decomposition in the char-bed. Res. Chem. Intermed. **20**, 13–28 (1994)
246. C.Y. Cha, Y. Kong, Enhancement of NO_x adsorption capacity and rate of char by microwaves. Carbon **33**, 1141–1146 (1995)
247. Y. Kong, C.Y. Cha, NO_x abatement with carbon adsorbents and microwave energy. Energy Fuels **9**, 971–975 (1995)
248. Y.-F. Chang, A. Sanjurjo, J. Mccarty, G. Krishnan, B. Woods, E. Wachsman, Microwave-assisted NO reduction by methane over Co-ZSM-5 zeolites. Catal. Lett. **57**, 187–191 (1999)
249. X. Wang, T. Zhang, C. Xu, X. Sun, D. Liang, L. Lin, Microwave effects on the selective reduction of NO by CH_4 over an In–Fe_2O_3/HZSM-5 catalyst. *Chem. Commun.* 279–280 (2000)
250. W.L. Perry, J.D. Katz, D. Rees, M.T. Paffett, A.K. Datye, Kinetics of the microwave-heated CO oxidation reaction over alumina-supported Pd and Pt catalysts. *J. Catal.* **171**, 431–438 (1997)
251. I. Silverwood, G. Mcdougall, G. Whittaker, Comparison of conventional versus microwave heating of the platinum catalysed oxidation of carbon monoxide over EUROPT-1 in a novel infrared microreactor cell. J. Mol. Catal. A Chem. **269**, 1–4 (2007)
252. I.P. Silverwood, G.S. Mcdougall, A.G. Whittaker, A microwave-heated infrared reaction cell for the in situ study of heterogeneous catalysts. Phys. Chem. Chem. Phys. **8**, 5412–5416 (2006)
253. H. Will, P. Scholz, B. Ondruschka, W. Burckhardt, Multimode microwave reactor for heterogeneous gas-phase catalysis. Chem. Eng. Technol. **26**, 1146–1149 (2003)
254. H. Will, P. Scholz, B. Ondruschka, Heterogeneous gas-phase catalysis under microwave irradiation—a new multi-mode microwave applicator. Topics Catal. **29**, 175–182 (2004)
255. J. Beckers, L.M.V.D. Zande, G. Rothenberg, Clean diesel power via microwave susceptible oxidation catalysts. ChemPhysChem **7**, 747–755 (2006)
256. D.O. Cooney, Z. Xi, Production of ethylene and propylene from ethane, propane, and n-butane mixed with steam in a microwave-irradiated silicon-carbide loaded reactor. Pet. Sci. Technol. **14**, 1315–1336 (1996)
257. I. Sinev, T. Kardash, N. Kramareva, M. Sinev, O. Tkachenko, A. Kucherov, L.M. Kustov, Interaction of vanadium containing catalysts with microwaves and their activation in oxidative dehydrogenation of ethane. Catal. Today **141**, 300–305 (2009)

258. Y. Liu, Y. Lu, S. Liu, Y. Yin, The effects of microwaves on the catalyst preparation and the oxidation of o-xylene over a V_2O_5/SiO_2 system. Catal. Today **51**, 147–151 (1999)
259. L. Seyfried, F. Garin, G. Maire, J.M. Thiebaut, G. Roussy, Microwave electromagnetic-field effects on reforming catalysts: 1. Higher selectivity in 2-methylpentane isomerization on alumina-supported Pt catalysts. J. Catal. **148**, 281–287 (1994)
260. J.M. Thiebaut, G. Roussy, M.S. Medjram, F. Garin, L. Seyfried, G. Maire, Durable changes of the catalytic properties of alumina-supported platinum induced by microwave irradiation. *Catal. Lett.* **21**, 133–138 (1993)
261. G. Roussy, S. Hilaire, J.M. Thiébaut, G. Maire, F. Garin, S. Ringler, Permanent change of catalytic properties induced by microwave activation on 0.3% $PtAl_2O_3$ (EuroPt-3) and on 0.3% Pt-0.3% $ReAl_2O_3$ (EuroPt-4). Appl. Catal. A Gen. **156**, 167–180 (1997)
262. X.-R. Zhang, L.-C. Wang, Y. Cao, W.-L. Dai, H.-Y. He, K.-N. Fan, A unique microwave effect on the microstructural modification of $Cu/ZnO/Al_2O_3$ catalysts for steam reforming of methanol. *Chem. Commun.* **32**, 4104–4106 (2005)
263. W.L. Perry, A.K. Datye, A.K. Prinja, L.F. Brown, J.D. Katz, Microwave heating of endothermic catalytic reactions: reforming of methanol. AIChE J. **48**, 820–831 (2002)
264. G. Bond, R.B. Moyes, I. Theaker, D.A. Whan, The effect of microwave heating on the reaction of propan-2-ol over alkalised carbon catalysts. Topics Catal. **1**, 177–182 (1994)
265. C.-J. Winter, Into the hydrogen energy economy—milestones. Int. J. Hydrogen Energy **30**, 681–685 (2005)
266. S. Horikoshi, A. Osawa, S. Sakamoto, N. Serpone, Control of microwave-generated hot spots. Part V. Mechanisms of hot-spot generation and aggregation of catalyst in a microwave-assisted reaction in toluene catalyzed by Pd-loaded AC particulates. Appl. Catal. A General **460–461**, 52–60 (2013)
267. S. Hodoshima, A. Shono, J. Kuwano, Y. Saito, Chemical recuperation of low-quality waste heats by catalytic dehydrogenation of organic chemical hydrides and its energy analysis. Energy Fuels **22**, 2559–2569 (2008)
268. S. Horikoshi, M. Kamata, T. Sumi, N. Serpone, Hydrogen evolution from organic hydrides through microwave selective heating in heterogeneous catalytic systems. Int. J. Hydrogen Energy **41**, 12029–12037 (2016)
269. S. Horikoshi, N. Serpone, On the influence of the microwaves' thermal and non-thermal effects in titania photoassisted reactions. Catal. Today **224**, 225–235 (2014)
270. A. Fujishima, K. Hashimoto, T. Watanabe, *TiO_2 Photocatalysis: Fundamentals and Applications* (BKC Incorporated, Tokyo, Japan, 1999)
271. S. Horikoshi, H. Hidaka, N. Serpone, Environmental remediation by an integrated microwave/UV illumination method. 1. Microwave-assisted degradation of rhodamine-B dye in aqueous TiO_2 dispersions. Environ. Sci. Technol. **36**, 1357–1366 (2002)
272. S. Horikoshi, N. Serpone, Photochemistry with microwaves: catalysts and environmental applications. J. Photochem. Photobiol. C Photochem. Rev. **10**, 96–110 (2009)
273. S. Horikoshi, H. Hidaka, N. Serpone, Hydroxyl radicals in microwave photocatalysis. Enhanced formation of ·OH radicals probed by ESR techniques in microwave-assisted photocatalysis in aqueous TiO_2 dispersions. Chem. Phys. Lett. **376**, 475–480 (2003)
274. S. Horikoshi, M. Abe, N. Serpone, Influence of alcoholic and carbonyl functions in microwave-assisted and photo-assisted oxidative mineralization. Appl. Catal. B Environ. **89**, 284–287 (2009)
275. S. Horikoshi, A. Matsubara, S. Takayama, M. Sato, F. Sakai, M. Kajitani, M. Abe, N. Serpone, Characterization of microwave effects on metal-oxide materials: zinc oxide and titanium dioxide. Appl. Catal. B Environ. **1**, 362–367 (2009)
276. S. Horikoshi, F. Sakai, M. Kajitani, M. Abe, N. Serpone, Microwave frequency effects on the photoactivity of TiO_2: dielectric properties and the degradation of 4-chlorophenol, bisphenol A and methylene blue. Chem. Phys. Lett. **470**, 304–307 (2009)
277. S. Horikoshi, M. Kajitani, N. Serpone, The microwave-/photo-assisted degradation of bisphenol-A in aqueous TiO_2 dispersions revisited. re-assessment of the microwave Non-thermal effect, J. Photochem. Photobiol. A Chem. **188**, 1–4 (2007)

278. S. Horikoshi, H. Tsutsumi, H. Matsuzaki, A. Furube, A. V. Emeline, N. Serpone, In situ picosecond transient diffuse reflectance spectroscopy of opaque TiO_2 systems under microwave irradiation and influence of oxygen vacancies on the UV-driven/microwave-assisted TiO_2 photocatalysis. J. Mater. Chem. C **3**, 5958–5969 (2015)
279. A.C. Metaxas, J.G.P. Binner, in *Advanced Ceramic Processing and Technology*, vol. 1 ed. by J.G.P. Binner (Noyes Publications, Park Ridge, NJ, USA, 1990)
280. P.S. Sai Prasad, N. Lingaiah, P. Kanta Rao, F.J. Berry, L.E. Smart, The influence of microwave heating on the morphology and benzene hydrogenation activity of alumina- and silica-supported palladium catalysts, *Catal. Lett.* **35**, 345–351 (1995)
281. E.J. van Wyk, S.M. Bradshaw, J.B. de Swardt, The dependence of microwave regeneration of activated carbon on time and temperature. J. Microwave Power Electromagn. Energy **33**, 151–157 (1998)
282. I. Sayago, E. Terrado, E. Lafuente, M.C. Horrillo, W.K. Maser, A.M. Benito, Hydrogen sensors based on carbon nanotubes thin films. Synth. Metals **148**, 15–19 (2005)
283. A. Ansón, E. Lafuente, E.P. Urriolabeitia, R. Navarro, A.M. Benito, W.K. Maser, M.T. Martínez, Hydrogen capacity of palladium-loaded carbon materials. J. Phys. Chem. B **110**, 6643–6648 (2006)
284. A. Ansón, E. Lafuente, E.P. Urriolabeitia, R. Navarro, A.M. Benito, W.K. Maser, M.T. Martínez, Preparation of palladium loaded carbon nanotubes and activated carbons for hydrogen sorption. J. Alloys Compd. **436**, 294–297 (2007)
285. M. Cano, A. Benito, W.K. Maser, E.P. Urriolabeitia, One-step microwave synthesis of palladium–carbon nanotube hybrids with improved catalytic performance. Carbon **49**, 652–658 (2011)
286. C. Linyuan, L. Yanlan, Z. Baohua, L. Lehui, In situ controllable growth of Prussian blue nanocubes on reduced graphene oxide: facile synthesis and their application as enhanced nanoelectrocatalyst for H_2O_2 reduction. ACS Appl. Mater. Interf. **2**, 2339–2346 (2010)
287. K.S. Subrahmanyam, K.M. Arun, K.P. Swapan, C.N.R. Rao, A study of graphene decorated with metal nanoparticles. Chem. Phys. Lett. **497**, 70–75 (2010)
288. H. Chang, G. Wang, A. Yang, X. Tao, X. Liu, Y. Shen, Z. Zheng, A transparent, flexible, low- temperature, and solution-processible graphene composite electrode. Adv. Funct. Mater. **20**, 2893–2902 (2010)
289. M. Lotya, Y. Hernandez, P.J. King, R.J. Smith, V. Nicolosi, L.S. Karlsson, F.M. Blighe, S. De, Z. Wang, I.T. McGovern, G.S. Duesberg, J.N. Coleman, Liquid phase production of graphene by exfoliation of graphite in surfactant/water solutions. J. Am. Chem. Soc. **131**, 3611–3620 (2009)
290. Y. Xina, J.-G. Liu, Y. Zhou, W. Liu, J. Gao, Y. Xie, Y. Yin, Z. Zou, Preparation and characterization of Pt supported on graphene with enhanced electrocatalytic activity in fuel cell. J. Power Sources **196**, 1012–1018 (2011)
291. K. Jasuja, J. Linn, S. Melton, V. Berry, Microwave-reduced uncapped metal nanoparticles on graphene: tuning catalytic, electrical, and Raman properties. J. Phys. Chem. Lett. **1**, 1853–1860 (2010)
292. F. Li, Y. Guo, Y. Liu, H. Qiu, X. Sun, W. Wang, Y. Liu, J. Gao, Fabrication of Pt–Cu/RGO hybrids and their electrochemical performance for the oxidation of methanol and formic acid in acid media. Carbon **64**, 11–19 (2013)
293. H. Zhang, X. Xu, P. Gu, C. Li, P. Wu, C. Cai, Microwave-assisted synthesis of graphene-supported $Pd1Pt_3$ nanostructures and their electrocatalytic activity for methanol oxidation. Electrochim. Acta **56**, 7064–7077 (2011)
294. N. Toshima, M. Harada, T. Yonezawa, K. Kushihashi, K. Asakura, A. Structural analysis of polymer-protected palladium/platinum bimetallic clusters as dispersed catalysts by using extended X-ray absorption fine structure spectroscopy. J. Phys. Chem. **95**, 7448–7453 (1991)
295. W. Qian, R. Hao, J. Zhou, M. Eastman, B.A. Manhat, Q. Sun, A.M. Goforth, J. Jiao, Exfoliated graphene-supported Pt and Pt-based alloys as electrocatalysis for direct methanol fuel cells. Carbon **52**, 595–604 (2013)

296. X. Chen, Z. Cai, X. Chen, M. Oyama, Synthesis of bimetallic Pt/Pd nanocubes on graphene with *N, N*-dimethylformamide and their direct use for methanol electrocatalytic oxidation. Carbon **66**, 387–394 (2014)
297. Y. Gogotsi, J. Libera, A.J. Koster van Groos, M. Yoshimura, Hydrothermal synthesis of carbon nanotubes, in *Proceedings Joint 6th International Symposium on Hydrothermal Reactions (ISHR) & 4th International Conference on Solvothermal Reactions (ICSTR)*, 2001, pp. 350–355
298. A.M.R. Galletti, C. Antonetti, I. Longo, G. Capannelli, A.M. Venezia, A novel microwave assisted process for the synthesis of nanostructured ruthenium catalysts active in the hydrogenation of phenol to cyclohexanone. Appl. Catal. A: General **350**, 46–52 (2008)
299. G. Glaspell, L. Fuoco, M.S. El-Shall, Microwave synthesis of supported Au and Pd nanoparticle catalysts for CO oxidation. J. Phys. Chem. B **109**, 17350–17355 (2005)
300. S. Shi, J.-Y. Hwang, Microwave-assisted wet chemical synthesis: advantages, significance, and steps to industrialization. J. Miner. Mater. Charact. Eng. **2**, 101–110 (2003)
301. Y. Ning, J. Wang, P. Hong, Influence of microwave radiation on properties and structure of Fe_2O_3/SO_4^{2-} solid superacid catalyst. J. Mater. Sci. Technol. **12**, 307–311 (1996)
302. C. Ragupathi, J.J. Vjaya, L.J. Kennedy, Preparation, characterization and catalytic properties of nickel aluminate nanoparticles: a comparison between conventional and microwave method. J. Saudi Chem. Soc. **21**, S231–S239 (2014)
303. N. Kumada, N. Kinomura, S. Komarneni, Microwave hydrothermal synthesis of ABi_2O_6 (A = Mg, Zn). Mater. Res. Bull. **33**, 1411–1414 (1998)
304. R. Trujillano, E. Rico, M.A. Vicente, V. Rives, L. Bergalaouis, S.B. Chaabene, A.D. Ghorbel, Microwave-assisted synthesis of Fe^{3+} saponites. Characterization by X-ray diffraction and FT-IR spectroscopy. Rev. Soc. Esp. Miner. **11**, 189–190 (2009)
305. Z. Derikvand, F. Farzaneh, Synthesis, characterization and catalytic behavior of special type of silico-alumino-phosphate with opal structure. J. Sci. (Iran) **14**, 235–238 (2003)
306. A.Y. Shan, T.I.M. Ghazi, S.A. Rashid, Immobilization of titanium dioxide onto supporting materials in heterogeneous photocatalysis: a review. Appl. Catal. A: General **389**, 1–8 (2010)
307. M. Horie, H. Kato, K. Fujita, S. Endoh, H. Iwahashi, In vitro evaluation of cellular response induced by manufactured nanoparticles. Chem. Res. Toxicol. **25**, 605–619 (2012)
308. A. Weir, P. Westerhoff, L. Fabricus, K. Hristovski, N. von Goetz, Titanium dioxide nanoparticles in food and personal care products. Environ. Sci. Technol. **46**, 2242–2250 (2012)
309. S. Horikoshi, S. Sakamoto, N. Serpone, Formation and efficacy of TiO_2/AC composites prepared under microwave irradiation in the photoinduced transformation of the 2-propanol VOC pollutant in air. Appl. Catal. B Environ. **140–141**, 646–651 (2013)
310. G. Gunawan, S. Bourdo, V. Saini, A.S. Biris, T. Viswanathan, Novel microwave-assisted synthesis of Nickel/Carbon (Ni/C) nanocomposite with tannin as the carbon source. J. Wood Chem. Technol. **31**, 345–356 (2011)
311. R.F. Schiffmann, Microwave and dielectric drying, in *Handbook of Industrial Drying*, 4th ed. (CRC Press, NY, USA, 2014)
312. P.J. Clark, Electromagnetic energy in food processing. *Food Technol.* **67**(4) (2013)

Chapter 10
Materials Processing by Microwave Heating

Abstract Attempts to use microwaves in materials processing are very old and are comparable to those of microwave heating of food. This chapter discusses the characteristics and advantages of the microwave processing of ceramics and rapid sintering of ceramics and introduces the sintering of inorganic materials with new structures through selective heating by microwaves. In addition, the refining of metals and drying technology of metal films and nano-inks is explained. Characteristics of the microwave method for the synthesis of metallic nanoparticles and quantum dots, together with their effects are described. In addition, the chapter introduces sintering of ceramics with structural features using microwave characteristic heating. The coffee break discusses future developments in microwave ovens.

Keywords Microwave sintering · Ceramics · Structural features
Metallic substrates · Metal metallurgy · Microwave iron manufacturing
Metal alloys · Drying · Nano-ink · Solid-phase diffusion · Spinel oxide
Nanoparticle synthesis · GaN semiconductor generator · Selective food heating

10.1 Processing of Solid-State Materials

10.1.1 Sintering and Drying of Ceramics

The calcination and sintering of ceramics by microwaves was reported as early as 1972 by Shimomura et al. [1]; fundamental and application studies followed thereafter around 1980. Even so, much of the theory and some of the relevant data on microwave-material interactions, which provided a valuable background for microwave heating and data of dielectric factors, had already been reported in 1954 in a book edited by von Hippel [2]. This book did, in part, accelerate the early studies into microwave research. von Hippel compiled the first tabulation of measured dielectric properties for numerous materials over a wide range of frequencies from 10 Hz to 10 GHz. Materials so characterized were both organic and

S. Horikoshi et al., *Microwave Chemical and Materials Processing*,
https://doi.org/10.1007/978-981-10-6466-1_10

inorganic in nature, and consisted of both liquids and solids. Subsequently, dielectric data were obtained on high-temperature solid-state materials that included oxides, nitrides, silicates, rocks, and minerals [3]. Since the early 1990s, four reports emanated from the Laboratory for Insulation Research of the Massachusetts Institute of Technology [3]. These reports contain a great deal of valuable information on the frequency and temperature dependence of the dielectric properties of a large number of ceramic oxides, and some non-oxide materials, together with various minerals, several other organic compounds, and some foodstuffs. The microwave frequency varied up to 25 GHz and the temperature up to 1000 °C [3].

Through the 1970s, considerable interest focused on the heating of ceramics (e.g., drying, calcining, and sintering); several other applications were also developed, including accelerated slip casting and acid digestion of ceramics and minerals. The essential role that microwaves play in the field of ceramics was identified in the mid-1980s, and is not different from what is currently known. Four categories of potential applications were identified; they are reported in Fig. 10.1 [4, 5]. To the extent that many ceramic materials are transparent to microwaves at ambient temperature, microwaves can be used to detect moisture, defects, and pores that have different absorption characteristics than the ceramics themselves. Such characteristics tend to be similar to a microwave process occurring in solid materials [6]. Thus, microwaves can be used for both process control and nondestructive testing (NDT). In the former case, there are several examples of moisture measurements, contact-free thickness gaging, flaw detection, and vibration measurements [7].

Microwave generated plasmas can be used in different methodologies: (i) microwave-plasma chemical vapor deposition (MPCVD), and (ii) microwave-plasma-induced sintering (MPIS). The MPCVD process assists in decomposing gaseous species to produce highly excited electronic states of the elements that recombine to form metastable elements. Application of microwave heating to liquids in the form of solutions or suspensions provides a new means to process, synthesize, and analyze materials. Pilot-scale microwaves have also been

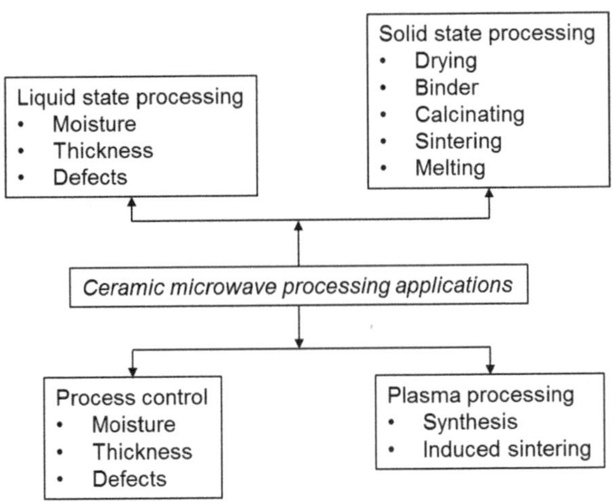

Fig. 10.1 Application of microwave energy for the processing of ceramics. Reprinted with permission from Ref. [4]; the flow chart was adapted from an original chart given by Roy et al. [5]. Copyright 1985 by the American Ceramic Society

used successfully in casting, setting, and drying slip-cast whiteware and refractory bodies.

The early industrial use of microwaves was as a heat source to remove water from inorganic materials. In the latter half of the 1970s, it was discovered [8] that ceramics could also be heated with microwaves, which led to the effective sintering of ceramic powders. Heat conduction and radiant heating are the conventional heating methods for such tasks. A sample that absorbs microwaves generates heat after irradiation for a short time and yields a sintered compact product at low temperature.

Studies of microwave processing of ceramics started in the late 1960s [9]. Since then, interest in the use of microwaves for heating and sintering of ceramics has grown steadily. However, the requirements for ceramic powder sintering make this technique one of the most challenging applications for microwave processing. These requirements often include, among other things, high temperatures, high heating rates, and uniform temperature distribution [10]. During microwave heating, the absorption of electromagnetic energy raises the temperature of the entire sample. However, heat loss from the surface causes the near-surface temperatures to be lower than the internal temperatures, hence large temperature gradients can develop in ceramics that possess poor thermal conductivity. The shape of the ceramic substrate and the microwave frequency strongly influence the temperature gradients, which can make it difficult to achieve uniform heating. Fast heating rates can lead to localized heating that imparts non-uniform properties (even cracking) to the sample, and even thermal runaway, in which the temperature of the sample increases uncontrollably and too rapidly [11].

A symposium held at a Conference on Materials Science by the Materials Research Society (MRS) revealed that materials processing by microwaves was a very hot topic that drew a great deal of interest [4]. Indications from the Conference also showed that microwaves could not only be used to determine the quantity of water in ceramics but could also be used as a technique to detect and analyze the formation of fissures in ceramics and metals [4]. The basic interactions between the microwaves and materials were also widely discussed; various types of ceramics to be treated by the microwave method were identified [See e.g. 12, 13]. Recent years have witnessed calcination of special ceramics with a specific microwave method [14].

It is therefore appropriate at this stage to query why microwaves are used for the sintering of ceramics. When absorbed, microwave energy is transformed into heat inside the material, resulting in significant energy savings and shorter sintering times for ceramic substrates when compared to other sintering methods. These two factors play a decisive role in most applications, and to date have gained considerable industrial acceptance especially in view of significant production efficiency. Ceramic sintering by microwave heating results from self-heating of the entire volume of material (i.e. volumetric heating). Through this high volume heat (calorific power per unit volume), it is possible to obtain a homogeneous and close-sintered product [4]. In addition to rapid processing, other reasons for the growing interest in the use of microwave energy include reduced production costs, reduced energy consumption, improved product quality, and uniformity, development of novel microstructures and properties, and synthesis of new materials. Table 10.1 summarizes these and other characteristics of the microwave process. Microwave processing of ceramics is a relatively new development and is fundamentally different from traditional heating processes used by the ceramic industry.

The typical microwave frequencies for materials processing are 915 MHz, 2.45, 5.8, 22.00, 24.12, 28 and 60 GHz; however, the most often used frequencies are 915 MHz and 2.45 GHz [15]. The advantages of higher frequencies are more uniform electric field distribution in the cavity and higher power dissipated in the dielectric materials. Unfortunately, the necessary microwave apparatus is often available on a very limited basis and at an extremely high cost. Also, at the higher frequencies, the penetration depth of the microwave energy is too small to be useful, except in special cases.

Many ceramic materials that are difficult to heat at room temperature possess electrical conductivity or dielectric loss factors that rapidly increase in magnitude as the temperature rises. Consequently, if they can be preheated to a suitable temperature by using another heat source, these materials will absorb microwave energy. This has led to the development of hybrid heating methods that use either an independent heat source (e.g., a gas or electric furnace) in combination with microwaves, or else use an external susceptor material (high-loss material) that couples with the microwaves to supply the required heat. In the latter case, the material is exposed simultaneously to both radiant heats produced by the susceptor material and to microwave radiation. (Note that a susceptor material is simply referred to as a *susceptor*; see Sect. 8.6). Currently, microwaves are used in the sintering of various ceramics. Table 10.2 illustrates examples of rapid sintering of ceramics using microwaves—the reported data have been updated from an original list given by Roy et al. [5].

10.1.2 Ceramics with Structural Features, Heated by Microwave Sintering

Sample morphology and crystallinity are also influenced by the microwave synthesis. In the synthesis of LiV_3O_8, a commonly investigated cathode material for lithium batteries, the morphology, the crystallite size, and the concentration of defects all depend upon irradiation time and power; in turn, this directly affects conductivity, discharge capacity, and cycle performance of the product [41]. Samples prepared at a relatively high irradiation power, and at very high temperatures, presented very stable cycle performance since as-prepared crystallites had a similar microstructure and narrow distribution of crystal size.

Microwave-induced sintering of a sample consisting of mixed precursors that display large differences in the absorption of microwave energy results in grain growth in short times and at low temperatures, compared with sintering the same

Table 10.1 Characteristics and advantages of the microwave processing of ceramics

Characteristics	Advantages of microwave heating over conventional heating
Direct coupling (absorbing) of microwave creates volumetric (bulk) heating	• Potential to heat large sections uniformly • Reversed thermal gradients: surface cooler than interior • Process materials at lower surface temperatures • Rapid removal of water, binders, and gases without rupture or cracking • Internal stresses reduction by lower thermal gradients • Heat in clean (pure) environment; air, controlled atmospheres, vacuum, or pressure • Control partial pressure of reactive gases for selective oxidation/reduction of certain phases • Improvement of product quality, uniformity, and yields • Instantaneous response to microwave power changes • Low thermal mass; precise and automated temperature control
Dielectric losses (and heating) accelerate very rapidly with increasing temperatures above critical temperature (T_{crit})	• Ability to heat "transparent" materials above critical temperature (T_{crit}) • Very rapid processing (2 to 50 times faster than conventional) • Densify materials rapidly with minimum grain growth (accelerated sintering) • Reduce process costs (time, energy, and labor) • Ability to heat ceramics well above 2000 °C (in air, vacuum, or controlled atmospheres)
Microwaves are polarized and coherent; location of maximum electric and magnetic fields can be controlled	• Capability of high-energy concentration in short times and in selected regions • Frequency and power level optimization for given material, size, and shape • Potential for process automation, flexibility, efficiency, and energy savings • Precise heating of selected regions, i.e., brazing or sealing of joints, fiber drawing, and plasma generation • Acceleration of sintering and diffusion due to high electrical fields; thus densification at lower temperatures

(continued)

Table 10.1 (continued)

Characteristics	Advantages of microwave heating over conventional heating
Differential micro-wave coupling of phases, additives, and constituents leads to selective heating	• Synthesis of new materials and microstructures • Heating of selected zones (brazing and sealing) • Enhanced coupling of microwave transparent materials • Use of fugitive coupling materials for preheating of otherwise transparent materials • Use of microwave-coupling materials as shapes or containers to heat the more transparent materials • Superior control over state of oxidation through selective heating of phases and control over oxygen partial pressure

Reprinted with permission from Ref. [4]. Copyright 1985 by the American Ceramic Society

sample in an electric furnace [42–44]. For instance, strontium carbonate ($SrCO_3$) and ferric oxide (γ-Fe_2O_3) are used as raw material powders to synthesize strontium ferrite ($SrFe_{12}O_{19}$) through a sintering reaction. The extent of microwave absorption is rather weak for $SrCO_3$, contrary to γ-Fe_2O_3 for which microwave absorption is much stronger. Consequently, microwaves heat γ-Fe_2O_3 selectively, which then becomes the source of conductive-heat for $SrCO_3$ to yield $SrFe_{12}O_{19}$ in short time and at low temperature, compared to the conventional method in an electric furnace. Microwave heating produces significantly greater particle growth than with conventional sintering method because of the microwave heating efficiency of this ferromagnetic material $SrFe_{12}O_{19}$ (Fig. 10.2).

A 2013 study by Endo et al. [45] reported an anisotropic grain growth by microwave heating. Figure 10.3 illustrates images of the difference in grain growth of $BaFe_{12}O_{19}$ when heated by a conventional electric furnace and microwave heating. The conventional heating method provides isotropic heat energy to a sample and so the growth of grains becomes isotropic; on the other hand, when using microwave heating that provides thermal energy anisotropically, the behavior of grain growth is clearly different (highly anisotropic) from the conventional heating method. For example, different physical properties are imparted to the structure of the vertical layers or the parallel layers of Na_xCoO_2, InO-ZnO, and $CuFeO_2$ as transparent conductive oxide layers, resulting an anisotropic grain growth is indicated by microwave processing. The $BaFe_{12}O_{19}$ also has c-axis anisotropic magnetization and displays an anisotropic grain growth by microwave processing.

Table 10.2 Examples of microwave-heated ceramics

Material	Sample size, quantity, and weight	Frequency/ Mode/Field	Atmosphere	Temperature/°C	Hold time/s	Theoretical density/%	Ref.
$LaCrO_3$	10 mm-dia. pellet, 2 mm thickness	2.45 GHz/S		1350	600	96	[16]
$LaCoO_3$	10 mm-dia. pellet, 2 mm thickness	2.45 GHz/S		1200	1800	99.6	[16]
$LaNiO_3$	10 mm-dia. pellet, 2 mm thickness	2.45 GHz/S		800	3000	95	[16]
ZnO-based varistors	8 mm-dia. pellet	2.45 GHz/S/E		1100	300	98.3 ± 0.5	[17]
ZnO-based varistors	8 mm-dia. pellet	2.45 GHz/S/H		1100	300	99.2 ± 0.5	[17]
PZT ceramics	10 mm-dia. pellets	2.45 GHz/M	Air	1310	300	Dense	[18]
PLZT ceramics	10 mm-dia. pellets	2.45 GHz/M	Air	1200	300	Dense	[18]
Mullite, SiO_2, Al_2O_3, gels	2–26 g	2.45 GHz/M	Air	1500+			[19]
$YBa_2Cu_3O_{7-x}$	10 mm rod	2.45 GHz/M	Air/O_2	1800	750/500		[20]
Ba-ferrite	12.7 mm disks	2.45 GHz/M	Air	1205	2700		[21]
Al_2O_3	~17 mm length, ~9 mm inner diameter, and ~11 mm outer diameter	2.45 GHz/M		1500	3600	92.2	[22]
Al_2O_3 spark plug insulator	186 insulators	2.45 GHz/M	Air	1600	1740	99.9	[23]
Al_2O_3 castable shapes	1 m (68 kg) long	2.45 GHz/M	Air	1200	2.9×10^{4a}		[24]
Portland cement	10 mm cylinders	2.45 GHz/S	Air	>1400	420^a		[25]
Al_2O_3	Pellet	2.45 GHz/M	Air	1500		95	[26]
99.995% pure Al_2O_3	10 mm pellet	60 GHz		1700	370^a	95.7	[27]
99% pure Al_2O_3	Pellet	2.45 GHz/M	Air			91.7	[28]
Al_2O_3 + 0.1MgO	100 g pellet	28 GHz	Vacuum	1200	3600	98.2	[29]
Al_2O_3 + 2SiC (powder)	Pellet	2.45 GHz/M	Air			96.9	[28]
Al_2O_3 + 10SiC (platelets)	Pellet	2.45 GHz/M	Air	1600		94	[30]
Al_2O_3 + 10SiC (whiskers)	Pellet	2.45 GHz/M	Air	1600		72	[30]

(continued)

Table 10.2 (continued)

Material	Sample size, quantity, and weight	Frequency/ Mode/Field	Atmosphere	Temperature/°C	Hold time/s	Theoretical density/%	Ref.
Al_2O_3 + 10SiC (whiskers)	10 mm pellet	60 GHz		1450	320[a]	77	[27]
(Al_2O_3 + 0.2MgO) + 30TiC	4 mm rod	2.45 GHz/M	1.1 MPa	1850		95	[31]
99% pure Y_2O_3	Pellet	2.45 GHz/M	Air			57.4	[28]
Y_2O_3 + $2ZrO_2$	Pellet	2.45 GHz/M	Air			93.3	[28]
Y_2O_3 +10Zr-oxynitrate	Pellet	2.45 GHz/M	Air		1800[a]	99.6	[28]
ZrO_2 + 4CaO	Pellet	2.45 GHz/M	Air			67.5	[28]
ZrO_2 + 8 Y_2O_3	Pellet	2.45 GHz/M	Air			97.6	[28]
ZrO_2 + 10Zr-oxynitrate	Pellet	2.45 GHz/M	Air			99.6	[28]
ZrO_2 + 10SiC (whiskers)	Pellet	2.45 GHz/M	Air		>1700		[32]
ZrO_2 + $10Si_3N_4$ (whiskers)	Pellet	2.45 GHz/M	Air		>1700		[32]
$SrTiO_3$	Pellet	2.45 GHz/M	Air	1300	430[a]	98	[26]
B_4C (pure)	10 mm cylinder	2.45 GHz/M	Air	2000	450	95	[33]
SiC + 0.6B + 4C	4.5 mm rod	2.45 GHz/M	0.8 MPa N_2	2200			[31]
Al_2O_3 + 0.25MgO	4.8 mm rod	2.45 GHz/S	4 kPa N_2	1600	<120	99.0	[34]
Al_2O_3 + 0.25MgO	4.8 mm rod	2.45 GHz/S	4 kPa N_2	1660	600	99.9	[34]
UO_2	103 g	2.45 GHz/M	Air	>2600	900[a]	Melted	[35]
U_3O_x	55 g	2.45 GHz/M	Air		1200[a]	Melted	[35]
Al_2O_3 + sealing glass	50*50*0.5 mm plates	2.45 GHz/M	Air	1400	1200		[36]
99% pure Al_2O_3 + flux	3 mm rod; butt join	2.45 GHz/S	Air	1800	180		[37]
92% or 96% pure Al_2O_3, no flux	3 mm rod; butt join	2.45 GHz/S	Air	1750	180		[37]

(continued)

Table 10.2 (continued)

Material	Sample size, quantity, and weight	Frequency/ Mode/Field	Atmosphere	Temperature/°C	Hold time/s	Theoretical density/%	Ref.
Mullite, no flux	9.5 mm rod; butt join	2.45 GHz/S	Air	1300	<600		[38]
Si_3N_4 + flux	3 mm rod; butt join	2.45 GHz/S	N_2	1720	180		[37]
SiO (pure)	Rod	2.45 GHz/S	Air	>2000			[39]
SiO_2 (ultrapure)	Tube/rod	2.45 GHz/S	Air	>2000			[40]

S Single mode, *M* Multi mode, *E* Maximum electric field, *H* Maximum magnetic field

[a]Heating time from room temperature to temperature indicated

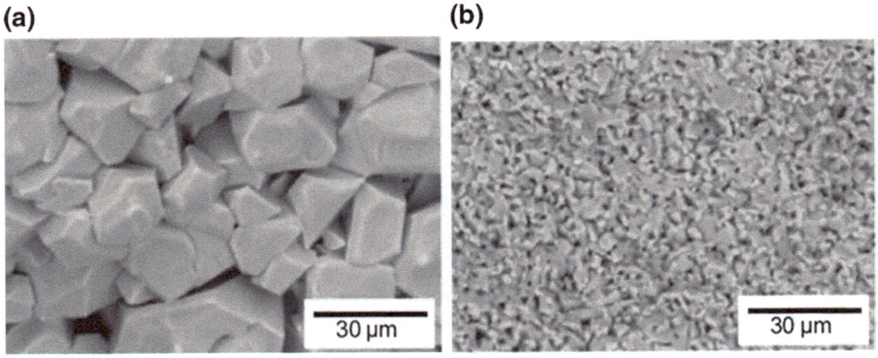

Fig. 10.2 Sintering of strontium ferrite ($SrFe_{12}O_{19}$) using (**a**) microwave heating (temperature rise for 10 s and keeping temperature constant at 1100 °C for 100 s), and (**b**) conventional electric furnace heating (temperature rise for 100 min and then keep temperature at 1200 °C for 120 min). Reproduced from Ref. [46]. Copyright 2013 by S. Horikoshi

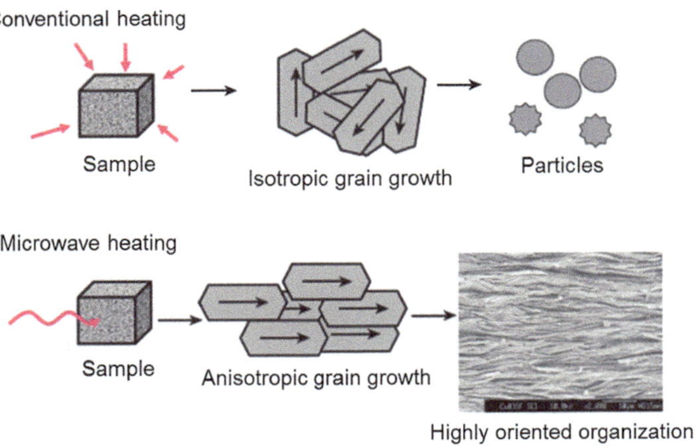

Fig. 10.3 Growth images of $BaFe_{12}O_{19}$ by conventional heating (e.g. electric furnace) resulting in an isotropous grain growth whereas microwave heating leads to an anisotropic grain growth. Reproduced from Ref. [46]. Copyright 2013 by S. Horikoshi

10.1.3 Metallic Substrates

Bulk metals typically reflect microwaves. However, in powdered form, metallic substrates absorb microwaves at room temperature and are subsequently heated rapidly and efficiently. Hence, microwave heating can be used to sinter various powdered metallic substrates. Initially, the success in microwave heating and sintering was confined primarily to oxide and some non-oxide ceramics, but recently the technique has been extended to carbide ceramics used in cutting tools [47]. Most of the early literature was mainly concerned with microwave-sintering of powdered metal compacts [48]. Virtually all metals, alloys, and intermetallics couple with microwaves and are heated efficiently and effectively, resulting in highly sintered substrates with improved mechanical properties. For example, Fe–Ni–C (FN208) and Fe–Cu–C (FC208) are two commonly reported commercial steel compositions that have been sintered in a controlled atmosphere furnace with multimode 2.45-GHz microwaves, and with typical cycle times of about 90 min; sintering temperatures ranged from 1100 to 1300 °C for soaking times of 5–60 min [49]. Mechanical properties, such as the module of rupture (MOR) and hardness of microwave-processed samples, were greater with microwaves than with conventionally processed samples. As an example, microwave sintering of copper steel (MPIF FC-0208 composition) displayed good sintered density, hardness, flexural strength, and near net dimensions (i.e. no shrinkage), thus yielding equivalent or even superior mechanical properties compared to conventional sintering (see Fig. 10.4).

An important distinction in the microstructures of conventionally sintered and microwave- sintered samples was that the pores in the microwave-sintered samples (FC208 steel) had more rounded edges than the conventionally sintered samples, resulting in improved ductility and strength [50]. The conventionally sintered sample failed at a load of 145 kg (320 lbs), while the microwave-sintered sample failed at 195 kg (430 lbs), indicating a 30% increase in strength. A more important observation was the manner in which the substrates failed after applying the maximum load: the conventionally sintered sample broke into four pieces, very typical of standard PM substrates, whereas the microwave-processed substrate broke into two flat pieces, indicating higher ductility. This difference was attributed to the porosity of the resulting substrates: the microwave-sintered samples exhibited round edges in contrast to the typical sharp-edged porosity found in conventionally sintered samples. Many commercial powdered metal components of various alloy compositions that include iron and steel, copper, aluminum, nickel, molybdenum, cobalt, tungsten, tungsten carbide, tin, and their alloys have been sintered using

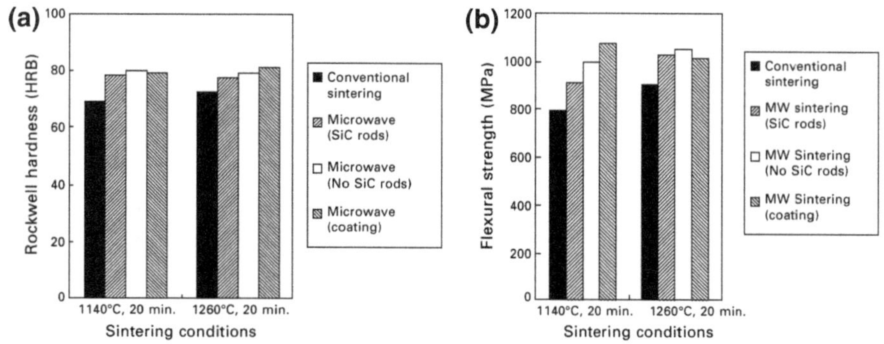

Fig. 10.4 Comparison of (**a**) Rockwell hardness and (**b**) flexural strength of microwave and conventionally sintered FC208 samples of copper steel (FC 208: 2% Cu, 0.08% C). Reprinted with permission from Ref. [49]. Copyright 2001 by Taylor & Francis

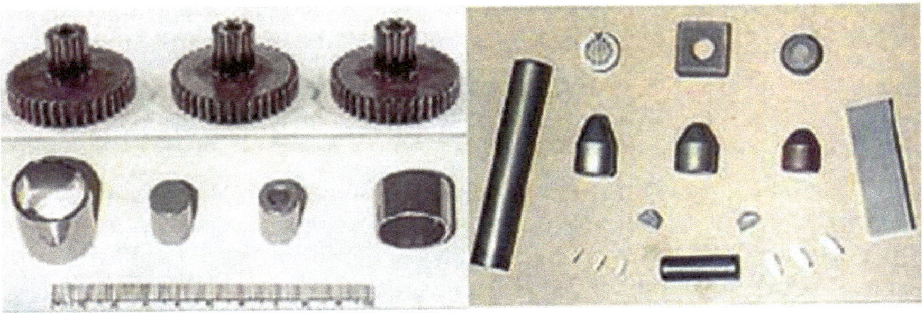

Fig. 10.5 Metallic parts produced by microwave sintering: gears cylinders, rods, and disks. Reproduced from [51]

microwaves producing essentially fully dense substrates. Figure 10.5 illustrates some of the metallurgical parts processed using the microwave technology. The largest commercial steel component that has been fully sintered so far is an automotive gear of 10 cm in diameter and about 2.5 cm in height.

Metallic copper powders can be sintered by 30-GHz microwave heating at 100 °C min^{-1} and then maintaining the temperature at 1000–1100 °C for 10 min [52]. A sintering density of 88% of theoretical was achieved in this case, while the formation density was 68% (Fig. 10.6), shrinkage was 7%, and no transformation of the shape occurred. In addition, a sintered copper object with sufficient sintering density and pull stress was obtained by the conventional heating method at nearly 100 °C from the cold.

Sintering of high melting-point metals (e.g., W, Mo, and others) is difficult; conventional heating requires high temperatures and long times to sinter the

Fig. 10.6 (top) Copper compact before sintering; (bottom) after sintering by 30-GHz microwave heating at 1000 °C. Reprinted with permission from Ref. [52]. Copyright 2006 by Taylor & Francis

corresponding powdered samples [53]. By contrast, microwaves can sinter such metal substrates at a lower sintering temperature and in a shorter time. For example, microwaves can sinter Mo powders with as high as 98% of theoretical density within 5 min compared to 10–20 h needed by conventional heating.

The implications of the microwave sintering of powder metals in the field of powder metallurgy are obvious. Metal powders are used in a diverse range of products and applications in various industries, such as the automotive and aerospace, and industries handling heavy machinery [54]. The challenging demands of new and improved processes and materials of high integrity for advanced engineering applications require innovation and new technologies. Finer microstructures and near-theoretical densities in special powder metal components are widely desired, yet they remain elusive. Increasing cost is also a concern for the industry. Microwave processing offers a new method to meet demands of producing fine microstructures and better properties at potentially lower costs.

10.1.4 Why Microwave Sintering?

There are two main reasons why the microwave process results in better mechanical properties, especially in the case of powdered metals: for instance, it produces a finer grain size and the shapes of porosity (if any) are quite different from the conventional process. In microwave-processed powdered metal components, round-edged porosities were observed producing higher ductility and toughness. Table 10.3 summarizes some of the data regarding microwave sintering of metals and alloys.

10.1.5 Drying of Monolithic Refractory Substrates

Refractory materials come in two fixed shapes: (1) brick-type refractories and (2) monolithic refractories. A brick refractory material is produced from high-temperature calcination of a refractory raw material. The construction of monolithic refractories occurs by binding a powder from the nano-size to the micro-size with aluminate cement. A refractory raw material typically contains ca. 5–10 mass% of water, which can be removed most effectively by microwaves through an aperture, especially when the pores size is small. Reducing the water content can decrease not only the number of pores but also the size of the pores. Thus, drying and sintering of monolithic refractories with microwaves are performed at an industrial scale. The rapid removal of water from the interior of monolithic refractories can be accomplished using either a gas-jet heating method or using microwave heating. For example, Nippon Steel & Sumitomo Metal Co. of Japan uses a blast furnace for drying large size monolithic refractories; however, because of the large size of the furnace, there is a large difference in temperature

Table 10.3 Summary of data on microwave-induced sintering of metallic substrates

Material	Sample size, quantity, or weight	Frequency/Mode/Field	Atmosphere	Temperature/°C	Hold time/s	Sintered density/g cm^{-3}	Ref.
Fe–Ni	Cold-pressed, 2–5 cm in the largest dimension	2.45 GHz	Forming gas ($N_2 + H_2$)	1100–1300	300–3600	7.15	[48]
Fe–Cu (industrial part)	Cold-pressed, 2–5 cm in the largest dimension	2.45 GHz	Forming gas ($N_2 + H_2$)	1100–1300	300–3600	7.17	[48]
Fe–Cu (Lab. sample)	Cold-pressed, 2–5 cm in the largest dimension	2.45 GHz	Forming gas ($N_2 + H_2$)	1100–1300	300–3600	6.96	[48]
Cobalt	Pellet	2.45 GHz	H_2	900	600	8.70	[48]
Cobalt	Pellet	2.45 GHz	H_2	1000–1050	600	8.88	[48]
Aluminum	Pellet	2.45 GHz/M	Ar		1800	2.70 ± 0.01	[55]
Magnesium	Pellet	2.45 GHz/M	Ar		1800	1.74 ± 0.01	[55]
lead-free Viromet 349 solder alloy (Sn: 91.4, Cu: 0.5, Ag: 4.1, In: 4)	Pellet	2.45 GHz/M	Ar		1800	7.39 ± 0.006	[55]
92.5W–6.4Ni–1.1Fe alloy	Pellet	2.45 GHz/M	H_2	1500	1200	>17.3	[56]
Magnesium	35 mm pellet	2.45 GHz/M		640	1500	1.738 ± 0.002	[57]
Mg/0.3Cu	35 mm pellet	2.45 GHz/M		640	1500	1.758 ± 0.002	[57]
Mg/0.6Cu	35 mm pellet	2.45 GHz/M		640	1500	1.776 ± 0.006	[57]
Mg/1.0Cu	35 mm pellet	2.45 GHz/M		640	1500	1.809 ± 0.007	[57]
FC-0208 (Fe–2Cu–0.8C composition steel powder)	Cylindrical tubular sample	2.45 GHz/M	95% N_2 + 5% H_2	760	1200	6.94	[58]

(continued)

Table 10.3 (continued)

Material	Sample size, quantity, or weight	Frequency/Mode/Field	Atmosphere	Temperature/°C	Hold time/s	Sintered density/g cm^{-3}	Ref.
FN-0208 (Fe–2Cu–0.8C composition steel powder)	Cylindrical tubular sample	2.45 GHz/M	95% N_2 + 5% H_2	760	1200	7.47	[58]
Austenitic stainless steel (316L: Fe–16.5Cr–12.97Ni–2.48Mo–0.93Si–0.21Mn–0.025C–0.008S (in wt%))	Cylindrical pellets (16 mm diameter and 6 mm average height)	2.45 GHz/M	H_2	1400	3600	6.82	[59]
Ferritic stainless steel (434L:Fe–17Cr–1Mo–0.2Mn–0.02C–0.02S–0.02P (in wt%))	Cylindrical pellets (16 mm diameter and 6 mm average height)	2.45 GHz/M	H_2	1400	3600	7.26	[59]
Cu-5 vol.% CNT (carbon nanotube)	Disk-like pellets (16 mm diameter and 6 mm average height)	2.45 GHz		800	300	>8.1 (>95% TD)	[60]
Cu-15 vol.% CNT (carbon nanotube)	Disk-like pellets (16 mm diameter and 6 mm average height)	2.45 GHz		800	300	>7.5 (>96% TD)	[60]
Titanium (Atlantic equipment engineering, <20 μm)	Cylindrical pellets (10 mm diameter and 10 mm height)	2.45 GHz	$\sim 7 \times 10^{-3}$ Pa vacuum	1200	7200	>96.30 ± 0.45% TD	[61]
Titanium (Sumitomo Co., 45–63 μm)	Cylindrical pellets (10 mm diameter and 10 mm height)	2.45 GHz	$\sim 7 \times 10^{-3}$ Pa vacuum	1200	7200	>91.77 ± 0.44% TD	[61]
Titanium (Cerac Inc., 100–150 μm)	Cylindrical pellets (10 mm diameter and 10 mm height, 800 MPa)	2.45 GHz	$\sim 7 \times 10^{-3}$ Pa vacuum	1200	7200	>93.30 ± 0.22% TD	[61]

(continued)

Table 10.3 (continued)

Material	Sample size, quantity, or weight	Frequency/Mode/Field	Atmosphere	Temperature/°C	Hold time/s	Sintered density/g cm^{-3}	Ref.
Copper	40*10*6 mm^3	2.45 GHz/M	Air	850	300	8.68	[62]
90W–7Ni–3Cu	Cylindrical pellets (16 mm diameter and 8 mm height)	2.45 GHz/M	H_2	1300	3600	>95% TD	[63, 64]
W–25Cu	Cylindrical pellets (10 mm diameter and 7 mm height)	2.45 GHz/M	H_2	1250	600	99.8% TD	[65]
W–24Cu–1Fe	Cylindrical pellets (10 mm diameter and 7 mm height)	2.45 GHz/M	H_2	1250	600	97.3% TD	[65]
Cu–12Sn	Cylindrical pellets (12.7 mm diameter and 10 mm height)	2.45 GHz/M	H_2	830	1800	78% TD	[66]
W-1wt%Ni	Cylindrical pellets (13 mm diameter and 3 mm height)	2.45 GHz/M	95% Ar + 5% H_2	1450	300	99% TD	[67]
Al-30wt%(Ti, W)C	Cuboid compact (23 × 7 × 7 mm)	2.45 GHz/M	Ar	560	2700	3.146 ± 0.001	[68]
Tungsten	Cylindrical pellets (13 mm diameter and 3 mm height)	2.45 GHz/M	95% Ar + 5% H_2	1500	1800	18.3 (95% TD)	[69]
Al–0.25Cu–1Mg–0.8Si	Cylindrical pellets (16 mm diameter and 6 mm height)	2.45 GHz/M	Vacuum (10^{-6} Torr)	630	3600	95% TD	[70]

(continued)

Table 10.3 (continued)

Material	Sample size, quantity, or weight	Frequency/Mode/Field	Atmosphere	Temperature/°C	Hold time/s	Sintered density/g cm^{-3}	Ref.
Cu (149 μm)	Cylindrical pellets (6.4 mm average diameter and 8–11 mm height)	30 GHz/M	92% N_2 + 8% H_2	1000	600	79.3% TD	[71]
Cu (149 μm)	Cylindrical pellets (6.4 mm average diameter and 8–11 mm height)	30 GHz/M	N_2	1000	600	79.8% TD	[71]
Cu (149 μm)	Cylindrical pellets (6.4 mm average diameter and 8–11 mm height)	30 GHz/M	Ar	1000	600	81.5% TD	[71]
Cu (149 μm)	Cylindrical pellets (6.4 mm average diameter and 8–11 mm height)	30 GHz/M	92% N_2 + 8% H_2	1000	7200	83.7% TD	[71]
Cu (149 μm)	Cylindrical pellets (6.4 mm average diameter and 8–11 mm height)	30 GHz/M	N_2	1000	7200	83.5% TD	[71]
Cu (149 μm)	Cylindrical pellets (6.4 mm average diameter and 8–11 mm height)	30 GHz/M	Ar	1000	7200	76.9% TD	[71]

(continued)

Table 10.3 (continued)

Material	Sample size, quantity, or weight	Frequency/Mode/Field	Atmosphere	Temperature/°C	Hold time/s	Sintered density/g cm^{-3}	Ref.
Cu (88 + 53 μm)	Cylindrical pellets (6.4 mm average diameter and 8–11 mm height)	30 GHz/M	92% N_2 + 8% H_2	1000	7200	86.4% TD	[71]
Cu (88 + 53 μm)	Cylindrical pellets (6.4 mm average diameter and 8–11 mm height)	30 GHz/M	N_2	1000	7200	86.6% TD	[71]
Cu (88 + 53 μm)	Cylindrical pellets (6.4 mm average diameter and 8–11 mm height)	30 GHz/M	Ar	1000	7200	90.2% TD	[71]
Cu (10 μm)	Cylindrical pellets (6.4 mm average diameter and 8–11 mm height)	30 GHz/M	92% N_2 + 8% H_2	1000	600	96.8% TD	[71]

% TD Percent theoretical density

from the outside to the interior of the substrate sample. To avoid this situation, hot air is continuously circulated to maintain a hot surface, while microwave generators of 20–150 kW are used for drying; the manufacturing process occurs in short time, is of low cost (energy saving), and yields high-quality products [72].

10.1.6 Drying of Transparent Conductive Films and Nano-Inks

Transparent conductive films are the materials that have been developed for the production of solar cells and liquid crystals for televisions. One typical transparent conductive film consists of tin-doped indium oxide coated on glass and referred to as ITO (indium tin oxide). ITOs are sintered under high-temperature conditions. Only inorganic materials can endure high-temperature conditions, and so are used in the fabrication of circuit boards. Clear plastic substrates can be used in the fabrication of circuit boards, making possible flexible transparent conductive films using the microwave selective heating method. For example, polyethylene terephthalate (PET) is a very poor microwave absorber and is used as a substrate for circuit boards. Hence, microwaves can selectively sinter an ITO powdered sample pasted on a PET film. However, when microwaves are applied to the same ITO acting as a plain metallic substrate, the microwaves are either reflected or cause sparking. Accordingly, precise control of the microwaves' magnetic and electric fields (mainly magnetic field irradiation) produced from a semiconductor generator and a resonator is necessary to heat the ITO layer. The production of an ITO/PET substrate unit by microwave magnetic field heating is illustrated in Fig. 10.7a. Under microwave irradiation, the ITO reaches the sintering temperature; neither transformation nor melting of the PET circuit board substrate occurs (Fig. 10.7b). Therefore, transparent conductive films can be calcined continuously using a roll-to-roll process under normal atmospheric conditions.

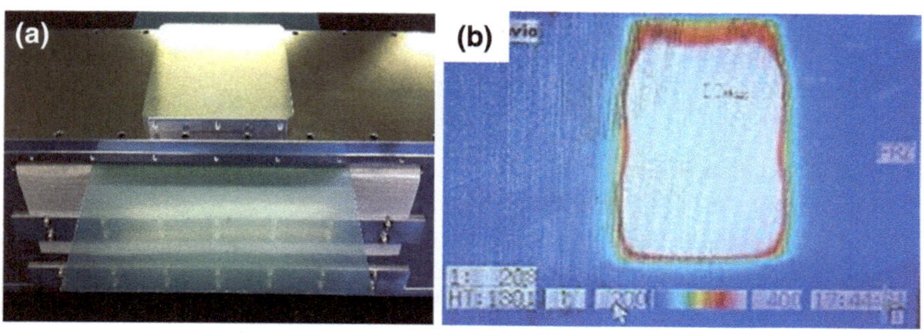

Fig. 10.7 (**a**) Photograph of the production of an ITO/PET film unit by microwave magnetic field heating, and (**b**) thermographic temperature measurement of the ITO/PET film under microwave magnetic field irradiation. Photograph provided by courtesy of Fuji Electronic Ind. Co

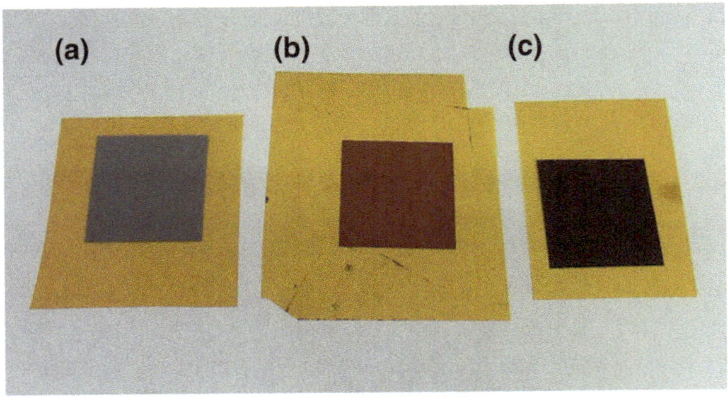

Fig. 10.8 Picture of the metallic nanoparticle ink on the PET substrate after microwave magnetic field sintering: (**a**) silver nanoparticle ink, (**b**) copper nanoparticle ink, (**c**) copper oxide nanoparticle ink. Photograph provided by courtesy of Fuji Electronic Ind. Co

Evidently, developing a printed circuit board requires simply pasting metallic nanoparticles onto some support material. It is well known that metal nanoparticles possess lower melting points than bulk metals. Therefore, after printing on an appropriate substrate support, it is possible, using microwaves, to use a lower temperature for the drying and/or sintering process. Copper produces the least expensive metal nanoparticles; unfortunately, these are subject to oxidation on heating under normal atmospheric conditions, and so require it be done under oxygen-free conditions. Since a resin coating is produced under normal atmospheric conditions, this would affect the quality of the ultimate product. These problems can be resolved through microwave selective heating and microwave-induced redox reactions. In this regard, the photographs illustrated in Fig. 10.8a–c display PET substrates onto which nanoparticles of silver (Fig. 10.8a), copper (Fig. 10.8b), and copper oxide (Fig. 10.8c) inks were sintered using microwave selective heating. Even though microwave heating was carried out in the presence of air, neither the silver nor the copper nanoparticles were oxidized, nor was resin coating necessary.

10.1.7 Features of Microwaves in Syntheses

In an ideal situation, where the precursors are good microwave absorbers, some or all of the samples in the synthesis of materials will exhibit high coupling efficiencies with microwaves at various frequencies. Such coupling will facilitate rapid heating of the samples in a microwave field, and lead to the successful rapid synthesis of the target material. In some cases, the materials used conventionally tend to be poor microwave absorbers; however, it is often possible for the synthesis to select alternative precursors that possess more favorable dielectric properties. The microwave selective heating method is most effective in a reaction of inorganic solid materials. For example, when microwave heating is performed on mixed raw materials, one of which is efficiently heated and the other is not, then a temperature gradient forms between the two raw materials in the microwave-assisted synthesis (Fig. 10.9). This temperature gradient may be caused by the size of the raw material powder, leading to a microscopic heat distribution in the sample. When considering substance diffusion, the diffusion of a solid-phase progresses from the hot substance to the cold substance, thereby creating a peculiar diffusion state at the microscopic level on microwave heating.

An example of the aforementioned peculiarity was reported by Takizawa and coworkers [46] in the synthesis of a mesoscopic-structured material involving TiO_2 and SnO_2, as the raw materials. Different heating efficiencies are illustrated by this mixture when heated by microwaves: high for SnO_2 and low for TiO_2. Figure 10.10 illustrates the synthesized mesostructured material at the nanoscale level. Microwave heating of the sample produces a supersaturated solid solution, a process that leads to microphase separation (spinodal decomposition) and generation of the mesoscopic structure. A material has been synthesized with minute striped layers of about 40 nm from a mixture of ZnO and Fe_2O_3 (Fig. 10.10). The state of two components that melted together on heating is referred to as a *solid solution* and under the microwave heating, promoting the solid solution of ZnO and Fe_2O_3 occurred and a supersaturated solid solution, resulting spinodal decomposition. Another synthetic example is that of Sn_2TiO_4 produced by microwave heating under non-equilibrium diffusion conditions the precursors SnO_2 (good microwave absorber) and TiO_2 (poor microwave absorber); the process occurs through a disproportionation reaction that involves the inorganic Sn^{2+} oxide (SnO) system at temperatures beyond ca. 300 °C. By contrast, conventional heating of the same precursors in an electric furnace does not produce Sn_2TiO_4 even if the raw precursors were heated to temperatures beyond 1000 °C.

Inorganic oxides possessing superlattice structures that mimic noble functional materials with dielectric and super-conductivity properties can be synthesized via a pulsed-laser irradiation method to form homologous compounds. One such homologous compound is the multiple oxide $MNO_3(ZnO)_m$ {M = Ga and/or Fe and/or In; N = Ga and/or Fe and/or In and/or Mn}. The synthesis of $Fe_2O_3(ZnO)_m$

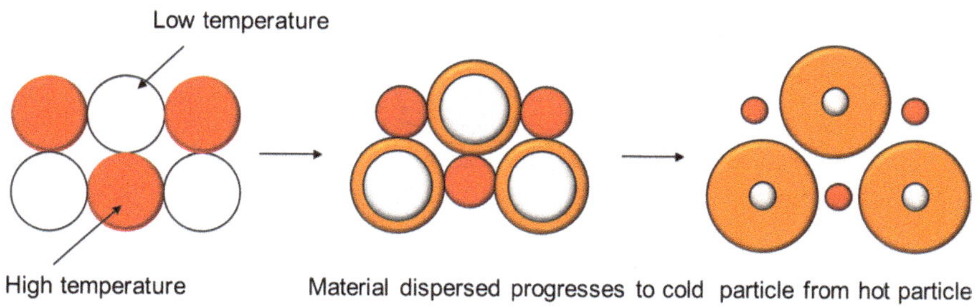

Fig. 10.9 Image of solid-phase diffusion by microwave selective heating. Reproduced from Ref. [46]. Copyright 2013 by S. Horikoshi

Fig. 10.10 Image of solid-phase diffusion by microwave selective heating. Reproduced from Ref. [46]. Copyright 2013 by S. Horikoshi

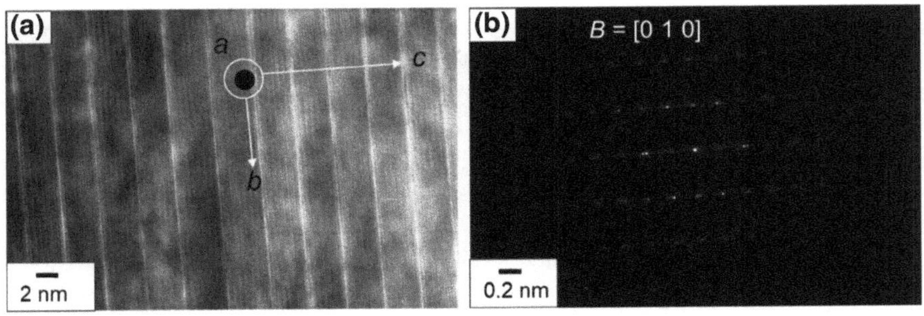

Fig. 10.11 (**a**) Transmission electron microscopy (TEM) images and (**b**) the electron diffraction patterns of samples of $Fe_2O_3(ZnO)_m$ irradiated ten times by 28 GHz microwave at 1200 °C. Reproduced from Ref. [74]. Copyright 2015 by Elsevier VB

necessitates heating to temperatures greater than 1350 °C for a period of 3 days to 3 weeks, resulting in a compound with various strain rate sensitivity indices and with various *m* numbers (Fig. 10.11). When synthesized in an electric furnace and heated to more than 1550 °C, rapid cooling is necessary; the product $Fe_2O_3(ZnO)_m$ has $m > 8$. By contrast, when microwave heating (2.45 GHz) is used, the product $Fe_2O_3(ZnO)_m$ has $m = 3$ [73]. The synthesis of $Fe_2O_3(ZnO)_m$ can also be achieved with a 28-GHz multimode applicator; the respective synthetic temperature is less than 1200 °C; a synthetic process different from the conventional method is suggested.

10.1.8 Control of Magnetic Properties of Spinel Oxide by Microwave Magnetic Field Irradiation

Valence control in transition-metal-containing materials is used for obtaining desired functional (e.g., conducting, magnetic, and superconducting) materials, which can be done through doping and through oxidation/reduction reactions. Nickel manganite ($NiMn_2O_4$) is an essential application material for negative-temperature-coefficient (NTC) thermistors [75], for use as a catalyst [76], and in future spintronic devices [77]. Nickel manganite crystallizes in the spinel structure in which manganese ions of several valences (Mn^{2+}, Mn^{3+}, and Mn^{4+}) occupy tetrahedral and octahedral sites. The magnetic properties of nickel manganite can be controlled by reduction, which brings about changes of the valences of manganese ions. The relationship between field strength ***H*** and magnetization ***M*** (***M-H*** loops) of microwave-irradiated $NiMn_2O_4$ (irradiated at different temperatures) are shown in Fig. 10.12 [78].

A higher processing temperature results in a decrease in the magnetization and an increase in the coercivity. Furthermore, emission of oxygen starts at a relatively low temperature (400 °C). Hence, the nonstoichiometry and magnetic properties of $NiMn_2O_4$, can be controlled by microwave magnetic field irradiation at different temperatures. $NiMn_2O_4$ irradiated with microwaves at the point of the maximum magnetic field has a lower magnetization and a higher coercivity than the sample heated conventionally in an electric furnace (Fig. 10.12). Only the sample irradiated with the microwave magnetic field shows the Mn^{2+}-derived peak in the ESR spectrum, suggesting that Mn^{3+} has been partially reduced. Different magnetic properties of $NiMn_2O_4$ can be obtained by microwave magnetic field irradiation, by

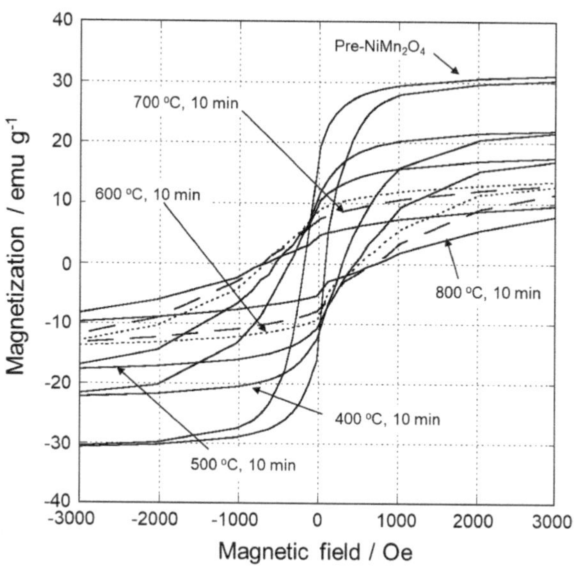

Fig. 10.12 Relationship between field strength ***H*** and magnetization ***M*** (***M-H*** loops) at 5 K of $NiMn_2O_4$. Irradiated with a microwave magnetic field at different temperatures. Magnification (−3000 to +3000 Oe). Reproduced from Ref. [78]. Copyright 2016 by Multidisciplinary Digital Publishing Institute (MDPI)

simply manipulating the temperature of the sample. The abovementioned effects are likely due to the microwave magnetic field's promotion effect on the reduction reaction. Clearly, with microwave magnetic field irradiation, one can control the stoichiometry of $NiMn_2O_4$, and thus control its magnetic properties.

10.1.9 Minerals Processing

As a relatively new energy source, microwave radiation has considerable potential in minerals processing and extractive metallurgy—that is hydrometallurgy and pyrometallurgy. Research activity in these areas has been growing significantly, as evidenced by the exponential growth in publications for both minerals and metals (see Fig. 10.13). The major advantages of utilizing microwaves as the energy source in extractive metallurgy systems can be summarized as follows [79]:

- The microwave energy is transported from the microwave source to the interior of the material wherein it is converted into thermal energy; there is very little heat deposited in or lost to the surroundings.
- In many cases, the temperature in the interior of the material is much higher than at the surface. Thus, significant improvements in heat transfer can be achieved for materials, such as oxides whose poor thermal conductivity limits conventional heat transfer from the outside to the inside. In addition, this improved temperature gradient minimizes the temperature of the refractory container (or crucible).
- The energy densities in microwave systems can be relatively high and can lead to very high internal heating rates, in combination with the low thermal conductivities of oxides and minimal heat absorption by the surroundings.
- The electrical energy source (microwaves) is relatively clean and easily controlled, thus facilitating continuous processing. Since there is no combustion of

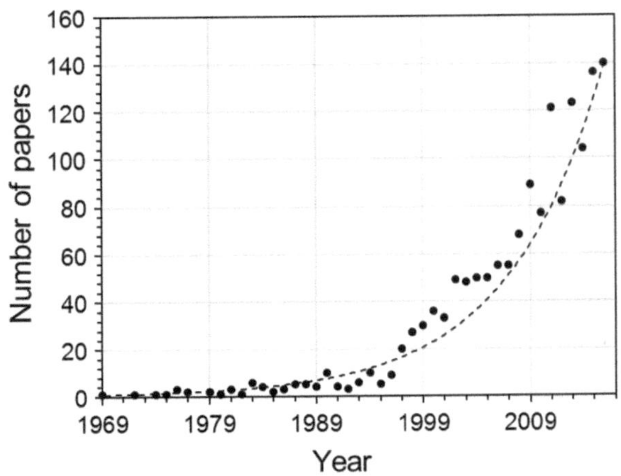

Fig. 10.13 Number of papers published on *microwave and metal and extraction* from 1940 to 2016 obtained through SciFinder Scholar™ 2016, American Chemical Society. Copyright 2017 by S. Horikoshi

carbonaceous or hydrocarbon fuels, the only gases generated are those produced as a result of the intended reactions, which minimizes the amount of off-gas and also the amount of entrained dust particles.

- The working conditions in microwave processes are far superior to those in conventional processes.
- The thermal energy is generated at the atomic or molecular levels, such that both endothermic and exothermic reactions can be promoted.
- In special cases, some degree of selective heating can be attained, although this is restricted by heat conduction, which limits temperature gradients.

Since traditional manganese oxide-containing ores are being depleted, manganese carbonate ores are becoming an increasingly important potential source of manganese oxide. The calcination and agglomeration of these ores by microwave irradiation has been investigated by Amankwah and Pickles [80], who found that the calcine was an excellent microwave absorber and that the addition of a small percentage of calcine to the feedstock resulted in improved microwave coupling. The decomposition rate of the ore was ascertained in both a conventional resistance furnace at a constant temperature of 1200 °C and in a microwave system. Since the resistance furnace was preheated to 1200 °C, the decomposition rate of the first 10 min was faster than the rate of microwave heating, following which the microwave calcination rate increased dramatically above that of conventional heating. After 27 min, the microwaved sample had been calcined and sintered with a nearly 34% mass loss. By comparison, complete calcination of the sample heated in the conventional furnace was achieved after 38 min, and no agglomeration occurred; the loss in mass was ca. 31%. Clearly, calcination and sintering by the microwave process were much faster than the conventional process. Figure 10.14a, b show, respectively, the differences in the morphological features of the calcines produced in the conventional resistance furnace and the microwave system. The conventionally calcined sample exhibited some fracturing, whereas fracturing was more pronounced under microwave heating. With the latter heating method, carbon dioxide evolved faster and resulted in higher internal pressures, which caused enhanced fracturing. In addition, as the microwave calcination process continued, thermal stresses originating from selective heating of the hyperactive manganese oxides could also promote cracking. Under optimal microwave conditions and for a sample mass of 100 g, sintering temperatures of over 1500 °C were achieved. The major phases in the calcine were manganosite (MnO) and hausmannite (Mn_3O_4). Sinter strengths seen were comparable to those of conventionally processed materials.

Microwave-assisted production of pig iron is an actively investigated area of research in microwave chemistry, more so now than in the 1990s. For iron manufacturing of blast furnace steel, hot CO (carbon monoxide gas) is typically blown into an iron ore (ferric oxide) causing a reduction reaction to occur, a process that has been performed for hundreds of years. However, CO is a deadly toxic gas; nonetheless, it is used in significant quantities generating CO_2 gas (global warming gas) through oxidation of CO. The iron and steel industries are perhaps the biggest

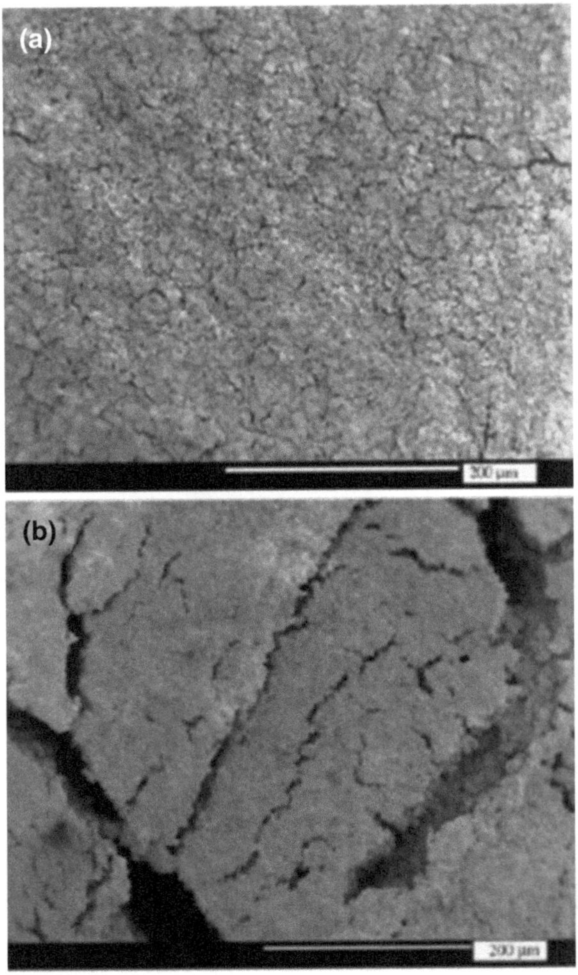

Fig. 10.14 Scanning electron micrographs of a manganese carbonate ore: (**a**) calcined in a conventional furnace at 1200 °C and (**b**) microwave calcined. Reproduced from Ref. [80]. Copyright 2005 by Taylor & Francis

producers of CO_2. The use of the microwave refining method has led to significant reduction in CO_2 emissions [81]. Figure 10.15 shows a continuous microwave iron manufacturing hearth apparatus (maximum output power at 20 kW) in which a mixed powder of iron ore and carbon are used as precursors. The role of carbon powder is as both the microwave absorber and as the reducing agent. The mixed sample is placed in the upper part of the microwave applicator system, where it is microwave irradiated. A photograph of the refined pig iron obtained by microwave heating is shown in Fig. 10.16. It is possible, using iron ore, to produce refined pig iron of the same mole content as the raw material. Microwave refining causes fewer impurities (such as, for example, phosphorus) compared to the usual shaft-furnace procedure, because of the higher heating efficiency and faster reduction reaction kinetics in the microwave hearth than in the conventional furnace.

Fig. 10.15 Picture of continuation type microwave iron manufacturing hearth (maximum output power, 20 kW). Photograph provided by courtesy of Professor Emeritus K. Nagata, Tokyo Institute of Technology, Japan

Fig. 10.16 Picture of fined iron by using continuation type microwave iron manufacturing hearth (maximum output power, 20 kW). Photograph provided by courtesy of Professor Emeritus K. Nagata, Tokyo Institute of Technology, Japan

Yoshikawa et al. [82] have investigated the microwave carbo-thermal reduction reaction of nickel oxide (NiO). When NiO particles (~7 μm) and graphite particles (5 μm; source of carbon) are used as a mixed powder, 100% metallic nickel is produced after only a few minutes by microwave heating. The reduction reaction is highly effective under microwave magnetic field irradiation. Also, Fukushima and Takizawa [83] reported using microwave magnetic field heating to reduce copper oxide in the presence of boron nitride (BN) as the reducing agent. A CuO pellet was placed in a quartz tube reactor positioned in a single-mode microwave applicator; the pellet was subsequently covered at the top and bottom with BN powder, and then irradiated with the microwaves' magnetic field. After heating for 2 min at 1000 °C, the surface of the sample was still the oxide; however, the oxide at the center of the pellet was reduced to metallic copper. Few reduction of CuO in the presence of BN occurred when the conventional electric furnace heating method was used to carry out the process under otherwise identical temperature conditions.

Transition metal nitrides are materials in which nitrogen atoms are located at interstitial sites within the metal lattice. Metal nitrides are typically very hard materials, and so are often used as coatings on metal alloys, in steel, in carbides, and aluminum components to improve the surface properties of these substrates. For example, titanium nitride (TiN)—a typical nitride material—thin films can be produced via the physical vapor deposition method {PVD, usually sputter deposition, cathodic arc deposition or electron beam heating, and by chemical vapor deposition (CVD}. In both PVD and CVD methods, pure titanium sublimes and then reacts with nitrogen in a high-energy vacuum environment. TiN films can also be produced on Ti workpieces by reactive growth (e.g., annealing) in a nitrogen atmosphere. Note that the production equipment of TiN may be expensive and the process complicated.

It is often difficult to coat a sample that is not flat and backside of the substrates. However, by microwave processing, TiN can be coated to not flat and back side of the substrates with simple procedures. Titanium is a metal that tends to be oxidized in air, but when it is microwave-heated in air (800 °C; 10 min) the surface of the titanium metal substrate changes to TiN (note that no formation of TiO_2 occurs under these conditions) [84], and no expensive equipment and vacuum conditions are necessary. The surface of alloys, ceramics, and plastics with various shapes can be coated with a TiN layer by microwave heating. This method has been used in various industrial applications such as, for example, in the abutment of dental implants (Fig. 10.17a). Although a titanium alloy is highly compatible with a living body, when used for dental implants the gums tend to appear black because of the inherent silvery luster of titanium metal (Fig. 10.17b). By contrast, the use of TiN for dental implants is more acceptable by patient since the gums looks a pink resulting from the gold color of the TiN as seen in Figure 10.17c.

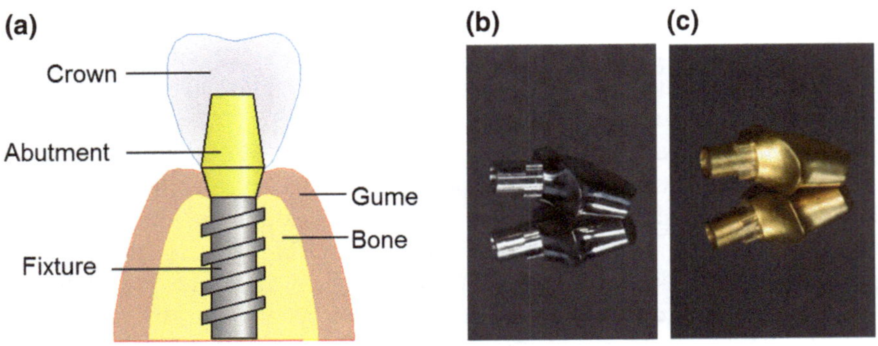

Fig. 10.17 (**a**) Schematic image of a dental implant, (**b**) conventional titanium alloy abutment and (**c**) titanium nitride abutment after microwave processing. Photograph provided by courtesy of Professor H. Takizawa, Tohoku University, Japan

10.2 Microwave Processing in the Liquid State

10.2.1 Syntheses in Liquid Media

The use of microwaves in syntheses involving the liquid state was reported in a seminal paper by Komarneni and Roy [85]. Unfortunately, the historical significance of this paper, titled *Titania gel spheres by a new sol-gel process,* received scant attention as no reference to microwave radiation was made. Nonetheless, the abstract did clearly point out that microwaves led to fast reactions; paraphrasing the abstract:

> A new technique is devised for the preparation of solution-derived titania gel spheres. The gel-sphere preparation process involved the dispersal of a polymerized titanium ethoxide solution into a two-phase liquid suspension in kerosene, followed by heating in a microwave oven.

A microwave-hydrothermal (M-H) process was also reported by Komarneni et al. [86] for reactions performed at temperatures above the boiling point of water (or of other solvents), and at pressures greater than 1 atm. In the field of microwave-assisted reactions in materials synthesis, and processing in water and other liquids, the microwave-hydrothermal/solvothermal process has been widely developed for the synthesis of nanosized oxides and metallic materials. The microwave-hydrothermal process possesses many advantages: (a) rapid heating to the temperature of the reaction, (b) increased reaction kinetics, (c) elimination of metastable phases, and (d) formation of novel phases.

Microwave-assisted reactions are being investigated worldwide, resulting in extraordinary new scientific developments in the processing of metal-oxide-semiconductors with metallic nanophases that find use in, for example, quantum dots, nanowires, nanorods, nanobelts, and several others. Two other processes for the synthesis of metals and other materials are worth noting: (i) the microwave-polyol process and (ii) the microwave-solvothermal process. The microwave-solvothermal synthesis (MSS) process was used by Lojkowskiet al. [87] to prepare nanoparticles of some binary and complex oxides: for instance, ZrO_2, ZnO, and yttrium-aluminum-garnet (YAG; formula $Y_3Al_5O_{12}$). The unique effect of microwave heating allowed the preparation of a very uniform nanosized powder, once the precise planning of a time-temperature regime was scheduled to ensure a uniform high supersaturation of the reagents in the reactor vessel. The preparation of functional oxides with complex crystalline structures has been reported by Prado-Gonjal et al. [88], who provided an explanation of how, by exploiting single-mode microwaves in microwave-assisted hydrothermal syntheses, a broad variety of different microwave-assisted synthesis techniques opens new opportunities in the preparation of inorganic nanoparticles and nanostructures, especially in relation to solid-state reactivity.

Microwave irradiation can lead to the morphology-controlled growth of nano- and microstructures of ZnO. Various basic ZnO structures have been obtained by Cho et al. [89] including nanorods, nanocandles, nanoneedles, nanodisks, nanonuts, microstars, microUFOs, and microballs (Fig. 10.18), all of which were prepared by a fast, simple, and reproducible method that required no template, no catalyst, and no surfactant, at relatively low temperatures (90 °C), using low power microwave-assisted heating ($\sim$50 W) and a subsequent aging process. Such results were obtained by changing the precursor chemicals, the capping agents, and the aging times. Even more complex ZnO structures, including ZnO bulky stars, cakes, and jellyfishes were constructed by microwave irradiation of a mixture of the as-prepared basic ZnO structures (Fig. 10.18). Mechanisms were proposed for the growth of shape-selective ZnO synthesis based on these results (Fig. 10.19).

The industrial importance of the use of microwaves in the preparation of catalysts is best exemplified by a 2013 US patent co-authored by El-Shall et al. [90]. These authors produced graphene using microwave irradiation of a solution phase of graphite oxide (GO) in the presence of a reducing agent; alternatively, it was also prepared by subjecting the GO to the effects of a microwave plasma that needed no reducing agent. The latter method was environmentally advantageous and cost effective. It could also encompass the simultaneous reduction of solution and solid-phase graphite oxide, together with a variety of metal (e.g., Pd, Co, Au, Ag, Cu, Pt, Ni, Fe, Mn, Cr, V, Ti, and Sc, among others) precursors, which resulted in the dispersion of metallic nanoparticles supported on the large surface area of

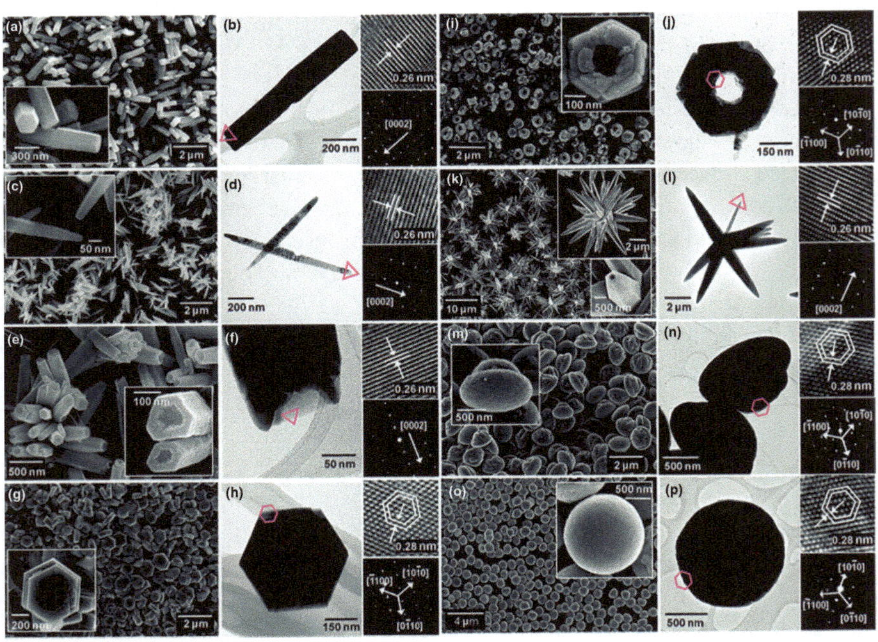

Fig. 10.18 SEM (left) and TEM (right) images of the basic ZnO structures synthesized by microwave irradiation: (**a**) and (**b**) nanorods, (**c**) and (**d**) nanoneedles, (**e**) and (**f**) nanocandles, (**g**) and (**h**) nanodisks, (**i**) and (**j**) nanonuts, (**k**) and (**l**) microstars, (**m**) and (**n**) microUFOs, (**o**) and (**p**) microballs. Reproduced from Cho et al. [89]. Copyright 2008 by the American Chemical Society

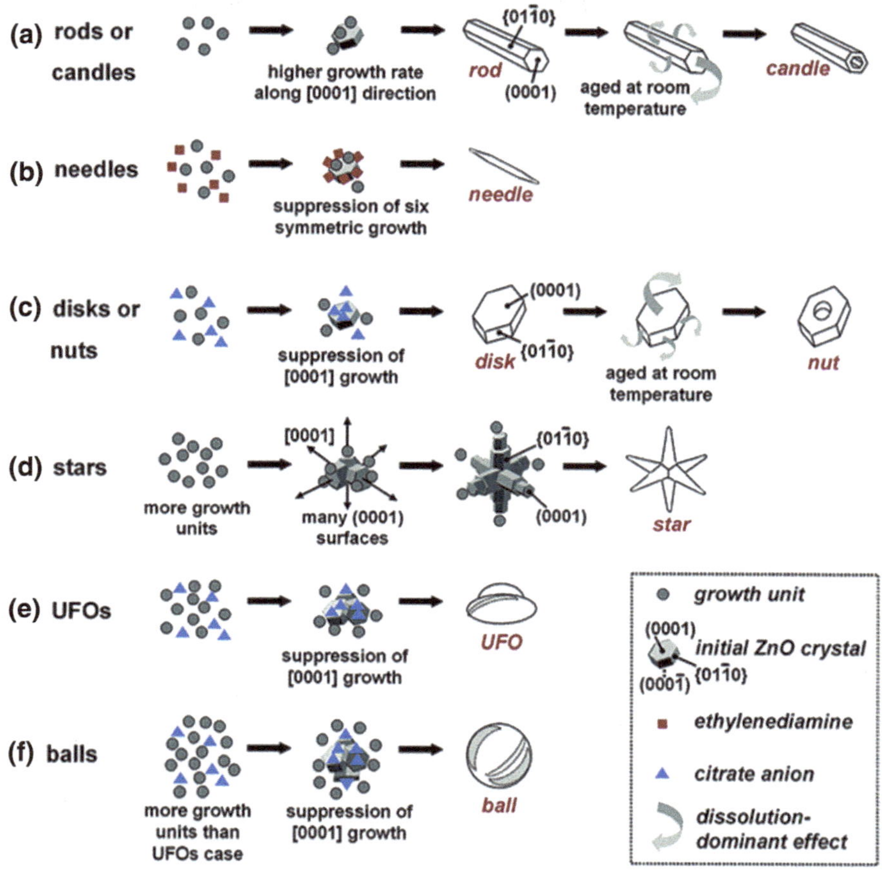

Fig. 10.19 Schematic diagram of the proposed formation processes of some basic ZnO structures: (**a**) nanorods or nanocandles, (**b**) nanoneedles, (**c**) nanodisks or nanonuts, (**d**) microstars, (**e**) microUFOs, (**f**) microballs. Reproduced from Cho et al. [89]. Copyright 2008 by the American Chemical Society

thermally stable two-dimensional graphene sheets. The graphene-supported metal nanoparticles have been used profitably as catalysts in various Suzuki-type reactions involving aryl bromides, and were free of contaminants from residual solvent and reducing agent.

The microwave method has also been used [91] to synthesize solids with the saponite structure, with Mg^{2+} or Ni^{2+} as octahedral cations, and Fe^{3+} in the tetrahedral sheet; the formulae of the resulting solids determined from elemental analysis were for SNiFe: $\{[Si_{7.136}Fe_{0.864}][Ni_{5.813}][Na_{0.810}]O_{20}(OH)_4{\cdot}13H_2O\}$ and for SMgFe $\{[Si_{7.200}Fe_{0.800}][Mg_{5.820}][Na_{0.7020}]O_{20}(OH)_4{\cdot}9H_2O\}$. A special type of silico-alumino-phosphate with an opal structure (Opal-SAPO) was prepared under microwave irradiation using tetraethylorthosilicate, triethanolamine, aluminum isopropoxide, and phosphoric acid (85%) [92]. The catalytic behavior of the

Opal-SAPO was examined in the regeneration of methylethylketone, benzaldehyde and cyclohexanone from the corresponding semicarbazones derivatives; results indicated that the OH groups on the Opal-SAPO surface provided the Bronsted acidity to the substrates. The first step in the regeneration of aldehydes or ketones from semicarbazone derivatives involved the chemisorption of semicarbazones on the Bronsted acid sites. Cleavage of the C–N bond was catalyzed by H^+ ions followed by hydrolysis with small amounts of water; no reaction occurred without the presence of water. The reactivity of Opal-SAPO toward the regeneration of carbonyls was methylethylketone > benzaldehyde > cyclohexanone [92].

Of particular relevance are the syntheses of solid-state nanomaterials taking place in liquid media using the microwave heating approach. Accordingly, the synthesis of nanoparticles in liquids is the focus of the next section, in which some of the features of the microwave method are explained in detail.

10.2.2 Nanoparticle Synthesis in Liquid Media

Recent years have witnessed an exponential growth in the application of microwaves in organic syntheses. Despite a growing number of publications in nanoscience and nanotechnology that have made effective use of dielectric heating from microwave irradiation [93], syntheses of inorganic nanoparticles (e.g., metallic and metal oxides, in particular, and metal chalcogenides, in general) and general microwave chemistry have yet to reach their full potential. Microwave heating significantly reduces not only the reaction times, as described previously, but also minimizes, if not suppresses side reactions, thereby leading to enhanced chemical yields and reproducible processes. The major features of microwaves in nanoparticle synthesis also include uniform heating, and short reaction times, two major points that are emphasized throughout this section.

A simple and straightforward microwave-assisted synthesis of bulk and shape-controlled prisms, cubes, hexagonal and spherical nanostructures of noble metals (e.g., Au, Ag, Pt, Pd) by reduction of their corresponding salts in aqueous glucose, sucrose, and maltose media, was reported by Mallikarjuna and Varma [94]. This is part of the growing need to develop ecofriendly processes that avoid the use of toxic chemicals in preparative protocols. Uniform nano and bulk particle sizes are typically obtained by uniform nuclei growth through specific microwave heating procedures. In this regard, there is a need for high-temperature-initiated nucleation, followed by controlled precursor addition to the reaction media. Synthetic methods that utilize conventional convective heating have relied on a conduction path to drive the synthetic process, for which the reactor acts as the intermediary in transferring thermal energy from the external heat source to the solvent and ultimately to the reactants [95]. Such a pathway typically leads to sharp thermal gradients throughout the bulk media, and to inefficient and non-uniform reactions. This can cause serious issues when attempting process scale-up and, more importantly, in nanomaterial syntheses where uniform nucleation and growth rates are essential in maintaining product quality, not to mention the formation of a metallic coating on the inner walls of the reactor. Unlike convective heating, microwave internal heating provides uniform temperatures, can prevent the formation of a metallic coating on the reactor walls in nanoparticle synthesis, and plays an important role in the wash-free continuous-flow production of metallic nanoparticles [96]. There is a clear need to clarify some of the features of microwaves in nanoparticle syntheses and develop novel microwave methodologies for possible process scale-up. Interested readers may wish to investigate a recent book [97] that deals with nanoparticle syntheses in much greater detail; it also describes various synthesis protocols.

Tsukahara and coworkers have noted that the size distribution of nucleating particles becomes uniform under microwave heating [98] as evidenced by employing *in-situ* Surface-Enhanced Raman Scattering (SERS) spectroscopy to observe the early stage(s) of the formation of Ag nanoparticles in solution (reaction 1).

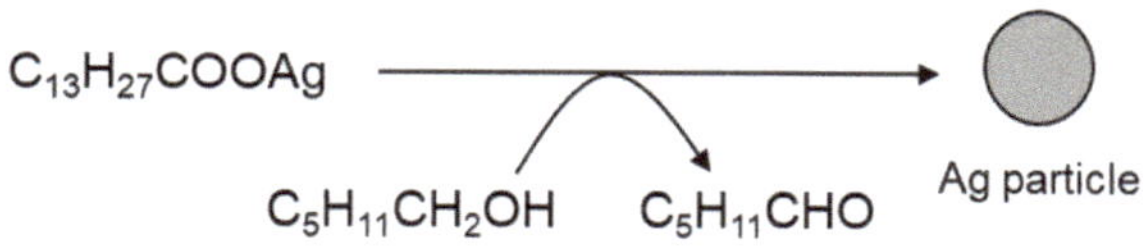

(reaction 1)

The subsequent appearance of the Raman spectrum of Rhodamine 6G, which was amplified by the surface-enhanced effect, is seen by the rise of the band intensities, which confirmed the homogeneous formation of silver nanoparticles. Experimental observations of the SERS scattering spectra of Rhodamine 6G (Fig. 10.20) showed that the spectra appeared simultaneously with the formation of silver particles under microwave irradiation, irrespective of whether the Raman

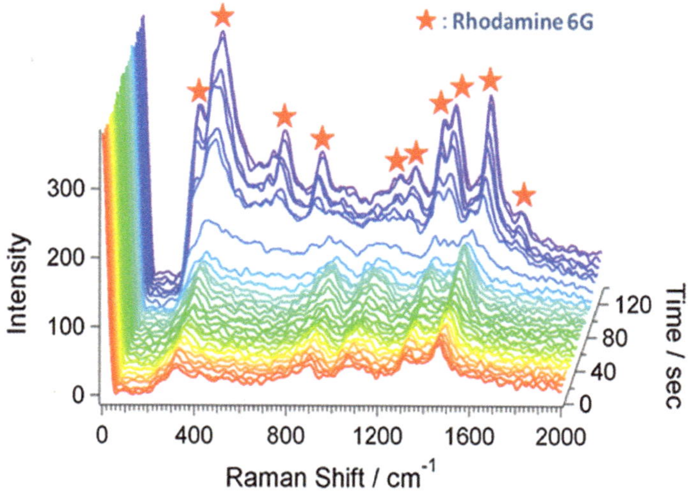

Fig. 10.20 In situ SERS spectra demonstrating the formation of Ag nanoparticles in solution. Reproduced with permission from Ref. [98]. Copyright 2006 by the Japan Chemical Society

probe was located at the center of the reaction vessel or beside the wall of the vessel.

The increase in the intensity of the Raman signal for the Rhodamine 6G adsorbed on silver particles occurred because of an enhancement in the electric field provided by the Ag particle surface when struck by the incident light, which causes localized surface *plasmons* to be excited. The enhancement is greatest when the plasmon frequency, ω_p, is in resonance with the radiation. For this scattering effect to occur, however, requires that the plasmon oscillations be perpendicular to the surface (note that no scattering occurs when oscillations are in-plane with the surface) [99]. When the reaction vessel was heated in a hot-oil bath, the Raman spectrum appeared sooner than was the case when the Raman probe was positioned at the center of the reactor. Tsukahara et al. [98] noted that the increase of the intensity of the Raman spectrum coincided completely with the increase of temperature when both microwave heating and oil-bath heating were used. These authors also demonstrated that the reaction solution was microwave-heated in the inner parts of the solution, resulting in homogeneous reaction conditions.

In general, a microwave-assisted synthesis of nanoparticles is characterized by rapid and homogeneous heating, in contrast to a conventional heat-assisted synthesis, even though the thermal effects are similar to those of other heating methods [100]. The observed rapid heating by microwaves can be controlled and monitored closely, so as to utilize the characteristic thermal features of the microwaves. Temperature conditions of both heating methods should be matched closely in nanoparticle synthesis. This can be achieved by soaking the cylindrical reactor in a preheated oil bath to reach the appropriate reaction temperature. The rise in temperature of the solution exposed to microwave radiation can be matched to the temperature rise in the oil-bath heating using suitable microwave power levels at 1 W increment, controlled by a proportional–integral–derivative (PID) device available in the microwave apparatus.

At this point, it is worth noting some of the thermal features of 2.45-GHz microwave heating used in the synthesis of silver nanoparticles, in comparison with conventional heating (oil bath), with special emphasis on the temperature effects and the characteristics (if any) of different microwave synthesizers [96]. The procedure used for the synthesis of silver nanoparticles involved introducing an aqueous solution of the diamminesilver(I) complex and carboxymethylcellulose (CMC) into a 150-mL Pyrex® glass batch-type cylindrical reactor, which was subsequently microwave irradiated (power, 64 W) continuously using a microwave single-mode resonance apparatus. To do this, the reactor was positioned such that it was exposed to the microwaves' electric field at the field's maximum density by suitable adjustments with the short plunger and the three-stub tuner. The conventional heating source was a silicone oil bath. After heating the sample solution, either with microwaves or conventionally, the reactor contents were cooled rapidly in an ice/water bath to arrest the synthesis of the Ag nanoparticles, otherwise, the reaction would have continued because of the residual heat from microwaves or the oil-bath heating.

Following the temperature match between microwave heating and oil-bath heating, the formation of silver nanoparticles was monitored at heating times of 0.5, 1, 2, 3, 4, and 5 min by the surface plasmon resonance band around 420 nm in the UV-visible absorption spectra. Silver nanoparticles formed after a 3-min heating period by both microwave and oil-bath methods. In both cases, the synthesis of Ag nanoparticles necessitated a temperature of 100 °C, reached only after this time period. Figure 10.21a, b show the corresponding transmission electron microscopic (TEM) images of the resulting silver nanoparticles after a 5-min heating time; they reveal a uniform particle size distribution of the colloids obtained from microwave heating under otherwise identical temperature conditions as with conventional heating. Results from light scattering tests showed that Ag nanoparticles prepared by microwave heating were in the range 1–2.3 nm (Fig. 10.21a inset), whereas the oil-bath heating method yielded polydispersed nanoparticles, mostly in the 3–5.7 nm range (Fig. 10.21b inset), with some up to 30 nm. Thus, even though the reaction temperature conditions for the microwave and oil-bath methods were identical, there was an otherwise noticeable difference in the rate of formation of silver nanoparticles and in their size distributions.

Differences in the effect of the two heating methods were evident in the results illustrated in Fig. 10.22. Five minutes into the microwave irradiation of the aqueous CMC/diaminesilver(I) solution led to the formation of a slightly yellow-colored sol in the reactor (Fig. 10.22a), whose concentration did not change even after 60 min of irradiation. Silver nanoparticles adsorbed on the reactor walls were easily removed by simply washing with water. By contrast, after 4 min by oil-bath heating, a silver film (mirror) coating formed in the inner reactor walls

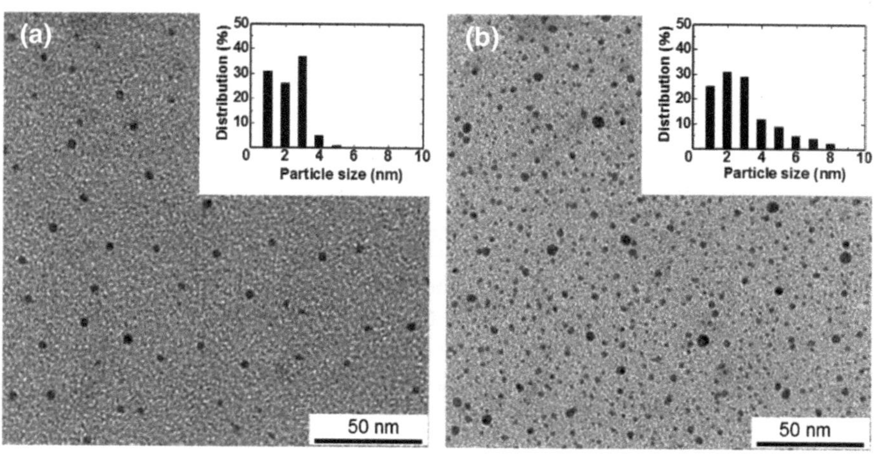

Fig. 10.21 TEM images of the silver nanoparticles produced by microwave and oil-bath heating methods (insets show the particle size distribution from light scattering experiments): (**a**) the microwave result, (**b**) the oil-bath result. Reproduced with permission from Ref. [96]. Copyright 2010 by the Royal Society of Chemistry

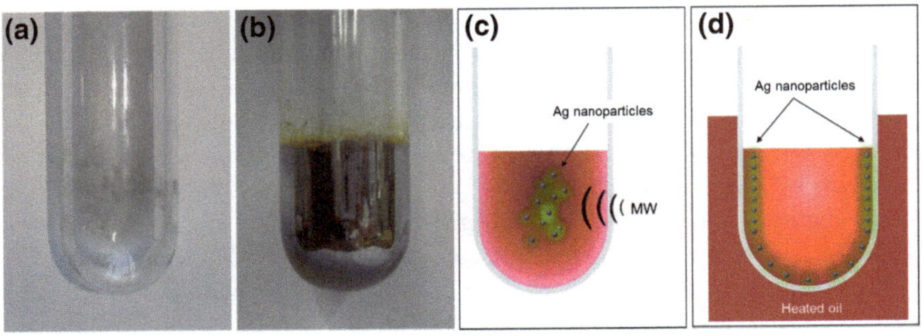

Fig. 10.22 Photographs of the reactor after the sample discharge: (**a**) after 4 min of microwave irradiation, and (**b**) after 5 min of oil-bath heating. Illustrations representing the temperature distribution in the reactor: (**c**) after 4 min of microwave heating and (**d**) after 4 min of oil-bath heating. Note that the photographs in (**a**) and (**b**) were taken immediately after the 4-min heating time. Experiments were carried out under non-stirring conditions. Reproduced with permission from Ref. [96]. Copyright 2010 by the Royal Society of Chemistry

(Fig. 10.22b) under temperature conditions identical to the microwave method; the coating resisted the water washings used for the microwave procedure.

The initial temperature distributions and formation of silver nanoparticles produced under non-stirring conditions by both microwave and oil-bath heating are illustrated, respectively, in Fig. 10.22c, d. In the former case, microwave radiation penetrated the CMC/diaminesilver(I) solution causing the temperature to rise by dielectric loss, and to some extent by conduction loss heating, followed by subsequent loss of heat to the surroundings through the reactor walls. As such, the temperature near the inner walls of the reactor tended to be lower than at the center of the reactor. That is, the synthesis of Ag nanoparticles by the microwave-assisted process progresses outward from the center of the reactor to the inner reactor wall, owing to a temperature gradient giving rise to a concentration gradient (thermophoretic migration, i.e. the Ludwig–Soret effect [101]). On the other hand, heat from the oil bath was most prominent at the reactor wall and was subsequently transmitted to the solution by thermal conduction and convection mechanisms. This ultimately led to the formation of a silver film at the reactor's inner wall as the concentration of the nanoparticles was greatest at this location. Related to the present discussion, Schanche [102] described the temperature distributions produced from microwave and oil-bath heating methods in various organic syntheses. In a later study, the Kappe group [100] noted that the temperature distribution in the microwave method might not always be uniform; that is, the location where heat is generated may be different from the position where the endothermic process occurs. This has called attention to the notion that even completely homogeneous solutions must be stirred/agitated when using single-mode microwave reactors so as to avoid temperature gradients from developing, a consequence of the inherent field inhomogeneities that exist inside the single-mode microwave cavity, as well as the effects of the penetration depth into the reactants.

10.2.3 Specific Microwave Synthesis

Kundu et al. [103] used a highly effective, rapid microwave method to synthesize shape-controlled gold nanoparticles in the presence of 2,7-dihydroxynaphthalene (2,7-DHN) as a new reducing agent and under microwave dielectric heating for 60–90 s. The growth of the particles with different shapes (spherical, polygonal, rod-like, and triangular/prismatic shapes) was governed by the surfactant-to-metal ion molar ratios and the concentration of 2,7-DHN. For example, experiments carried out at different microwave irradiation times led to the formation and growth of triangularly shaped particles. Figure 10.23 displays TEM images of Au nanoprisms at various stages of microwave heating; Fig. 10.23a, b show, respectively, the TEM images of the formation of Au nanoprisms after 30 and 45 s of microwave heating. There were about 40 particles (10 nm in diameter) having no defined shape, but they started growing by adopting a triangular shape. Figure 10.23c shows the low-magnification image after 60 s of microwave heating with particles perfectly grown into triangular shapes, whereas Fig. 10.23d shows different magnified images of nanoprisms after 90 s of microwave heating. The latter image contains more than 95% of prism-shaped particles with a few other-shaped particles (ca. 65) with an average particle size around 10 nm. The contrasting changes compared to the particles of Fig. 10.23c are likely due to the internal stress of the growing crystal layers. The thickness of the prisms was around 3.5–4 nm (height profile) determined from atomic force microscopy (AFM). Because the sample could be heated uniformly by microwave irradiation, seed particles may have formed first followed by crystal growth.

Shape control may be facilitated in the presence of dispersing agents. When a microwave-assisted synthesis of nanomaterials is performed in the presence of a dispersing agent with a high dielectric loss factor, the dispersing agent becomes the local heat source in the solution. For example, Tsuji et al. [104] noted that when polyvinylpyrrolidone (PVP) was the dispersing agent, it was heated selectively in the solution by the microwave radiation, followed rapidly by reduction of the metallic ions and production of seed particles. Also, Yamamoto et al. [105] reported that local microwave heating occurred at the junction between Ag^+ ions and the PVP dispersing agent. Formation of prism-like particles did not, however, occur at the stage of seed particle development, as portrayed in Fig. 10.24; it occurred later with the generation of globular particles and subsequent to aggregation produced the prism-shaped nanoparticles.

Tsuji et al. [106] also used the microwave polyol method to synthesize Au@Ag core-shell nanoparticles in gold nanoparticle dispersions in the presence of a silver precursor. The shape of the core-shell particles was controlled by the shape of the Au core and by the ratio of the gold ($HAuCl_4{\cdot}4H_2O$) to the silver ($AgNO_3$) precursor. Figure 10.25 displays TEM photographs of the Au core (Fig. 10.25a) and the Au@Ag core-shell nanocrystals (Fig. 10.25b–h), overgrown over truncated-triangular-bipyramidal-like (TTBP-like) Au cores in a $[AgNO_3]/[HAuCl_4]$ molar ratio ranging from 5 to 30. At the low $[AgNO_3]/[HAuCl_4]$ molar

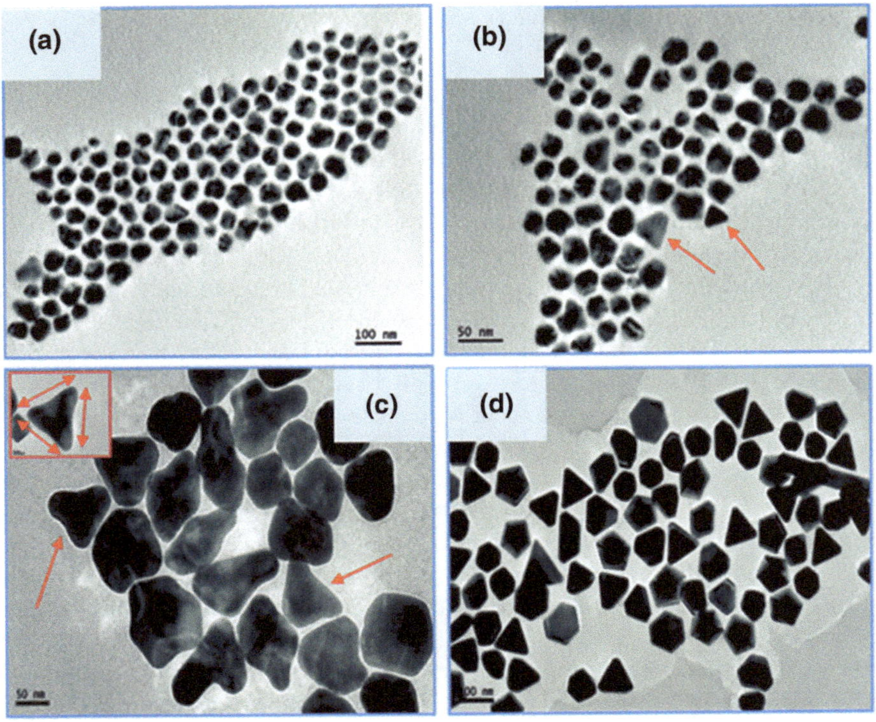

Fig. 10.23 TEM images for the synthesis of gold nanoprisms under different microwave irradiation times: (**a**) 30 s, (**b**) 45 s, (**c**) 60 s, and (**d**) 90 s. Reproduced with permission from Ref. [103] Copyright 2008 by the American Chemical Society

ratios of 5:9, TTBP crystals developed until the Ag edge lengths observed from the top view became twice those of the Au core plates (Fig. 10.25b–e). The same triangular-bipyramidal shells formed at the higher [$AgNO_3$]/[$HAuCl_4$] molar ratios of 11:30 (Fig. 10.25f–h).

The edge length of the Ag shells increased upon increasing the [$AgNO_3$]/[$HAuCl_4$] molar ratio; at the highest [$AgNO_3$]/[$HAuCl_4$] molar ratio of 30:1 it was about 7 times larger than that of the TTBP-like Au core (~50 nm) [106]. When the total sizes of Au@Ag crystals were large, it was difficult to see the exact position of the Au cores. However, better contrast TEM images unambiguously demonstrated that all the cores located in the middle, and consisted of inverted outer larger triangular-bipyramidal Ag shells. The position of the Au core was also confirmed by etching Ag shells using $AuCl_4^-$ ions. Owing to the displacement reaction between Ag and Au (reaction 2), the Ag shells could be partially broken and an Au core could then be observed at the center of the Au/Ag alloy shells.

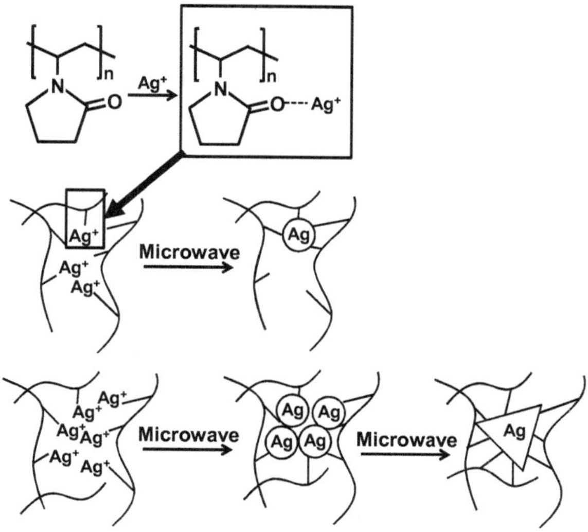

Fig. 10.24 Mechanistic view in the formation of Ag nanoparticles in a PVP network under microwave irradiation. Reproduced with permission from Ref. [105]. Copyright 2004 by the Japan Chemical Society

$$3Ag_{(s)} + AuCl_4{}^{-}{}_{(aq)} \rightarrow Au_{(s)} + 3Ag^{+}{}_{(aq)} + 4Cl^{-}{}_{aq} \qquad \text{(reaction 2)}$$

The synthesis of metal chalcogenides with microwave radiation has been accomplished either via the hydrothermal method or the polyol method (see e.g., Ref. [97] for more details). As an example, a novel microwave synthesis of water-dispersed and highly luminescent CdTe/CdS core-shell nanocrystals were reported nearly a decade ago by He et al. [107]. Their method allowed for the rapid synthesis of high-quality CdTe/CdS core-shell nanocrystals of various colors in aqueous phase under moderate conditions (see Fig. 10.26). The *as-prepared* CdTe/CdS core-shell nanocrystals required no post-preparative treatment and revealed a high photoluminescence quantum yield (up to 75%) and a narrow size distribution ($\sim$35 nm). More importantly, their investigations demonstrated the suitability and the advantages of microwave radiation for accelerating the epitaxial growth of the CdS shell; only 5 min was needed to form an optimal thickness of the CdS shell in the microwave synthesis, compared with several days expended using an illumination method with ambient light. In this regard, several successful examples of the formation of core-shell structures in aqueous systems have been reported; for instance, citrate-stabilized CdSe/CdS nanocrystals by Wang et al. [108] and TGA-stabilized (TGA = thioglycolic acid) CdTe/CdS nanocrystals by Bao et al. [109]. Both groups prepared the systems by exposing citrate-stabilized CdSe nanocrystals in the presence of cadmium and sulfide ions [108] or TGA-stabilized CdTe nanocrystals [109] to ambient light in order to achieve the formation of the CdS shell through the photodegradation of the respective stabilizers. Both of the as-prepared core-shell nanocrystals also demonstrated high quantum yields of photoluminescence (QYPL) owing to the CdS shell, which passivated (make unreactive) the core CdSe and CdTe nanocrystal surfaces by effectively smoothing

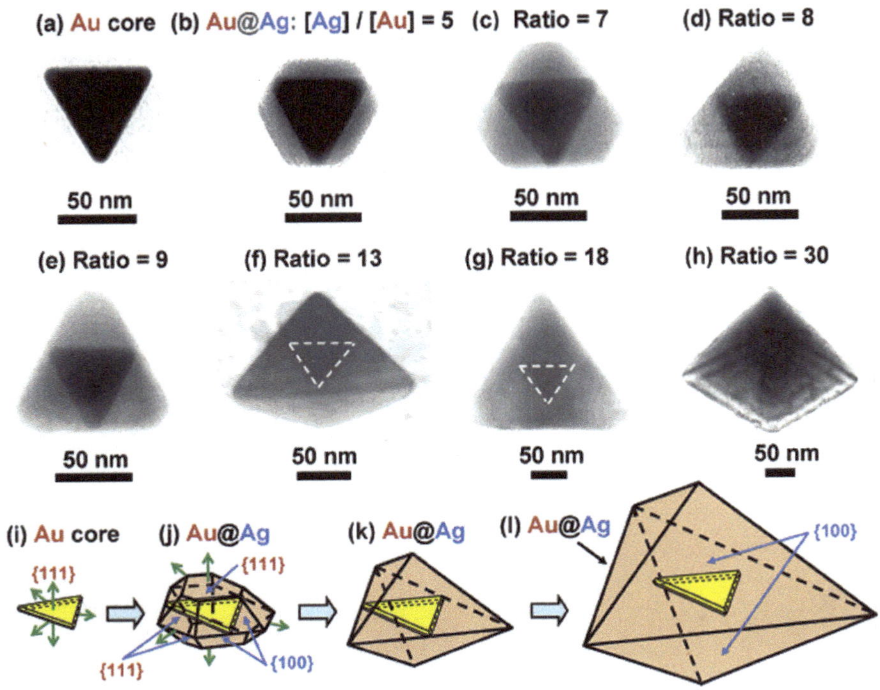

Fig. 10.25 TEM photographs of (**a**) the triangular twin Au core and (**b**)–(**h**) Au@Ag nanocrystals prepared by addition of various amounts of $AgNO_3$ to Au cores with microwave heating for 2 min. Dotted lines denote the triangular twin Au core plates, which can be observed using photographs with better contrast. (**i**)–(**l**) Growth mechanism of triangular-bipyramidal Au@Ag crystals from triangular twin Au cores. Reproduced with permission from Ref. [106]. Copyright 2006 by the American Chemical Society

out the surfaces and reducing the surface traps of these core nanocrystals. Enhancement of the quantum yields of photoluminescence also resulted from the destruction of the stabilizers through their photodegradation.

However, the illumination methods were time-consuming requiring 5 or 20 days, respectively, to obtain citrate-stabilized CdSe/CdS nanocrystals [108] and TGA-stabilized CdTe/CdS nanocrystals [109] with the highest QYPL; in the latter case, it required a gradual and slow photooxidation of TGA to form the CdS shell. In addition, the long reaction times generated a wide range of sizes distributed in these systems.

In comparison to the thioglycolic acid method for synthesizing CdTe/CdS core-shell nanocrystals by the illumination method under ambient light, the microwave synthesis can also be applied to synthesize core-shell nanocrystals capped with other thiols (e.g., with 3-mercaptopropionic acid) [107]. Clearly, the microwave synthesis has proven to be an extremely simple and convenient method, free of complicated vacuum manipulation and expensive chemical reagents, and it

Fig. 10.26 Photograph of the wide spectral range of bright luminescence from a sample of microwave-assisted synthesis of CdTe/CdS core-shell nanocolloids in aqueous media without any post-preparative treatment under irradiation with 365-nm ultraviolet light from a UV lamp [107]. Copyright 2006 by the American Chemical Society

is also environmentally and biologically friendly since the whole synthesis was performed in aqueous media.

10.2.4 Nanoparticle Syntheses in Continuous-Flow Reactors

The synthesis of Ag@SiO_2 core-shell particles has been performed using a two-step process with microwave systems, [110] in which a polytetrafluoroethylene (PTFE) tube (internal diameter 1 mm) was mounted coaxially in the center of the TM_{010} single-mode cavity for use as a flow-type reactor (Fig. 10.27a). Since the microwave system was a unit type microwave heating device, the interesting feature of the equipment was homogeneous heating, and it could be connected perpendicularly.

A large-scale ultrasonic atomization-type microwave nanoparticle continuous-flow synthesis equipment system (Kojundo Chemical Lab. Co. Ltd) is illustrated in Fig. 10.27b. A 3.5-matter reactor is contained at the center of the equipment; the sample solution is ultrasonically sprayed into the apparatus, after which resistance heating plus microwave heating leads to the formation of nanoparticles. This type of system has been used for the synthesis of ferrite, an excellent absorber of electromagnetic waves [97].

A microwave-solvothermal synthesis apparatus with a stop-flow reactor has also been proposed by Lojkowski et al. [87]. It is an automatically operating system equipped with a 250-mL chamber that can be used at a maximum pressure of 4 MPa and at a temperature of 240 °C (Fig. 10.27c); the reaction chamber consists of a Teflon® tube (3 cm, diameter). The microwave energy is supplied by means of two magnetrons with 1 kW each output power. The magnetrons are placed at different levels in the chamber, opposite to each other. The temperature of the PTFE

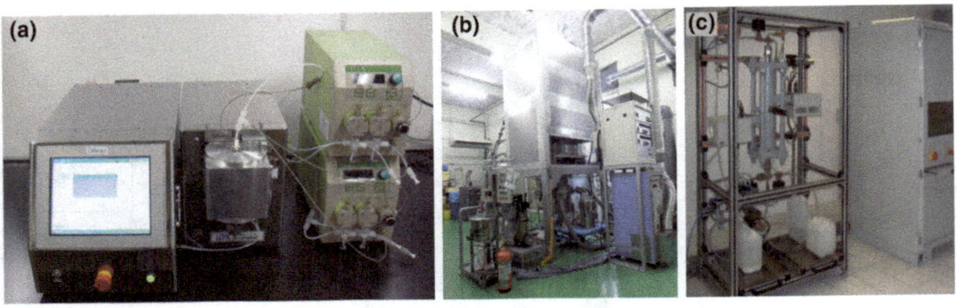

Fig. 10.27 Continuous-flow microreactor nanoparticle synthesis systems. (**a**) Unit type microwave heating device; (**b**) industrial scale device comprising ultrasonic atomization-type microwave nanoparticle continuous-flow synthesis equipment. Reproduced with permission from Ref. [97]. Copyright 2013 by the Wiley-VCH Verlag GmbH. (**c**) Microwave-solvothermal synthesis with stop-flow reactor operating automatically, with a 250-mL chamber, maximum operating pressure was 4 MPa and temperature 240 °C. Reproduced with permission from Ref. [87]. Copyright 2016 by Multidisciplinary Digital Publishing Institute (MDPI)

chamber is measured with the use of a pyrometer (TXP7, Raytek, Santa Cruz, CA, USA) placed between the magnetrons, while the pressure is measured with a pressure sensor with a capillary and a separation membrane (SEN-3291, Kobold, Pittsburgh, PA, USA). A circulator prevents the reflected microwave power from damaging the magnetrons. The use of two magnetrons allows for a uniform field distribution; if necessary, in principle it is possible to use only one magnetron, resulting in a temperature gradient and/or microwave stirring effect.

The features of another microwave nanoparticle synthesis procedure are worth describing. A continuous-flow reactor system was proposed a few years ago by Horikoshi et al. [96] that consisted of a Pyrex tube (length = 135 mm; internal diameter, 8 mm) placed horizontally in a microwave waveguide through which an aqueous CMC/diaminesilver(I) solution was circulated by means of a peristaltic pump. Irradiation of the reactor contents was performed with 1200 W microwaves from a microwave generator (maximum power, 3000 W). In the photograph of Fig. 10.28a, the microwaves emanated from the generator toward the reactor, with the maximum of the microwaves' electric field positioned at the center of the reactor. A metal mesh closed the waveguide to prevent microwave leakage, and to allow safe observation of progress in the reactor. For maximum heating efficiency, the flow rate of the peristaltic pump was set at 600 mL min^{-1}. Under microwave irradiation, the solution temperature increased rapidly to 100 °C to yield a colloidal sol of Ag nanoparticles; these were collected in the receiver flask and rapidly cooled by an ice-water bath to arrest any further reaction.

The transmission electron microscopic image of the nanoparticles displayed in Fig. 10.28a confirmed the generation of a narrow size distribution of silver nanoparticles (range, 1–4 nm) with the reactor setup. The graphic in Fig. 10.28b illustrates the reactor system and the observations reported [96]. Interestingly, the internal reactor walls showed no visible evidence of a yellow stain resulting from this continuous-flow synthesis of silver nanoparticles, even after 7 min of

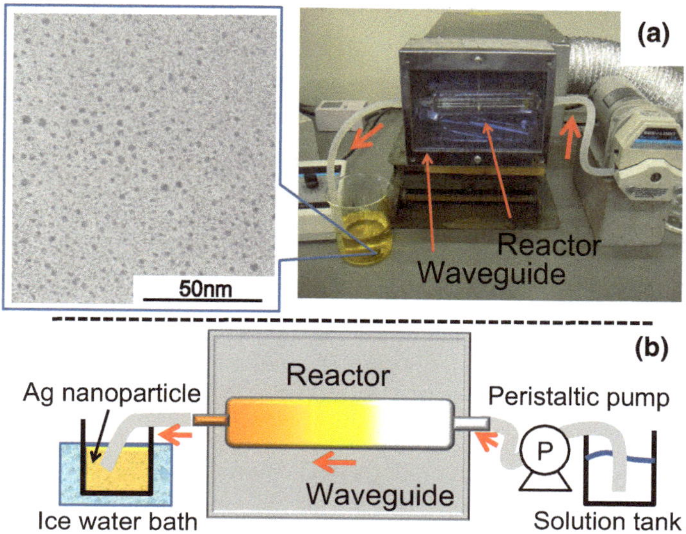

Fig. 10.28 (**a**) Photograph of the continuous-flow reactor system for the microwave-assisted synthesis of silver nanoparticles; the TEM image of the resulting silver colloids is also displayed. (**b**) Schematic image of the overall experimental setup. Reproduced with permission from Ref. [96]. Copyright 2010 by the Royal Society of Chemistry

microwave irradiation. While microwave power levels greater than 1200 W were available, there was some difficulty in controlling the temperature at these higher wattages, which could (in principle) be corrected by cooling the reactor externally with a cooling device and by a faster solution flow rate to enhance synthesis efficiency. However, the addition of any cooling device (e.g. cooling jacket or cold air) would have unnecessarily complicated the reactor system. In the present instance, with the 1200-W microwaves, there was no necessity for cooling; this microwave power level was maintained throughout the synthesis. Understandably, a similar continuous-flow rapid synthesis of silver nanoparticles of similar size distribution (i.e., 1–4 nm), would have been somewhat difficult to achieve in a reactor system that relied on a conventional heating method.

Microfluidic reactors have been used in the synthesis of various organic, inorganic, and biological materials [111]. Such microreactors have also been used successfully in the synthesis of colloidal nanoparticles [112]. The microreactor method and the microwave method can lead to nanoparticles of uniform size. A uniform seed crystal is the principal feature of the microreactor; by combining the microreactor and the microwave heating method, in which the microreactor can generate a uniform seed crystal and a uniform crystal growth is possible, resulted in nanoparticles with a narrower size distribution. As a result, a novel hybrid system was proposed that combined a microreactor and microwave radiation for the continuous nanoparticle synthesis, as exemplified by the formation of Ag nanoparticles [113]. The synthesis of silver nanoparticulates using microwaves and a

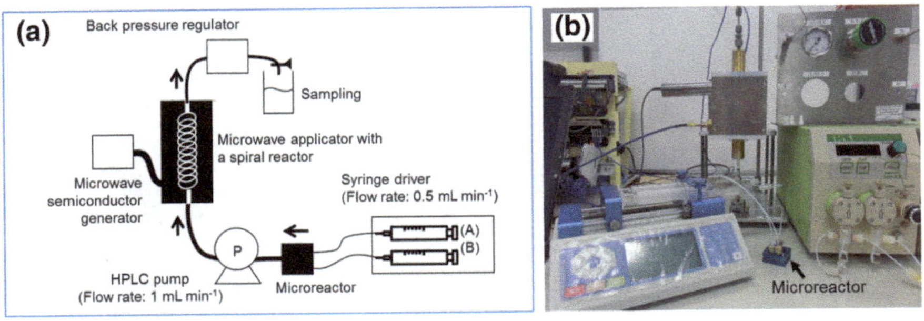

Fig. 10.29 Experimental setup of a hybrid microreactor/microwave system: (**a**) detailed schematic; (**b**) whole photograph. (Microwave equipment: Saida UMS Co. Ltd.) Reproduced with permission from Ref. [112]. Copyright 2013 by Elsevier B.V.

microreactor involved the addition of polyvinylpyrrolidone (PVP) as the dispersing agent and glucose (the reducing agent) in aqueous media (**solution A**) (Fig. 10.29); **solution B** consisted of an aqueous solution of the diaminesilver(I) complex. The two solutions were then mixed in the microreactor (YMC. Co. Ltd Deneb mixer) using a syringe driver, and subsequently added to a quartz spiral reactor using an HPLC pump. The reactor was heated rapidly using a microwave resonator (Saida FDS Inc); the temperature of the sample solution was adjusted to 162 °C and the pressure was regulated to 1 MPa pressure using a back-pressure regulator.

Heating the mixed solution with microwaves in the microreactor generated Ag nanoparticles of uniform size ($\sim$5 nm) (Fig. 10.30a). The maximum absorption of the surface plasmon resonance band of the resulting silver colloids occurred at ca. 420 nm (Fig. 10.30b). Control experiments were also performed, in which both **solutions A** and **B** were mixed in the microreactor system, but without microwave radiation. No silver nanoparticles formed under these conditions; also, none formed in the presence of glucose as the reducing agent, no doubt due to a very slow rate of the reduction process in the absence of microwaves [112]. No silver nanoparticles were produced when the mixed solution **A** + **B** was heated by microwaves but without the microreactor, in spite of identical temperature conditions that occurred under microwave irradiation. However, silver nanoparticles did form by the hydrothermal synthesis method, utilizing a continuous-flow system and a synthesis apparatus that combined both microwaves and the microreactor.

10.2.5 Compendium of Microwave-Assisted Nanoparticle Syntheses

Research into the synthesis of nanoparticles of various types and shapes through the use of microwave radiation has had some remarkable, albeit limited success as reported in Table 10.4, which lists several examples of metallic, metal-oxide and

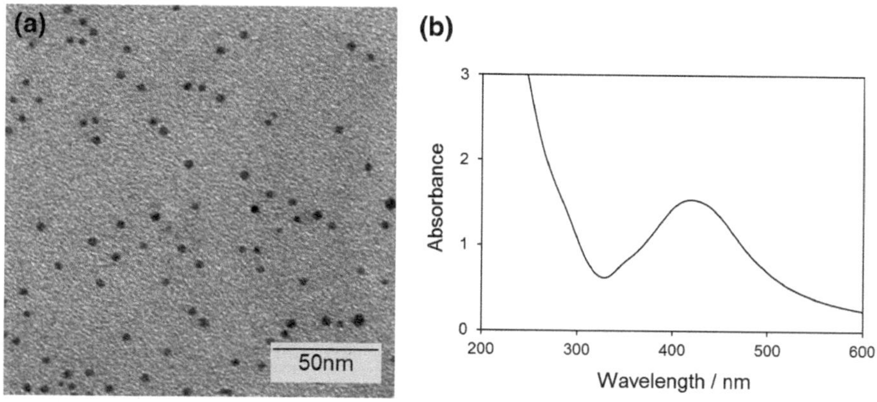

Fig. 10.30 (**a**) TEM images and (**b**) Plasmon resonance absorption of the silver nanoparticles produced by the hybrid microreactor/microwave system [112]. Copyright 2013 by Elsevier B.V

metal-chalcogenide nanoparticles formed from their corresponding precursor salts, together with the dispersing agents, reducing agents, solvent media and reaction temperature conditions [1–75].

10.3 Coffee Break 10: Future Developments in Microwave Ovens

What are some of the future developments in microwave ovens? Over the next few years, we can expect some improvements in cooking performance in microwave ovens. Three things will likely dominate these developments.

(1) ***Semiconductor generation of microwave energy***: the dream of replacing magnetrons and their power supplies with semiconductors is becoming a reality. In early 2016, NXP Semiconductors NV and Wave Technologies Ltd. co-developed a portable microwave oven in which the conventional magnetron/power supply was replaced by an NXP power transistor system with the heating output of 200 W, capable of heating 100–200 mL of food. It is equipped with a rechargeable battery pack. NXP and several of the major companies have formed the RF Energy Alliance for the development and utilization of semiconductor technology in various industries, and have targeted microwave ovens as a potential for this technology. Currently, the semiconductor transistor/power supply is too expensive to replace conventional magnetrons and power supplies. Material costs for magnetron systems including their power supplies is about $18 (USA), or 1970 Japanese Yen, or about €17.25. A solid-state system for 250–300 W of microwave energy output currently costs about €200 to €300, or about $208 to $213 or 22,750 to 23,300 Japanese Yens. The hope is

Table 10.4 Synthesis of nanoparticles under a microwave heating method

Particles	Shape (size)	Precursor	Dispersing agent	Reducing agent/solvent	Reaction temp. and time	Ref.
Ag	Spheres (50 nm)	$AgNO_3$	PVP	EG	150 °C, 15 min	[114]
Ag	Rods (10–20 nm × 50–200 nm)	$AgNO_3$	Na_3Citrate	H_2O	100 °C, 10 min	[115, 116]
Ag	spheres and cubes(5–10 nm)	$AgNO_3$	Dodecylthiol	EG, toluene, Thiourea	160–170 °C, 3 h	[117]
Ag	Spheres (26 ± 3 nm)	$AgNO_3$	Starch	L-lysine, L-arginine, H_2O	150 °C, 10 s	[118]
Ag	Spheres, cubes, rods, wires (30–100 nm)	$AgNO_3$	PVP	Na_2S, EG	130–170 °C, 20 s	[119]
Ag	Rods (40–120 nm × 1–8 μm)	Ag_2O	1,2-ethanedithiol	1,2-ethanedithiol	80–140 °C, 10 min	[120]
Ag	Spheres (1–2.3 nm)	$Ag(NH_3)_2$	Carboxymethylcellulose	Carboxymethyl-cellulose, H_2O	100 °C, flow rates = 600 mL min^{-1}	[96]
Ag	Spheres (5 nm)	$Ag(NH_3)_2$	PVP	Glucose, H_2O	100 °C, flow rates = 1 mL min^{-1}, 1 MPa	[112]
Au	Wires (20 nm × 1–3 μm)	$HAuCl_4$	DNA	DNA, EDTA, H_2O	<100 °C, 3 min	[121]
Au	Spheres, Triangles, Rods	$HAuCl_4$	CTAB	2,7-DHN, NaOH, H_2O	30–90 s	[103]
Au	Plates (30 nm)	$HAuCl_4$	PVP	EG	196 °C, 1 min	[122]
Au	Spheres (5–25 nm)	$HAuCl_4 \cdot 3H_2O$	PVP	HCl, NaOH	1000 W, 5–60 s	[123]
Au	Spheres (9 ± 2 nm)	$HAuCl_4$	Oleylamine	Oleylamine	<15 min, 2.45 or 5.8 GHz	[124]
Ni	Spheres (6 ± 3 nm)	$Ni(OH)_2$	PVP	EG, H_2O	150 °C, 5 min	[125]

(continued)

Table 10.4 (continued)

Particles	Shape (size)	Precursor	Dispersing agent	Reducing agent/solvent	Reaction temp. and time	Ref.
Ni	Polyhedrons (40–100 nm)	$Ni(CH_3CO_2)_2$	PVP, DDA	EG	195 °C, 45 min	[126]
Pt	Spheres (1.5–6.5 nm)	$H_2[PtCl_6]\cdot 6H_2O$	PVP	EG, glycerol, or 1,3-propanediol	>160 °C, flow rates = 3–200 mL h^{-1}	[127]
Pt	Spheres (5 nm)	H_2PtCl_6	PVP	EG	150 °C, 15 min	[128]
Pd	10–30 nm	$PdCl_2$	PEG	D-glucose	800 W, 20 s	[129]
Pd	Cubes (25 nm)	$PdCl_2$	PVP, CTAB, CTAC	TEG	900 W, 80 s	[130]
Te	Rods, wires (15–40 nm, 710 nm)	TeO_2	PVP	(Bpy)[BF_4], $NaBH_4$	180 °C, 10 min	[131]
Cu	Spheres (10 nm)	$CuSO_4$	PVP	NaH_2PO_2, EG	5 min	[132]
Cu	Spheres (90–240 nm)	$Cu(ac)_2$	PVP	Ascorbic acid, DEG	<170 °C	[133]
Cu	Spheres (2–6 nm)	$Cu(octa)_2$, Cu $(myri)_2$, $Cu(AOT)_2$, $Cu(EHP)_2$	–	1-propanol, 1-hexanol, 1-heptanol	<170 °C, 20–120 min	[134]
Cr, Mo, W, Mn, Re, Fe, Ru, Os, Co, Rh, Ir	Spheres (<13 nm)	Mn(CO)m (n = 1–6, m = 6–16)	–	[Bmim][BF_4]	250 °C, 3 min	[135]
CuO, Cu_2O	Spheres (30 nm), cubes (300 nm)	$Cu(OAc)_2\cdot H_2O$	–	EG, H_2O	15 min	[136]
CuO	Spheres (4 nm)	$Cu(OAc)_2$	PEG	NaOH, EtOH	10 min	[137]
Cu_2O	Ellipsoids (300 nm)	$Cu(OAc)_2$	–	Glucose, H_2O	70 °C, 30 min	[138]
CuS	5–10 nm	$Cu(OAc)_2$, Thioacetamide	–	Formaldehyde, H_2O	Reflux, 20 min	[139]

(continued)

Table 10.4 (continued)

Particles	Shape (size)	Precursor	Dispersing agent	Reducing agent/solvent	Reaction temp. and time	Ref.
TiO_2	Cubes (5 nm)	TTIP	–	EtOH, [bmim][BF_4]	3–40 min	[140]
TiO_2	Cubes (25 nm), rods (4 × 17 nm), spheres (8 nm)	$TiCl_4$	–	NH_3 or NaCl or NH_4Cl, H_2O	20–60 min	[141]
TiO_2	Spheres (10 nm), rods 10 × 100 nm)	$TiOCl_2$	–	H_2O	195 °C, 5–60 min	[142]
Fe_3O_4	3–10 nm	$FeCl_3$	Polyacrylic acid	NaOH, Diethylene glycol	10 min	[143]
Fe_2O_3	Cubes (240 nm), ellipsoids (300 nm), spheres (300 nm)	$FeCl_3$	–	$NH_4H_2PO_4$, H_2O	220 °C	[144]
Fe_2O_3	<20 nm	$FeCl_3$	–	EtOH, AcOH	150 °C, 15 min	[145]
Fe_3O_4	Wires (30–50 nm × 1 μm)	$FeSO_4$	PEG	H_2O	180 °C, 8 min	[146, 147]
Fe_3O_4	Spheres (6 nm)	$Fe(acac)_3$	Oleylamine, 1,2-Hexadecanediol	[bmin][BF_4], Phenyl ether	250 °C, 5–10 min	[148]
Tl_2O_3	Spheres (20 nm)	$TlCl_3$	–	NH_3/H_2O	60 min	[149]
SnO_2	Spheres (5 nm)	$SnCl_4$	–	HCl, H_2O	180 °C, 45–120 min	[150]
CeO_2	Cubes (50 nm)	$Ce(NO_3)_3$	Oleic acid	EG, t-butylamine, H_2O	60 min	[151]
CeO_2	Spheres (3 nm)	$Ce(NO_3)_3$	Hexamine	Diethylene glycol	60 min, 10 bar	[152]

(continued)

Table 10.4 (continued)

Particles	Shape (size)	Precursor	Dispersing agent	Reducing agent/solvent	Reaction temp. and time	Ref.
$PbTiO_3$	50 nm	$Pb(Ac)_2$, $Ti(OPr^i)_4$	–	EG	1 min	[153]
BiO_2	Rods (2 × 30 μm)	$Bi(NO_3)_3$	PVP	NaOH, H_2O	6 min reflux	[154]
$Cd(OH)_2$	Wires (5–30 nm × 0.3 μm)	$Cd(NO_3)_2$	–	NH_3, H_2O	5 min	[155]
$BaTiO_3$	50–100 nm	$BaCl_2$, $Ti(OPr^i)_4$	–	EG	5 min	[156]
$BaTiO_3$	Cube particles	$BaCl_2{\cdot}2H_2O$, Titanium (IV) Isopropoxide	–	NaOH, HCl, H_2O	2–40 h 2.45–5 GHz	[157]
CdSe	Spheres (200 nm)	Cd(Ac), Se(powder)	–	EG	60 min	[158]
CdS	Spheres (<5 nm)	$Cd(OAc)_2$	–	Thiourea, DMF	18 s	[159]
CdS	Spheres (2.8 nm)	$Cd(CH_3COO)_2$	1-thioglycerol	DMF, Thiourea	900 W, 25–50 s	[160]
CdTe	Spheres (4.5 nm)	CdO, TeTBP	–	ODE, TDPA	180–280 °C, 1–90 s	[161]
CdTe	Spheres (5 nm)	$CdCl_2$, NaHTe	MPA	H_2O	300 W, 10–150 min	[162]
PbS	Spheres (10 nm)	$Pb(OAc)_2$, S	–	NaOH, EtOH	Reflux, 20 min	[163]
PbSe	Cubes (5–15 nm)	$Pb(OAc)_2$, selenourea	Oleylamine	Phenyl ether, DMF	160 °C, 1 min, 3 bar	[164]
ZnS, CdS	6 nm	$ZnSO_4$, $CdSO_4$	Sodium citrate	Thioacetamide, KOH	905 W, 1 min	[165]
ZnO	Rods, wires, spindles	$Zn(NO_3)_2{\cdot}6H_2O$	–	NaOH, H_2O	100–200 °C, 10–20 min, 0.5–4.0 MPa	[166]
ZnO	Sphere (57–210 nm)	$Zn(OAc)_2{\cdot}2H_2O$	–	DEG	180 °C, 10 min	[167]

(continued)

Table 10.4 (continued)

Particles	Shape (size)	Precursor	Dispersing agent	Reducing agent/solvent	Reaction temp. and time	Ref.
ZnS	Spheres (7 nm)	$Zn(OAc)_2$, Thiourea	PVP	DMF	1–12 min	[168]
ZrO_2	spheres(2 nm)	$Zr(NO_3)_4 \cdot 5H_2O$	–	PVA, NaOH	6 min	[169]
ZnSe	Spheres (2.5–4 nm)	$ZnCl_2$, Se(powder)	MPA	$NaBH_4$, H_2O	140 °C, 5 min	[170]
WO_3	Wires (20–30 nm × >1 μm)	$Na_2WO_4 \cdot 2H_2O$	$(NH_4)_2SO_4$	HCl, H_2O	150 °C, 20–180 min	[171]
WO_3	Aggregates	WCl_6	–	$C_6H_5CH_2OH$	210 °C, 5–15 min	[172]
$ZnWO_4$	Aggregates	H_2WO_4, $Zn(OAc)_2$	–	Citric acid, H_2O	30 min	[173]
Bi_2Se_3	50 × 300 nm	$BiONO_3$, Se	–	NaOH, EG	Reflux, 30 min	[174]
$MoSe_2$	Aggregates	$Mo(CO)_6$, Se	–	EG	Reflux, 60 min	[175]
$NiFe_2O_4$ $CoFe_2O_4$ $MnFe_2O_4$	Spheres (5 ± 2 nm)	$Fe(NO_3)_3 \cdot 9H_2O$ $Ni(NO_3)_2 \cdot 6H_2O$ $Co(NO_3)_2 \cdot 6H_2O$ $Mn(NO_3)_2 \cdot 4H_2O$	–	Oleic acid Toluene	500 W, 160 °C, 1 h	[176]
$MxZn_{1-x}O$ (M = V, Co, Fe, Ni, Mn)	7–19 nm	$Zn(CH_3COO)_3$ $VO[OCH(CH_3)_2]_3$ $Co(CH_3COO)_2$ $Fe(CH_3COO)_2$ Ni $(CH_3COO)_2 \cdot 4H_2O$ $Mn(CH_3COO)_2$	–	Benzyl alcohol	160 °C, 3 min	[177]
$MnZnFe_2O_4$	Aggregates	$Fe(NO_3)_3$, Mn $(NO_3)_2$, $Zn(NO_3)_2$	–	NaOH, H_2O	100 °C, 5–30 min	[178]
CdZnSe	Aggregates	$Cd(OAc)_2$, Zn $(OAc)_2$, Se	–	NaOH, EG	Reflux, 60 min	[179]

(continued)

Table 10.4 (continued)

Particles	Shape (size)	Precursor	Dispersing agent	Reducing agent/solvent	Reaction temp. and time	Ref.
GdF_3:Eu^{3+}	Aggregates	$Gd(TFA)_3$, $Eu(TFA)_3$	–	$[bmin][BF_4]$, EG	120 °C, 15 min	[180]
Cu-Ag	Spheres (20 ± 5 nm)	$Cu(NO_3-)_2 \cdot 3H_2O$, $AgNO_3$	–	Ascorbic acid, H_2O	90 s	[181]
Au@Pd	12 nm	$HAuCl_4 \cdot 6H_2O$, $PdCl_2$	–	EG	10–60 s	[182]
Pd@Pt	Cubes (10–30 nm)	$K_2(PtCl_4)$, $PdCl_2$	CTAB	Ascorbic acid, H_2O	100 °C, 3 min	[183]
CdTe@CdS@ZnS	Spheres (3–5 nm)	$CdCl_2$, NaHTe, $ZnCl_2$	MPA	Na_2S, H_2O	100 °C, 1–5 min	[184]
CdSe@ZnS	Aggregates	$CdCl_2$, Se, $ZnCl_2$	MPA	NaOH, NH_4OH, $NaBH_4$, H_2O	140–170 °C, 45–120 min	[185]

EG Ethylene glycol; *FA* Formaldehyde; *DHN* Dihydroxynaphthalene; *DDA* Dodecylamine; *MPA* 3-mercaptopropionic acid; *ODE* 1-octadecane; *PVP* Polyvinylpyrrolidone; *TDPA* Tetradecylphosphonic acid; *TTIP* Titanium tetraisopropoxide; *PVA* Polyvinyl alcohol that, once the price falls dramatically, this will lead to microwave- assisted cooking ovens that will provide excellent cooking but in much shorter times.

(2) ***Sensors***: Several developments are underway to utilize such things as optical scanning of UPC codes on food packages, IR scanning of foods inside the microwave oven while heating, and other technologies to take more or all the guesswork out of the preparation of foods in the microwave oven. However, the variability of prepared foods, especially of frozen foods, is so great that it may be a long while before any of these technologies become practical in Japan or in the United States.

(3) ***Internet connection***: The near-universality of ownership of smart phones has led to the potential for controlling many devices including air-conditioning and lighting in the home; by extension, this could be utilized in controlling the cooking or reheating of foods in a microwave oven. For example, it may be possible to maintain a library of recipes and cooking instructions on the smart phone and then deliver these instructions to the microwave oven along with the appropriate food, thus taking the guesswork out of cooking and reheating.

An innovative combination: Fig. 10.31a shows a prototype microwave oven and other items that illustrate the benefits of a semiconductor generator, sensors, and connection to the Internet, all of which automatically provide the best cooking performance for any purchased meal. This novel microwave oven includes a charge-coupled device (CCD) and a low-cost thermographic camera located at the top of the microwave cavity. Here, for example, a consumer places a box lunch (Fig. 10.31b) or a frozen food inside the microwave cavity and the barcode on the packages is automatically read by the CCD; this information is transferred to the Internet which responds by providing the best microwave irradiation time, power and heating pattern to cook that food item. Meanwhile, the thermographic camera and other sensors monitor the temperature (Fig. 10.31c) for each of the ingredients

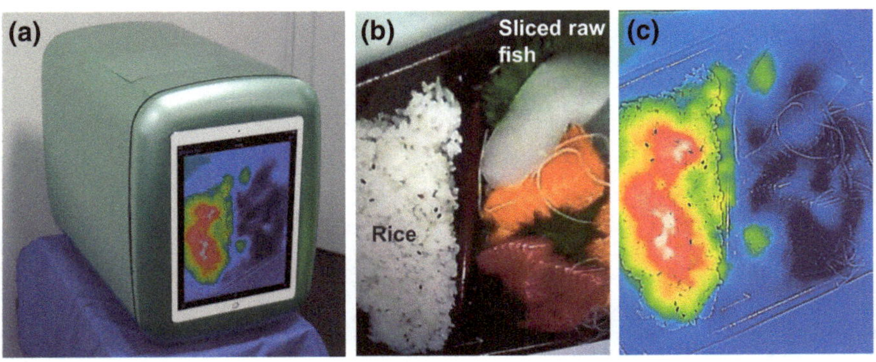

Fig. 10.31 (**a**) A novel microwave oven with a GaN semiconductor generator (total power, 1000 W), a CCD and an IR camera, all connected to an internet device. (**b**) Picture of a Sushi box lunch (rice and sliced raw fish), and (**c**) temperature monitoring by a thermographic camera. Photographs courtesy of S. Horikoshi. Copyright 2017 by S. Horikoshi

as it is microwaved. At this point, the oven *knows* the appropriate end temperature for the best eating quality for each of the components and provides the best microwave heating pattern and power to achieve those temperatures. In order to achieve this, all of the heating information data must first be uploaded to a data calculation server to the Internet, which also includes *failed heating* information, i.e. conditions that will not give the best quality. An artificial intelligence device analyzes this information and uploads the best heating information for any situation to the server.

References

1. K. Shimomura, T. Miyazaki, N. Taniguchi, A. Tutiya, Study of ceramics processing. *Annual Report, RIKEN*, Japan, vol. 48, pp. 11–19
2. A. von Hippel (ed.), *Dielectric Materials and Applications* (MIT Technology Press, New York, 1954)
3. A.C. Metaxas, J.G.P. Binner, Microwave processing of ceramics, in *Advanced Ceramic Processing and Technology*, vol 1, ed. by J.G.P. Binner (Noyes Publications, Park Ridge, NJ, USA, 1990), pp. 285–362
4. W.H. Sutton, Microwave processing of ceramic materials. Am. Ceram. Soc. Bull. **68**, 376–386 (1989)
5. H.R. Roy, S. Komarneni, L.J. Yang, Controlled microwave heating and melting of gels. J. Am. Ceram. Soc. **68**, 392–395 (1985)
6. C. Leonelli, S. Komarneni, Inorganic syntheses assisted by microwave heating. Inorganics **3**, 388–391 (2015)
7. W.B. Campbell, J.V. Shivers, Microwaves for ceramic process control. Am. Ceram. Soc. Bul. **52**, 260–262 (1973)
8. H.M. Kingston, S.J. Haswell (eds.), *Microwave-Enhanced Chemistry* (American Chemical Society, Washington D.C., 1997)
9. W.R. Tinga, A.G. Voss, *Microwave Power Engineering* (Academic, New York, 1968)
10. R.R. Menezes, P.M. Souto, R.H.G.A. Kiminami, Microwave hybrid fast sintering of porcelain bodies. J. Mater. Process. Technol. **190**, 223–229 (2007)
11. M.N. Rahaman, *Ceramic Processing and Sintering* (Marcel Dekker Inc., 0-8247-9573-3, New York, 1995)
12. Materials Research Advisory Board, *Microwave Processing of Materials*, National Research Council, Publication NMAB-473 (National Academy Press, Washington, D. C., 1994)
13. D.C. Folz, J.H. Booske, D.E. Clark, J.F. Gerling (eds), Microwave and radio frequency applications, in *Proceedings from the Third World Congress on Microwave and RF Processing* (American Ceramic Society, Westerville, OH, 2003)
14. H. Takizawa, Microwave non-equilibrium chemistry in inorganic chemistry and materials chemistry. Inst. Elect. Eng. Jpn. **132**, 17–19 (2012)
15. J.D. Katz, Microwave sintering of ceramics. Annu. Rev. Mater. Sci. **22**, 153–170 (1992)
16. M. Panneerselvam, K.J. Rao, Microwave preparation and sintering of industrially important perovskite oxides: $LaMO_3$ (M = Cr Co, Ni). J. Mater. Chem. **13**, 596–601 (2003)
17. A. Badev, S. Marinel, R. Heuguet, E. Savary, D. Agrawal, Sintering behavior and non-linear properties of ZnO varistors processed in microwave electric and magnetic fields at 2.45 GHz. Acta Mater. **61**, 7849–7858 (2013)
18. W.B. Harrison, M.R.B. Hanson, B.G. Koepke, Microwave processing and sintering of PZT and PLZT ceramics. Microwave Process. Mater. **124**, 279–286 (1988)

19. S. Komarneni, E. Breval, R. Roy, Microwave preparation of mullite powders, in *Microwave Processing of Materials*, ed. by W.H. Sutton, M.H. Brooks, I.J. Chabinsky (Materials Research Society, Pittsburgh, PA, 124, 1988), pp. 235–238
20. I. Ahmad, G.T. Chandler, D.E. Clark, Processing of superconducting ceramics using microwave energy, in *Microwave Processing of Materials*, vol. 124, ed. by W. H. Sutton, M. H. Brooks, I.J. Chabinsky (Materials Research Society, Pittsburgh, PA, 1988), pp. 239–246
21. M.K. Krage, Microwave sintering of ferrites. Am. Ceram. Soc. Bull. **60**, 1232–1234 (1981)
22. K.H. Brosnan, G.L. Messing, D.K. Agrawal, Microwave sintering of alumina at 2.45 GHz. J. Am. Ceram. Soc. **86**, 1307–1312 (2003)
23. N.W. Schubring, Microwave sintering of alumina spark plug insulators. *Technical Report GMR-4471* (General Motors Research Laboratory, 1983), pp. 1–27
24. W.H. Sutton, Microwave firing of high alumina ceramics, in *Microwave Processing of Materials*, ed. by W.H. Sutton, M.H. Brooks, I.J. Chabinsky (Materials Research Society, Pittsburgh, PA, 124, 1988), pp. 287–295
25. L. Quemeneur, J. Choisnet, B. Raveau, J.M. Thiebaut, G. Roussy, Microwave clinkering with a grooved resonant applicator. J. Am. Ceram. Soc. **66**, 855–859 (1983)
26. V.K. Varadan, Y. Ma, A. Lakhtakia, W. Varadan, Microwave sintering of ceramics, in *Microwave Processing of Materials*, ed. by W.H. Sutton, M.H. Brooks, I.J. Chabinsky (Materials Research Society, Pittsburgh, PA, 124, 1988), pp. 45–57
27. T.T. Meek, R.D. Blake, J.J. Petrovic, Microwave sintering of Al_2O_3 and Al_2O_3–SiC-whisker composites. Ceram. Eng. Sci. Proc. **8**, 861–871 (1987)
28. T.T. Meek, C.E. Holcomb, N. Dykes, Microwave sintering of some oxide materials using sintering aids. J. Mater. Sci. Lett. **6**, 1060–1062 (1987)
29. M.A. Janney, H.D. Kimrey, Microwave sintering of alumina at 28 GHz, in *Ceramic Transactions, Ceramic Powder Science II B,* ed. by G.L. Messing, E.R. Fuller Jr., H. Hausner (American Ceramic Society, Westerville, OH, 1988),pp. 919–924
30. J.D. Katz, R.D. Blake, J.J. Petrovic, Microwave sintering of alumina-silicon carbide composites at 2.45 and 60 GHz. Ceram. Eng. Sci. Proc. **9**, 725–734 (1988)
31. Y.L. Tian, M.E. Brodwin, D.L. Johnson, Microwave sintering of ceramics under high gas pressure, in *Microwave Processing of Materials*, ed. by W.H. Sutton, M.H. Brooks, I. J. Chabinsky (Materials Research Society, Pittsburgh, PA, 124, 1988), pp. 213–218
32. R.D. Blake, T.T. Meek, Microwave processed composites. J. Mater. Sci. Lett. **5**, 1097–1098 (1986)
33. J.D. Katz, R.D. Blake, J.J. Petrovic, H. Scheinberg, Microwave sintering of boron carbide, in *Microwave Processing of Materials*, ed. by W.H. Sutton, M.H. Brooks, I.J. Chabinsky (Materials Research Society, Pittsburgh, PA, 124, 1988), pp. 219–226
34. E.L. Kerner, D.L. Johnson, Microwave plasma sintering of alumina. Am. Ceram. Soc. Bull. **64**, 1132–1136 (1985)
35. P.A. Haas, Heating of uranium oxides in a microwave oven. Am. Ceram. Soc. Bull. **58**, 873 (1979)
36. T.T. Meek, R.D. Blake, Ceramic-ceramic seals by microwave heating. J. Mater. Sci. Lett. **5**, 270–274 (1986)
37. H. Fukushima, T. Yamanka, M. Matsui, Microwave heating of ceramics and its application to joining, in *Microwave Processing of Materials*, ed. by W.H. Sutton, M.H. Brooks, I. J. Chabinsky (Materials Research Society, Pittsburgh, PA, 124, 1988), pp. 267–272
38. D. Palaith, R. Silberglitt, C.C.M. Wu, R. Kleiner, E.L. Libelo, Microwave joining of ceramics, in *Microwave Processing of Materials*, ed. by W.H. Sutton, M.H. Brooks, I. J. Chabinsky (Materials Research Society, Pittsburgh, PA, 124, 1988), pp. 255–266
39. A.J. Berteaud, J.C. Badot, High temperature microwave heating in refractory materials. J. Microwave Power **11**, 315–320 (1976)
40. Y. Hassler, L. Johansen, Microwave heating of fused quartz to high temperatures in the fabrication process of optical fibers, in *Microwave Processing of Materials*, ed. by W. H. Sutton, M.H. Brooks, I.J. Chabinsky (Materials Research Society, Pittsburgh, PA, 124, 1988), pp. 273–278

41. G. Yang, G. Wang, W.H. Hou, Microwave solid-state synthesis of LiV_3O_8 as cathode material for lithium batteries. J. Phys. Chem. B **109**, 11186–11196 (2005)
42. Z.J. Wang, H. Kokawa, H. Takizawa, M. Ichiki, R. Maeda, Low-temperature growth of high-quality lead zirconate titanate thin films by 28 GHz microwave irradiation. Appl. Phys. Lett. **86**, 212903–212906 (2005)
43. K.I. Rybakov, E.A. Olevsky, E.V. Krikun, Microwave sintering: fundamentals and modeling. J. Am. Ceram. Soc. **94**, 1003–1020 (2013)
44. W.H. Sutton, Microwave processing of ceramics materials. Am. Ceram. Soc. Bull. **68**, 376–388 (1987)
45. T. Endo, J. Fukushima, Y. Hayashi, H. Takizawa, Fabrication of $(Zn_{1-x}Al_xO)_5In_2O_3$ by microwave irradiation and thermoelectric characterization. J. Ceram. Soc. Jpn. **121**, 416–421 (2013)
46. S. Horikoshi, N. Shinohara, H. Takizawa, J. Fukushima, *Microwave Chemistry—Reaction, Processing and Application* (Sankyo Publishing Co. Ltd., 2013)
47. R. Roy, D. Agrawal, J.P. Cheng, M. Mathis, Microwave processing: triumph of applications-driven science in WC-composites and ferroic titanates. Ceram. Trans. **80**, 3–26 (1997)
48. R. Roy, D. Agrawal, J. Cheng, S. Gedevanishvili, Full sintering of powdered-metal bosies in a microwave field. Nature **399**, 668–670 (1999)
49. R.M. Anklekar, D.K. Agrawal, R. Roy, Microwave sintering and mechanical properties of P/M steel. Powder Metall. **44**, 355–362 (2001)
50. R.M. Anklekar, K. Bauer, D. Agrawal, R. Roy, Improved mechanical properties and microstructural development of microwave sintered copper and nickel steel PM parts'. Powder Metall. **48**, 39–46 (2005)
51. http://www.azom.com/article.aspx?ArticleID=937
52. S. Takayama, G. Link, S. Miksch, M. Sato, J. Ichikawa, M. Thumm, Millimetre wave effects on sintering behavior of metal powder compacts. Powder Metall. **46**, 274–280 (2006)
53. D. Demirskyi, A. Ragulya, D. Agrawal, Initial stage sintering of binderless tungsten carbide powder under microwave radiation. Ceramic Intl. **37**, 505–512 (2011)
54. D. Agrawal, Metal parts from microwaves. Mater. World **7**, 672–673 (1999)
55. M. Gupta, W.L.E. Wong, Enhancing overall mechanical performance of metallic materials using two-directional microwave assisted rapid sintering. Scripta Mater. **52**, 479–483 (2005)
56. A. Upadhyaya, S.K. Tiwari, P. Mishra, Microwave sintering of W–Ni–Fe alloy. Scripta Mater. **56**, 5–8 (2007)
57. W.L.E. Wong, M. Gupta, Development of Mg/Cu nanocomposites using microwave assisted rapid heating. Composites Sci. Technol. **67**, 1541–1552 (2007)
58. R.M. Anklekar, K. Bauer, D. Agrawal, R. Roy, Improved mechanical properties and microstructural development of microwave sintered copper and nickel steel PM parts. Powder Metall. **48**(1), 39–46 (2005)
59. S.S. Panda, V. Singh, A. Upadhyaya, D. Agrawal, Sintering response of austenitic (316L) and ferritic (434L) stainless steel consolidated in conventional and microwave furnaces. Scripta Mater. **54**, 2179–2183 (2006)
60. K. Rajkumar, S. Aravindan, Tribological studies on microwave sintered copper-carbon nanotube composites. Wear **270**, 613–621 (2011)
61. S.D. Luo, M. Yan, G.B. Schaffer, M. Qian, Sintering of titanium in vacuum by microwave radiation. Metall. Mater. Trans. A **42**, 2466–2474 (2011)
62. S. Takayama, Y. Saito, M. Sato, T. Nagasaka, T. Muroga, Y. Ninomiya, Sinteirng behavior of metal powders involving microwave-enhanced chemical reaction. J. Appl. Phys. **45**, 1816–1822 (2006)
63. A. Mondal, A. Upadhyaya, D. Agrawal, Microwave and conventional sintering of premixed and prealloyed tungsten heavy alloys, in *Materials Science and Technology (MS&T) 2008*, Pittsburgh, Pennsylvania, 5–9 October 2008, pp. 2502–2515

64. A. Mondal, A. Upadhyaya, D. Agrawal, Microwave and conventional sintering of 90W-7Ni-3Cu alloys with premixed and prealloyed binder phase. Mater. Sci. Eng. A **527**, 6870–6878 (2010)
65. S.-D. Luo, J.-H. Yi, Y.-L. Guo, Y.-D Peng, L.-Y. Li, J.-M. Ran, Microwave sintering W–Cu composites: analyses of densification and microstructural homogenization. J. Alloy Compd. **473**, L5–L9 (2009)
66. A. Upadhyaya, G. Sethi, Effect of heating mode on the densification and microstructural homogenization response of premixed bronze. Scripta Mater. **56**, 469–472 (2007)
67. Y. Zhou, K. Wang, R. Liu, X.P. Wang, C.S. Liu, Q.F. Fang, High performance tungsten synthesized by microwave sintering method. Int. J. Refract. Metals Hard Mater. **34**, 13–17 (2012)
68. R.R. Zheng, Y. Wu, S.L. Liao, A.H. Wang, Microstructure and mechanical properties of Al/(Ti, W)C composites prepared by microwave sintering. J. Alloy. Compd. **590**, 168–175 (2014)
69. K. Wang, X.P. Wang, R. Liu, T. Hao, T. zhang, C.S. Liu, Q.F. Fang, The study on the microwave sintering of tungsten at relatively low temperature, J. Nuclear Mater. **431**, 206–211 (2012)
70. C. Padmavathi, A. Upadhyaya, D. Agrawal, Effect of microwave and conventional heating on sintering behavior and properties of Al–Mg–Si–Cu alloy. Mater. Chem. Phys. **130**, 449–457 (2011)
71. M.M. Mahmoud, G. Link, M. Thumm, The role of the native oxide shell on the microwave sintering of copper metal powder compacts. J. Alloy. Compd. **627**, 231–237 (2015)
72. H. Taira, Y. Saito, K. Sawano, K. Goda, K. Asano, I. Takita, Microwave drying of monolithic refractories-1. TAIKABUTSU **60**, 512–521 (2007)
73. S. Katayose, T. Miyazaki, Y. Hayashi, H. Takizawa, Synthesis of natural superlattice structure in the binary ZnO–Fe_2O_3 system by microwave irradiation. J. Ceram. Soc. Jpn. **118**, 387–389 (2010)
74. D. Nagao, J. Fukushima, Y. Hayashi, H. Takizawa, Synthesis of homologous compounds $Fe_2O_3(ZnO)_m$ (m = 6, 8, 34) by various selective microwave heating conditions. Ceram. Int. **41**, 14021–14028 (2015)
75. A. Feteira, Negative temperature coefficient resistance (NTCR) ceramic thermistors: an industrial perspective. J. Am. Ceram. Soc. **92**, 967–983 (2009)
76. D. Mehandjiev, A. Naydenov, G. Ivanov, Ozone decomposition, benzene and CO oxidation over $NiMnO_3$-ilmenite and $NiMn_2O_4$-spinel catalysts. Appl. Catal. A **206**, 13–18 (2001)
77. B.B. Nelson-Cheeseman, R.V. Chopdekar, J.M. Iwata, M.F. Toney, E. Arenholz, Y. Suzuki, Modified magnetic ground state in $NiMn_2O_4$ thin films. Phys. Rev. B **82**, 144419–144426 (2010)
78. H. Goto, J. Fukushima, H. Takizawa, Control of magnetic properties of $NiMn_2O_4$ by a microwave magnetic field under air. Materials **9**, 169–176 (2016)
79. C.A. Pickles, Microwaves in extractive metallurgy: part 2—a review of applications. Mineral. Eng. **22**, 1112–1118 (2009)
80. R.K. Amankwah, C.A. Pickles, Microwave calcination and sintering of manganese carbonate ore. Can. Metall. Q. **44**, 239–248 (2005)
81. K. Hara, M. Hayashi, M. Sato, K. Nagata, Continuous pig iron making by microwave heating with 12.5 kW at 2.45 GHz. J. Microwave Power Electromagn. Energy **45**, 137–147 (2011)
82. N. Yoshikawa, E. Ishizuka, K. Mashiko, S. Taniguchi, Carbon reduction kinetics of NiO by microwave heating of the separated electric and magnetic fields. Metall. Mater. Trans. B Process Metall. Mater. Proc. Sci. **38**, 863–868 (2007)
83. J. Fukushima, H. Takizawa, Enhanced reduction of copper oxides via internal heating, selective heating, and cleavage of Cu–O bond by microwave magnetic-field irradiation. Mater. Chem. Phys. **172**, 47–53 (2016)
84. H. Takizawa, J. Fukushima, Titanium nitride coating by microwave. Kinzoku **83**, 694–699 (2013)

85. S. Komarneni, R. Roy, Titania gel spheres by a new sol-gel process. Mater. Lett. **3**, 165–167 (1985)
86. S. Komarneni, R. Roy, Q.H. Li, Microwave- hydrothermal synthesis of ceramic powders. Mater. Res. Bull. **27**, 1393–1405 (1992)
87. W. Lojkowski, C. Leonelli, T. Chudoba, J. Wojnarowicz, A. Majcher, A. Mazurkiewicz, High-energy low-temperature technologies for the synthesis of nanoparticles: microwaves and high pressure. Inorganics **2**, 606–619 (2014)
88. J. Prado-Gonjal, R. Schmidt, E. Morán, Microwave-assisted routes for the synthesis of complex functional oxides. Inorganics **3**, 101–117 (2015)
89. S. Cho, S.-H. Jung, K.-H. Lee, Morphology-controlled growth of ZnO nanostructures using microwave irradiation: from basic to complex structures. J. Phys. Chem. C **112**, 12769–12776 (2008)
90. M.S. El-Shall, V. Abdelsayed, H.M.A. Hassan, S.K. Abd El Rahman, K.M. Abouzeid, Q. Dai, P. Afshani, F. Gupton, A.R. Siamaki, Z. Abdullah, M. Alothman, H.Z. Alkhathlan, Production of graphene and nanoparticle catalysts supported on graphene using microwave radiation. US Patent Application, US2013/0211106 A1, 15 August 2013
91. R. Trujillano, E. Rico, M.A. Vicente, V. Rives, L. Bergalaouis, S.B. Chaabene, A.D. Ghorbel, Microwave-assisted synthesis of Fe^{3+} saponites. Characterization by x-ray diffraction and FT-IR spectroscopy. Rev. Soc. Esp. Miner. **11**, 189–190 (2009)
92. Z. Derikvand, F. Farzaneh, Synthesis, characterization and catalytic behavior of special type of silico-alumino-phosphate with opal structure. J. Sci. (Iran) **14**, 235–238 (2003)
93. I. Bilecka, M. Niederberger, Microwave chemistry for inorganic nanomaterials synthesis. Nanoscale **2**, 1358–1374 (2010)
94. N.N. Mallikarjuna, R.S. Varma, Microwave-assisted shape-controlled bulk synthesis of noble nanocrystals and their catalytic properties. Cryst. Growth Des. **7**, 686–690 (2007)
95. J.A. Gerbec, D. Magana, A. Washington, G.F. Strouse, Microwave-enhanced reaction rates for nanoparticle synthesis. J. Am. Chem. Soc. **127**, 15791–15800 (2005)
96. S. Horikoshi, H. Abe, K. Torigoe, M. Abe, N. Serpone, Access to small size distributions of nanoparticles by microwave-assisted synthesis. Formation of Ag nanoparticles in aqueous carboxymethylcellulose solutions in batch and continuous-flow reactors. Nanoscale **2**, 1441–1447 (2010)
97. S. Horikoshi, N. Serpone (eds.), *Microwaves in Nanoparticle Synthesis: Fundamentals and Applications* (Wiley-VCH Verlag GmbH & Co., Germany, 2013)
98. Y. Tsukahara, T. Nakamura, T. Kobayashi, Y. Wada, Homogeneous Ag particle formation confirmed by real-time in situ surface-enhanced Raman scattering measurements under microwave irradiation. Chem. Lett. **35**, 1396–1397 (2006)
99. E. Smith, G. Dent, *Modern Raman Spectroscopy: A Practical Approach* (John Wiley and Sons Ltd., Chichester, UK, 2005), p. 210
100. M.A. Herrero, J.M. Kremsner, C.O. Kappe, Nonthermal microwave effects revisited: on the importance of internal temperature monitoring and agitation in microwave chemistry. J. Org. Chem. **73**, 36–47 (2008)
101. S. Wiegand, Thermal diffusion in liquid mixtures and polymer solutions. J. Phys. Condens. Matter **16**, R357–R379 (2004)
102. J.-S. Schanche, Microwave synthesis solutions from personal chemistry. Molec. Diversity **7**, 291–298 (2003)
103. S. Kundu, L. Peng, H. Liang, A new route to obtain high-yield multiple-shaped gold nanoparticles in aqueous solution using microwave irradiation. Inorg. Chem. **47**, 6344–6352 (2008)
104. M. Tsuji, M. Hashimoto, Y. Nishizawa, M. Kubokawa, T. Tsuji, Microwave-assisted synthesis of metallic nanostructures in solution. Chem. Eur. J. **11**, 440–452 (2005)
105. T. Yamamoto, H. Yin, Y. Wada, T. Kitamura, T. Sakata, H. Mori, S. Yanagida, Morphology-control in microwave-assisted synthesis of silver particles in aqueous solutions. Bull. Chem. Soc. Jpn. **77**, 757–761 (2004)

106. M. Tsuji, N. Miyamae, S. Lim, K. Kimura, X. Zhang, S. Hikino, M. Nishio, Crystal structures and growth mechanisms of Au@Ag core—shell nanoparticles prepared by the microwave-polyol method. Cryst. Growth Des. **6**, 1801–1807 (2006)
107. Y. He, H.-T. Lu, L.-M. Sai, W.-Y. Lai, Q.-L. Fan, L.-H. Wang, W. Huang, Microwave-assisted growth and characterization of water-dispersed CdTe/CdS core-shell nanocrystals with high photoluminescence. J. Phys. Chem. B **110**, 13370–13374 (2006)
108. Y. Wang, Z. Tang, M.A. Correa-Duarte, I. Pastoriza-Santos, M. Giersig, N.A. Kotov, L.M. Liz-Marzan, Mechanism of strong luminescence photoactivation of citrate-stabilized water-soluble nanoparticles with CdSe cores. J. Phys. Chem. B **108**, 15461–15469 (2004)
109. H. Bao, Y. Gong, Z. Li, M. Gao, Enhancement effect of illumination on the photoluminescence of water-soluble CdTe nanocrystals: toward highly fluorescent CdTe/CdS core—shell structure. Chem. Mater. **16**, 3853–3859 (2004)
110. M. Nishioka, M. Miyakawa, H. Kataoka, H. Koda, K. Sato, T.M. Suzuki, Facile and continuous synthesis of Ag@SiO_2 core–shell nanoparticles by a flow reactor system assisted with homogeneous microwave heating. Chem. Lett. **40**, 1204–1206 (2011)
111. W. Ehrfeld, V. Hessel, H. Lowe, *Microreactor* (Wiley-VCH Verlag, GmbH, Weinheim, Germany, 2001)
112. S. Horikoshi, T. Sumi, N. Serpone, A hybrid microreactor/microwave high-pressure flow system of a novel concept design and its application to the synthesis of silver nanoparticles. Chem. Eng. Proc. **73**, 59–66 (2013)
113. National Research Council, *Plasma Processing of Materials: Scientific Opportunities and Technological Challenges* (National Academies Press, Washington, D.C., 1991)
114. M. Tsuji, M. Hashimoto, T. Tsuji, Fast preparation of nano-sized nickel particles under microwave irradiation without using catalyst for nucleation. Chem. Lett. **31**, 1232–1233 (2002)
115. E.K. Liu, P.W. Huang, Y.C. Chang, C.J. Ko, T.C. Chu, Formation of silver nanorods by microwave heating in the presence of gold seeds. J. Cryst. Growth **273**, 439–445 (2005)
116. F.K. Liu, P.W. Huang, T.C. Chu, C.J. Ko, Gold seed-assisted synthesis of silver nanomaterials under microwave heating. Mater. Lett. **59**, 940–944 (2005)
117. F. Gao, Q. Lu, S. Komarneni, Interface reaction for the self-assembly of silver nanocrystals under microwave-assisted solvothermal conditions. Chem. Mater. **17**, 856–860 (2005)
118. B. Hu, S.B. Wang, K. Wang, M. Zhang, S.H. Yu, Microwave-assisted rapid facile "green" synthesis of uniform silver nanoparticles: self-assembly into multilayered films and their optical properties. J. Phys. Chem. C **112**, 11169–11174 (2008)
119. T. Zhao, J.B. Fan, J. Cui, J.H. Liu, X.B. Xu, M.Q. Zhu, Microwave-controlled ultrafast synthesis of uniform silver nanocubes and nanowires. Chem. Phys. Lett. **501**, 414–418 (2011)
120. Y.J. Zhu, X.L. Hu, Microwave-assisted polythiol reduction method: a new solid-liquid route to fast preparation of silver nanowires. Mater. Lett. **58**, 1517–1519 (2004)
121. S. Kundu, H. Liang, Microwave synthesis of electrically conductive gold nanowires on DNA scaffolds. Langmuir **24**, 9668–9674 (2008)
122. F.K. Liu, C.J. Ker, Y.C. Chang, F.H. Ko, T.C. Chu, B.T. Dai, Synthesis of nanoparticles and one-dimensional nanomaterials. J. Appl. Phys. Part **1**(42), 4152–4158 (2003)
123. S. Kundu, K. Wang, H. Liang, Rapid continuous microwave-assisted synthesis of silver nanoparticles to achieve very high productivity and full yield: from mechanistic study to optimal fabrication strategy. J. Phys. Chem. C **113**, 5157–5163 (2009)
124. S. Horikoshi, H. Abe, T. Sumi, K. Torigoe, H. Sakai, N. Serpone, M. Abe, Microwave frequency effect in the formation of Au nanocolloids in polar and non-polar solvents. Nanoscale **3**, 169–172 (2011)
125. Y.J. Zhu, X.L. Hu, Microwave-polyol preparation of single-crystalline gold nanorods and nanowires. Chem. Lett. **32**, 1140–1141 (2003)
126. M. Fernandez-Garcia, A. Martinez- Arias, J.C. Hanson, J.A. Rodriguez, Certain aspects of the formation and identification of nanosized oxide components in heterogeneous catalysts prepared by different methods. Chem. Rev. **104**, 4063–4104 (2004)

127. M. Nishioka, M. Miyakawa, Y. Daino, H. kataoka, H. Koda, K. Sato, T.M. Suzuki, Characterization of Pt nanoparticles encapsulated in Al_2O_3 and their catalytic efficiency in propene hydrogenation. Chem. Lett. **40**, 1327–1329 (2011)
128. S. Komarneni, D. Li, B. Newalkar, H. Katsuki, A.S. Bhalla, Microwave- and conventional-hydrothermal synthesis of CuS, SnS and ZnS: Optical properties. Langmuir **18**, 5959–5962 (2002)
129. S.K. Mehta, S. Gupta, Microwave-assisted shape-controlled bulk synthesis of noble nanocrystals and their catalytic properties. J. Appl. Electrochem. **41**, 1407–1417 (2011)
130. Y. Yu, Y. Zhao, T. Huang, H. Liu, Microwave-assisted synthesis of palladium nanocubes and nanobars. Mater. Res. Bull. **45**, 159–164 (2010)
131. Y.J. Zhu, W.W. Wang, R.J. Qi, X.L. Hu, Microwave-assisted synthesis and magnetic property of magnetite and hematite nanoparticles. Angew. Chem. Int. Ed. **43**, 1410–1414 (2004)
132. H. Zhu, C. Zhang, Y. Yin, Electrochemical preparation of biomolecule stabilized copper nanoparticles decorated reduced graphene oxide for the sensitive and selective determination of hydrogen peroxide. J. Cryst. Growth **270**, 722–728 (2004)
133. M. Blosi, S. Albonetti, M. Dondi, C. Martelli, G. Baldi, Microwave-assisted polyol synthesis of Cu nanoparticles. J. Nanopart. Res. **13**, 127–138 (2011)
134. T. Nakamura, Y. Tsukahara, T. Sakata, H. Mori, Y. Kanbe, H. Bessho, Y. Wada, Facile synthesis of bimetallic Cu–Ag nanoparticles under microwave irradiation and their oxidation resistance. Bull. Chem. Soc. Jpn. **80**, 224–232 (2007)
135. C. Vollimer, E. Redel, K.A. Shandi, R. Thomann, H. Manyar, C. Hardacre, C. Janiak, Microwave Irradiation for the Facile Synthesis of Transition-Metal Nanoparticles (NPs) in Ionic Liquids (ILs) from Metal–Carbonyl Precursors and Ru-, Rh-, and Ir-NP/IL Dispersions as Biphasic Liquid–Liquid Hydrogenation. Chem. Eur. J. **16**, 3849–3858 (2010)
136. Y. Zhao, J.J. Zhu, J.M. Hong, N. Bian, H.Y. Chen, Microwave-induced polyol-process synthesis of copper and copper oxide nanocrystals with controllable morphology. Eur. J. Inorg. Chem. **20**, 4072–4080 (2004)
137. H. Wang, J.-Z. Xu, J.-J. Zhu, H.-Y. Chen, Preparation of CuO nanoparticles by microwave irradiation. J. Cryst. Growth **244**, 88–94 (2002)
138. D.P. Volanti, M.O. Orlandi, J. Andres, E. Longo, Microwave hydrothermal and solvothermal processing of materials and compounds. Cryst. Eng. Commun. **12**, 1696–1699 (2010)
139. X.-H. Liao, J.-J. Zhu, H.-Y. Chen, Microwave synthesis of nanocrystalline metal sulfides in formaldehyde solution. Mater. Sci. Eng. B **85**, 85–89 (2001)
140. K. Ding, Z. Miao, Z. Liu, Z. Zhang, B. Han, G. An, S. Miao, Y. Xie, Ionic liquid-assisted synthesis of carbon nanotube/platinum nanocomposites. J. Am. Chem. Soc. **129**, 6362–6363 (2007)
141. T. Suprabha, H. Roy, J. Thomas, K.P. Kumar, S. Mathew, Microwave-assisted synthesis of titania nanocubes, nanospheres and nanorods for photocatalytic dye degradation. Nanoscale Res. Lett. **4**, 144–152 (2009)
142. A.B. Corradi, F. Bondioli, B. Focher, A.M. Ferrari, C. Grippo, E. Mariani, C. Villa, Green synthesis and dye-sensitized solar cell application of rutile and anatase TiO_2 nanorods. J. Am. Ceram. Soc. **88**, 2639–2641 (2005)
143. W. Xiao, H. Gu, D. Li, D. Chen, X. Deng, Z. Jiao, J. Lin, Microwave-assisted synthesis of magnetite nanoparticles for MR blood pool contrast agents. J. Mag. Mag. Mater. **324**, 488–494 (2012)
144. X. Hu, J.C. Yu, Phase transformation and shape evolution of iron oxide nanocrystals synthesized in the ethylene glycol-water system. Adv. Funct. Mater. **18**, 880–887 (2008)
145. X. Li, H. Li, R. Cao, Facile fabrication of pure α-Fe2O3 nanoparticles via forced hydrolysis using microwave-assisted esterification and their sensing property. J. Am.Ceram. Soc. **92**, 2188–2191 (2009)
146. T. Muraliganth, A.V. Murugan, A. Manthiram, Facile synthesis of porous Fe3O4@C nanospheres as high-performance anode for lithium-ion battery. Chem. Commun. **47**, 7360–7362 (2009)

147. X. Hu, J. Yu, Microwave-assisted snthesis of a superparamagnetic surface-functionalized porous Fe_3O_4/C nanocomposite. Chem. Asian J. **1**, 605–610 (2006)
148. H. Hu, H. Yang, P. Huang, D. Cui, Y. Peng, J. Zhang, F. Lu, J. Lian, D. Shi, Unique role of ionic liquid in microwave-assisted synthesis of monodisperse magnetite nanoparticles. Chem. Commun. **46**, 3866–3868 (2010)
149. C.R. Patra, A. Gedanken, Rapid synthesis of nanoparticles of hexagonal type In_2O_3 and spherical type Tl_2O_3 by microwave irradiation. New J. Chem. **28**, 1060–1065 (2004)
150. J. Jouhannaud, J. Rossignol, D. Stuerga, Rapid synthesis of tin (IV) oxide nanoparticles by microwave induced thermohydrolysis. J. Solid State Chem. **181**, 1439–1444 (2008)
151. Y. Tao, F.H. Gong, H.P. Wu, G.L. Tao, Preparation and characterization of monodisperse cerium oxide nanoparticles in hydrocarbon solvents. Mater. Chem. Phys. **112**, 973–976 (2008)
152. M. Zawadzki, Preparation and characterization of ceria nanoparticles by microwave-assisted solvothermal process. J. Alloys Compd. **454**, 347–351 (2008)
153. O. Palchik, J.-J. Zhu, A. Gedanken, Microwave assisted preparation of binary oxide nanoparticles. J. Mater. Chem. **10**, 1251–1254 (2000)
154. S. Anandan, J.J. Wu, Photocatalytic activity of enlarged microrods of α-Bi_2O_3 produced using ethylenediamine-solvent. Mater. Lett. **63**, 2387–2389 (2009)
155. D.S. Raj, T. Krishnakumar, R. Jayaprakash, N. Donato, M. Latino, G. Neri, Microwave-assisted synthesis of metal oxide nanostructures for sensing applications. Sci. Adv. Mater. **2**, 432–437 (2010)
156. W.W. Wang, Microwave-induced polyol-process synthesis of $M^{II}Fe_2O_4$ (M = Mn, Co) nanoparticles and magnetic property. Mater. Chem. Phys. **108**, 227–231 (2008)
157. E.K. Nyutu, C.H. Chen, P.K. Dutta, S.L. Suib, Synthesis and characterization of Barium Titanate, Calcium Titanate and Strontium Titanate thin films. J. Phys. Chem. C **112**, 9659–9667 (2008)
158. O. Palchik, R. Kerner, A. Gedanken, A.M. Weiss, M.A. Slifkin, V. Palchik, Microwave-assisted polyol method for the preparationof CdSe “nanoballs”. J. Mater. Chem. **11**, 874–878 (2001)
159. Y. Wada, H. Kuramoto, J. Anand, T. Kitamura, T. Sakata, H. Mori, S. Yanagida, Microwave-assisted size control of CdS nanocrystallites. J. Mater. Chem. **11**, 1936–1940 (2001)
160. S. Karan, B. Mallik, Tunable visible-light emission from CdS nanocrystallites prepared under microwave irradiation. J. Phys. Chem. C **111**, 16734–16741 (2007)
161. Q. Song, X. Ai, T. Topuria, P.M. Rice, F.H. Alharbi, A. Bagabas, M. Bahattab, J.D. Bass, H.-C. Kim, J.C. Scott, R.D. Miller, Microwave-assisted synthesis of monodispersed CdTe nanocrystals. Chem. Commun. **46**, 4971–4973 (2010)
162. L. Li, H. Qian, J. Ren, Rapid synthesis of highly luminescent CdTe nanocrystals in the aqueous phase by microwave irradiation with controllable temperature. Chem. Commun. **4**, 528–530 (2005)
163. T. Ding, J.-J. Zhu, Microwave heating synthesis of HgS and PbS nanocrystals in ethanol solvent. Mater. Sci. Eng. B **100**, 307–313 (2003)
164. M.A. Sliem, A. Chemseddine, U. Bloeck, R.A. Fischer, PbSe nanocrystal shape development: oriented attachment at mild conditions and microwave assisted growth of nanocubes. Cryst. Eng. Commun. **13**, 483–488 (2011)
165. S. Ortíz, I. Gómez, P. Elizondo, J. Cavazos, Highly luminescent nanostructures of CdS and ZnS prepared by microwaves heating: effect of sulphide concentration. Phys. Status Solid C **7**, 2683–2687 (2010)
166. J. Huang, C. Xia, L. Cao, X. Zeng,One-pot synthesis of Copper−Indium Sulfide nanocrystal heterostructures with acorn, bottle, and larva shapes. Mater. Sci. Eng. B **150**, 187–193 (2008)
167. X. Hu, J. Gong, L. Zhang, J.C. Yu, Humidity sensing properties of CeO2–NiO nanocomposite materials. Adv. Mater. **20**, 4845–4850 (2008)

168. R. He, X.-F. Qian, J. Yin, H.-A. Xi, L.-J. Bian, Z.-K. Zhu, Colloids Surf. A **220**, 151–157 (2003)
169. J. Liang, Z. Deng, X. Jiang, F. Li, Y. Li, Photoluminescence of tetragonal ZrO_2 nanoparticles synthesized by microwave irradiation. Inorg. Chem. **41**, 3602–3604 (2002)
170. H.F. Qian, X. Qiu, L. Li, J.C. Ren, Hydrothermal synthesis of nanocrystalline ZnSe: An in situ. J. Phys. Chem. B **110**, 9034–9040 (2006)
171. A. Phuruangrat, D.J. Ham, S.J. Hong, S. Thongtem, J.S. Lee, Synthesis of hexagonal WO_3. J. Mater. Chem. **20**, 1683–1690 (2010)
172. N.L. Houx, G. Pourroy, F. Camerel, M. Comet, D. Spitzer, Comparison of the morphology and structure of WO_3. J. Phys. Chem. C **114**, 155–161 (2010)
173. J.H. Ryu, C.S. Lima, W.C. Oh, K.B. Shim, J. Ceram. Process. Res. **5**, 316–320 (2004)
174. R. Harpeness, A. Gedanken, A simple fast microwave-assisted synthesis of thermoelectric bismuth telluride nanoparticles from homogeneous reaction-mixture. New J. Chem. **27**, 1191–1193 (2003)
175. R. Harpeness, A. Gedanken, A.M. Weiss, M.A. Slifkin, Microwave-assisted synthesis of nanosized MoSe. J. Mater. Chem. **13**, 2603–2606 (2003)
176. B. Babita, M.N. Nadagouda, R.S. Varma, Nano-catalysts with magnetic core: Sustainable options for greener synthesis. J. Phys. Chem. C **112**, 18399–18404 (2008)
177. I. Bilecka, L. Luo, I. Djerdj, M.D. Rossell, M. Jagodic, Z. Jaglicic, Y. Masubuchi, S. Kikkawa, M. Niederberger, Microwave-assisted nonaqueous sol–gel chemistry for highly concentrated ZnO-based magnetic semiconductor nanocrystals. J. Phys. Chem. C **115**, 1484–1495 (2011)
178. L. Zhenyu, X. Guangliang, Z. Yalin, Rapid synthesis of dittmarite by microwave-assisted hydrothermal method. Nanoscale Res. Lett. **2**, 40–43 (2007)
179. H. Grisaru, O. Palchik, A. Gedanken, V. Palchik, M.A. Slifkin, A. M. Weiss, Y. R. Hacohen, Preparation of $Cd_{1-x}Zn_xSe$ Using Microwave-Assisted Polyol Synthesis. Inorg. Chem. **40**, 4814–4815 (2001)
180. C. Lorbeer, J. Cybinska, A.V. Mudring, Facile preparation of quantum cutting $GdF_3 : Eu^{3+}$. Chem. Commun. **46**, 571–573 (2010)
181. M. Valodkar, S. Modi, A. Pal, S. Thakore, Synthesis and anti-bacterial activity of Cu, Ag and Cu–Ag alloy nanoparticles: A green approach. Mater. Res. Bul. **46**, 384–389 (2011)
182. R. Harpeness, A. Gedanken, Preparation of bimetallic palladium-platinum colloids in organic solvent by solvent extraction-reduction. Langmuir **20**, 3431–3434 (2004)
183. H. Zhang, Y. Yin, Y. Hu, C. Li, P. Wu, S. Wei, C. Cai, J. Phys. Chem. C **114**, Pd@Pt core−shell nanostructures with controllable composition synthesized by a microwave method and their enhanced electrocatalytic activity toward oxygen reduction and methanol oxidation. 11861–11867 (2010)
184. Y. He, H.T. Lu, L.M. Sai, Y.Y. Su, M. Hu, C.H. Fan, W. Huang, L.H. Wang, Microwave synthesis of water-dispersed CdTe/CdS/ZnS core-shell-shell quantum dots with excellent photostability and biocompatibility. Adv. Mater. **20**, 3416–3421 (2008)
185. W. Schumacher, A. Nagy, W.J. Waldman, P.K. Dutta, Direct synthesis of aqueous CdSe/ZnS-based quantum dots using microwave irradiation. J. Phys. Chem. C **113**, 12132–12139 (2009)

Appendix A

Abstract This chapter provides Appendices that report some basic physical properties of various substances from which one can extract useful information needed to conduct experiments.

Keywords Microwave heating effect • Physical properties • Thermal conductivities • Emissivity

See Tables A.1, A.2, A.3 and A.4.

S. Horikoshi et al., *Microwave Chemical and Materials Processing*,
https://doi.org/10.1007/978-981-10-6466-1

A.1 Microwave heating effect on solids indicating the time needed to reach the indicated temperatures (particle size: 60 mm mesh)

Substance	Temp./°C	Heating time/min	Substance	Temp./°C	Heating time/min
AgBr	39	1.5	CCl_4	34	4.0
AgCl	45	4.0	$CdCl_2$	37	4.0
Agl	64	4.0	$CdCl_2 \cdot 2.5H_2O$	144	2.5
Ag_2S	652	5.3	CdS	87	5.0
Al	577	6.0	$CeCl_3$	48	2.8
$AlCl_3$	41	4.0	$CeCl_3 \cdot 7H_2O$	38	1.5
$AlCl_3 \cdot 6H_2O$	46	4.0	Co	697	3.0
$Al(ClO_3)_3 \cdot 6H_2O$	72	5.0	$CoCl_2$	36	2.0
$Al_2(SO_4)_3 \cdot 18H_2O$	206	3.8	$CoCl_2 \cdot 6H_2O$	153	4.0
$Al_2(SO_4)_3 \cdot K_2SO_4 \cdot 24H_2O$	56	4.5	Co_3O_4	1290	3.0
$BaCl_2$	45	4.0	CoS	158	7.0
$BaCl_2 \cdot 2H_2O$	50	0.8	$CrCl_3$	82	6.0
BaS	160	5.0	$CrCl_3 \cdot 6H_2O$	181	5.0
C (amorphous, <1μm)	1283	1.0	CrO_3	112	9.0
C (graphite, <1μm)	1073	1.8	Cr_2O_3	130	7.0
C (graphite, 200mesh)	780	6.0	Cu	228	7.0
$CaBr_2 \cdot 2H_2O$	47	4.0	CuBr	722	11.0
$CaCl_2$	32	1.8	$CuBr_2$	193	3.25
$CaCl_2 \cdot 2H_2O$	45	0.8	CuCl	619	13.0
$CaCO_3$	61	7.0	$CuCl_2 \cdot 2H_2O$	171	2.8
CaF_2	81	4.3	$CuF_2 \cdot 2H_2O$	78	6.0
CaS	102	4.5	CuO	1012	6.3
Cus	440	4.8	Cu_2O	89	7.0
Fe	768	7.0	Ni	384	1.0
$FeBr_2 \cdot 6H_2O$	180	4.0	$NiBr_2 \cdot 2H_2O$	159	3.0
$FeCl_2$	33	1.5	$NiCl_2$	51	2.8
$FeCl_2 \cdot 2H_2O$	41	4.0	$NiCl_2 \cdot 6H_2O$	168	1.8
$FeCl_2 \cdot 4H_2O$	47	4.0	$NiF_2 \cdot 4H_2O$	50	5.0
$FeCl_3$	41	4.0	NiO	1305	6.3
$FeCl_3 \cdot 6H_2O$	220	4.5	NiS	251	7.0
$Fe_2(SO_4)_3 \cdot 9H_2O$	154	6.0	$PbBr_2$	56	2.8
Hg	40	6.0	$PbCl_2$	51	2.0
$HgCl_2$	112	7.0	PbO	91	2.3
HgS	105	7.0	PbS	1024	1.3
KBr	46	0.3	$PrCl_3$	30	1.3
KCl	31	1.0	S	163	6.0
$LaCl_3$	38	1.8	Sb	390	1.0
$LaCl_3 \cdot 7H_2O$	34	2.0	$SbCl_3$	224	1.8

(continued)

A.1 (continued)

Substance	Temp./°C	Heating time/min	Substance	Temp./°C	Heating time/min
$LiBr \cdot H_2O$	45	1.0	Sn	297	6.0
LiCl	35	0.5	$SnBr_2$	472	10.0
Mg	120	7.0	$SnCl_2$	476	2.0
$MgBr_2 \cdot 6H_2O$	322	9.0	$SnCl_2 \cdot 2H_2O$	252	5.3
$MgCl_2 \cdot 6H_2O$	254	4.0	$SnCl_4$	49	8.0
MgO	203	5.5	$SnCl_4 \cdot 5H_2O$	151	6.8
$MgSO_4 \cdot 7H_2O$	80	4.5	$SrCl_2 \cdot 6H_2O$	32	3.0
$MnBr_2 \cdot 4H_2O$	176	2.3	Ta	177	7.0
$MnCl_2$	53	1.8	TiB_2	843	7.0
$MnCl_2 \cdot 4H_2O$	170	4.0	$TiCl_4$	31	4.0
MnO	113	6.0	TiO_2	79	8.5
MnO_2	1287	6.0	V	557	1.0
$MnSO_4 \cdot H_2O$	47	5.0	V_2O_5	714	11.0
Mo	660	4.0	YCl_3	40	1.75
MoO_2	69	5.5	Y_2O_3	115	7.0
MoS_3	1106	7.0	W	690	6.3
NaBr	40	4.0	WCl_6	45	3.0
NaCl	83	7.0	WO_3	1270	6.0
Na_2MoO_4	53	4.3	Zn	581	3.0
Na_2SO_4	67	6.0	$ZnBr_2$	574	7.0
Nb	358	6.0	$ZnCl_2$	609	7.0
Nb_2O_3	114	6.0	ZnO	326	5.5
NH_4Cl	31	3.5	ZnS	67	3.0
$5(NH_4)_2 \cdot 12WO_2$	68	4.0	Zr	462	6.0
Pb	277	7.0	ZrO_2	63	4.0

Table A.2 Physical properties of some metals and alloys and the penetration depth of microwaves (particle size: 60 mm mesh)

Metal	Conductivity $\delta / \times 10^7$ S m^{-1}	Resistivity $/ \times 10^{-8}$ Ω m^{-1}	Relative magnetic permeability/μ_r'	Penetration depth/μm
SUS304	0.14	71.4	1	8.60
Titanium	0.18	55.0	1	7.58
Brass	2.56	3.91	1	2.01
Aluminum	3.63	2.75	1.00002	1.69
Gold	4.66	2.15	1	1.49
Copper	5.81	1.72	0.99999	1.33
Silver	6.17	1.62	0.99998	1.3
Nickel	1.38	7.24	600	0.11
Soft iron	1.02	9.80	2000	0.07
Steel	0.67	15.0	5000	0.06

Table A.3 Thermal conductivities of selected solids and liquids

Solid substances	Thermal conductivity λ/W m^{-1} K^{-1}	Temperature/K
Au	319	273
Ag	428	273
Al	236	273
Cu	403	273
C (graphite)	50–130	300
C (diamond)	1000	273
Fe	83.5	273
Ni	94	273
Si	168	300
Ti	22	273
Ti-6Al-4V	7.6	300
W	177	273
α-Al_2O_3	36	300
BaO	55	490
BeO	210	373
CaO	17	300
CaO (porosity 8.75)	13.9	373
CeO_2	11.7	320
Cu_2O	5.58	299
Fe_2O_3	6.4	300
Fe_3O_4	3.8	300
MgO	4.2	Room temperature
MnO	3.5	573
NaCl	6.4	273
NiO	82	194
SiO_2 (single crystal)	10.4 (parallel to the *c*-axis)	300
SiO_2 (single crystal)	6.21 (perpendicular to the *c*-axis)	300
SiO_2 (glass)	1.37	298
SnO_2	30.4	328
TiO_2	8.4	300
TiO_2	10.4 (parallel to the *c*-axis)	300
TiO_2	7.4 (perpendicular to the *c*-axis)	300
UO_2	8.2	330
Y_2O_3	27	290
Y_2O_3(Nd 1 mol% doped)	14	290
Y_2O_3 (black)	18.9	293
Y_2O_3 (colorless)	12.5	293
ZnO	54	300
ZrO_2	3.2	298
ZrO_2(7.9wt%Y_2O_3)	1.3	293

(continued)

Table A.3 (continued)

Solid substances	Thermal conductivity λ/W m^{-1} K^{-1}	Temperature/K
$BaTiO_3$	1.9	Room temperature
$BaTiO_3$	6.2	300
$CaTiO_3$	4.7	Room temperature
$MgAl_2O_4$	20	290
$PbTiO_3$	3.8	330
$SrTiO_3$	11.2	300
Air	0.02	298
$YBa_2Cu_3O_{7-\delta}$	3	300
$NaCo_2O_4$	3.9	300
$Y_3Al_5O_{12}$	13	300
$Y_3Al_5O_{12}$ (Nd 1mol% doped)	14	300
YPO_4	7	373
$LaPO_4$	2.5	573
$SrZrO_3$	2.9	573
$Sr_2Nb_2O_7$	1	573
$Yb_2Sn_2O_7$	2.7	573
$Gd_2Zr_2O_7$	1.3	673
$La_2Zr_2O_7$	1.8	673
$Nd_2Zr_2O_7$	1.4	673
$La_2Mo_2O_9$	0.7	573
$Zr_2Y_4O_{12}$	2	673
$Bi_4Ti_3O_{12}$	0.9	573
$W_3Nb_{14}O_4$	1.9	573
$LaCuO_4$	3	300
SiC (Polycrystalline)	107	300
SiC (Single crystal)	89.5	330
SiC (sintered compact, 94%TD)	31.9	290
Si_3N_4	39.2	298
GaN	130	298
BN	60	298
LaF_3	5.1	300
AgCl	1.3	273
BaF_2	10.9	305
Bi_2Te_3	3.6	303
CaF_2	10	273
Asbestos	0.09	273
Liquid and gaseous substances	**Thermal conductivity λ/W m^{-1} K^{-1}**	**Temperature/K**
Aniline	0.172	293
Ethanol	0.189	233

(continued)

Table A.3 (continued)

Liquid and gaseous substances	Thermal conductivity λ/W m^{-1} K^{-1}	Temperature/K
Ethanol	0.15	353
Toluene	0.159	193
Toluene	0.119	353
Water	0.561	273
Water	0.673	353
Methanol	0.223	233
Methanol	0.186	333
Heptane	0.1223	300
Boron	0.122	300
Carbon tetrachloride	0.104	300
Carbon dioxide	0.074	300
Chlorine	0.134	300
Ammonia	0.479	300
dinitrogen tetroxide	0.13	300
Sulfur dioxide	0.195	300
Acetone	0.159	300
Ethylene	0.077	280
Ethylene glycol	0.258	300
Methyl chloride	0.156	300
Octane	0.132	300
Glycerin	0.288	300
Chloroform	0.117	300
Diethylether	0.129	300
Decane	0.138	300
Nonane	0.136	300
Dimethyl phthalate	0.1472	300
Butane	0.106	300
Hexane	0.123	300
Benzene	0.144	300
Pentane	0.118	300
Trichlorofluoromethane	0.0926	300
Dichlorodifluoromethane	0.069	300
Chlorodifluoromethane	0.0858	300
Mercury	8.25	298
Trichloromethane	0.117	298
Dibromomethane	0.108	298
Methanol	0.2	298
Tetrachlorethylene	0.11	298
1,1,1-trichlorethylene	0.101	298
Acetonitrile	0.188	298

(continued)

Table A.3 (continued)

Liquid and gaseous substances	Thermal conductivity λ/W m^{-1} K^{-1}	Temperature/K
Acetic acid	0.158	298
Chloroethane	0.119	298
N-methylformamide	0.203	298
Ethylene glycol	0.256	298
Ethanolamine	0.299	298
Epichlorohydrin	0.131	298
Acetone	0.161	298
Methyl acetate	0.153	298
N, N-dimethylformamide	0.184	298
1-propanol	0.154	298
2-propanol	0.135	298
1,2-propanediol	0.2	298
Glycerol	0.292	298
Triethylamine	0.133	273
Franc	0.126	298
Thiophene	0.199	298
2-butyne	0.121	298
2-butane	0.145	298
Tetrahydrofuran	0.12	298
1,4-dioxane	0.159	298
Ethyl acetate	0.144	298
1-butanol	0.154	298
Diethylether	0.13	298
Pyridine	0.165	298
Cyclopentene	0.129	298
1-pentene	0.116	298
Cyclopentane	0.126	298
Pentane	0.113	298
1-pentanol	0.153	298
Cyclobenzene	0.127	298
Benzene	0.1411	298
Phenol	0.156	323
Cyclohexane	0.13	298
Mesityl oxide	0.156	298
Cyclohexane	0.123	298
1-hexane	0.121	298
Cyclohexanol	0.134	298
2-hexanone	0.139	298
Hexane	0.12	298
1-hexanol	0.15	298

(continued)

Table A.3 (continued)

Liquid and gaseous substances	Thermal conductivity λ/W m^{-1} K^{-1}	Temperature/K
Benzaldehyde	0.151	298
Toluene	0.1311	298
Anisole	0.156	298
Heptane	0.1228	298
1-heptanol	0.159	298
Styrene	0.137	298
Ethylbenzene	0.13	298
O-xylene	0.131	298
M-xylene	0.13	298
P-xylene	0.13	298
Octane	0.128	298
1-octanol	0.161	298
Cumene	0.128	298
Mesitylene	0.136	298
Nonane	0.131	298
1-nonanol	0.161	298
Dimethyl phthalate	0.1473	298
P-cymene	0.122	298
Decane	0.132	298
1-decanol	0.162	298
Undecane	0.14	298
Diphenyl ether	0.139	323
Dodecane	0.152	298
1-dodecanol	0.146	298
Tridecane	0.137	298
Tetradecane	0.136	298
1-tetradecanol	0.167	323
Dibutyl phthalate	0.136	298
Hexadecane	0.14	298
Octadecane	0.146	298

Table A.4 Emissivity of solids at indicated temperatures

	Material	Total emissivity	Temperature/°C
Metal	Aluminum (nonoxidizing)	0.02	25
		0.03	100
		0.06	500
	Aluminum (oxidation)	0.11	199
		0.16	599
	Aluminum (coarse gloss)	0.18	Room temperature
	Bismuth (nonoxidizing)	0.05	25
		0.06	100
	Cadmium	0.02	25
	Carbon	0.81	25
		0.81	100
		0.79	500
	Graphite	0.76	100
		0.75	300
		0.71	500
	Chromium	0.08	38
		0.26	538
	Chromium oxide	0.81	Room temperature
	Cobalt	0.13	500
		0.23	1000
	Copper oxide (Cu_2O)	0.87	38
		0.83	260
		0.77	538
	Copper oxide (CuO)	0.78	38
	Copper (after polishing)	0.03	38
	Iron (oxidized surface)	0.74	100
		0.84	499
		0.89	1199
	Iron (non-oxidized surface)	0.05	100
	Lead (polished surface)	0.06–0.08	38–260
	Lead (oxidized surface)	0.43	38
	Magnesium	0.07–0.13	38–260
	Magnesium oxide	0.16–0.20	1027–1727
	Molybdenum	0.06	38
		0.08	260
		0.11	538
		0.18	1093
	Molybdenum (oxidized surface)	0.8	316
		0.84	371
		0.84	427
		0.83	482
		0.82	538

(continued)

Table A.4 (continued)

	Material	Total emissivity	Temperature/°C
Nickel (polished surface)	0.05	38	
Nickel (oxidized surface)	0.31–0.46	38–260	
Nickel oxide	0.59–0.86	538–1093	
Platinum	0.05	38	
Silver (polished surface)	0.01	38	
	0.02	260	
	0.03	538	
	0.03	1093	
Tin (non-oxidized surface)	0.04	25	
	0.05	100	
Tungsten (non-oxidized surface)	0.02	25	
	0.03	100	
	0.07	500	
	0.15	1000	
	0.23	1500	
	0.28	2000	
Uranium oxide	0.79	1027	
Zinc oxide	0.11	260–538	
Zinc (polished surface)	0.02	38	
	0.03	260	
	0.04	538	
	0.06	1093	
Nonmetal	A brick	0.9	20
	Asbestos (board)	0.96	38
	Paper	0.93	38–371
	Slate	0.97	20
	Aluminum oxide	0.9	93
		0.76	Room temperature
	Silicon carbide	0.83–0.96	149–649
	Glass	0.94	Room temperature
	Carbon resistance	0.85	Room temperature
	Steel (oxidation)	0.78	Room temperature

(continued)

Table A.4 (continued)

	Material	Total emissivity	Temperature/°C
	Cast iron (oxidation)	0.78	Room temperature
	Stainless steel (non-glossy)	0.74	Room temperature
	Copper (oxidation)	0.568	Room temperature
	Ceramics	0.90–0.94	Room temperature
	Firefly stones	0.30–0.40	Room temperature
	Alumina ceramics	0.93–0.44	260–816
	Magnesia ceramics	0.55–0.18	260–816

Kolibri 360 GmbH
Ferdinand-Jühlke-Straße 7
99095 Erfurt, Deutschland
produktsicherheit@kolibri360.de